Discontinuous Fiber Composites, Volume II

Discontinuous Fiber Composites, Volume II

Editors

Tim Osswald
Christoph Kuhn

MDPI • Basel • Beijing • Wuhan • Barcelona • Belgrade • Manchester • Tokyo • Cluj • Tianjin

Editors

Tim Osswald
Mechanical Engineering
Department
University of
Wisconsin-Madison
Madison, WI
United States

Christoph Kuhn
Engineering & Planning Center
Volkswagen Group of America
Chattanooga, TN
United States

Editorial Office
MDPI
St. Alban-Anlage 66
4052 Basel, Switzerland

This is a reprint of articles from the Special Issue published online in the open access journal *Journal of Composites Science* (ISSN 2504-477X) (available at: https://www.mdpi.com/journal/jcs/ special_issues/Discontinuous_Fibers_v2).

For citation purposes, cite each article independently as indicated on the article page online and as indicated below:

LastName, A.A.; LastName, B.B.; LastName, C.C. Article Title. *Journal Name* **Year**, *Volume Number*, Page Range.

ISBN 978-3-0365-1292-1 (Hbk)
ISBN 978-3-0365-1291-4 (PDF)

Contents

About the Editors . . . . **vii**

Christoph Kuhn and Tim A. Osswald
Editorial for the Special Issue on Discontinuous Fiber Composites, Volume II
Reprinted from: *J. Compos. Sci.* **2021**, *5*, 71, doi:10.3390/jcs5030071 . . . . **1**

Susanne Katrin Kugler, Armin Kech, Camilo Cruz, Tim Osswald
Fiber Orientation Predictions—A Review of Existing Models
Reprinted from: *J. Compos. Sci.* **2020**, *4*, 69, doi:10.3390/jcs4020069 . . . . **5**

Sara Andrea Simon, Abrahán Bechara Senior and Tim Osswald
Experimental Validation of a Direct Fiber Model for Orientation Prediction
Reprinted from: *J. Compos. Sci.* **2020**, *4*, 59, doi:10.3390/jcs4020059 . . . . **27**

Abrahán Bechara Senior and Tim Osswald
Measuring Fiber Length in the Core and Shell Regions of Injection Molded Long
Fiber-Reinforced Thermoplastic Plaques
Reprinted from: *J. Compos. Sci.* **2020**, *4*, 104, doi:10.3390/jcs4030104 . . . . **45**

Tzu-Chuan Chang, Abrahán Bechara Senior, Hakan Celik, Dave Brands, Angel Yanev and Tim Osswald
Validation of Fiber Breakage in Simple Shear Flow with Direct Fiber Simulation
Reprinted from: *J. Compos. Sci.* **2020**, *4*, 134, doi:10.3390/jcs4030134 . . . . **61**

Nils Meyer, Oleg Saburow, Martin Hohberg, Andrew N. Hrymak, Frank Henning and Luise Kärger
Parameter Identification of Fiber Orientation Models Based on Direct Fiber Simulation with
Smoothed Particle Hydrodynamics
Reprinted from: *J. Compos. Sci.* **2020**, *4*, 77, doi:10.3390/jcs4020077 . . . . **77**

Fabian Willems, Philip Reitinger and Christian Bonten
Calibration of Fiber Orientation Simulations for LFT—A New Approach
Reprinted from: *J. Compos. Sci.* **2020**, *4*, 163, doi:10.3390/jcs4040163 . . . . **97**

Susanne Katrin Kugler, Argha Protim Dey, Sandra Saad, Camilo Cruz, Armin Kech and Tim Osswald
A Flow-Dependent Fiber Orientation Model
Reprinted from: *J. Compos. Sci.* **2020**, *4*, 96, doi:10.3390/jcs4030096 . . . . **127**

Christian Hopmann, Jonas Neuhaus, Kai Fischer, Daniel Schneider and René Laschak Pinto Gonçalves
Metamodelling of the Correlations of Preform and Part Performance for Preform Optimisation
in Sheet Moulding Compound Processing
Reprinted from: *J. Compos. Sci.* **2020**, *4*, 122, doi:10.3390/jcs4030122 . . . . **149**

Christoph Kuhn and Simon Wehler
A Force-Balanced Fiber Retardation Model to Predict Fiber-Matrix-Separation during
Polymer Processing
Reprinted from: *J. Compos. Sci.* **2020**, *4*, 165, doi:10.3390/jcs4040165 . . . . **167**

Armin Kech, Susanne Kugler and Tim Osswald
Significance of Model Parameter Variations in the pARD-RSC Model
Reprinted from: *J. Compos. Sci.* **2020**, *4*, 109, doi:10.3390/jcs4030109 . . . . **179**

Jan Teuwsen, Stephan K. Hohn and Tim A. Osswald
Direct Fiber Simulation of a Compression Molded Ribbed Structure Made of a Sheet Molding
Compound with Randomly Oriented Carbon/Epoxy Prepreg Strands—A Comparison of
Predicted Fiber Orientations with Computed Tomography Analyses
Reprinted from: *J. Compos. Sci.* **2020**, *4*, 164, doi:10.3390/jcs4040164 . . . . **189**

Manuel V. C. Morais, Marco Marcellan, Nadine Sohn, Christof Hübner and Frank Henning
Process Chain Optimization for SWCNT/Epoxy Nanocomposite Parts with Improved
Electrical Properties
Reprinted from: *J. Compos. Sci.* **2020**, *4*, 114, doi:10.3390/jcs4030114 . . . . **239**

Lukas Knorr, Robert Setter, Dominik Rietzel, Katrin Wudy and Tim Osswald
Comparative Analysis of the Impact of Additively Manufactured Polymer Tools on the Fiber
Configuration of Injection Molded Long-Fiber-Reinforced Thermoplastics
Reprinted from: *J. Compos. Sci.* **2020**, *4*, 136, doi:10.3390/jcs4030136 . . . . **251**

Tobias Heckner, Michael Seitz, Sven Robert Raisch, Gerrit Huelder and Peter Middendorf
Selective Laser Sintering of PA6: Effect of Powder Recoating on Fibre Orientation
Reprinted from: *J. Compos. Sci.* **2020**, *4*, 108, doi:10.3390/jcs4030108 . . . . **283**

Jochen Wellekötter, Julia Resch, Stephan Baz, Götz Theo Gresser and Christian Bonten
Insights into the Processing of Recycled Carbon Fibers via Injection Molding Compounding
Reprinted from: *J. Compos. Sci.* **2020**, *4*, 161, doi:10.3390/jcs4040161 . . . . **299**

Hicham Ghossein, Ahmed Arabi Hassen, Seokpum Kim, Jesse Ault and Uday K. Vaidya
Characterization of Mechanical Performance of Composites Fabricated Using Innovative
Carbon Fiber Wet Laid Process
Reprinted from: *J. Compos. Sci.* **2020**, *4*, 124, doi:10.3390/jcs4030124 . . . . **319**

Tianran Chen, Dana Kazerooni, Lin Ju, David A. Okonski and Donald G. Baird
Development of Recyclable and High-Performance In Situ Hybrid TLCP/Glass
Fiber Composites
Reprinted from: *J. Compos. Sci.* **2020**, *4*, 125, doi:10.3390/jcs4030125 . . . . **337**

Daniel P. Pulipati and David A. Jack
Strength Prediction Sensitivity of Foamed Recycled Polymer Composite Structures due to the
Localized Variability of the Cell Density Distribution
Reprinted from: *J. Compos. Sci.* **2020**, *4*, 93, doi:10.3390/jcs4030093 . . . . **353**

About the Editors

Tim Osswald

Tim Osswald is a professor of mechanical engineering and the director of the Polymer Engineering Center at the University of Wisconsin-Madison, and he is originally from Cúcuta, Colombia. He received the National Science Foundation's Presidential Young Investigator Award as well as the 2001 VDI-K Dr-Richard-Escales-Preis. In 2006, he was named an honorary professor at the University of Erlangen-Nuremberg in Germany, and in 2011, he was named an honorary professor at the National University of Colombia. Professor Osswald teaches polymer and polymer composites processing, and designing with polymers and polymer composites and conducts research in the same areas, in particular in the areas of fiber orientation, fiber density and fiber length distributions. Professor Osswald has published over 300 papers and numerous books.

Christoph Kuhn

Christoph Kuhn currently supports the introduction of VW's electric fleet in the US as a project manager at the Engineering and Planning Center (EPC) in Chattanooga. Prior to this, Dr. Kuhn was a research project manager with the Materials Division at Volkswagen Group Research in Germany, focusing on Functional Battery Materials, Lightweight Design and Composite Material Simulation. As project lead in a publicly funded research project on a lightweight composite center tunnel ("LehoMit-Hybrid") with Porsche AG and university partners, Dr. Kuhn was awarded the JEC's Innovation Award in 2020. Dr. Kuhn graduated with a PhD in mechanical engineering from the FAU Erlangen-Nuremberg, with a MSc in plastics processing from RWTH Aachen, and with a MSc in mechanical engineering from the UW-Madison. Dr. Kuhn conducted his research at the Polymer Engineering Center (PEC) in Madison, WI, and the Institute for Plastics Processing (IKV) in Aachen, Germany.

Journal of
Composites Science

Editorial

Editorial for the Special Issue on Discontinuous Fiber Composites, Volume II

Christoph Kuhn [1,*] and Tim A. Osswald [2,*]

1   Volkswagen Group of America, Chattanooga, TN 37416, USA
2   Department of Mechanical Engineering, University of Wisconsin Madison, Madison, WI 53706, USA
*   Correspondence: christoph.kuhn@vw.com (C.K.); tosswald@wisc.edu (T.A.O.)

Citation: Kuhn, C.; Osswald, T.A. Editorial for the Special Issue on Discontinuous Fiber Composites, Volume II. *J. Compos. Sci.* **2021**, *5*, 71. https://doi.org/10.3390/jcs5030071

Received: 25 February 2021
Accepted: 26 February 2021
Published: 5 March 2021

Publisher's Note: MDPI stays neutral with regard to jurisdictional claims in published maps and institutional affiliations.

This Special Issue on discontinuous fiber composites and its published papers, like its predecessor, give the polymer engineer and scientist an insight into challenges and research topics in the field of discontinuous fiber-reinforced composites. This Special Issue addresses a number of current topics from industry and academia, displays the latest achievements in the field and further introduces novel research topics as inspiration for future work.

In various industrial applications, discontinuous fiber-reinforced composites have been successful due to their wide range of properties and advantages regarding, for example, their mechanical property to weight ratio, cost-efficient and flexible production and recyclability. These advantages over traditional materials like steel and aluminum have been continuously improved by academic and industry research.

A prominent research topic with discontinuous fiber-reinforced composites is the understanding and prediction of the process-related fiber microstructure inside the composite. The introduction of discontinuous fibers, such as glass, carbon or natural fibers into a polymer resin, synergistically creates a composite material superior to the individual components. Next to the fiber's material properties, the final product's properties rely strongly on the inner fiber microstructure, e.g., orientation, content and length. Due to process-induced effects, phenomena like fiber reorientation, attrition and even fiber-matrix phase separation occur. Examples of continued work on changes in fiber microstructure inside a component, presented in this special edition, are shown by the work of Simon et al. [1] on direct fiber simulation of orientation development during flow, and the work of Chang et al. [2] on the local fiber structure with their investigations into fiber length distribution inside injection molded plaques.

For decades, particle orientation has been the major focus due to the importance of the process-induced microstructure on mechanical component behavior. Fiber alignment during processing leads to regional part anisotropy, which has created great opportunities for mechanical applications if these effects are understood and accounted for. Hence understanding and predicting the final microstructure is of significant importance for successful application.

With the introduction of computers, the first simulative models were created to predict the changes of fiber orientation during processing with the help of phenomenological models [3]. With the increase in computational capacity and improved models, simulations have increased both in accuracy and computational speed with great improvements for part development. A comprehensive review of the history of these phenomenological fiber orientation models by Kugler et al. [4] is included in this Special Issue. Current research is still focused on further understanding and improving the prediction of fiber properties during various established and novel processes, from injection and compression molding to extrusion and additive manufacturing. On the side of simulative models, improved prediction accuracy and faster development times are the focus of research. A fitting example of faster development times is Willems et al.'s new method to calibrate the fiber orientation model parameter for injection molding simulations [5]. Improvements in accuracy

1

continually lead to newer simulation models. Kugler et al. present their latest approach with a novel flow-dependent fiber orientation model [6]. Another example is shown by Hopmann et al., where trained neuronal networks are applied in a metamodeling approach to optimize part performance with a compression molded sheet molding compound (SMC) in the preforming stage [7]. Further, Kuhn and Wehler present a phenomenological model on fiber-matrix-phase separation during compression molding, a model to predict changes in fiber content distribution inside a component [8].

With increasing computational power in recent years, detailed simulations on particle level with fiber models inside a polymer flow became feasible, thus enabling detailed studies of fiber behavior, regarding orientation, attrition and fiber-matrix phase separation [9,10]. The latest advances with this direct fiber simulation approach are presented in this Special Issue [1,2]. Kech et al. present their investigations on the significance of fiber parameter variations for the pARD-RSC orientation model in a shear cell simulation with a similar model [11]. Further, Teuwsen et al. [12] successfully compare fiber orientation predictions from a mechanistic fiber model with compression molded carbon fiber epoxy composites via computed tomography. Chang et al. [12] apply the detailed fiber simulation to validate fiber breakage experiments inside a simple shear flow. Meyer et al. [13] applied a direct fiber simulation approach to successfully identify fitting parameters for phenomenological fiber orientation models. Morais et al. [14] present their research on carbon nanotube (CNTs) composites and investigate the influence of local nanofiber microstructure on electrical conductivity. They show that with optimized microstructure, polymers can be designed as electrically conductive and hence, can be suitable for metal substitution and implementation in complex component design.

Next to the traditional processing techniques like injection and compression molding, research on fiber-induced properties within the field of additive manufacturing have increased in recent years. Heckner et al. present their study of fiber-filled powders in selective laser-sintering processing [15]. While this process is comparably different to melt-based processing techniques, a process-induced fiber microstructure during the recoating process, and hence an anisotropic component behavior is also found in this process. Knorr et al. show the application of additive manufactured injection molding tools for small series manufacturing and compare them to steel and aluminum tooling. In this application, both the tools and the later parts are manufactured with different fiber-reinforced composite materials [16]. This area of additive tooling shows how discontinuous fiber microstructure is affected by the heat transfer during flow and cooling of the reinforced melt [16].

The latest industry trends with fiber-reinforced composites have been increasingly focused on the fields of recycling where the reuse of both polymer and matrix materials, and the understanding of recycled material behavior are key factors for successful product application. The reduction of carbon emissions through the reuse and recycling of materials is a key factor for a more sustainable future. Wellenkoetter et al. [17] present a novel approach to incorporate recycled fiber and matrix material in an injection molding process via direct fiber feed. Ghossein et al. present a novel wet laying process for the efficient reuse of recycled carbon fibers [18]. Chen et al. further present their latest developments on recyclable thermotropic liquid crystalline polymer and glass fiber composites [19]. Pulipati and Jack present their work on polymer foam structures made from recycled polyethylene and polypropylene [20].

This Special Issue relates not only to a variety of materials, but also to different processes, such as injection and compression molding as well as additive manufacturing. The set of papers in this issue can help advance discontinuous fiber-reinforced composites by contributing to the overall understanding of microstructure development during processing.

References

1. Simon, S.A.; Bechara Senior, A.; Osswald, T. Experimental Validation of a Direct Fiber Model for Orientation Prediction. *J. Compos. Sci.* **2020**, *4*, 59. [CrossRef]
2. Chang, T.-C.; Bechara Senior, A.; Celik, H.; Brands, D.; Yanev, A.; Osswald, T. Validation of Fiber Breakage in Simple Shear Flow with Direct Fiber Simulation. *J. Compos. Sci.* **2020**, *4*, 134. [CrossRef]
3. Folgar, F.; Tucker, C.L. Orientation Behavior of Fibers in Concentrated Suspensions. *J. Reinf. Plast. Compos.* **1984**, *3*, 98–119. [CrossRef]
4. Kugler, S.K.; Kech, A.; Cruz, C.; Osswald, T. Fiber Orientation Predictions—A Review of Existing Models. *J. Compos. Sci.* **2020**, *4*, 69. [CrossRef]
5. Willems, F.; Reitinger, P.; Bonten, C. Calibration of Fiber Orientation Simulations for LFT—A New Approach. *J. Compos. Sci.* **2020**, *4*, 163. [CrossRef]
6. Kugler, S.K.; Dey, A.P.; Saad, S.; Cruz, C.; Kech, A.; Osswald, T. A Flow-Dependent Fiber Orientation Model. *J. Compos. Sci.* **2020**, *4*, 96. [CrossRef]
7. Hopmann, C.; Neuhaus, J.; Fischer, K.; Schneider, D.; Gonçalves, R.L.P. Metamodelling of the Correlations of Preform and Part Performance for Preform Optimisation in Sheet Moulding Compound Processing. *J. Compos. Sci.* **2020**, *4*, 122. [CrossRef]
8. Kuhn, C.; Wehler, S. A Force-Balanced Fiber Retardation Model to Predict Fiber-Matrix-Separation during Polymer Processing. *J. Compos. Sci.* **2020**, *4*, 165. [CrossRef]
9. Kuhn, C. Analysis and Prediction of Fiber Matrix Separation during Compression Molding of Fiber Reinforced Plastics. Ph.D. Thesis, Friedrich-Alexander-Universität Erlangen-Nürnberg, Erlangen, Germany, 2018.
10. Londoño-Hurtado, A. *Mechanistic Model for Fiber Flow*; University of Wisconsin-Madison: Madison, WI, USA, 2009.
11. Kech, A.; Kugler, S.; Osswald, T. Significance of Model Parameter Variations in the pARD-RSC Model. *J. Compos. Sci.* **2020**, *4*, 109. [CrossRef]
12. Teuwsen, J.; Hohn, S.; Osswald, T.A. Direct Fiber Simulation of a Compression Molded Ribbed Structure Made of a Sheet Molding Compound with Randomly Oriented Carbon/Epoxy Prepreg Strands—A Comparison of Predicted Fiber Orientations with Computed Tomography Analyses. *J. Compos. Sci.* **2020**, *4*, 164. [CrossRef]
13. Meyer, N.; Saburow, O.; Hohberg, M.; Hrymak, A.N.; Henning, F.; Kärger, L. Parameter Identification of Fiber Orientation Models Based on Direct Fiber Simulation with Smoothed Particle Hydrodynamics. *J. Compos. Sci.* **2020**, *4*, 77. [CrossRef]
14. Morais, M.V.C.; Marcellan, M.; Sohn, N.; Hübner, C.; Henning, F. Process Chain Optimization for SWCNT/Epoxy Nanocomposite Parts with Improved Electrical Properties. *J. Compos. Sci.* **2020**, *4*, 114. [CrossRef]
15. Heckner, T.; Seitz, M.; Raisch, S.R.; Huelder, G.; Middendorf, P. Selective Laser Sintering of PA6: Effect of Powder Recoating on Fibre Orientation. *J. Compos. Sci.* **2020**, *4*, 108. [CrossRef]
16. Knorr, L.; Setter, R.; Rietzel, D.; Wudy, K.; Osswald, T. Comparative Analysis of the Impact of Additively Manufactured Polymer Tools on the Fiber Configuration of Injection Molded Long-Fiber-Reinforced Thermoplastics. *J. Compos. Sci.* **2020**, *4*, 136. [CrossRef]
17. Wellekötter, J.; Resch, J.; Baz, S.; Gresser, G.T.; Bonten, C. Insights into the Processing of Recycled Carbon Fibers via Injection Molding Compounding. *J. Compos. Sci.* **2020**, *4*, 161. [CrossRef]
18. Ghossein, H.; Hassen, A.A.; Kim, S.; Ault, J.; Vaidya, U.K. Characterization of Mechanical Performance of Composites Fabricated Using Innovative Carbon Fiber Wet Laid Process. *J. Compos. Sci.* **2020**, *4*, 124. [CrossRef]
19. Chen, T.; Kazerooni, D.; Ju, L.; Okonski, D.A.; Baird, D.G. Development of Recyclable and High-Performance In Situ Hybrid TLCP/Glass Fiber Composites. *J. Compos. Sci.* **2020**, *4*, 125. [CrossRef]
20. Pulipati, D.P.; Jack, D.A. Strength Prediction Sensitivity of Foamed Recycled Polymer Composite Structures due to the Localized Variability of the Cell Density Distribution. *J. Compos. Sci.* **2020**, *4*, 93. [CrossRef]

Journal of
Composites Science

Review

Fiber Orientation Predictions—A Review of Existing Models

Susanne Katrin Kugler [1,2,*], **Armin Kech** [1], **Camilo Cruz** [1] and **Tim Osswald** [2]

[1] Corporate Sector Research and Advance Engineering, Robert Bosch GmbH, 71272 Renningen, Germany; Armin.Kech@de.bosch.com (A.K.); Camilo.Cruz@de.bosch.com (C.C.)

[2] Polymer Engineering Center, University of Wisconsin-Madison, Madison, WI 53706, USA; tosswald@wisc.edu

[*] Correspondence: Susanne.Kugler2@de.bosch.com

Received: 11 May 2020; Accepted: 2 June 2020; Published: 8 June 2020

Abstract: Fiber reinforced polymers are key materials across different industries. The manufacturing processes of those materials have typically strong impact on their final microstructure, which at the same time controls the mechanical performance of the part. A reliable virtual engineering design of fiber-reinforced polymers requires therefore considering the simulation of the process-induced microstructure. One relevant microstructure descriptor in fiber-reinforced polymers is the fiber orientation. This work focuses on the modeling of the fiber orientation phenomenon and presents a historical review of the different modelling approaches. In this context, the article describes different macroscopic fiber orientation models such as the Folgar-Tucker, nematic, reduced strain closure (RSC), retarding principal rate (RPR), anisotropic rotary diffusion (ARD), principal anisotropic rotary diffusion (pARD), and Moldflow rotary diffusion (MRD) model. We discuss briefly about closure approximations, which are a common mathematical element of those macroscopic fiber orientation models. In the last section, we introduce some micro-scale numerical methods for simulating the fiber orientation phenomenon, such as the discrete element method (DEM), the smoothed particle hydrodynamics (SPH) method and the moving particle semi-implicit (MPS) method.

Keywords: fiber orientation; fiber reinforced thermoplastics; modeling

---

1. Introduction

Fiber reinforced polymers are key materials across different industries. For example short fiber reinforced thermoplastics are wildly used in the automotive industry to reduce weight. The manufacturing processes of those materials have typically strong impact on their final microstructure, which at the same time controls the mechanical performance of the part. A reliable virtual engineering design of fiber-reinforced polymers therefore requires considering the simulation of the process-induced microstructure.

One relevant microstructure descriptor in fiber-reinforced polymers is the fiber orientation. This work focuses on the modeling of the fiber orientation phenomenon and presents a historical review of the different modeling approaches. In this context, the article describes different modeling approaches such as the addition of a scalar diffusion by Folgar and Tucker [1], the nematic potential approach [2], the modeling of a retarding rate in the reduced strain closure (RSC) [3] and the retarding principle rate (RPR) model [4], and lastly the anisotropic rotary diffusion approach (ARD) by Phelps et al. [5]. Additionally, reduced parameters models like the improved anisotropic rotatory diffusion (iARD) [4], principal anisotropic rotary diffusion (pARD) [6], and Moldflow rotary diffusion (MRD) [7] will be introduced. The mentioned models are provided in commercial injection molding software such as Autodesk Moldflow®, Moldex 3D®, Sigmasoft® and Cadmould®. For example,

Autodesk Moldflow® provides the Folgar-Tucker, RSC and ARD-RSC model and Moldex 3D® the Folgar-Tucker, ARD, and iARD-RPR model.

Furthermore, we briefly discuss closure approximations, which are a common mathematical element of those macroscopic fiber orientation models. Simple closure approximation like the linear, quadratic, and hybrid closure [8,9] will be introduced. Then, we focus on exact closure approximations and fitted closure approximations. In the field of fitted closures we distinguish between orthotropic and invariant based closures.

Afterwards, we introduce micro-scale numerical methods for simulating the fiber orientation phenomenon, such as the discrete element method (DEM), smoothed particle hydrodynamics (SPH) and moving particle semi- implicit MPS. The focus will be on DEM based method and existing approaches will be looked at under the points of fiber discretization, imposed flow fields, fluid-fiber interaction, and fiber-fiber interaction.

The last section focuses on combining the advantages of both scales, for example through parameter fitting or machine learning approaches.

2. Fiber Orientation

The orientation of a single fiber can be characterized by a unit vector $\mathbf{p} \in \mathbb{S}^3 := \{\mathbf{p} \in \mathbb{R}^3 : ||\mathbf{p}|| = 1\}$ along the fiber axis. All $\mathbf{p} \in \mathbb{S}^3$ can be defined by two angles (ϕ, θ) (Figure 1)

$$
\begin{aligned}
p_1 &= \sin\theta\cos\phi & (1)\\
p_2 &= \sin\theta\sin\phi & (2)\\
p_3 &= \cos\theta & (3)
\end{aligned}
$$

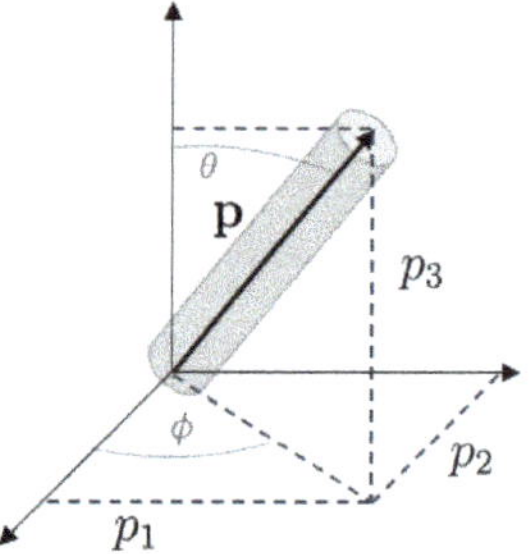

Figure 1. Orientation of a single rigid fiber $\mathbf{p}$.

To describe the orientation of many fibers statistical methods are useful. The probability density function (PDF) $\psi()$ [1] is defined, such that $\psi(\mathbf{p}, t)\mathrm{d}\mathbf{p}$ is the probability that a fiber is directed between $\mathbf{p}$ and $\mathbf{p} + \mathrm{d}\mathbf{p}$ at time t. The PDF has the following distinct properties.

$$
\begin{aligned}
\mathcal{B}(\psi) &= [0,1] & (4)\\
\oint \psi(\mathbf{p}, t)\mathrm{d}\mathbf{p} &= 1 & (5)\\
\psi(\mathbf{p}) &= \psi(-\mathbf{p}) & (6)\\
\frac{D\psi}{Dt} &= -\nabla_s \cdot (\psi\dot{\mathbf{p}}). & (7)
\end{aligned}
$$

Equation (6) is valid under the assumption that the fibers are cylindrical and have no preferred end. Equation (7) is called the continuity equation. The operator ∇_s represent the gradient operator on the surface of the unit sphere.

Since the PDF is defined on the unit sphere, computation is expensive and numerically difficult. For that reason, moments of the PDF are commonly used for computational efficiency [8]. The orientation tensors are defined by

$$a_{ij} = \int p_i p_j \psi(\mathbf{p}) d\mathbf{p} \tag{8}$$

$$a_{ijkl} = \int p_i p_j p_k p_l \psi(\mathbf{p}) d\mathbf{p}. \tag{9}$$

$$a_{i...n} = \int p_i \cdots p_n \psi(\mathbf{p}) d\mathbf{p}. \tag{10}$$

Since the PDF is even (Equation (6)) all odd-ordered orientation tensors are zero. To simplify notation we introduce $\mathbf{A} = a_{ij}, \mathbb{A} = a_{ijkl}$. The orientation tensors have important properties [8], which are described here for the second and fourth moment only. The second order tensor is symmetric and has a unit trace.

$$\mathbf{A}_{ij} = \mathbf{A}_{ji} \tag{11}$$

$$\mathrm{tr}\mathbf{A} = 1. \tag{12}$$

The fourth order tensor is symmetric with respect to any pair of indices.

$$\mathbb{A}_{ijkl} = \mathbb{A}_{jikl} = \mathbb{A}_{ijkl} = \mathbb{A}_{kjil} = \mathbb{A}_{ljki} = \mathbb{A}_{ikjl} = \mathbb{A}_{ilkj} \tag{13}$$

and all information of the second order orientation tensor can be retrieved from the fourth order tensor

$$\mathbf{A}_{ij} = \mathbb{A}_{ijkk}. \tag{14}$$

The use of orientation tensors simplifies the computation because no discretization of the unit sphere is necessary, but it is impossible to distinguish between certain orientations. For example, the orientation tensor for a bipolar and planar random orientation is identical with $\mathbf{A} = \begin{pmatrix} 0.5 & 0 & 0 \\ 0 & 0.5 & 0 \\ 0 & 0 & 0 \end{pmatrix}$.
An advantage is the objectivity, that means the equation is independent of the coordinate system.

3. Macroscopic Fiber Orientation Models

Macroscopic fiber orientation models are used to predict fiber orientation in parts. The models are integrated in commercial software such as Autodesk Moldflow®, Moldex 3D®, Sigmasoft® and Cadmould®. Additionally open source software such as OpenFoam can be used to implement the models.

3.1. Macroscopic Fiber Orientation Models in the Dilute Regime

The first description of the motion of a single fiber was developed by Jeffrey [10]. This description is based on the following assumptions: The fluid is Newtonian and has no turbulences (rot $\mathbf{u} = 0$). The particle is an ellipsoid and perfectly rigid, such that no bending or breaking occurs. The velocity field of the fluid is not influenced by the particle and there exists a perfect contact between the particle and the fluid. Under these assumptions the model is accurate up to order two. This has been proven in a more general way by Junk and Illner [11] under the same assumptions. Jeffrey's model has the form:

$$\dot{\mathbf{p}} = \mathbf{W} \cdot \mathbf{p} + \xi(\mathbf{D} \cdot \mathbf{p} - \mathbf{D} : \mathbf{ppp}), \tag{15}$$

where $\mathbf{W} = \frac{1}{2}((\nabla \mathbf{u})^T - \nabla \mathbf{u})$ is the vorticity tensor, $\mathbf{D} = \frac{1}{2}((\nabla \mathbf{u})^T + \nabla \mathbf{u})$ the rate of strain tensor, $\xi = \frac{r_e^2 - 1}{r_e^2 + 1}$ the particle shape function and ∇ the nabla operator. Applying it to injection molding simulation, it has to be highlighted that the polymer melt is non-Newtonian and fibers are not perfectly rigid. In fact they can break and bend. Furthermore for highly filled polymers, the fluid is influenced by the fiber [2]. Coupling between fluid and fiber orientation will not be considered in this review.

3.2. Macroscopic Fiber Orientation Models in the Concentrated Regime

In the concentrated regime fiber interaction is dominant for fiber orientation. All macroscopic modeling approaches published up to today, accounting for interaction, are phenomenological.

Folgar and Tucker [1] added a diffusion term to account for fiber interaction to predict the orientation in semi-dilute and concentrated solutions. The model is valid under the following additional assumptions: The fibers are rigid cylinders, uniform in length and diameter. Moreover, they are sufficient large such that Brownian motion is negligible. The matrix is incompressible and sufficient viscous, such that particle inertia and buoyancy is negligible. The center of mass of the fibers are randomly distributed and no external forces or torques act on the suspension.

The fiber motion of a single fiber can then be described by

$$\dot{\mathbf{p}} = \mathbf{W} \cdot \mathbf{p} + \xi(\mathbf{D} \cdot \mathbf{p} - \mathbf{D} : \mathbf{ppp}) + C_I \dot{\gamma} \cdot \nabla_s \psi, \tag{16}$$

where C_I describes the interaction coefficient and $\dot{\gamma} = \sqrt{2\mathbf{D} : \mathbf{D}}$ the scalar magnitude of the rate of strain tensor. The equation is, in contrast to Jeffrey's equation (15), not reversible. The interaction coefficient C_I is empirically determined and describes the rate of interaction. Setting $C_I = 0$ retrieves Jeffrey's equation (15). Interactions of fibers cause random orientation. The Fokker-Planck equation (17) [1] expresses the rate of change for the PDF, using the Folgar-Tucker equation for a single fiber (16) and the continuity equation (7). Fibers are modeled as independent, identically distributed, random variables with zero mean. Each interaction causes an orientation change in both fibers

$$\begin{aligned}
\frac{D\psi}{Dt} &= -\nabla_s \cdot (\psi(\mathbf{W} \cdot \mathbf{p} + \xi(\mathbf{D} \cdot \mathbf{p} - \mathbf{D} : \mathbf{ppp}) + C_I \dot{\gamma} \cdot \nabla_s \psi) \\
&= -\nabla_s \cdot (\psi(\mathbf{W} \cdot \mathbf{p} + \xi(\mathbf{D} \cdot \mathbf{p} - \mathbf{D} : \mathbf{ppp}))) + C_I \dot{\gamma} \cdot \nabla_s^2 \psi.
\end{aligned} \tag{17}$$

The concept introduced by Folgar and Tucker offers advantages in the prediction of fiber orientation but is difficult to solve numerically and can only be solved with high computational effort. Based on the Fokker-Planck equation (17), Advani and Tucker [8] developed an equation for the rate of change of the second order orientation tensor

$$\begin{aligned}
\frac{D\mathbf{A}}{Dt} = \dot{\mathbf{A}} &= \dot{\mathbf{A}}^h + \dot{\mathbf{A}}^d \tag{18} \\
\dot{\mathbf{A}}^h &= (\mathbf{W} \cdot \mathbf{A} - \mathbf{A} \cdot \mathbf{W}) + \xi(\mathbf{D} \cdot \mathbf{A} + \mathbf{A} \cdot \mathbf{D} - 2\mathbb{A} : \mathbf{D}) \tag{19} \\
\dot{\mathbf{A}}^d &= 2C_I \dot{\gamma}(\mathbf{I} - 3\mathbf{A}).. \tag{20}
\end{aligned}$$

The influence of the phenomenological parameter C_I is displayed in Figure 2. With decreasing diffusion (small C_I) fibers are more aligned.

Evolution equations of the second order orientation tensor $\mathbf{A}$ contain the fourth order fiber orientation tensor $\mathbb{A}$. It is possible to derive an equation for the fourth order orientation tensor, but it will contain the sixth order orientation tensor. This leads to an infinite series of evolution equations. Therefore, it is common to truncate the series at the level of the second order orientation tensor. Since the fourth order orientation tensor is then not explicitly computed, a so-called closure approximation is necessary for calculating the fourth order fiber orientation tensor. In Section 3.3 different closure approximations will be introduced.

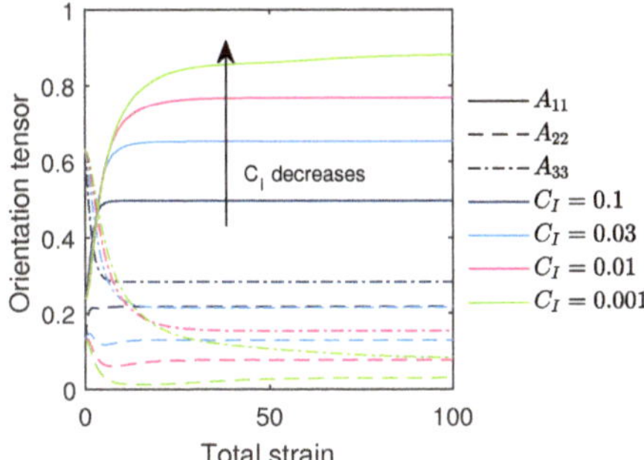

Figure 2. Influence of the phenomenological parameter C_I on fiber orientation evolution.

Three model enhancements have been made since the simulated fiber orientation shows deviation to fiber orientation determined experimentally.

To slow down the evolution speed, Huynh [12] introduced the strain reduction factor (SRF) $\frac{1}{\alpha}$ by multiplying equation (18) with $\alpha \in [0, 1]$

$$\dot{\mathbf{A}} = \alpha(\dot{\mathbf{A}}^{\mathrm{h}} + \dot{\mathbf{A}}^{\mathrm{d}}). \tag{21}$$

The SRF model violates the material objectivity, so Wang et al. [3] developed the reduced strain closure (RSC), based on the eigenvalue λ and eigenvector $\mathbf{e}$ decomposition, that is, $\mathbf{A} = \sum_{i=1}^{3} \lambda_i \mathbf{e}_i \mathbf{e}_i$ with orthonormal eigenvectors. The modified growth rate has the following form

$$\dot{\lambda}_i^{\mathrm{RSC}} = \kappa \dot{\lambda}_i \tag{22}$$

$$\dot{\mathbf{e}}_i^{\mathrm{RSC}} = \dot{\mathbf{e}}_i, \tag{23}$$

with the constant $\kappa \in [0, 1]$. In the differential equation form, this equals

$$\dot{\mathbf{A}} = \dot{\mathbf{A}}^{\mathrm{RSC}} + \kappa \dot{\mathbf{A}}^{\mathrm{d}} \tag{24}$$

$$\dot{\mathbf{A}}^{\mathrm{RSC}} = \mathbf{W} \cdot \mathbf{A} - \mathbf{A} \cdot \mathbf{W} + \zeta[\mathbf{D} \cdot \mathbf{A} + \mathbf{A} \cdot \mathbf{D} - 2(\mathbb{A} + (1 - \kappa)(\mathbb{L} - \mathbb{M} : \mathbb{A})) : \mathbf{D}] \tag{25}$$

$$\mathbb{L} = \sum_{i=1}^{3} \lambda_i \mathbf{e}_i \mathbf{e}_i \mathbf{e}_i \mathbf{e}_i \tag{26}$$

$$\mathbb{M} = \sum_{i=1}^{3} \mathbf{e}_i \mathbf{e}_i \mathbf{e}_i \mathbf{e}_i. \tag{27}$$

Figure 3 displays the influence of the phenomenological constant κ with a smaller retarding rate (smaller κ) a slower orientation evolution is reached. The steady state is unchanged by κ. Tseng et al. [4] introduced at the same time the retarding principal rate (RPR), modifying the growth rate of the eigenvalue by

$$\dot{\lambda}_i^{\mathrm{RPR}} = \kappa \dot{\lambda}_i + (1 - \kappa)\beta(\dot{\lambda}_i^2 + \dot{\lambda}_j \dot{\lambda}_k) \tag{28}$$

$$\dot{\mathbf{e}}_i^{\mathrm{RPR}} = \dot{\mathbf{e}}_i, \tag{29}$$

with a fine-tuning parameter β. For $\beta = 0$ this equals the RSC model equation (22). The differential equation form (similar to (24) with PRR) can be developed based on equations (28) and (29).

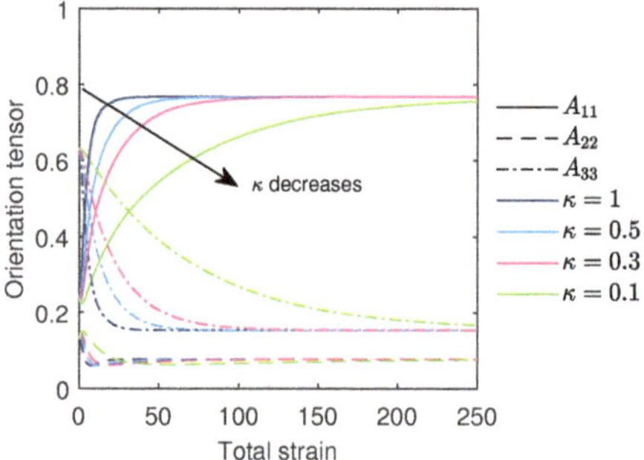

Figure 3. Influence of the phenomenological parameter κ on fiber orientation evolution. The diffusion constant is set to $C_I = 0.01$.

The second model approach, added by Phelps and Tucker [5], introduced an anisotropic rotary diffusion (ARD) term by replacing C_I with a rotary diffusion tensor $\mathbf{C}$. This allows spatially non-uniform rotary diffusion, which makes the rotary diffusion effect a function of the orientation state

$$\dot{\mathbf{A}} = \dot{\mathbf{A}}^{\mathrm{h}} + \dot{\mathbf{A}}^{\mathrm{ARD}} \tag{30}$$

$$\dot{\mathbf{A}}^{\mathrm{ARD}} = \dot{\gamma}[2\mathbf{C} - 2\mathrm{tr}(\mathbf{C})\mathbf{A} - 5(\mathbf{C}\cdot\mathbf{A} + \mathbf{A}\cdot\mathbf{C}) + 10\mathbb{A}:\mathbf{C}]. \tag{31}$$

The different modeling approaches for the anisotropic rotary diffusion are listed below

$$\mathbf{C} = \mathbf{C}(\mathbf{D},\mathbf{A}) = b_1\mathbf{I} + b_2\mathbf{A} + b_3\mathbf{A}^2 + \frac{b_4}{\dot{\gamma}}\mathbf{D} + \frac{b_5}{\dot{\gamma}^2}\mathbf{D}^2. \tag{32}$$

ARD model [5] with constants $b_i, i = 1, ..., 5$

$$\mathbf{C} = C_I\left(\mathbf{I} - 4C_M\frac{\mathbf{D}^2}{\dot{\gamma}}\right) \tag{33}$$

iARD model [4] with constants C_I, C_M

$$\mathbf{C} = C_I\mathbf{R}_A \begin{pmatrix} D_1 & 0 & 0 \\ 0 & D_2 & 0 \\ 0 & 0 & D_3 \end{pmatrix} \mathbf{R}_A^\top \tag{34}$$

pARD model [6] with constants C_I, D_1, D_2, D_3.

The eigenmatrix $\mathbf{R}_A$ is defined by

$$\mathbf{A} = \mathbf{R}_A \begin{pmatrix} \lambda_1 & 0 & 0 \\ 0 & \lambda_2 & 0 \\ 0 & 0 & \lambda_3 \end{pmatrix} \mathbf{R}_A^\top. \tag{35}$$

Tseng et al. [6] chose

$$D_1 = 1, D_2 = c, D_3 = 1 - c \tag{36}$$

to reduce the needed amount of parameters. This implies that the rotary diffusion factor in the first principal fiber orientation direction is defined by C_I. The second principal fiber orientation direction is scaled with $c\cdot C_I$ and in the third principal fiber orientation direction the smallest rotary diffusion factor with $(1 - c)C_I$ is applied. The influence od D_2 is displayed in Figure 4.

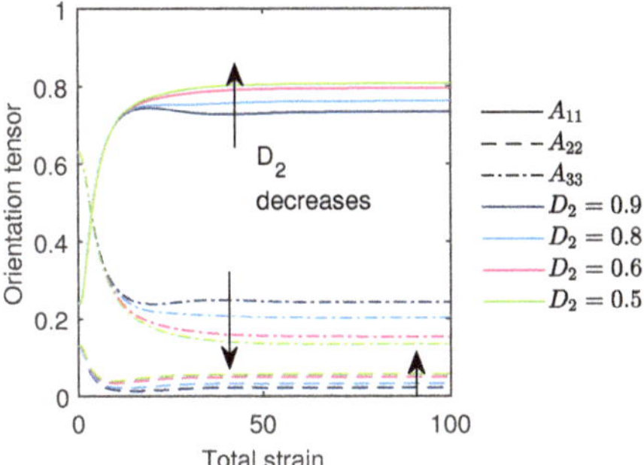

Figure 4. Influence of the phenomenological parameter D_2 on fiber orientation evolution. The diffusion is et to $C_I = 0.01$.

A slightly different approach to add anisotropic rotary diffusion is defined in the MRD model [7]. The rotary diffusion is modeled according to equation (34), but not the full ARD equation (31) is used. The reduced form is given by

$$\dot{\mathbf{A}} = \dot{\mathbf{A}}^{\mathsf{h}} + \dot{\mathbf{A}}^{\mathsf{mARD}} \tag{37}$$

$$\dot{\mathbf{A}}^{\mathsf{mARD}} = 2\dot{\gamma}(\mathbf{C} - \mathrm{tr}(\mathbf{C})\mathbf{A}). \tag{38}$$

Recently, Favaloro and Tucker [13] published a framework to compare anisotropic rotary diffusion approaches and compared the mentioned approaches in shear and elongation flows. They added a suggestion for a more general but stable model by using equation (34) but making D_i a function of λ_i or using equation (32) with $b_4 = b_5 = 0$ and $b_i, i = 1, 2, 3$ as scalar functions of $\mathrm{tr}\mathbf{A}^2$ and $\mathrm{tr}\mathbf{A}^3$.

Latz et al. [2] added a nematic potential to the Folgar-Tucker model to account for excluded volume effects. So far, there is no clear dependency of C_I on the volume fraction and aspect ratio of the fillers. Different experiments in different regimes even showed contradictory results [14–16]. Latz et al. [2] stated that the excluded volume mechanism may be a possible explanation for the observed effects. The integration of a second constant U_0 could decouple this effects and perhaps give a clear dependence on physical descriptors. The model is defined by

$$\dot{\mathbf{A}} = \dot{\mathbf{A}}^{\mathsf{h}} + \dot{\mathbf{A}}^{\mathsf{nem}} \tag{39}$$

$$\dot{\mathbf{A}}^{\mathsf{nem}} = (C_{\mathsf{I}}(\mathbf{I} - 3\mathbf{A}) + U_0(\mathbf{A} \cdot \mathbf{A} - \mathbf{A} : \mathbb{A}))\dot{\gamma}, \tag{40}$$

with one additional constant U_0.

In injection molding simulation combination of the methods are used. Combining Equations (24) and (31) leads to the following (p)ARD-RSC model

$$\dot{\mathbf{A}} = \dot{\mathbf{A}}^{RSC} + \dot{\mathbf{A}}^{(p)ARD-RSC} \tag{41}$$

$$\dot{\mathbf{A}}^{(p)ARD-RSC} = \dot{\gamma}[2[\mathbf{C} - (1-\kappa)\mathbb{M} : \mathbf{C}] - 2\kappa(\mathrm{tr}\mathbf{C})\mathbf{A} - 5(\mathbf{C} \cdot \mathbf{A} + \mathbf{A} \cdot \mathbf{C})$$
$$+ 10[\mathbb{A} - (1-\kappa)(\mathbb{L} - \mathbb{M} : \mathbb{A})] : \mathbf{C}]. \tag{42}$$

3.3. Closure Approximations

Since the fourth order orientation tensor is used in all models, closure approximations are needed. Any fourth order tensors with the symmetric properties stated in equation (13) can be represented by a 6×6 matrix with at most 36 independent entries [17]

$$\mathbb{A} = \begin{pmatrix} \mathbb{A}_{11} & \mathbb{A}_{12} & \mathbb{A}_{13} & \mathbb{A}_{14} & \mathbb{A}_{15} & \mathbb{A}_{16} \\ \mathbb{A}_{21} & \mathbb{A}_{22} & \mathbb{A}_{23} & \mathbb{A}_{24} & \mathbb{A}_{25} & \mathbb{A}_{26} \\ \mathbb{A}_{31} & \mathbb{A}_{32} & \mathbb{A}_{33} & \mathbb{A}_{34} & \mathbb{A}_{35} & \mathbb{A}_{36} \\ \mathbb{A}_{41} & \mathbb{A}_{42} & \mathbb{A}_{43} & \mathbb{A}_{44} & \mathbb{A}_{45} & \mathbb{A}_{46} \\ \mathbb{A}_{51} & \mathbb{A}_{52} & \mathbb{A}_{53} & \mathbb{A}_{54} & \mathbb{A}_{55} & \mathbb{A}_{56} \\ \mathbb{A}_{61} & \mathbb{A}_{62} & \mathbb{A}_{63} & \mathbb{A}_{64} & \mathbb{A}_{65} & \mathbb{A}_{66} \end{pmatrix}, \tag{43}$$

where $\mathbb{A}_{\alpha\beta} = \mathbb{A}_{ijkl}$ for $\alpha, \beta \in \{1,2,3,4,5,6\}$ the indices α and β represent an index pair ij or kl in the following way:

$$\alpha, \beta : \begin{cases} 1 & \rightarrow & 11 \\ 2 & \rightarrow & 22 \\ 3 & \rightarrow & 33 \\ 4 & \rightarrow & 23 \\ 5 & \rightarrow & 31 \\ 6 & \rightarrow & 12. \end{cases} \tag{44}$$

The first simple closure approaches have been introduced by Advani and Tucker [8]. The linear approach [8] is a summation of all products of a_{ij} and δ_{ij}. After applying symmetry and normalization condition the following linear approximation occurs

$$\begin{aligned} \mathbb{A}^{lin}_{ijkl} &= -\tfrac{1}{35}(\delta_{ij}\delta_{kl} + \delta_{ik}\delta_{jl} + \delta_{il}\delta_{jk}) \\ &\quad + \tfrac{1}{7}(\mathbf{A}_{ij}\delta_{kl} + \mathbf{A}_{ik}\delta_{jl} + \mathbf{A}_{il}\delta_{jk} + \mathbf{A}_{kl}\delta_{ij} + \mathbf{A}_{jl}\delta_{ik} + \mathbf{A}_{jk}\delta_{il}). \end{aligned} \tag{45}$$

The quadratic closure [8] omits all linear terms

$$\mathbb{A}^{quad}_{ijkl} = \mathbf{A}_{ij}\mathbf{A}_{kl}. \tag{46}$$

The quadratic closure does not preserve the symmetry of the fourth order tensor (equation (8)), but applying it to equation (19), it does preserve the symmetry of the second order tensor.

It is also possible to combine both approaches. This leads to the hybrid closure [8]

$$\mathbb{A}^{hyb}_{ijkl} = (1-f)\mathbb{A}^{lin}_{ijkl} + f\mathbb{A}^{quad}_{ijkl} \tag{47}$$

$$f = \frac{3}{2}\mathbf{A}_{ij}\mathbf{A}_{ji} - \frac{1}{2}. \tag{48}$$

Another approach to determine the function for the hybrid closure was determined by Advani and Tucker [9]

$$f = 1 - 27 \cdot \det \mathbf{A}. \tag{49}$$

A more advanced approach is the natural closure (NAT). It is based on the relationship $\mathbb{A} = f(\mathbf{A})$ when no diffusion occurs ($C_I = 0$).

Verley and Dupret [18] stated that there exists an exact closure in the case that the orientation is at one time isotropic and used a numerical approximation in the 3-D case to obtain manageable computational cost.

Montgomery-Smith et al. [19] determined the exact formulation in Cartesian coordinates on the sphere using the Carlson form of elliptic integrals. They used the analytic solution of the Jeffrey equation presented by Dinh and Armstrong [20]. The approach is based on the assumption of isotropic orientation at $t = 0$. Then the second order orientation tensor can be expressed by

$$\mathbf{A} = \int_S \frac{\mathbf{pp}}{4\pi(\mathbf{B}:\mathbf{pp})^{\frac{3}{2}}}d\mathbf{p} \tag{50}$$

$$\mathbf{B} = -\mathbf{B}\cdot(\mathbf{W}+\xi\mathbf{D})-(-\mathbf{W}+\xi\mathbf{D})\cdot\mathbf{B},\ \mathbf{B}=\mathbf{I}\ \text{at}\ t=0. \tag{51}$$

The method presented uses high computational effort, so Montgomery-Smith et al. [19] introduced the fast exact closure (FEC). The FEC introduced a computationally efficient way to compute the closure. Instead of computing $\mathbf{B}$ by inverting the integral, equations (18) and (51) are solved simultaneously. If the initial data is not isotropic, $\mathbf{B}$ has to be computed for the initial condition.

Later Montgomery-Smith et al. [21] also include the anisotropic rotary diffusion of Phelps and Tucker [5]. In this case $(\mathbf{C}\neq 0)$ the closure is not exact. The key idea is to introduce a matrix $\mathbb{B}$ and define two conversion tensors $\mathbb{C}$ and $\mathbb{D}$ such that

$$\frac{D\mathbf{A}}{Dt}=-\mathbb{C}:\frac{D\mathbf{B}}{Dt},\ \frac{D\mathbf{B}}{Dt}=-\mathbb{D}:\frac{D\mathbf{A}}{Dt}. \tag{52}$$

hold true. The ordinary differential equations (ODEs) are solved simultaneously, which can be computed very efficiently. The special form used for this approach is

$$\frac{D\mathbf{A}}{Dt}=-\mathbb{C}:F(\mathbf{B})+G(\mathbf{A}),\ \frac{D\mathbf{B}}{Dt}=F(\mathbf{B})-\mathbb{D}:G(\mathbf{A}). \tag{53}$$

They showed that it can also by applied to the reduced strain closure in equation (24) and proved that the solution stays physical for example, that $\mathbf{A}$ stays positive definite with $\operatorname{tr}\mathbf{A}=1$.

A large family of closure are fitted closures. Chung and Kwon [22] stated two ways to develop fitted closure. The method depends on the coordinate system, global or eigenspace, and the representation of the approximation, invariants or eigenvalues, of the orientation tensor. The orthotropic closures use the eigenspace coordinate system and eigenvalues as representatives. They are the most used family of closures. Cintra and Tucker [23] stated that an objective closure has to be orthotropic. Orthotropic means that the principal axes must match those of the second order tensor and each principal fourth order component is a function of just two principal values of the second order tensor.

The second order orientation tensor can be transformed in the eigenraum representation

$$\hat{\mathbf{A}}=\begin{pmatrix}\lambda_1 & 0 & 0\\ 0 & \lambda_2 & 0\\ 0 & 0 & \lambda_3\end{pmatrix}=\mathbf{R}_A^T\mathbf{A}\mathbf{R}_A, \tag{54}$$

with non-negative eigenvalues $\lambda_1 \geq \lambda_2 \geq \lambda_3 \geq 0$ and $\lambda_1+\lambda_2+\lambda_3=1$. Figure 5 shows the reference triangle for orthotropic closures.

Kuzmin [17] showed that regarding all symmetries of the fourth order orientation tensor the orthotropic representation has the form

$$\underline{\hat{\mathbb{A}}}=\begin{pmatrix}\hat{\mathbb{A}}_{11} & \hat{\mathbb{A}}_{12} & \hat{\mathbb{A}}_{13} & 0 & 0 & 0\\ \hat{\mathbb{A}}_{12} & \hat{\mathbb{A}}_{22} & \hat{\mathbb{A}}_{23} & 0 & 0 & 0\\ \hat{\mathbb{A}}_{13} & \hat{\mathbb{A}}_{23} & \hat{\mathbb{A}}_{33} & 0 & 0 & 0\\ 0 & 0 & 0 & \hat{\mathbb{A}}_{44} & 0 & 0\\ 0 & 0 & 0 & 0 & \hat{\mathbb{A}}_{55} & 0\\ 0 & 0 & 0 & 0 & 0 & \hat{\mathbb{A}}_{66}\end{pmatrix}. \tag{55}$$

It has been proven by Cintra and Tucker [23] and Kuzmin [17] that the fourth order orientation tensor can be represented by three independent components. This is due to the orthotropic properties, the symmetries of the second order tensor, the symmetry with respect to any pair of indices for the fourth order tensor and the normalization condition. The three components can be expressed as functions of the two largest eigenvalues of the second order tensor

$$\hat{\mathbb{A}}_{11} = f_1(\lambda_1, \lambda_2) \tag{56}$$
$$\hat{\mathbb{A}}_{22} = f_2(\lambda_1, \lambda_2) \tag{57}$$
$$\hat{\mathbb{A}}_{33} = f_3(\lambda_1, \lambda_2). \tag{58}$$

The remaining entries are defined by symmetry and normalization condition

$$\hat{\mathbb{A}}_{12} = \hat{\mathbb{A}}_{66} \tag{59}$$
$$\hat{\mathbb{A}}_{23} = \hat{\mathbb{A}}_{44} \tag{60}$$
$$\hat{\mathbb{A}}_{13} = \hat{\mathbb{A}}_{55} \tag{61}$$
$$\hat{\mathbb{A}}_{55} + \hat{\mathbb{A}}_{66} = \lambda_1 - \hat{\mathbb{A}}_{11} \tag{62}$$
$$\hat{\mathbb{A}}_{44} + \hat{\mathbb{A}}_{66} = \lambda_2 - \hat{\mathbb{A}}_{22} \tag{63}$$
$$\hat{\mathbb{A}}_{44} + \hat{\mathbb{A}}_{55} = \lambda_3 - \hat{\mathbb{A}}_{33}. \tag{64}$$

Cintra and Tucker [23] used the eigenspace coordinate system and eigenvalues as representatives to fit the closure (EBOF). Complete second order polynomials were used to approximate the components (ORF). The orthotropic fitted (ORF) closure and the NAT closure are based on identical assumptions. In conclusion they can be transferred to each other. In fact they are mathematically equivalent, but the ORF closure is numerically more stable for repeated eigenvalues [19] The three remaining components are fitted by

$$\hat{\mathbb{A}}_{ii} = f_i(\lambda_1, \lambda_2) = C_i^1 + C_i^2 \lambda_1 + C_i^3 \lambda_1^2 + C_i^4 \lambda_2 + C_i^5 \lambda_2^2 + C_i^6 \lambda_1 \lambda_2 \tag{65}$$

for for $i = 1, 2, 3$.

The components are fitted with different flow types (simple shear, two shearing/stretching flows, uniaxial elongation, biaxial elongation) using a least squares routine. The closure approximation shows good agreement with the distribution function and the NAT and better performance than the hybrid closure. For small C_I the ORF closure shows oscillating behavior.

Chung and Kwon [24] improved the ORF closure to overcome the oscillation for small C_I values and introduced the orthotropic fitted closure approximation for wide interaction coefficients (OWE) and orthotropic fitted closure approximation for wide interaction coefficients with third order polynomial approximations (OWE3) closure. The OWE closure is fitted with two additional flow fields (shear/planar elongation, balanced shear/biaxial elongation) to cover the orientation triangle $\hat{K}$ (Figure 5) more closely. Consequently the only difference between the ORF and OWE closure are the values of the fitted parameters. The OWE3 closure uses the same flows for the parameter fit but approximates the coefficients values by a third order polynomial expression

$$\begin{aligned}\hat{\mathbb{A}}_{ii} &= f_i(\lambda_1, \lambda_2) = C_i^1 + C_i^2 \lambda_1 + C_i^3 \lambda_1^2 + C_i^4 \lambda_2 + C_i^5 \lambda_2^2 + C_i^6 \lambda_1 \lambda_2 \\ &\quad + C_i^7 \lambda_1^2 \lambda_2 + C_i^8 \lambda_1 \lambda_2^2 + C_i^9 \lambda_1^3 + C_i^{10} \lambda_2^3\end{aligned} \tag{66}$$

for for $i = 1, 2, 3$. The closures show stable behavior for a wide range of C_I but tend to over predict the orientation.

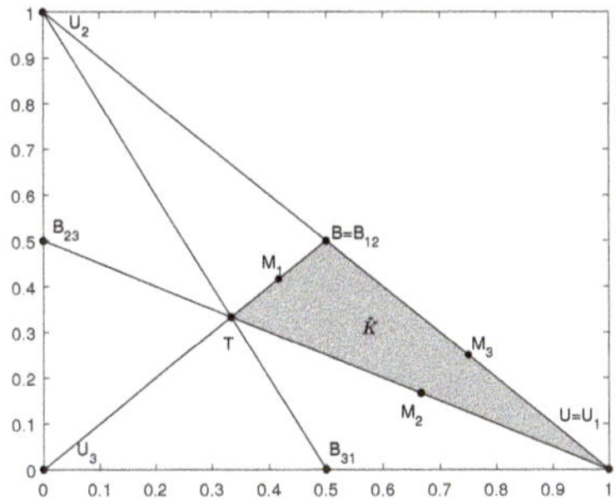

Figure 5. Reference triangle for orthotropic closure.

Kuzmin [17] introduced a mathematical concept to develop orthotropic closures. A concept for planar orientations was developed and extended to the 3D case. In this work only the 3D case is explained, for the planar case refer to Reference [17] The linear and smooth closures were stated in the orthotropic state [17]. An orthotropic version of the quadratic closure was developed

$$\hat{\mathbb{A}}_{11} = f_1(\lambda_1, \lambda_2) = \lambda_1^2 \tag{67}$$

$$\hat{\mathbb{A}}_{22} = f_2(\lambda_1, \lambda_2) = \lambda_2^2 \tag{68}$$

$$\hat{\mathbb{A}}_{33} = f_3(\lambda_1, \lambda_2) = (1 - \lambda_1 - \lambda_2)^2. \tag{69}$$

Since there does not exist an analytical form of the standard NAT, natural closures based on extended quadratic and piecewise linear interpolation have been developed.

The extended quadratic fit is fitted on cubic polynomials of the form

$$\begin{aligned}
\hat{\mathbb{A}}_{ii} = \ & f_i(\lambda_1, \lambda_2) = C_i^1 + C_i^2 \lambda_1 + C_i^3 \lambda_1^2 + C_i^4 \lambda_2 + C_i^5 \lambda_2^2 + C_i^6 \lambda_1 \lambda_2 \\
& + C_i^7 \lambda_1 \lambda_2 (1 - \lambda_1 - \lambda_2).
\end{aligned} \tag{70}$$

The data points U_i, B_{ij}, $i = 1, 2, 3$ $j = i + 1$ for $i = 1, 2$, $j = 1$ for $i = 3$ and T are used, using the planar natural closure and the triaxial orientation state at T. For extrapolation the extended quadratic finite element method is used.

For the exact midpoint fit quadratic interpolation polynomials are used

$$\hat{\mathbb{A}}_{ii} = f_i(\lambda_1, \lambda_2) = C_i^1 + C_i^2 \lambda_1 + C_i^3 \lambda_1^2 + C_i^4 \lambda_2 + C_i^5 \lambda_2^2 + C_i^6 \lambda_1 \lambda_2. \tag{71}$$

The values at points U, B, T and the midpoints M_1, M_2, M_3 are validated using the exact closure.

In contrast to all eigenspace based closures, Reference [22] introduced the invariants-based optimal fitted (IBOF) closure in the global coordinate system with invariants as representatives. This closure is computational more efficient than the orthotropic closures.

4. Microscopic Fiber Simulation

Simulation methods on the microscopic level can be used to simulate fiber movement for discretized fibers. In contrast to the phenomenological macroscopic fiber orientation models, modeling approaches on the microscopic scale enable the approximation of the physical behavior more accurately. However, they are computationally very expensive and are, in most cases, not suitable for the simulation of actual parts. Mostly, they are used on reference elements to investigate physical effects and quantities. Different modeling approaches can be used on the micro level. They can coarsely be divided in particle based methods, where the polymer matrix and the fibers are treated as particles for example the smoothed particle hydrodynamics (SPH) and moving particle semi-implicit

(MPS) method, or element based methods, which treat fibers as particles and the matrix is treated as a continuous media.

One big advantage of particle based method, is the computationally cheap coupling between fluid motion and fibers in two ways. Yashiro et al. [25,26] developed a method based on MPS to predict fiber movement in injection molding. They simulated complex flow fields and investigated orientation in a T-shaped bifurcation.

The SPH method is also applied to polymer composites. He et al. [27] simulated the 3D injection molding process for short-fiber reinforced polymers; Bertevas et al. [28] simulated the 3D printing process of fiber-reinforced polymer and Yamagata and Ichimiya [29] used the method to simulate the solidification process for injection molded short-fiber reinforced parts.

The SPH method can also be combined with the discrete element method (DEM) [30], or the element bending group (EBG) method [31,32]. The fluid is then model by the SPH method and the fibers by the respective method.

Element based methods are mostly developed on the principle of the DEM. In contrast to particle based methods, which are per se two-way coupled, DEM based simulations are often solved one-way coupled. The backcoupling from the fiber motion on the fluid is computationally expensive, but can be integrated in a coarse or refined way. The resulting fluid equation can be solved by multiple approaches such as the direct numerical simulation (DNS), the Latice-Bolzmann or the particle finite element analysis (pFEA). In the DEM, fibers are considered as particles and the movement of each particle is calculated by the solution of the force and torque balance acting on each particle. Fibers are either discretized as chain of spheres, prolate spheroids or chain of rods. The forces acting on a fiber are

- hydrodynamic forces
- fiber fiber interaction forces
- elastic and bending forces (intra fiber forces)

Hydrodynamic forces are exerted from the fluid on the fiber. The fluid motion can be considered undisturbed by the fiber motion or disturbed. In the second case a backcoupling is necessary. The interaction forces can be divided in two cases: long-range hydrodynamic interaction and short range interaction. The short range interactions can then be divided in three regimes: short range lubrication forces, transition and mechanical contact. Fibers can be modeled flexible by using chains of beads or rods connected by joints. Breaking can also be incorporated at the defined joints. The modeling of the forces, the discretization of the fibers and the fluid motion varies between the different approaches and the following Table 1 gives an overview of published approaches, without claim for completeness. Only inertia free models are considered. Examples for models incorporating inertia are References [33–35].

Table 1. Literature overview of element based simulations for fiber reinforced polymers in chronological order.

	Discretization of Fibers	Flow Fields	Fluid-Fiber Interaction	Fiber-Fiber Interaction	Flexibility	Regarded Quantities
Yamamoto and Matsuoka 1993 [36]	chain of beads	shear	one-way coupled	-	flexible	single fiber movement
Yamamoto and Matsuoka 1994 [37]	chain of beads	shear	one-way coupled	-	flexible	viscosity of dilute solutions
Yamane et al., 1994 [38]	rods	shear	one-way coupled	lubrication	-	semi dilute suspensions, orientation evolution, diffusion constant, shear viscosity
Yamane et al., 1995 [39]	rods	shear	one-way coupled	lubrication	-	semi dilute suspensions, bounded and unbounded system
Yamamoto and Matsuoka 1995 [40]	chain of beads	shear	one-way coupled	lubrication	flexible	concentrated suspension, viscosity, stresses
Thomasset et al., 1997 [41]	rigid rods	varies flow fields	one-way coupled	lubrication, mechanical and hydrodynamical contact, no friction	-	2D,effects of fiber motion and orientation
Sundararajakumar and Koch 1997 [42]	rods	shear	one-way coupled	lubrication, mechanical and hydrodynamical contact, no friction	-	dilute: hydrodynamical contact most important, approaching higher concentration fiber contact
Skjetne et al., 1997 [43]	prolate spheroids	shear	one-way coupled	-	flexible, rigid	single fiber movement
Ross and Klingenberg 1997 [44]	prolate spheroids	shear	one-way coupled	repulsive interactions	flexible and rigid	single fiber movement, viscosity
Fan et al., 1998 [45]	rods	shear	one-way coupled	lubrication, no friction, long range hydrodynamic interactions by slender body theory	-	orientation, viscosity, stresses, all regimes
Harlen et al., 1999 [46]	rods	no imposed flow	-	mechanical contact, friction, long range hydrodynamic interactions by slender body theory	-	sphere settling through suspension of neutrally buoyant fibers, fiber contact has significant influence

Table 1. *Cont.*

	Discretization of Fibers	Flow Fields	Fluid-Fiber Interaction	Fiber-Fiber Interaction	Flexibility	Regarded Quantities
Phan-Thien et al., [14]	rods	shear	one-way coupled	lubrication, no friction, long range hydrodynamic interactions by slender body theory	-	FT constant, dilute and semi dilute
Joung et al., 2001 [47]	chain of spherical beads	shear and extensional flows	one-way coupled	lubrication, preventing from overlapping, long range hydrodynamic interactions	rigid, flexible	viscosity, orientation
Joung et al., 2002 [48]	chain of spherical beads	shear and complex flows	one-way coupled	lubrication, preventing from overlapping, long range hydrodynamic interactions	rigid, curved	viscosity for curved fibers
Joung et al., 2003 [49]	chain of spherical beads	shear and complex flows	one-way coupled	lubrication, preventing from overlapping, long range hydrodynamic interactions	flexible	Jeffrey orbits for rigid and flexible fibers, relationship between fiber stiffness, and bulk viscosity, arbitrary particle shapes, dilute regime
Switzer and Klingenberg 2003 [50]	chain of rods	shear	one-way coupled	mechanical interaction, friction	flexible	effects of shape, friction, aspect ratio and stiffness, yield stress, rheology in flocculated systems
Kromkamp et al., 2005 [51]	rods	shear	coupled, Lattice Bolzmann for fluid forces,particles as boundary surfaces	lubrication correction,mechanical interaction, no friction	-	2D, effects of shear rate on flow behavior and micro structure, shear-induced self diffusion
Ausias et al., 2006 [52]	rigid prolate spheroids	shear	one-way coupled	lubrication and interaction in normal direction, no friction,no long range interactions	-	orientation, viscosity, stresses, up to $\Phi = 11.5$
Wang et al., 2006 [53]	rod-chain	shear	one-way coupled		flexible, rigid	optimal rod length for high accuracy and efficient calculation

Table 1. *Cont.*

	Discretization of Fibers	Flow Fields	Fluid-Fiber Interaction	Fiber-Fiber Interaction	Flexibility	Regarded Quantities
Lindström and Uesaka 2007 [54]	rod-chain model	shear	coarse two-way coupling	lubrication and interaction in normal direction, friction	flexible	Jeffrey orbits, curvature, regimes of motions for flexible fibers
Lindström and Uesaka 2008 [55]	rod-chain model	shear	coarse two-way coupling	lubrication and interaction in normal direction, friction	flexible	orientation, viscosity, dilute and semidilute regime, rheological properties
Lindström and Uesaka 2009 [56]	rod-chain model	shear	coarse two-way coupling	lubrication and interaction in normal direction, friction	flexible	rheological properties
Yamanoi and Maia 2010 [57]	chain of beads	shear	one-way coupling	lubrication, mechanical contact, long range hydrodynamic interactions	-	rheological properties, orientation
Yamanoi et al., 2010 [58]	chain of beads	shear	one-way coupling	lubrication, mechanical contact, long range hydrodynamic interactions	flexible	nylon fiber, rheological properties, orientation, effect of flexibility
Yamanoi and Maia 2010 [59]	chain of beads	uniaxel elongation flow	one-way coupling	lubrication, mechanical contact, long range hydrodynamic interactions	flexible	rheological properties, orientation, orientation tensor independent of aspect ratio, volume fraction
Yamanoi and Maia 2011 [60]	chain of beads	shear	two way coupling	single fiber	rigid and flexible	hydrodynamic interaction in single fiber movement in shear
Andrić et al., 2013 [61]	rod-chain model	turbulent flow	two way coupling, DNS for fluid motion	single fiber	rigid and flexible	fiber-flow interaction for a single fiber
Andrić et al., 2014 [62]	rod-chain model	shear	two way coupling, DNS for fluid motion	-	rigid and flexible	dilute solution, no interaction, rheological properties, orbit drifts
Do-Quang et al., 2014 [63]	rod-chain model	turbulent flow	two way coupling, entropy lattice Boltzmann for fluid, external boundary force method	lubrication and mechanical contact	rigid	cellulose fibers in water, accumulation effects

Table 1. *Cont.*

	Discretization of Fibers	Flow Fields	Fluid-Fiber Interaction	Fiber-Fiber Interaction	Flexibility	Regarded Quantities
Mezher et al., 2015 [64]	prolate spheroids	shear	one-way coupling	lubrication and interaction in normal direction,no friction, no long range hydrodynamics	flexible	concentrated, orientation, normalized stresses, interactions, elastic energy
Mezher et al., 2016[65]	prolate spheroids	shear	one-way coupling	lubrication and interaction in normal direction,no friction, no long range hydrodynamics	flexible	concentrated $\Phi = 7 - 18.2$, orientation, diffusion constants, confinement effects
Wang et al., 2016 [66]	rod-chain model	shear	one-way coupled	-	flexible	new rod chain model, optimal rod length
Perez et al., 2016 [67]	rod	shear	one-way coupling	only wall interaction	-	dilute, confinement effects
Sasayama and Inagaki 2017 [68]	simplified bead-chain model	shear	one-way	mechanical, lubrication	flexible	simplified bead-chain model for hydrodynamic calculations
Kuhn et al., 2017 [69]	rod-chain model	complex flow fields	one-way coupling	mechanical, friction, no long-range hydrodynamic	flexible	fiber matrix separation in compression molding, LFRT
Kuhn et al., 2018 [70]	rod-chain model	complex flow fields	one-way coupling	mechanical, friction, no long-range hydrodynamic	flexible	rib filling
Meirson and Hrymak 2018 [71]	rod-chain model	squeeze	one-way	-	flexible	2D, fiber orientation and deformation
Wu et al., 2018 [30]	bounded spheres	complex flow	2-way coupled,SPH for fluid motion	linear contact	-	2D, fiber orientation, accumulation during injection molding
Sasayama and Inagaki 2019 [72]	efficient bead-chain model	shear	one-way	mechanical, lubrication, friction	flexible	efficient bead-chain model for hydrodynamic calculations
Sasayama et al., 2019 [73]	efficient bead-chain mode	shear	one-way	mechanical, lubrication, friction	flexible, breakage	fiber breakage
Laurentcin et al., 2019 [74]	sphero-cylinder	squeeze flow (lubricated, non-lubricated)	one-way	-	rigid	non-Newtonian fluid, dilute regime, comparison between numerical analytical and experimental results

The listed approaches show, that depending on the evaluated quantities, the matrix and fiber material and the volume fraction, different modeling approaches show more promising results. In case of rigid fibers, long range hydrodynamic interactions show the highest influence in semidilute solution, whereas mechanical interaction gets dominant in the concentrated regime and long-range hydrodynamic interaction can be neglected [42,46,57]. The backcoupling has neglectable influence on rigid single fiber movement [60,75]. Once flexibility of the fiber is higher the backcoupling cannot be neglected [60]. Due to the high computational effort it has not be used in the concentrated regimen. If there is a significant influence in the concentrated regime can not be answered.

Based on the movement of single fibers, orientation evolution curves can be calculated. With the discrete position of each fiber the second order orientation tensor can be calculated in each time step by

$$A_{ij}(t) = \sum_{n=1}^{N} \frac{p_{n,j}(t)p_{n,i}(t)}{N}, \tag{72}$$

where N denotes the number of fibers, t the actual time, and $p_n(t)$ the position of fiber n at time t.

5. Using Microscopic Models for an Enhanced Prediction on the Macroscopic Scale

A combination of two scales can enhance the prediction on the macroscopic scale, while being computationally cheap. A straight forward approach is to use the microscopic simulation for parameter definition of existing macroscopic model. This has been done by many authors, for example Reference [65,75]. Microscopic fiber orientation evolution results can also lead to new macroscopic models [67]. A different approach is to create orientation data with a microscopic simulation and use a machine learning based approach on the macroscopic level [76].

6. Summary

The fiber orientation phenomenon can be simulated either on a macro- or micro-scale by using current numerical techniques and state-of-the-art modelling approaches. Macroscopic fiber orientation models are computationally efficient and are the preferred solution for estimating fiber orientation in large industrial applications. The limitation of those models is their phenomenological nature and the dependency on a set of fitting parameters. The challenge is typically the correct choice of parameters for a specific material. On the other hand, microscopic fiber models have a larger physical basis, but since fibers are explicitly discretized they are not suitable for large-scale simulations, because they are computational expensive. A combination of the models from the two scales can relieve the individual shortcomings and provides an interesting numerical solution for the virtual engineering design of fiber-reinforced polymer parts. One possibility is the prediction of optimal macroscopic parameters by a microscopic simulation, or the derivation of a macroscopic data-driven model using microscopic simulation results as input data.

Author Contributions: Conceptualization, S.K.K.; Writing—Original Draft Preparation, S.K.K.; Writing—Review and Editing, T.O., A.K. and C.C.; Visualization, S.K.K.; Supervision, T.O., A.K. and C.C. All authors have read and agreed to the published version of the manuscript.

Funding: This research received no external funding.

Conflicts of Interest: The authors declare no conflict of interest.

Abbreviations

The following abbreviations are used in this manuscript:

PDF	probability density function
ODE	ordinary differential equation
FT	Folgar-Tucker

nem	nematic
SRF	strain reduction factor
RSC	reduced strain closure
RPR	retarding principle rate
ARD	anisotropic rotary diffusion
iARD	improved anisotropic rotary diffusion
pARD	principal anisotropic rotary diffusion
MRD	Moldflow rotary diffusion
NAT	natural closure
FEC	fast exact closure
IBOF	invariant-based optimal fitted
EBOF	eigenvalue-based optimal fitted
ORF	orthotropic fitted
OWE	orthotropic fitted closure approximation for wide interaction coefficients
OWE3	orthotropic fitted closure approximation for wide interaction coefficients with third order polynomial approximation
SPH	smoothed particle hydrodynamic
DEM	discrete element method
pFEA	particle finite element analysis
DNS	direct numerical simulation
MPS	moving particle semi-implicit
EBG	element bending group

References

1. Folgar, F.; Tucker, C.L. Orientation Behavior of Fibers in Concentrated Suspensions. *J. Reinf. Plast. Compos.* **1984**, *3*, 98–119, doi:10.1177/073168448400300201. [CrossRef]
2. Latz, A.; Strautins, U.; Niedziela, D. Comparative numerical study of two concentrated fiber suspension models. *J. Non-Newton. Fluid Mech.* **2010**, *165*, 764–781, doi:10.1016/j.jnnfm.2010.04.001. [CrossRef]
3. Wang, J.; O'Gara, J.F.; Tucker, C.L. An objective model for slow orientation kinetics in concentrated fiber suspensions: Theory and rheological evidence. *J. Rheol.* **2008**, *52*, 1179–1200, doi:10.1122/1.2946437. [CrossRef]
4. Tseng, H.C.; Chang, R.Y.; Hsu, C.H. Phenomenological improvements to predictive models of fiber orientation in concentrated suspensions. *J. Rheol.* **2013**, *57*, 1597–1631, doi:10.1122/1.4821038. [CrossRef]
5. Phelps, J.H.; Tucker, C.L., III. An anisotropic rotary diffusion model for fiber orientation in short- and long-fiber thermoplastics. *J. Non-Newton. Fluid Mech.* **2009**, *156*, 165–176, doi:10.1016/j.jnnfm.2008.08.002. [CrossRef]
6. Tseng, H.C.; Chang, R.Y.; Hsu, C.H. The use of principal spatial tensor to predict anisotropic fiber orientation in concentrated fiber suspensions. *J. Rheol.* **2017**, *62*, 313–320, doi:10.1122/1.4998520. [CrossRef]
7. Bakharev, A.; Yu, H.; Ray, S.; Speight, R.; Wang, J. *Using New Anisotropic Rotational Diffusion Model to Improve Prediction of Short Fibers in Thermoplastic Injection Molding*; ANTEC: Orlando, FL, USA, 2018.
8. Advani, S.G.; Tucker, C.L. The Use of Tensors to Describe and Predict Fiber Orientation in Short Fiber Composites. *J. Rheol.* **1987**, *31*, 751–784, doi:10.1122/1.549945. [CrossRef]
9. Advani, S.G.; Tucker, C.L. Closure approximations for three-dimensional structure tensors. *J. Rheol.* **1990**, *34*, 367–386, doi:10.1122/1.550133. [CrossRef]
10. Jeffery, G.B. The motion of ellipsoidal particles immersed in a viscous fluid. *Proc. R. Soc. Lond. A Math. Phys. Eng. Sci.* **1922**, *102*, 161–179.
11. Junk, M.; Illner, R. A New Derivation of Jeffery's Equation. *J. Math. Fluid Mech.* **2007**, *9*, 455–488. [CrossRef]
12. Huynh, H. M. Improved Fiber Orientation Prediction For Injection-Molded Composites. Master's Thesis, University of Illinois Urbana-Champaign, Champaign County, IL, USA, 2001.
13. Favaloro, A.J.; Tucker, C.L. Analysis of anisotropic rotary diffusion models for fiber orientation. *Compos. Part A Appl. Sci. Manuf.* **2019**, *126*, 105605, doi:10.1016/j.compositesa.2019.105605. [CrossRef]
14. Phan-Thien, N.; Fan, X.J.; Tanner, R.; Zheng, R. Folgar–Tucker constant for a fibre suspension in a Newtonian fluid. *J. Non-Newton. Fluid Mech.* **2002**, *103*, 251–260, doi:10.1016/S0377-0257(02)00006-X. [CrossRef]

15.	Bay, R.S. Fiber Orientation in Injection-Molded Composites: A Comparison of Theory and Experiment. Ph.D. Thesis, Mechanical Science and Engineering, University of Illinois at Urbana-Champaign, Champaign County, IL, USA, 1991.

16.	Petrich, M.P.; Koch, D.L.; Cohen, C. An experimental determination of the stress–microstructure relationship in semi-concentrated fiber suspensions. *J. Non-Newton. Fluid Mech.* **2000**, *95*, 101–133, doi:10.1016/S0377-0257(00)00172-5. [CrossRef]

17.	Kuzmin, D. Planar and orthotropic closures for orientation tensors in fiber suspension flow models. *SIAM J. Appl. Math.* **2018**, *78*, 3040–3059. [CrossRef]

18.	Verley, V.; Dupret, F. Numerical prediction of the fiber orientation in complex injection molded parts. *Trans. Eng. Sci.* **1994**, doi:10.2495/CP940341. [CrossRef]

19.	Montgomery-Smith, S.; He, W.; Jack, D.; Smith, D. Exact tensor closures for the three-dimensional Jeffery's equation. *J. Fluid Mech.* **2011**, *680*, 321–335, doi:10.1017/jfm.2011.165. [CrossRef]

20.	Dinh, S.M.; Armstrong, R.C. A Rheological Equation of State for Semiconcentrated Fiber Suspensions. *J. Rheol.* **1984**, *28*, 207–227, doi:10.1122/1.549748. [CrossRef]

21.	Montgomery-Smith, S.; Jack, D.; Smith, D.E. The Fast Exact Closure for Jeffery's equation with diffusion. *J. Non-Newton. Fluid Mech.* **2011**, *166*, 343–353, doi:10.1016/j.jnnfm.2010.12.010. [CrossRef]

22.	Chung, D.H.; Kwon, T.H. Invariant-based optimal fitting closure approximation for the numerical prediction of flow-induced fiber orientation. *J. Rheol.* **2002**, *46*, 169–194, doi:10.1122/1.1423312. [CrossRef]

23.	Cintra, J.S.; Tucker, C.L. Orthotropic closure approximations for flow-induced fiber orientation. *J. Rheol.* **1995**, *39*, 1095–1122, doi:10.1122/1.550630. [CrossRef]

24.	Chung, D.H.; Kwon, T.H. Improved model of orthotropic closure approximation for flow induced fiber orientation. *Polym. Compos.* **2001**, *22*, 636–649, doi:10.1002/pc.10566. [CrossRef]

25.	Yashiro, S.; Okabe, T.; Matsushima, K. A Numerical Approach for Injection Molding of Short-Fiber-Reinforced Plastics Using a Particle Method. *Adv. Compos. Mater.* **2011**, *20*, 503–517, doi:10.1163/092430411X584423. [CrossRef]

26.	Yashiro, S.; Sasaki, H.; Sakaida, Y. Particle simulation for predicting fiber motion in injection molding of short-fiber-reinforced composites. *Compos. Part A Appl. Sci. Manuf.* **2012**, *43*, 1754–1764, doi:10.1016/j.compositesa.2012.05.002. [CrossRef]

27.	He, L.; Lu, G.; Chen, D.; Li, W.; Lu, C. Three-dimensional smoothed particle hydrodynamics simulation for injection molding flow of short fiber-reinforced polymer composites. *Model. Simul. Mater. Sci. Eng.* **2017**, *25*, 055007, doi:10.1088/1361-651X/aa6dc9. [CrossRef]

28.	Bertevas, E.; Férec, J.; Khoo, B.C.; Ausias, G.; Phan-Thien, N. Smoothed particle hydrodynamics (SPH) modeling of fiber orientation in a 3D printing process. *Phys. Fluids* **2018**, *30*, 103103, doi:10.1063/1.5047088. [CrossRef]

29.	Yamagata, N.; Ichimiya, M. Numerical Approach of Viscous Flow Containing Short Fiber by SPH Method. In *Computational and Experimental Simulations in Engineering*; Okada, H., Atluri, S.N., Eds.; Springer International Publishing: Cham, Switzerland, 2020; Volume 75, pp. 301–307, doi:10.1007/978-3-030-27053-7_28. [CrossRef]

30.	Wu, K.; Wan, L.; Zhang, H.; Yang, D. Numerical simulation of the injection molding process of short fiber composites by an integrated particle approach. *Int. J. Adv. Manuf. Technol.* **2018**, *97*, 3479–3491, doi:10.1007/s00170-018-2204-6. [CrossRef]

31.	Yang, X.; Liu, M.; Peng, S. Smoothed particle hydrodynamics and element bending group modeling of flexible fibers interacting with viscous fluids. *Phys. Rev. E* **2014**, *90*, doi:10.1103/PhysRevE.90.063011. [CrossRef]

32.	Yang, X.; Liu, M.B. Bending modes and transition criteria for a flexible fiber in viscous flows. *J. Hydrodyn.* **2016**, *28*, 1043–1048, doi:10.1016/S1001-6058(16)60709-6. [CrossRef]

33.	Challabotla, N.R.; Zhao, L.; Andersson, H.I. On fiber behavior in turbulent vertical channel flow. *Chem. Eng. Sci.* **2016**, *153*, 75–86, doi:10.1016/j.ces.2016.07.002. [CrossRef]

34.	Dotto, D.; Marchioli, C. Orientation, distribution, and deformation of inertial flexible fibers in turbulent channel flow. *Acta Mech.* **2019**, *230*, 597–621, doi:10.1007/s00707-018-2355-4. [CrossRef]

35.	Njobuenwu, D.O.; Fairweather, M. Simulation of inertial fibre orientation in turbulent flow. *Phys. Fluids* **2016**, *28*, 063307, doi:10.1063/1.4954214. [CrossRef]

36.	Yamamoto, S.; Matsuoka, T. A method for dynamic simulation of rigid and flexible fibers in a flow field. *J. Chem. Phys.* **1993**, *98*, 644–650, doi:10.1063/1.464607. [CrossRef]

37. Yamamoto, S.; Matsuoka, T. Viscosity of dilute suspensions of rodlike particles: A numerical simulation method. *J. Chem. Phys.* **1994**, *100*, 3317–3324, doi:10.1063/1.466423. [CrossRef]

38. Yamane, Y.; Kaneda, Y.; Dio, M. Numerical simulation of semi-dilute suspensions of rodlike particles in shear flow. *J. Non-Newton. Fluid Mech.* **1994**, *54*, 405–421, doi:10.1016/0377-0257(94)80033-2. [CrossRef]

39. Yamane, Y.; Kaneda, Y.; Doi, M. The Effect of Interaction of Rodlike Particles in Semi-Dilute Suspensions under Shear Flow. *J. Phys. Soc. Jpn* **1995**, *64*, 3265–3274, doi:10.1143/JPSJ.64.3265. [CrossRef]

40. Yamamoto, S.; Matsuoka, T. Dynamic simulation of fiber suspensions in shear flow. *J. Chem. Phys.* **1995**, *102*, 2254–2260, doi:10.1063/1.468746. [CrossRef]

41. Thomasset, J.; Grmela, M.; Carreau, P.J. Microstructure and rheology of polymer melts reinforced by long glass fibres: direct simulations. *J. Non-Newton. Fluid Mech.* **1997**, *73*, 195–203. [CrossRef]

42. Sundararajakumar, R.R.; Koch, D.L. Structure and properties of sheared fiber suspensions with mechanical contacts. *J. Non-Newton. Fluid Mech.* **1997**, *73*, 205–239. [CrossRef]

43. Skjetne, P.; Ross, R.F.; Klingenberg, D.J. Simulation of single fiber dynamics. *J. Chem. Phys.* **1997**, *107*, 2108–2121, doi:10.1063/1.474561. [CrossRef]

44. Ross, R.F.; Klingenberg, D.J. Dynamic simulation of flexible fibers composed of linked rigid bodies. *J. Chem. Phys.* **1997**, *106*, 2949–2960, doi:10.1063/1.473067. [CrossRef]

45. Fan, X.; Phan-Thien, N.; Zheng, R. A direct simulation of fibre suspensions. *J. Non-Newton. Fluid Mech.* **1998**, *74*, 113–135. [CrossRef]

46. Harlen, O.G.; Sundararajakumar, R.R.; Koch, D.L. Numerical simulations of a sphere settling through a suspension of neutrally buoyant fibres. *J. Fluid Mech.* **1999**, *388*, 355–388, doi:10.1017/S0022112099004929. [CrossRef]

47. Joung, C.G.; Phan-Thien, N.; Fan, X.J. Direct simulation of flexible fibers. *J. Non-Newton. Fluid Mech.* **2001**, *99*, 1–36. [CrossRef]

48. Joung, C.G.; Phan-Thien, N.; Fan, X.J. Viscosity of curved fibers in suspension. *J. Non-Newton. Fluid Mech.* **2002**, *102*, 1–17. [CrossRef]

49. Joung, C.G. Direct Simulation Studies of Suspended Particles and Fibre-Filled Suspensions. Ph.D. Thesis, School of Aerospace, Mechanical and Mechatronic Engineering The University of Sydney, Sydney, Australia, 2003.

50. Switzer, L.H.; Klingenberg, D.J. Rheology of sheared flexible fiber suspensions via fiber-level simulations. *J. Rheol.* **2003**, *47*, 759–778, doi:10.1122/1.1566034. [CrossRef]

51. Kromkamp, J.; Van Den Ende, D.T.M.; Kandhai, D.; Van Der Sman, R.G.M.; Boom, R.M. Shear-induced self-diffusion and microstructure in non-Brownian suspensions at non-zero Reynolds numbers. *J. Fluid Mech.* **2005**, *529*, 253–278, doi:10.1017/S0022112005003551. [CrossRef]

52. Ausias, G.; Fan, X.; Tanner, R. Direct simulation for concentrated fibre suspensions in transient and steady state shear flows. *J. Non-Newton. Fluid Mech.* **2006**, *135*, 46–57, doi:10.1016/j.jnnfm.2005.12.009. [CrossRef]

53. Wang, G.; Yu, W.; Zhou, C. Optimization of the rod chain model to simulate the motions of a long flexible fiber in simple shear flows. *Eur. J. Mech.-B/Fluids* **2006**, *25*, 337–347, doi:10.1016/j.euromechflu.2005.09.004. [CrossRef]

54. Lindström, S.B.; Uesaka, T. Simulation of the motion of flexible fibers in viscous fluid flow. *Phys. Fluids* **2007**, *19*, 113307, doi:10.1063/1.2778937. [CrossRef]

55. Lindström, S.B.; Uesaka, T. Simulation of semidilute suspensions of non-Brownian fibers in shear flow. *J. Chem. Phys.* **2008**, *128*, 024901, doi:10.1063/1.2815766. [CrossRef]

56. Lindström, S.B.; Uesaka, T. A numerical investigation of the rheology of sheared fiber suspensions. *Phys. Fluids* **2009**, *21*, 083301, doi:10.1063/1.3195456. [CrossRef]

57. Yamanoi, M.; Maia, J.M. Analysis of rheological properties of fibre suspensions in a Newtonian fluid by direct fibre simulation. Part1: Rigid fibre suspensions. *J. Non-Newton. Fluid Mech.* **2010**, *165*, 1055–1063, doi:10.1016/j.jnnfm.2010.05.003. [CrossRef]

58. Yamanoi, M.; Maia, J.; Kwak, T.s. Analysis of rheological properties of fibre suspensions in a Newtonian fluid by direct fibre simulation. Part 2: Flexible fibre suspensions. *J. Non-Newton. Fluid Mech.* **2010**, *165*, 1064–1071, doi:10.1016/j.jnnfm.2010.05.004. [CrossRef]

59. Yamanoi, M.; Maia, J.M. Analysis of rheological properties of fiber suspensions in a Newtonian fluid by direct fiber simulation. Part 3: Behavior in uniaxial extensional flows. *J. Non-Newton. Fluid Mech.* **2010**, *165*, 1682–1687, doi:10.1016/j.jnnfm.2010.09.006. [CrossRef]

60. Yamanoi, M.; Maia, J.M. Stokesian dynamics simulation of the role of hydrodynamic interactions on the behavior of a single particle suspending in a Newtonian fluid. Part 1. 1D flexible and rigid fibers. *J. Non-Newton. Fluid Mech.* **2011**, *166*, 457–468, doi:10.1016/j.jnnfm.2011.02.001. [CrossRef]

61. Andrić, J.; Fredriksson, S.T.; Lindström, S.B.; Sasic, S.; Nilsson, H. A study of a flexible fiber model and its behavior in DNS of turbulent channel flow. *Acta Mech.* **2013**, *224*, 2359–2374, doi:10.1007/s00707-013-0918-y. [CrossRef]

62. Andrić, J.; Lindström, S.; Sasic, S.; Nilsson, H. Rheological properties of dilute suspensions of rigid and flexible fibers. *J. Non-Newton. Fluid Mech.* **2014**, *212*, 36–46, doi:10.1016/j.jnnfm.2014.08.002. [CrossRef]

63. Do-Quang, M.; Amberg, G.; Brethouwer, G.; Johansson, A.V. Simulation of finite-size fibers in turbulent channel flows. *Phys. Rev. E* **2014**, *89*, doi:10.1103/PhysRevE.89.013006. [CrossRef]

64. Mezher, R.; Abisset-Chavanne, E.; Férec, J.; Ausias, G.; Chinesta, F. Direct simulation of concentrated fiber suspensions subjected to bending effects. *Model. Simul. Mater. Sci. Eng.* **2015**, *23*, 055007, doi:10.1088/0965-0393/23/5/055007. [CrossRef]

65. Mezher, R.; Perez, M.; Scheuer, A.; Abisset-Chavanne, E.; Chinesta, F.; Keunings, R. Analysis of the Folgar & Tucker model for concentrated fibre suspensions in unconfined and confined shear flows via direct numerical simulation. *Compos. Part A Appl. Sci. Manuf.* **2016**, *91*, 388–397, doi:10.1016/j.compositesa.2016.10.023. [CrossRef]

66. Wang, J.; Cook, P.; Bakharev, A.; Costa, F.; Astbury, D. Prediction of fiber orientation in injection-molded parts using three-dimensional simulations. *AIP Conf. Proc.* **2016**, *1713*, 040007, doi:10.1063/1.4942272. [CrossRef]

67. Perez, M.; Scheuer, A.; Abisset-Chavanne, E.; Chinesta, F.; Keunings, R. A multi-scale description of orientation in simple shear flows of confined rod suspensions. *J. Non-Newton. Fluid Mech.* **2016**, *233*, 61–74, doi:10.1016/j.jnnfm.2016.01.011. [CrossRef]

68. Sasayama, T.; Inagaki, M. Simplified bead-chain model for direct fiber simulation in viscous flow. *J. Non-Newton. Fluid Mech.* **2017**, *250*, 52–58, doi:10.1016/j.jnnfm.2017.11.001. [CrossRef]

69. Kuhn, C.; Walter, I.; Taeger, O.; Osswald, T. Experimental and Numerical Analysis of Fiber Matrix Separation during Compression Molding of Long Fiber Reinforced Thermoplastics. *J. Compos. Sci.* **2017**, *1*, 2, doi:10.3390/jcs1010002. [CrossRef]

70. Kuhn, C. Analysis and Prediction of Fiber Matrix Separation during Compression Molding of Fiber Reinforced Plastics. Ph.D. Thesis, Friedrich-Alexander Universität Erlangen-Nürnberg, Erlangen, Germany, 2018.

71. Meirson, G.; Hrymak, A.N. Two dimensional long-flexible fiber orientation simulation in squeeze flow. *Polym. Compos.* **2018**, *39*, 4656–4665. doi:10.1002/pc.24580. [CrossRef]

72. Sasayama, T.; Inagaki, M. Efficient bead-chain model for predicting fiber motion during molding of fiber-reinforced thermoplastics. *J. Non-Newton. Fluid Mech.* **2019**, *264*, 135–143, doi:10.1016/j.jnnfm.2018.10.008. [CrossRef]

73. Sasayama, T.; Inagaki, M.; Sato, N. Direct simulation of glass fiber breakage in simple shear flow considering fiber-fiber interaction. *Compos. Part A Appl. Sci. Manuf.* **2019**, *124*, 105514, doi:10.1016/j.compositesa.2019.105514. [CrossRef]

74. Laurencin, T.; Laure, P.; Orgéas, L.; Dumont, P.; Silva, L.; Rolland du Roscoat, S. Fibre kinematics in dilute non-Newtonian fibre suspensions during confined and lubricated squeeze flow: Direct numerical simulation and analytical modelling. *J. Non-Newton. Fluid Mech.* **2019**, *273*, 104187, doi:10.1016/j.jnnfm.2019.104187. [CrossRef]

75. Pérez, C. The Use of a Direct Particle Simulation to Predict Fiber Motion in Polymer Processing. Ph.D. Thesis, University of Wisconsin-Madison, Madison, WI, USA, 2017.

76. Yun, M.; Argerich Martin, C.; Giormini, P.; Chinesta, F.; Advani, S. Learning the Macroscopic Flow Model of Short Fiber Suspensions from Fine-Scale Simulated Data. *Entropy* **2019**, *22*, 30, doi:10.3390/e22010030. [CrossRef]

 Journal of
Composites Science

Article

Experimental Validation of a Direct Fiber Model for Orientation Prediction

Sara Andrea Simon *, Abrahán Bechara Senior and Tim Osswald

Polymer Engineering Center (PEC), University of Wisconsin-Madison, 1513 University Ave, Madison, WI 53706, USA; bechara@wisc.edu (A.B.S.); tosswald@wisc.edu (T.O.)
* Correspondence: ssimon9@wisc.edu; Tel.: +1-608-265-2405

Received: 24 April 2020; Accepted: 19 May 2020; Published: 25 May 2020

Abstract: Predicting the fiber orientation of reinforced molded components is required to improve their performance and safety. Continuum-based models for fiber orientation are computationally very efficient; however, they lack in a linked theory between fiber attrition, fiber–matrix separation and fiber alignment. This work, therefore, employs a particle level simulation which was used to simulate the fiber orientation evolution within a sliding plate rheometer. In the model, each fiber is accounted for and represented as a chain of linked rigid segments. Fibers experience hydrodynamic forces, elastic forces, and interaction forces. To validate this fundamental modeling approach, injection and compression molded reinforced polypropylene samples were subjected to a simple shear flow using a sliding plate rheometer. Microcomputed tomography was used to measure the orientation tensor up to 60 shear strain units. The fully characterized microstructure at zero shear strain was used to reproduce the initial conditions in the particle level simulation. Fibers were placed in a periodic boundary cell, and an idealized simple shear flow field was applied. The model showed a faster orientation evolution at the start of the shearing process. However, agreement with the steady-state aligned orientation for compression molded samples was found.

Keywords: fiber reinforced plastics; long fiber reinforced thermoplastics (LFT); sliding plate rheometer; fiber microstructure; fiber orientation; direct fiber simulation; mechanistic model

---

1. Introduction

Over the past decade, the use of long fiber-reinforced thermoplastics (LFTs) has gained wide acceptance in the automotive industry in order to meet the increasingly tightened corporate average fuel efficiency standards [1,2]. As more efficient engines and electric powertrains will not carry the whole load, the automotive industry is pressured into finding alternative solutions. One of those solutions has been light weighting car components by retrofitting existing structures made out of steel [3–7]. LFTs are being considered as a substitute because they show high performance in terms of mechanical properties, are lightweight, noncorrosive, and can be tailored to satisfy performance requirements [3,5,6,8,9].

Although the mechanical advantages of LFTs are widely reported in literature, the uncertainty remains being able to accurately model this heterogenous material class. It has been shown by numerous researchers that LFTs' mechanical performance and dimensional stability are a direct function of their microstructure, and therefore depend on fiber orientation (FO), fiber length (FL), and fiber concentration (FC) [7]. Parts are stronger in the direction of the fiber alignment and if fiber length and volume fraction (φ) are increased [10]. Being able to accurately predict the microstructure of molded components is a key factor in the automotive industry, not only for design calculations and a successful process, but also for an optimized part's performance and for guaranteeing parts are safely introduced into vehicles.

Continuum-based models for fiber orientation [11–14], length distribution [15,16], and fiber content prediction [17–19] have been developed in the last decades. These models employ a probabilistic approach to obtain the fiber configuration evolution during processing which is based on the input parameters, such as fiber characteristics and flow conditions. These models are computationally very efficient and have been implemented into commercial software. However, they can only describe the mechanism of fiber–matrix separation, fiber-attrition, and fiber alignment individually. They lack a linked theory between the three mechanisms and their interdependency [7]. Additionally, these models require experimentally determined fitting parameters. Conducting these experiments is costly, lengthy, and their provided information is limited.

Particle level simulations (PLS) have been used in the past for microstructure predictions in the plastics industry to learn about the real mechanisms present in processing discontinuous fibers [20–22]. PLS provide a fundamental modeling approach, as they solve the forces and moments acting on individual fibers during processing by accounting for the complex interactions of fibers and matrix. Compared to continuum models, the state and motion of fibers are not described as averaged and homogenized properties, but rather by solving the governing equations of each fiber explicitly. Hydrodynamic forces, fiber flexibility, excluded volume forces, fiber–fiber, and fiber–wall contacts are taken into account, leading to more accurate simulation results [7,23]. Additionally, PLS can be used to determine the fitting parameters of continuum models numerically. This has advantages as all parameters can be accurately controlled, detailed information is always available, and simulations are relatively inexpensive to perform.

PLS of semi-concentrated to concentrated fiber suspensions must address two central problems: First, the formulation of the equation of motion for individual fibers. Second, the particle–particle interactions, which can affect the flow behavior. To do this, various approaches have been taken in the past. Yamane et al. [24] represented fibers a rigid rods and calculated motion based on Jeffery's theory. They considered only short range interactions and modeled them with lubrication theory. Fan et al. [20] extended Yamane's work with rigid fibers. They accounted for long range interactions by employing a slender body approximation. Fan and co-workers later used a chain of beads joined with connectors to model flexible fibers, improving their viscosity predictions [25]. Schmid et al. [26] used PLS to study fiber flocculation. They modeled fibers as chains of rigid rods interconnected with hinges. Rods could rotate and twist about the hinges, replicating fiber bending and twisting deformations. They modeled interactions by considering repulsive forces acting normal to fiber surfaces to represent the fiber's excluded volume, and frictional fiber forces which prevented fibers from sliding over one another.

The objective of this work is to generate reliable FO evolution data in a well-defined simple shear flow to aid in the validation and development of our PLS. Simple shear was chosen as it is one of the fundamental flow conditions present in most polymer processes, such as injection molding (IM). This allows us to directly correlate the rate of deformation with the filler's behavior. IM and compression molded (CM) glass fiber-reinforced polypropylene samples were sheared in a Sliding Plate Rheometer (SPR) following Cieslinski et al. [27]. As has been shown by the same author, CM has no control over the planar orientation of the fibers, and is therefore not a suitable sample preparation method. This work will present a CM technique which ensures a controlled and repeatable initial FO for shear experiments. Results from both simulation and experiment are presented in this work.

2. Direct Fiber Model

The direct fiber or mechanistic model, used in this work, was developed at the Polymer Engineering Center, University of Wisconsin-Madison, USA, and is based on Schmid et al. [26]. Every single fiber is discretized as a chain of rigid elements with circular cross sections of diameter D, interconnected with spherical joints. The joints elastically couple the segments of a fiber as shown in Figure 1.

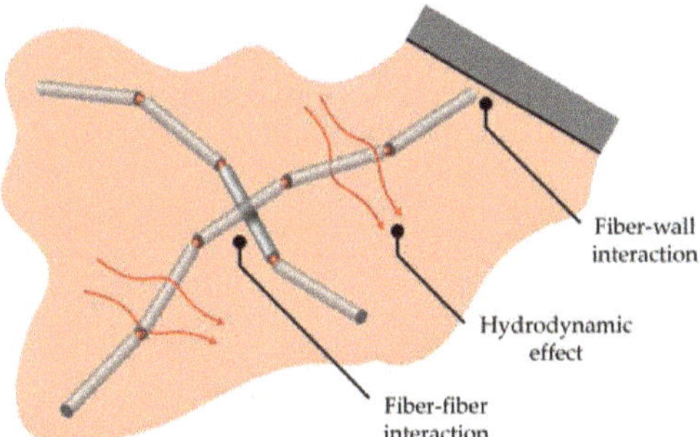

Figure 1. Particle-level simulation: modeling single fiber and macroscale interactions [7].

In the model, inertial effects are neglected due to the low Reynolds numbers (Re) that result from the high viscosity of the polymer matrix. It has been shown experimentally by Hoffman [28] and Barnes [29] that inertial effects become important at Re $\geq 10^{-3}$. In our work, the Re was calculated to be in a range of 10^{-9}–10^{-4}. Long-range hydrodynamic interactions are neglected as well due to the fluid's high viscosity [30]. Additionally, extensional and torsional deformations are neglected as fluid forces are not sufficiently high to cause fiber stretching or torsional deformation. Only bending deformation is included in the model. Brownian motion can be neglected as the Peclet number (Pe) is well above 10^3 and at this point Brownian interactions are disrupted by hydrodynamic forces [31]. Buoyant effects are neglected in the model as well [7,32]. The presence of fillers in any suspension modifies the flow field, especially in the semi-diluted and more concentrated regimes. However, the coupling between particle and fluid is not considered in the present formulation due to its high computational cost [32].

The force calculation for a segment i includes the drag force F_i^H, the interaction force with an adjacent segment j F_{ij}^C, and intra-fiber forces X_i exerted by internal connection loads at the nodes. The translation equation of motion is

$$0 = F_i^H + \sum_j F_{ij}^C + X_i - X_{i+1} \tag{1}$$

The rotational equation of motion is equally derived and includes elastic recovery terms M^b and a hydrodynamic torque T_i^H

$$0 = T_i^H - r_i \times X_{i+1} + \sum_j r_{ij} \times F_{ij}^C + M_i^b - M_{i+1}^b \tag{2}$$

where r_{ij} describes the shortest distance vector between two segments. Some of these quantities are presented in Figure 2. A fiber divided into more than one element requires an extra constraint that enforces connectivity between the different elements

$$0 = v_i + \omega_i \times (x_{i+1} - x_i) - v_{i+1} \tag{3}$$

Hydrodynamic effects are approximated by modeling each fiber segment as a chain of beads. The hydrodynamic force F_i^H is the sum of forces experienced by the beads F_k^H and is described as

$$F_i^H = \sum_{k=1}^m F_k^H = \sum_{k=1}^m 6\pi\mu a \left(U_k^\infty - u_k \right) \tag{4}$$

F_k^H is given by Stokes law, where k describes the number of beads, a the bead radius, μ the polymer matrix viscosity, U_k^∞ the surrounding fluid velocity, and u_k the velocity of the bead k ($u_k = u_i + \omega_i \times r_k$).

F_i^H is then be written as

$$F_i^H = \sum_{k=1}^{m} 6\pi\mu a U_k^\infty - m u_i - \omega_i \times \sum_{k=1}^{m} r_k \tag{5}$$

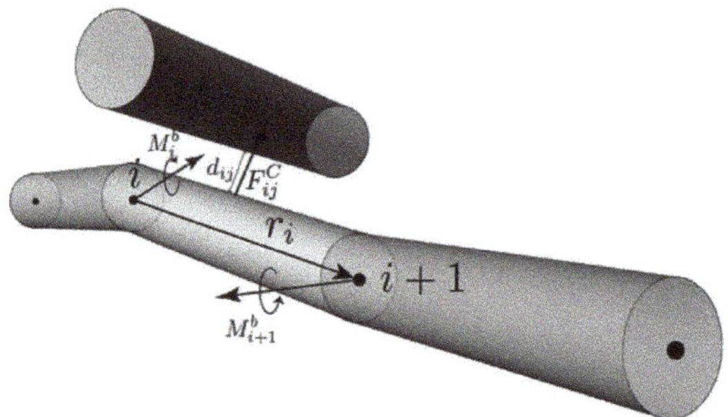

Figure 2. Fiber represented as a chain of segments. Positions are stored at the segment nodes. A neighboring fiber is depicted. The contact force $F_{ij}^C \neq 0$ when the distance between two segments is $d_{ij} < D$ [32].

Due to the fluid's vorticity each bead is also subjected to a hydrodynamic torque T_i^H

$$T_i^H = \sum_{k=1}^{m} T_k^H \tag{6}$$

where the hydrodynamic contribution T_k^H of bead k is

$$T_k^H = 8\pi\mu a^3 \left(\Omega_k^\infty - \omega_k \right) \tag{7}$$

Ω_k^∞ is the vorticity of the surrounding fluid and ω_k is the angular velocity of the bead k. Substituting the expression of T_k^H and the expression of u_k into the fluid's vorticity we can write

$$T_i^H = \sum_{k=1}^{m} \left(8\pi\mu a^3 \Omega_k^\infty \right) - 8\pi\mu a^3 m\omega_i + \sum_{k=1}^{m} 6\pi\mu a \left(r_k \times U_k^\infty \right) + u_i \times \sum_{k=1}^{m} \left(6\pi\mu a r_k \right) - \sum_{k=1}^{m} \left(r_k \times (\omega_i \times r_k) \right) \tag{8}$$

The fiber–fiber interaction force F_{ij}^C is decomposed in a normal force F_{ij}^N and a tangential force F_{ij}^T

$$F_{ij}^C = F_{ij}^N + F_{ij}^T \text{ for } d_{ij} < d_{threshold} \tag{9}$$

F_{ij}^C describes the friction between the segments and starts acting when the distance between adjacent fibers, d_{ij}, is below a defined threshold $d_{threshold}$. F_{ij}^N is an excluded volume force which is implemented as a discrete penalty method [7,32,33]. F_{ij}^N usually has a function of the following type [26,34,35],

$$F_{ij}^N = A\, exp\left[-B\left(\frac{2d_{ij}}{D} - 2 \right) \right] n_{ij} \tag{10}$$

where d_{ij} is the shortest distance between rod i and rod j, D the fiber diameter, and n_{ij} is the vector along the closest distance between the rods. A and B are parameters for which different values have been presented in the literature [26,34]. In this work, A has been chosen empirically to avoid the overlapping of fibers. A value of 1000 N was the minimum constant value at which penetrations cease to occur in the simulation [32]. B was set to 2.

The friction force F_{ij}^T between segment i and j is computed as

$$F_{ij}^T = \mu_f \left| F_{ij}^N \right| \frac{\Delta u_{ij}}{\left| \Delta u_{ij} \right|} \tag{11}$$

μ_f is the Coulomb coefficient between fibers, Δu_{ij} is the vector of the relative velocity between elements.

Bending of a fiber was approximated by using elastic beam theory

$$M_i^b = \frac{(\pi - \alpha)IE}{l} \left(\frac{r_i \times r_{i-1}}{|r_i \times r_{i-1}|} \right) \tag{12}$$

with M_i^b as the bending moment, α as the angle between segments, and l as the segment length.

A linear system of equations was assembled and solved to find the velocities and connective forces in each fiber. These velocities were integrated over time using an explicit Euler scheme to determine the fiber trajectory during the simulation [7].

3. Test Setup and Experimental Validation

3.1. Sample Preparation

The fiber suspension used in this work was a 10 and 20 wt % glass fiber in a polypropylene (PP) matrix. The 20 wt % (STAMAX PPGF20) was commercially available and provided by SABIC^{TM}. The material was supplied in the form of coated pellets with a nominal length of 15 mm, which also represents the initial and uniform length of the glass fibers. The used E-glass fibers ($\varrho = 2.55$ g/cm^3) are chemically coupled to the PP matrix ($\varrho = 0.91$ g/cm^3). The fiber diameter was measured to be 19 ± 1 μm using an optical microscope. The 10 wt % was achieved by mixing higher FCs with neat PP (SABIC^{TM} PP579S) in a cement mixer. The neat PP is identic to the matrix material for the commercial available STAMAX pellets. The matrix was considered as a generalized Newtonian fluid under the test conditions presented. The authors of [27,36] showed that the presence of polymer matrix eliminates any concerns regarding fiber sedimentation due to gravity.

Two separate processes were used to prepare samples that were tested in the SPR. The first method combined extrusion to disperse fibers and compression molding to form proper sample dimensions. The second approach used IM to produce plates.

The first sample preparation method used an Extrudex Kunstoffmaschinen single-screw extruder to process the pultruded material. The 45 mm 30 L/D extruder was equipped with a gradually tapering screw and fitted with a 3 mm die. The temperature zones of the extruder were set to 210, 210, 220, 220, 230, 230, and 230 °C and the die temperature was set to 230 °C. The composite was extruded at 5 rpm. Due to the low processing speed, most of the initial FL was maintained in the extruded strand ($L_N = 6.7$ mm, $L_W = 11.4$ mm). Computational time for the mechanistic model simulation is geometrically proportional to the maximum detected FL. To reduce computation, the strands were pelletized to 3.2 mm. Initial compression molding trials showed that manual alignment of pellets in the mold did not yield a repeatable initial FO, because pellets could move easily and rotate. Therefore, pellets were re-extruded and strands were cut to fit mold dimensions. The mold geometry was a rectangular prism (14 mm × 14 mm × 2.1 mm). Analysis of the re-extruded strands showed a L_N of 0.83 mm and L_W of 1.53 mm. The mold was coated and placed on a heating plate to cause partial melting of the aligned strands. This facilitated space reduction between irregular shaped strands. The plate was extracted, flipped 180°, and re-molten to remove trapped air bubbles between strands. Failure to remove caught air bubbles would alter matrix rheology. Plates were compression molded at 210 °C and a load of 1000 lbs. Final fiber properties are summarized in Table 1.

Table 1. Fiber properties. Average FLs, a_{11}, a_{22}, and a_{33} as the orientation tensors. a_{11} representing extrusion direction in compression molded samples and the melt flow direction in injection molded samples.

Material Property	CM Plates	IM Plates
L_N [mm]	0.83	1.28
L_W [mm]	1.53	2.92
a_{11} [-]	0.86	0.60
a_{22} [-]	0.11	0.37
a_{33} [-]	0.03	0.03

The second sample preparation procedure involved IM a simple plate geometry ($102 \times 305 \times 2.85$ mm^3). The cavity was filled through a 20 mm edge-gate with the same thickness as the plate and was fed through a 17 mm full-round runner. Parts were molded on a 130 ton Supermac Machinery SM-130 IM machine. The melt temperature was set to 250 °C, and back and holding pressure to 5 and 300 bar, respectively. Injection and holding time were set to 2 and 22 s, respectively. A full microstructure analysis was conducted. Fiber properties are summarized in Table 1. Samples for the SPR were only extracted from regions which showed a clear developed shell–core structure and identic FCs [4].

For both sample preparation methods, the SPR sample sizes were calculated by adding 20 mm to the stroke applied to be able to extract pure shear samples for FO analysis.

3.2. Sliding Plate Rheometer

FO evolution data was obtained in a SPR by shearing samples in a controlled simple shear flow (Figure 3). The rheometer was based on the design of [37]. The SPR was contained in a convection oven and the rectilinear plate displacement was generated by an Interlaken 3300 universal testing instrument. The rheometer had an effective surface of 100×300 mm^2 and a maximum stroke of 120 mm. The SPR gap thickness could be altered. However, it was chosen to be 2 mm, as this work yields to aid in the understanding of how fibers flow and orient themselves during the IM process. Typically, IM parts show a thickness of 2 mm. The maximum possible displacement and the chosen gap size, limited the deformation that could be imposed on the sample to a shear strain of 60. Both, shear rate and shear strain, were programmable through the Wintest$^®$ Software (Bose Corporation, Eden Prairie, MN, USA).

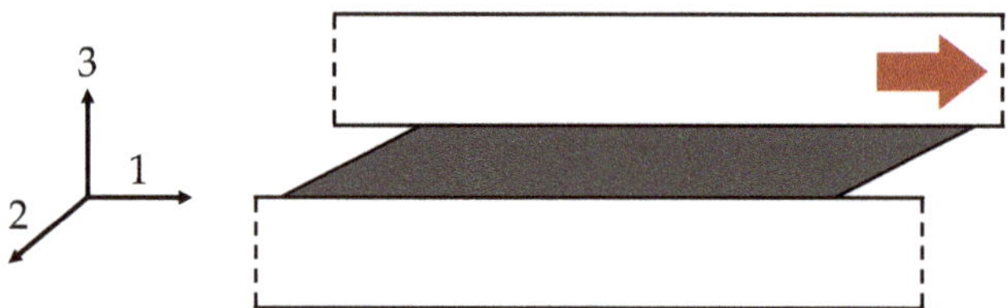

Figure 3. Sliding plate rheometer and defined coordinate system.

The experimental procedure was based on the SPR experiment conducted by [27]. The rheometer was heated to 260 °C for 2 h prior to sample loading. Upon loading, the test specimen was rotated 90° with respect to the extrusion/filling direction, effectively swapping the a_{11} and the a_{22} components of the orientation tensor. This allowed for a low alignment in the shearing direction, so a larger change in orientation could be observed. The sample was secured between the rheometer plates and allowed to melt evenly before the plates were tightened to a final gap of 2 mm (=initial condition). As the initial thickness of the test specimen was larger than the final gap thickness, samples were slightly compressed when tightening the plates to guarantee full contact. After an additional 10 min of heating, the sample was sheared at a rate of 1 s^{-1}. Forced convection was used to cool the sample to preserve its shape and FO for further analysis. Five repetitions per testing condition were used to ensure accuracy and repeatability of results.

3.3. Measurement of Fiber Microstructure

A fully characterized microstructure was required to accurately reproduce the initial conditions in the direct fiber simulation (Figure 4). FO was determined by using the X-ray microcomputed tomography (μCT) approach. A sample of dimensions $10 \times 10 \times 2$ mm^3 was loaded into a sample holder and then placed on a rotating platform. The X-rays penetrated the sample and were absorbed differently depending on the configuration of the sample's constituents. The detector recorded the attenuated X-rays as radiographs at incremental angles during the rotation of the sample to achieve a full scan of the sample. After completion of the 3D reconstruction, the μCT data set was processed with the VG StudioMAX (Volume Graphics) software to quantify the FO distribution using the structure tensor approach [4,38,39]. Samples were scanned with an industrial μCT system (Metrotom 800, Carl Zeiss AG, Oberkochen, Germany). Throughout the reported studies, the voltage was set to 80 V, the current to 100 A, the integration time to 1000 ms, the gain to 8, the number of projections to 2200, and the voxel size was set to 5 μm.

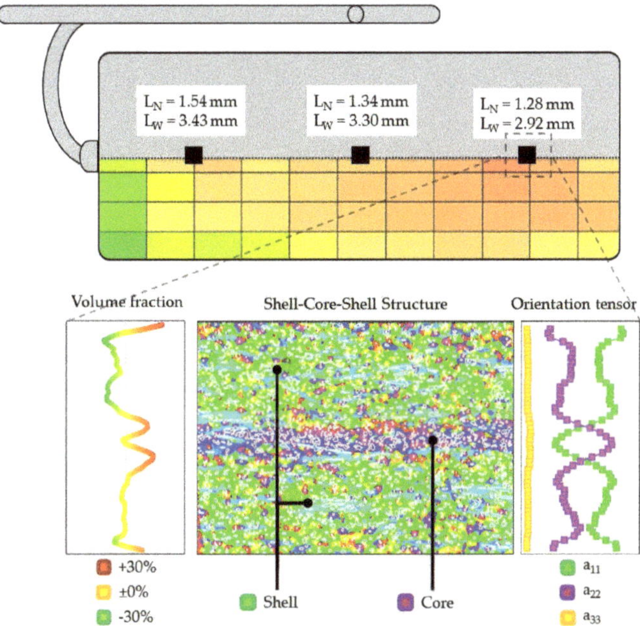

Figure 4. Fully characterized microstructure of an injection molded plaque.

FC was quantified with VG StudioMAX as well. The μCT data set was converted into a stack of 2D cross-sectional images aligned normal to the thickness direction. The grayscale images were transformed into binary images by thresholding, which separated the image into black (matrix) and white (fibers) pixels. Subsequently, the fiber volume fraction through the thickness was calculated [4].

The FL measurement technique presented in [40] was employed in this work. This technique consists of fiber dispersion and a fully automated image processing algorithm to quantify the fiber length distribution (FLD). It has been shown in [41] that downsampling methods yield to a preferentially capture of long fibers, and thus skew the real FLD. In this work, a correction was therefore applied to all results as published by [41].

4. Direct Fiber Simulation

4.1. Simulation

To set up a simulation that matches the SPR experiment, a number of fibers matching the experimental volume fraction was placed inside a shear cell. To guarantee a constant fiber volume fraction, Lees–Edwards periodic boundaries [42] were assigned to the cell faces perpendicular to the shearing direction (Figure 5). Mechanical properties were assigned to the fibers while rheological properties were assigned to the matrix. A simple shear flow field was applied to the matrix phase which translates into the hydrodynamic force term in the force balance calculation explained in Section 2.

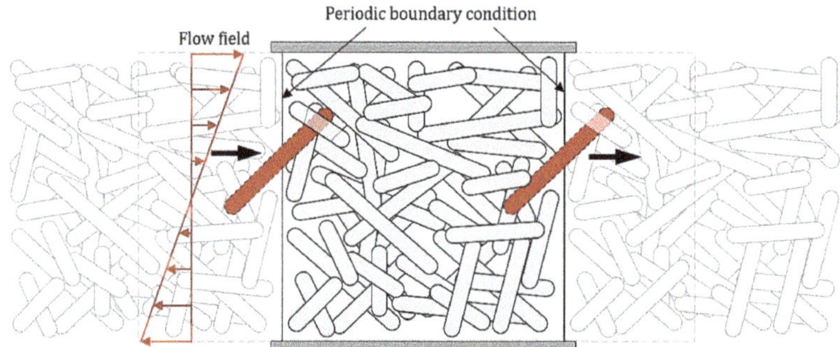

Figure 5. Shear cell with periodic boundary conditions.

4.2. Simulation Setup

In order to obtain reliable rheological information from discontinuous fiber composites, accurate and repeatable initial conditions are needed [27]. To create the initial cluster of fibers for the simulation, the microstructure of the experimental samples was carefully characterized. The experimental values of orientation (a_{ij}) and fiber density (vol %) were reproduced by discretizing each through-thickness profile into a set number of layers as shown in Figure 6. For each layer the average orientation tensor components were used to reconstruct the fiber orientation state by using a Fourier Series expansion. It is worth noticing that as the thickness of the individual layers is decreased, to obtain better resolution, the target off-plane orientation becomes harder to achieve.

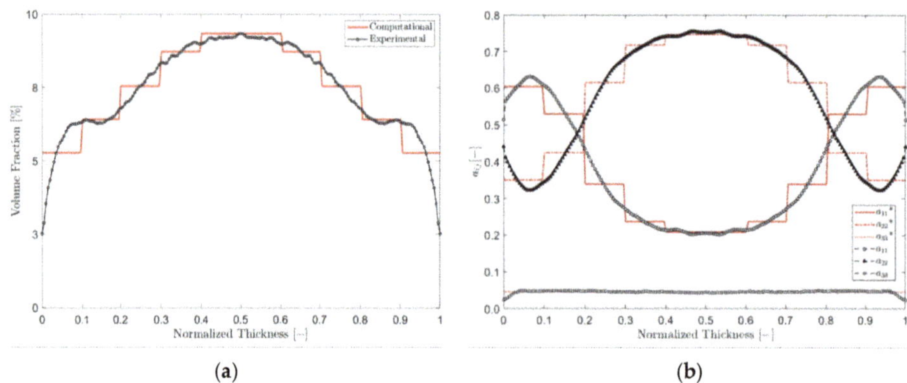

Figure 6. Through-thickness discretization for **(a)** fiber concentration and **(b)** fiber orientation of a compression molded sample. Experimental (black) and discretized, computational (red) data is shown [43].

The average FLs (L_N, L_W) were assigned as a global length distribution to the complete cluster of fibers. This distribution was obtained by fitting a Weibull probability distribution function to the experimental length measurement. The result was a heterogeneous cluster of fibers closely resembling the experimental condition (Figure 7). Prior to imposing the shear flow, a short simulation was conducted with no velocity field and only interaction forces present. This was applied as a relaxation step, so fibers are not overlapping or in extreme proximity at the start of the shearing simulation. This step does not modify the initial orientation state in a significant way.

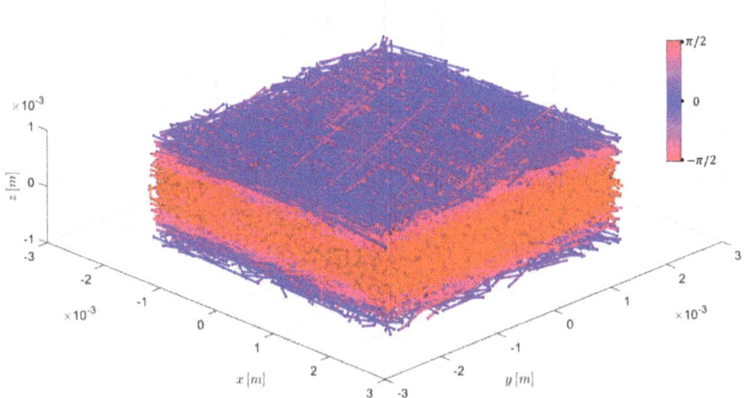

Figure 7. Computational compression molded cluster, with x as the shearing direction (a_{11}). Fibers are colored as a function of their XY (=12) planar orientation [43].

A simple shear flow with a rate of deformation of $1\ \text{s}^{-1}$ was applied to the shear cell. The viscosity was set to 110 Pas using the properties of the neat PP. The dimension of the cell is dictated by the SPR gap in Z direction of 2 mm. To allow free rotation, the dimension in X and Y were set to 1.1 times the length of the longest fiber. The cluster properties corresponding to both sample preparation methods are listed in Table 2.

Table 2. Cluster properties for injection- and compression molded samples.

Cluster Property	IM Plates	CM Plates
vol %	4	8.5
Longest fiber [mm]	5	4
L_N [mm]	0.71	0.83
L_W [mm]	1.32	1.53
a_{11} [-]	0.6	0.36
a_{22} [-]	0.39	0.62
a_{33} [-]	0.002	0.02

The simulation time was set to 60 s to obtain a total strain of 60. Cell walls parallel to the YZ plane have a periodic boundary condition. A tight array of fixed fibers is placed on the upper and lower boundaries to represent the SPR walls. The additional input parameters for the simulation are listed in Table 3. The simulation results were the coordinates of each fiber at every time step. With this information orientation tensors were calculated as a function of time and strain.

Table 3. Shear cell properties for injection- and compression molded samples.

Parameter	Value
E [GPa]	73
Fiber diameter [μm]	19
η [Pas]	110
$\dot{\gamma}$ [s^{-1}]	1
Time step [s]	5×10^{-5}
Integrations [-]	1,200,000

5. Results and Discussion

5.1. Injection Molded Samples

IM provides a repeatable flow history that will yield to FO that is consistent among multiple samples [4,27]. The global diagonal components of the orientation tensor at zero shear strain are shown in Figure 8. The common core–shell structure associated with molded composites can be seen. The core layer consists of fibers predominantly aligned in cross-flow direction (a_{22}), while fibers in the shell are oriented along the flow direction (a_{11}) due to the fountain flow effect. The orientation in thickness direction a_{33} is uniform and generally low with average values less than 0.06 [4].

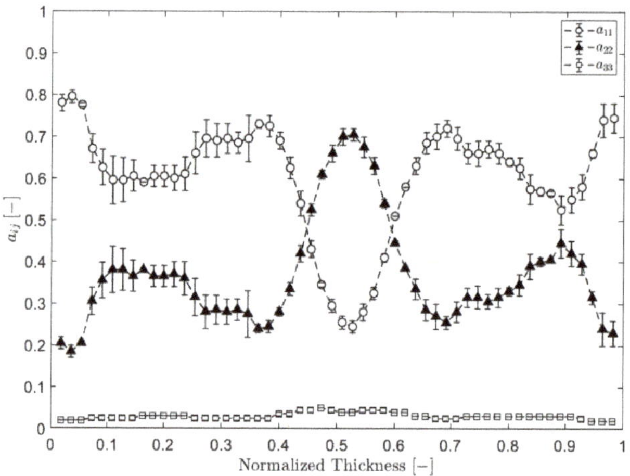

Figure 8. Initial fiber orientation for injection molded samples for simple shear flow tests.

The average orientation tensor as a function of shear strain is shown in Figure 9. For both simulation and experiment fibers gradually align in the flow direction over 60 shear strain units. The general trend of a_{11} for the experimental data begins at 0.53 and evolves to a value of 0.63. In contrast, the simulation starts at a value for a_{11} at 0.60 and evolves to a value of 0.74. The computational initial condition differs from the experimental one due to two main reasons: First, the complex core–shell structure present in the IM sample is difficult to discretize (Figure 8). There will be a loss of accuracy by averaging the orientation data into seven layers. Second, the FL of the experimental sample is very high which increases the shear cell size and therefore computational time. Therefore, the FL was truncated in the simulation. As the initial conditions were not accurately reproduced, a different rate of orientation evolution was obtained in the PLS. However, experiment and simulation still show similarities in their behavior. Both show a slight drop in a_{11} before fibers start aligning in shearing direction. This drop is less pronounced in the PLS, as fibers start off with a higher a_{11} alignment. An initial decrease in a_{11} was also found in [27]. In their work, this phenomenon was explained as follows. Fibers oriented slightly out of the 1–2 plane need to flip first to become oriented in flow direction, leading to an increase in a_{13}, thus a decrease in a_{11}. This increase in a_{13} could be seen in the experimental data. Depending on the initial value of a_{13}, the alignment in flow direction is different. If a_{13} is positive, the fiber will continue to align in flow direction. However, an initial negative a_{13} orientation would require the fiber to flip before becoming aligned in flow direction. It is expected that a fiber with a negative a_{13} would take longer to align in flow direction [27]. A slight initial decrease in a_{11} was also reported by [44]. The used SPR was limited to 60 strain units and a steady state response associated with fiber suspensions could not be obtained.

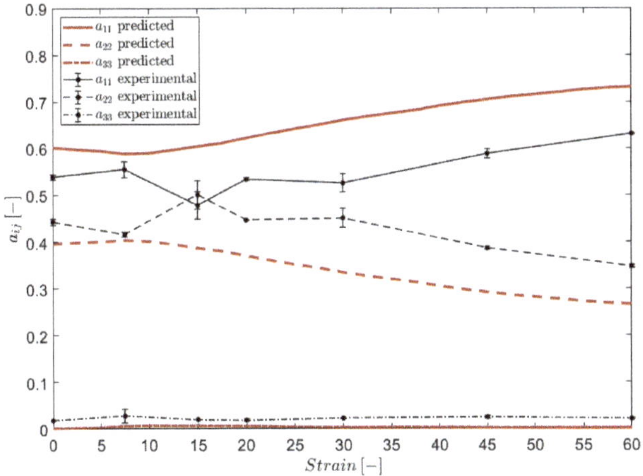

Figure 9. Experimental (black) and predicted (red) fiber orientation evolution of injection molded samples.

The simulation results for a_{11} through thickness were smoothed and plotted for different strains in Figure 10. Under a constant shear, the complex heterogeneous orientation profile transitions into a more homogeneous profile.

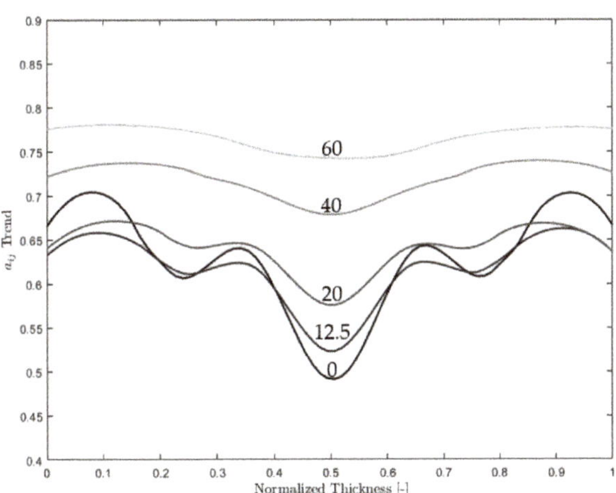

Figure 10. Smoothed computational a_{11} evolution of injection molded samples at varying shear strains.

5.2. Compression Molded Samples

It has been shown that IM samples provide a repeatable initial FO; however, IM specimens show a complex core–shell structure, which is difficult to reproduce computationally (Figure 8). In addition, samples showed a high alignment in shearing direction at zero strain; therefore, only a low further alignment of a_{11} could be observed (Figure 9). CM samples with a controlled FL, FC, homogenous structure, and with a low a_{11} alignment were consequently manufactured to accurately reproduce the initial conditions (Figure 6).

The average orientation through sample thickness as a function of shear strain is shown in Figure 11. Initial conditions could be accurately reproduced computationally and fit the experimental data. The standard deviation for experimental values at zero strain is low, which demonstrates a repeatable initial FO, validating the CM sample preparation method.

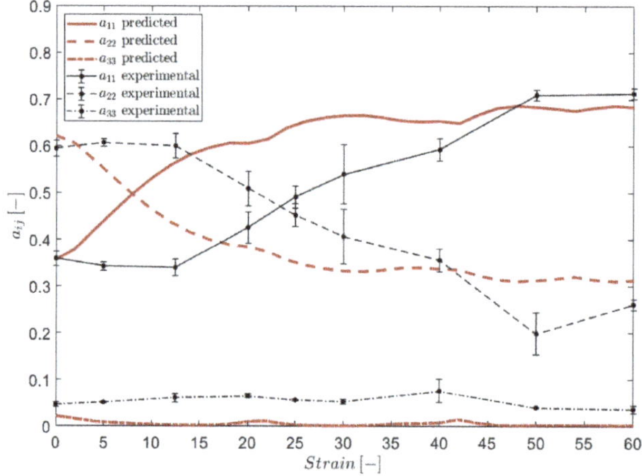

Figure 11. Experimental (black) and predicted (red) fiber orientation evolution of compression molded samples [43].

The a_{11} component starts at 0.36 and transitions to a steady-state value of approximately 0.7 for both the experiment and the PLS. A steady state is reached at a total strain of 50. A slight initial drop in a_{11} could be seen in the experimental data, but was not present for the PLS. The simulation shows faster orientation evolution than the experiment. This phenomenon has also been reported in literature [45,46] for other diffusion models. It has been established that Jeffrey's hydrodynamic model and Folgar–Tucker's model predict a faster transient orientation rate when compared to related experiments [46]. Therefore, recent models have been developed with slow fiber orientation kinetics. Wang et al. [12] introduced the reduced strain closure model. This model employs a scalar factor κ to slow the orientation dynamics. Results still showed a quicker initial rise of flow-direction orientation, but achieved nearly the same steady state orientation as obtained with experiments [45]. In an attempt to slow the orientation kinetics predicted, Phelps et al. [13] derived the anisotropic rotary diffusion model.

The distortion of the flow field due to the presence of particles gives rise to the particular rheology of suspensions. As the aspect ratio (r_p) of the particles increases, this effect becomes dependent on the orientation state, as described by Dinh et al. [47]. Lindstrom et al. [34] conducted a PLS with individual fibers in simple shear flow. They observed orbit periods were overestimated when not including two-way coupling. They concluded that particle–fluid interaction is essential to the fiber's dynamic behavior in shear flow. In that regard, neglecting the two-way coupling between the fiber orientation state and the underlying flow field in a PLS, can be a considerable source of error. However, various authors have shown that including two-way coupling has marginal effect on the orientation state for flow in common geometries. Mezi et al. [48] pointed out that accounting for the coupling effect had little impact on the fiber orientation distribution for the fluid flow in a planar channel, while having a considerable effect in the pressure drop. They also observed coupling increased the size of corner vortex in a contraction geometry. This last effect was also noted by VerWeyst et al. [49], who observed moderate impact of coupling on the fiber orientation in a center-gated disk. They showed effects of coupling decayed rapidly with increasing distance from the center of the disk. The difficulty with these approaches lies in the large computational domain required to resolve both the spatial and orientation domains. With PLS the computational cost increases rapidly as the two-way coupling must be calculated for each fiber.

In a PLS study for SFTs under simple shear, Kugler et al. [50] also observed a faster alignment in the simulation with respect to the experimental data. They suggested a reduced shear strain to correct for the lack of two-way coupling. Similarly, in an experimental study of fiber attrition employing a Couette rheometer, Moritzer et al. [51] use the scale factor $(1 - \phi)$ to account for the reduced rate of deformations experienced by fibers in the suspension.

Published work indicates that the coupling effect might have a greater impact on the fiber orientation in the dynamic portion of the orientation evolution, and that aspect ratio and volume fraction are scaling factors of such effect. A PLS study to determine quantitatively the impact of coupling under simple shear flow remains to be done; one that compares the predicted orientation to the complete microstructural data determined experimentally.

The a_{13} and a_{12} components of orientation are shown in Figure 12. In both prediction and experiment a_{13} and a_{12} show close agreement with the final steady state value. The measured a_{13} component shows small oscillations. This behavior indicates some fibers orbit in the plane of shear. The predicted a_{13} component also shows these orbits, however, with a shorter period. The initial measured a_{13} component has a negative value of -6×10^{-4}, while the computational cluster started with a very small, yet positive, value of 6×10^{-5}. Literature suggests that the initial value of a_{13} impacts whether components of orientation evolve monotonically to a steady state or present an overshoot and/or undershoot prior to reaching a steady state [52]. The fact that the initial computational cluster differed from the experimental one in this aspect, is a factor in why the a_{11} component oriented directly to steady state in the simulation, while the experimental a_{11} did show a slight undershoot (Figure 11). It can be seen that a_{12}, for the simulation, increases continuously with shear strain. As fibers oriented in the 1–2 plane orient themselves in 1, the value of a_{12} should approach zero.

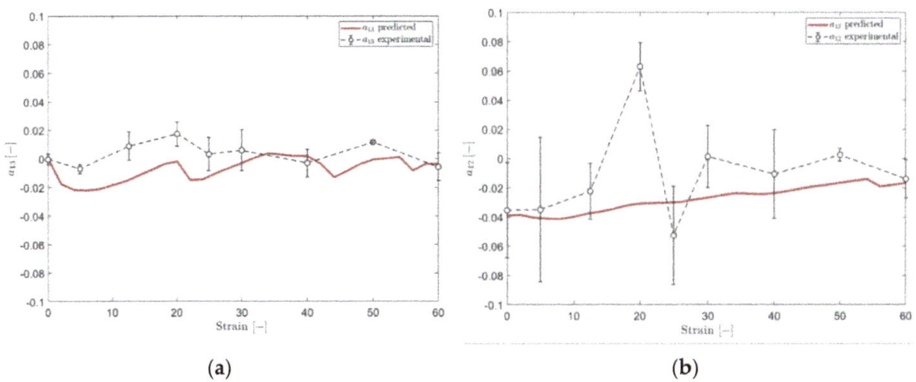

(a)   (b)

Figure 12. Experimental (black) and predicted (red) off-diagonal orientation components (a) a_{12} and (b) a_{13} as a function of shear strain.

The mentioned faster orientation kinetics in the PLS are also shown in Figure 13. Compared to the initial orientation, the experimental core–shell structure is largely unchanged at 12.5 strain units. At the same applied strain, the simulation already shows a large transition to a rather homogenous structure. At 60 strain units, the core–shell structure disappears, a homogeneous profile is reached and a good match between experiment and the PLS was observed. It is assumed that shearing beyond 60 strain units would lead to a steady state FO through thickness [27]. The smoothed computational

a_{11} evolution as a function of shear strain is shown Figure 14. With an increase in shear strain, the heterogeneous orientation profile transitions into a homogeneous steady state. The most significant change in orientation occurs in the core of the sample. The rate of change slows down as the orientation approaches steady state.

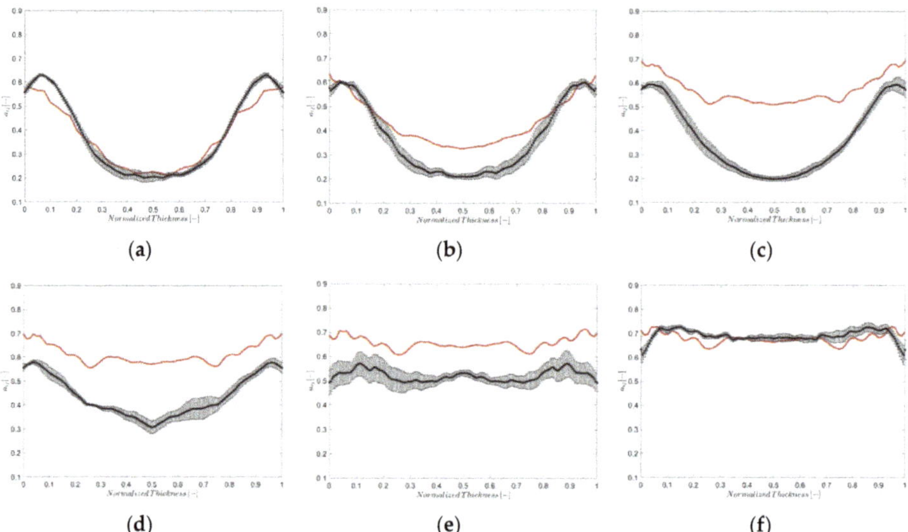

Figure 13. Experimental (black) and predicted (red) a_{11} values averaged through sample thickness at varying shear strains for compression molded samples. (**a**) Initial fiber orientation distribution at 0 shear strain, (**b**) fiber orientation distribution after 5 strain units, (**c**) 12.5, (**d**) 20, (**e**) 40, and (**f**) 60 shear strain units [43].

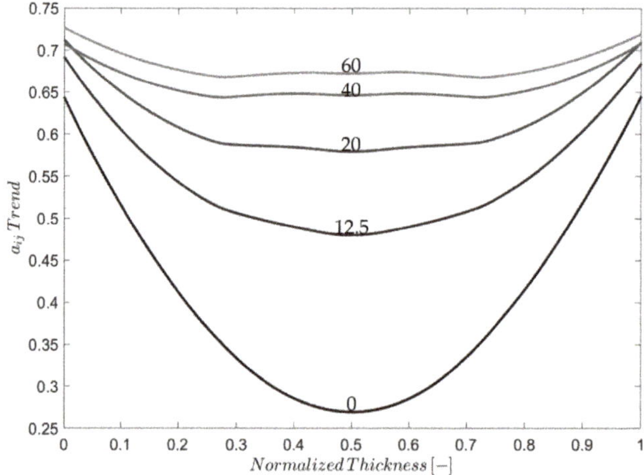

Figure 14. Smoothed computational a11 evolution of compression molded samples [43].

With the proposed CM technique, it was possible to manufacture samples with low alignment in shearing direction, which allowed a larger orientation transition. Higher alignment and faster orientation evolution was observed for the CM samples when compared to the IM ones. The faster alignment can be attributed to the lower FL of the CM samples. The long fibers present in the IM samples require more energy to rotate and also hinder the motion of adjacent fibers.

6. Conclusions and Outlook

A PLS was used to simulate the FO evolution of glass fiber-reinforced PP plates sheared in a SPR. The PLS results showed a faster orientation evolution at the beginning of the shearing process compared to the experimental data. However, an agreement with the final orientation state was observed for the CM plates after fine-tuning the initial condition discretization method. In the experiment, an early decrease in a_{11} was observed for IM and CM samples. The same behavior has been reported in literature by [27]. However, the author used shorter fibers compared to the FL in this work. It appears that this phenomenon is more easily observed with longer fibers due to a slower rate of FO.

A reliable sample preparation method for suspension rheology was developed. Repeatable and controlled initial orientation can be achieved through the presented CM technique. Even though it has been shown in [27] that CM samples do not exhibit repeatable orientation evolution data during the startup of simple shear flow in a SPR, the CM technique employed in this work differs significantly by manually depositing individual and aligned strands.

As the dynamics of fiber suspensions are not yet well captured by the PLS, future work will focus on implementing the effect of fiber fluid coupling and studying its impact under different combinations of the dimensionless number ϕr_p.

Author Contributions: S.A.S. and A.B.S. conceived and designed the experiments. S.A.S. was responsible for performing the experimental studies and the microstructure analysis. A.B.S. performed the simulations. S.A.S. and A.B.S. analyzed the data and wrote the paper. T.O. supervised the project and was involved in all stages of the research. All authors have read and agreed to the published version of the manuscript.

Funding: This research was supported by the National Science Foundation under Grant No. 1633967.

Acknowledgments: The authors would like to thank SABIC® for their ongoing support and collaboration with our research group and their material supply. Additionally, we would also like to thank our colleagues from the Polymer Engineering Center who provided insight and expertise that greatly assisted the research.

Conflicts of Interest: The authors declare no conflicts of interest.

References

1. NHTSA. *Corporate Average Fuel Economy (CAFE) Standards*; NHTSA: Washington, DC, USA, 2020.
2. Ning, H.; Lu, N.; Hassen, A.A.; Chawla, K.; Selim, M.; Pillay, S. A review of Long fibre thermoplastic (LFT) composites. *Int. Mater. Rev.* **2019**, *65*, 164–188. [CrossRef]
3. Jain, R.; Lee, L. *Fiber Reinforced Polymer (FRP). Composites for Infrastructure Applications*; Springer: Berlin/Heidelberg, Germany, 2012.
4. Goris, S. Characterization of the Process-Induced Fiber Configuration of Long Glass Fiber-Reinforced Thermoplastics. Ph.D. Thesis, University of Wisconsin-Madison, Madison, WI, USA, 2017.
5. Thomason, J.L. The influence of fibre length and concentration on the properties of glass fibre reinforced polypropylene. 6. the properties of injection moulded long fibre PP at high fibre content. *Compos. Part A Appl. Sci. Manuf.* **2005**, *36*, 995–1003. [CrossRef]
6. Nghiep Nguyen, B.; Kunc, V. An elastic-plastic damage model for long-fiber thermoplastics. *Int. J. Damage Mech.* **2010**, *19*, 691–725. [CrossRef]
7. Osswald, T.; Ghandi, U.; Goris, S. *Discontinous Fiber Reinforced Composites*, 1st ed.; Carl-Hanser Verlag: Munich, Germany, 2020.
8. Zhang, G.; Thompson, M.R. Reduced fibre breakage in a glass-fibre reinforced thermoplastic through foaming. *Compos. Sci. Technol.* **2005**, *65*, 2240–2249. [CrossRef]
9. Wang, J.; Geng, C.; Luo, F.; Liu, Y.; Wang, K.; Fu, Q.; He, B. Shear induced fiber orientation, fiber breakage and matrix molecular orientation in long glass fiber reinforced polypropylene composites. *Mater. Sci. Eng. A* **2011**, *528*, 3169–3176. [CrossRef]
10. Fu, S.Y.; Hu, X.; Yue, C.Y. Effects of fiber length and orientation distributions on the mechanical properties of short-fiber-reinforced polymers: A Review. *Int. J. Mater. Res.* **1999**, *5*, 74–83. [CrossRef]

11. Advani, S.G.; Tucker, C.L. The Use of Tensors to Describe and Predict Fiber Orientation in Short Fiber Composites. *J. Rheol.* **1987**, *31*, 751–784. [CrossRef]

12. Wang, J.; O'Gara, J.F.; Tucker, C.L. An objective model for slow orientation kinetics in concentrated fiber suspensions: Theory and rheological evidence. *J. Rheol.* **2008**, *52*, 1179–1200. [CrossRef]

13. Phelps, J.H.; Tucker, C.L. An anisotropic rotary diffusion model for fiber orientation in short- and long-fiber thermoplastics. *J. Nonnewton. Fluid Mech.* **2009**, *156*, 165–176. [CrossRef]

14. Tseng, H.C.; Chang, R.Y.; Hsu, C.H. Phenomenological improvements to predictive models of fiber orientation in concentrated suspensions. *J. Rheol.* **2013**, *57*, 1597–1631. [CrossRef]

15. Phelps, J.H.; Abd El-Rahman, A.I.; Kunc, V.; Tucker, C.L. A model for fiber length attrition in injection-molded long-fiber composites. *Compos. Part A Appl. Sci. Manuf.* **2013**, *51*, 11–21. [CrossRef]

16. Durin, A.; De Micheli, P.; Ville, J.; Inceoglu, F.; Valette, R.; Vergnes, B. A matricial approach of fibre breakage in twin-screw extrusion of glass fibres reinforced thermoplastics. *Compos. Part A Appl. Sci. Manuf.* **2013**, *48*, 47–56. [CrossRef]

17. Nott, P.R.; Brady, J.F. Pressure-driven flow of suspensions: Simulation and theory. *J. Fluid Mech.* **1994**, *275*, 157–199. [CrossRef]

18. Morris, J.F.; Boulay, F. Curvilinear flows of noncolloidal suspensions: The role of normal stresses. *J. Rheol.* **1999**, *43*, 1213–1237. [CrossRef]

19. Miller, R.M.; Morris, J.F. Normal stress-driven migration and axial development in pressure-driven flow of concentrated suspensions. *J. Non-Newton. Fluid Mech.* **2006**, *135*, 149–165. [CrossRef]

20. Fan, X.; Phan-Thien, N.; Zheng, R. A direct simulation of fibre suspensions. *J. Non-Newton. Fluid Mech.* **1998**, *74*, 113–135. [CrossRef]

21. Londoño-Hurtado, A.; Osswald, T.; Hernandez-Ortíz, J.P. Modeling the behavior of fiber suspensions in the molding of polymer composites. *J. Reinf. Plast. Compos.* **2011**, *30*, 781–790. [CrossRef]

22. Yashiro, S.; Sasaki, H.; Sakaida, Y. Particle simulation for predicting fiber motion in injection molding of short-fiber-reinforced composites. *Compos. Part A Appl. Sci. Manuf.* **2012**, *43*, 1754–1764. [CrossRef]

23. Strautins, U. Flow-driven orientation dynamics in two classes of fibre suspensions. Ph.D. Thesis, University of Kaiserslautern, Kaiserslautern, Germany, 2008.

24. Yamane, Y.; Kaneda, Y.; Doi, M. Numerical simulation of a concentrated suspension of rod-like particles in shear flow. *J. Non-Newton. Fluid Mech.* **1994**, *54*, 405–421. [CrossRef]

25. Joung, C.G.; Phan-Thien, N.; Fan, X.J. Direct simulations of flexible fibers. *J. Non-Newton. Fluid Mech.* **2001**, *99*, 1–36. [CrossRef]

26. Schmid, C.F.; Switzer, L.H.; Klingenberg, D.J. Simulations of fiber flocculation: Effects of fiber properties and interfiber friction. *J. Rheol.* **2000**, *44*, 781–809. [CrossRef]

27. Cieslinski, M.J.; Baird, D.G.; Wapperom, P. Obtaining repeatable initial fiber orientation for the transient rheology of fiber suspensions in simple shear flow. *J. Rheol.* **2016**, *60*, 161–174. [CrossRef]

28. Hoffman, R.L. Discontinuous and dilatant viscosity behavior in concentrated suspensions. 1. Observation of a flow instability. *T. Soc. Rheol.* **1972**, *16*, 155–173. [CrossRef]

29. Barnes, H.A. Shear-thickening (dilatancy) in suspension of nonaggregating solid particles dispersed in Newtonian liquids. *J. Rheol.* **1989**, *33*, 329–366. [CrossRef]

30. Sundararajakumar, R.R.; Koch, D.L. Structure and properties of sheared fiber suspensions with mechanical contacts. *J. Non-Newton. Fluid Mech.* **1997**, *73*, 205–239. [CrossRef]

31. Stickel, J.J.; Powell, R.L. Fluid mechanics and rheology of dense suspensions. *Annu. Rev. Fluid Mech.* **2005**, *37*, 129–149. [CrossRef]

32. Pérez, C. The Use of a Direct Particle Simulation to Predict Fiber Motion in Polymer Processing. Ph.D. Thesis, University of Wisconsin-Madison, Madison, WI, USA, 2016.

33. Tang, M.; Manochay, D.; Otaduyz, M.A.; Tongx, R. Continuous penalty forces. *ACM Trans. Graph.* **2012**, *31*, 1–9. [CrossRef]

34. Lindström, S.B.; Uesaka, T. Simulation of the motion of flexible fibers in viscous fluid flow. *Phys. Fluids* **2007**, *19*. [CrossRef]

35. Switzer, L.H.; Klingenberg, D.J. Rheology of sheared flexible fiber suspensions via fiber-level simulations. *J. Rheol.* **2003**, *47*, 759–778. [CrossRef]

36. Chaouche, M.; Koch, D.L. Rheology of non-Brownian rigid fiber suspensions with adhesive contacts. *J. Rheol.* **2001**, *45*, 369–382. [CrossRef]

37. Giacomin, A.J.; Samurkas, T.; Dealy, J.M. A Novel Sliding Plate Rheometer for Molten Plastics. *Polym. Eng. Sci.* **1989**, *29*, 499–504. [CrossRef]

38. Krause, M.; Hausherr, J.M.; Burgeth, B.; Herrmann, C.; Krenkel, W. Determination of the fibre orientation in composites using the structure tensor and local X-ray transform. *J. Mater. Sci.* **2010**, *45*, 888–896. [CrossRef]

39. Nguyen, B.N.; Bapanapalli, S.K.; Kunc, V.; Frame, B.J.; Phelps, J.H.; Tucker, C.L. Fiber length and orientation in long-fiber injection-molded thermoplastics—Part I: Modeling of microstructure and elastic properties. *J. Compos. Mater.* **2008**, *42*, 1003–1029. [CrossRef]

40. Goris, S.; Back, T.; Yanev, A.; Brands, D.; Drummer, D.; Osswald, T. A novel fiber length measurement technique for discontinuous fiber-reinforced composites: A comparative study with existing methods. *Polym. Compos.* **2018**, *39*, 4058–4070. [CrossRef]

41. Kunc, V.; Frame, B.; Nguyen, B.N.; Tucker, C.L.; Velez-Garcia, G. Fiber length distribution measurement for long glass and carbon fiber reinforced injection molded thermoplastics. In Proceedings of the SPE Automotive Composites Conference and Exhibition, Troy, MI, USA, 11 September–13 November 2007.

42. Evans, D.J.; Morriss, G.J. Non-Newtonian molecular dynamics. *Comput. Phys. Rep.* **1984**, *1*, 297–343. [CrossRef]

43. Simon, S.A.; Bechara, A.; Osswald, T. Direct fiber model validation: Orientation evolution in simple shear flow. In Proceedings of the SPE Automotive Composites Conference and Exhibition, Novi, MI, USA, 4–6 September 2019.

44. Ortman, K.; Baird, D.; Wapperom, P.; Whittington, A. Using startup of steady shear flow in a sliding plate rheometer to determine material parameters for the purpose of predicting long fiber orientation. *J. Rheol.* **2012**, *56*, 955–981. [CrossRef]

45. Wang, M.L.; Chang, R.Y.; Hsu, C.H. *Molding Simulation: Theory and Practice*; Carl-Hanser Verlag: Munich, Germany, 2018.

46. Tseng, H.C.; Chang, R.Y.; Hsu, C.H. Numerical investigations of fiber orientation models for injection molded long fiber composites. *Int. Polym. Process.* **2018**, *33*, 543–552. [CrossRef]

47. Dinh, S.M.; Armstrong, R.C. A rheological equation of state for semiconcentrated fiber suspension. *J. Rheol.* **1984**, *28*, 207–227. [CrossRef]

48. Mezi, D.; Ausias, G.; Advani, S.G.; Férec, J. Fiber suspension in 2D nonhomogeneous flow: The effects of flow/fiber coupling for Newtonian and power-law suspending fluids. *J. Rheol.* **2019**, *63*, 405–418. [CrossRef]

49. VerWeyst, B.E.; Tucker, C.L. Fiber suspension in complex geometries: Flow/orientation coupling. *Can. J. Chem. Eng.* **2002**, *80*, 1093–1106. [CrossRef]

50. Kugler, S.K.; Lambert, G.M.; Cruz, C.; Kech, A.; Osswald, T.A.; Baird, D.G. Efficient parameter identification for macroscopic fiber orientation models with experimental data and a mechanistic fiber simulation. *AIP Conf. Proc.* **2020**, *2205*, 3–8.

51. Moritzer, E.; Heiderich, G. *Fiber Length Degradation of Glass Fiber Reinforced Polypropylene during Shearing*; ANTEC: Fremont, CA, USA, 2016.

52. Cieslinski, M.J.; Wapperom, P.; Baird, D.G. Fiber orientation evolution in simple shear flow from a repeatable initial fiber orientation. *J. Non-Newton. Fluid Mech.* **2016**, *237*, 65–75. [CrossRef]

Journal of
Composites Science

Article

Measuring Fiber Length in the Core and Shell Regions of Injection Molded Long Fiber-Reinforced Thermoplastic Plaques

Abrahán Bechara * Senior and Tim Osswald

Polymer Engineering Center (PEC), University of Wisconsin-Madison, 1513 University Ave, Madison, WI 53706, USA; tosswald@wisc.edu
* Correspondence: bechara@wisc.edu; Tel.: +1-608-265-2405

Received: 29 June 2020; Accepted: 29 July 2020; Published: 31 July 2020

Abstract: Long fiber-reinforced thermoplastics are an attractive design option for many industries due to their excellent mechanical properties and processability. Processing of these materials has a significant influence on their microstructure, which controls the properties of the final part. The microstructure is characterized by the fibers' orientation, length, and concentration. Many characterization methods can capture the fiber orientation and concentration changes through the thickness in injection molded parts, but not the changes in fiber length. In this study, a technique for measuring fiber length in the core and shell regions of molded parts was proposed, experimentally verified, and used on injection molded 20 wt.% glass fiber-reinforced polypropylene plaques. The measured fiber length in the core was 50% higher than in the shell region. Comparison with simulation results shows disagreement in the shape of the through-thickness fiber length profile. Stiffness predictions show that the through-thickness changes in fiber length have little impact on the longitudinal and transverse Young's modulus.

Keywords: long fiber-reinforced thermoplastics (LFTs); core region; shell region; fiber length distribution (FLD)

---

1. Introduction

Long fiber-reinforced thermoplastics (LFTs) are increasingly being used in a number of industries and applications, mainly in the transportation industry, but also in electronics, durable consumer appliances, sporting goods, and even health care [1]. LFTs have become an attractive design option due to their improved mechanical properties over short fiber-reinforced thermoplastics (SFTs) while still being suitable for injection molding (IM) [2].

IM of discontinuous fiber composites imparts a microstructure on the molded material. This underlying structure controls the mechanical properties of the finished part [2–4]. Von Bradsky et al. stated that there are three important microstructural variables for discontinuous fiber composites which control the mechanical properties: fiber orientation distribution (FOD), fiber length distribution (FLD), and fiber content (FC) [5]. The characteristic flow pattern during mold filling and the no-slip condition on the mold walls cause fibers to re-orient, producing a distinctive configuration known as the core–shell structure [5,6]. In this structure, fibers near the mid-plane do not experience strong shearing deformations and usually align transverse to the flow direction (core); large shear strains in the regions adjacent to the core cause fibers to have strong alignment in the flow direction (shells). As the mechanical and physical properties of the final part highly depend on the microstructural variations along its thickness [7,8], great efforts have been made to accurately measure each important microstructural variable and its correlation with processing conditions [9–14].

Parallel to characterization work, many researchers have proposed mathematical models to predict the final fiber configuration in molded components. For example, attempts to characterize FOD date back as early as 1922, when Jeffery described the periodic motion of an ellipsoidal particle under the action of a simple shear flow [15]. From then, complex models such as the Folgar–Tucker model [16], the reduced strain closure (RSC), and the anisotropic rotary diffusion (ARD) [17,18] have evolved to better account for material characteristics such as anisotropy and fiber volume fraction effects. Comparatively fewer models have attempted predicting the process induced changes in FLD or FC. However, models such as the Phelps–Tucker model for fiber attrition [19] and Morris–Boulay model for fiber migration have been successfully used in mold filling simulations [20]. Much of the simulation efforts in IM of LFTs aim to provide mappable data that can be use in finite element analysis (FEA) for making mechanical property and dimensional stability predictions [21,22].

Modern measurement techniques such as image analysis of polished micrographs and micro-computed tomography (μCT) can accurately capture the local changes in FOD and FC through the thickness of molded parts [23,24]. However, current techniques to measure FLD for LFTs are limited to reporting the global fiber length over the whole part's thickness. Various studies have reported FLD measurements via μCT (Table 1); however, as high resolution is needed to differentiate individual fibers (four voxels per fiber diameter [25]), the size of the evaluated volume is limited to a few millimeters. However, parts molded with LFTs can still have fibers in the 10–15 mm range [14,19,26], far longer than what can be capture with μCT.

Table 1. Overview of fiber length measurement via micro-computed tomography (μCT) in recently published studies.

Material	Sampled Size	Voxel Size (μm)	Max Fiber Length Detected (μm)	Reference
PP-GF	$138 \times 413 \times 129\ \mu\text{m}^3$	8.73	7000	Teßmann et al. [27]
PP-GF20	$\varnothing\,4 \times 1.5\ \text{mm}^3$	3	4000	Pinter et al. [28]
PP-GF1	$4 \times 2 \times 2\ \text{mm}^3$	2	1650	Salaberger et al. [29]
Wood fiber-lignin	$4 \times 2 \times 2\ \text{mm}^3$	2.4	4000	Miettinen et al. [30]
PP-GF24	1.5 mm thick	2	1000	Köpplmayr et al. [31]
PA66-GF35	$1255 \times 1343 \times 1883\ \mu\text{m}^3$	1	1000	Hessman et al. [32]
PP-GF10-60	-	3	2000	Kastner et al. [33]

This work aims to determine FLD for the core and shell regions independently, by expanding on a currently used fiber length measurement technique [26]. Mechanical design software can benefit from having through-thickness measurements of FLD, as this additional information means having a more accurate representation of the material. Process simulation software can also benefit, as detailed fiber length data provide a better point of comparison and validation for models predicting fiber damage.

This paper presents an approach for determining FLD in the core and shell regions of IM components. First, the reasoning behind the concept is explained. Second, the characterization methods are described, and an experimental validation of the new approach is presented. Finally, the proposed technique is used for an LFT injection molded plaque and the results are compared with simulation predictions.

2. Rationale

In moldings with 50% weight fraction (wt.%) long fiber reinforced PA66, Bailey and Kraft observed significantly higher fiber length in the core compared to the shell region ($L_{N(\text{core})}$ = 1.46 mm, $L_{N(\text{shell})}$ = 0.55 mm) [13]. O'Regan and Akay also identified longer fibers in the core region ($L_{N(\text{core})}$ = 0.86 mm, $L_{N(\text{shell})}$ = 0.7 mm) for 60 wt.% long-fiber reinforced PA66 samples [34]. The standard technique for sample isolation used in these studies involved selecting a small amount of fibers with tweezers after matrix removal. Aside from the risk of short fibers being dropped or fibers breaking, the fiber population in these studies was very low (800–3500 fibers). However, to have statistical confidence, large fiber populations are required, specifically when characterizing LFTs for which the fibers' aspect ratio can vary over two orders of magnitude.

Since matrix removal is usually achieved through pyrolysis, the shortest fibers tend to fall towards the bottom as the matrix melts and burns-off [14]. Therefore, to measure FLD in either the core or the shell, such region should be isolated from the complete sample before the pyrolysis step. As the shell is generally thicker than the core and more accessible [24], we propose measuring fiber length in the shell and indirectly calculating the fiber length in the core. The extraction of the shell is addressed in Section 3.3. Fiber length in the core can be determined as follows.

FLD data are often given as an average value. However, to properly describe this type of distributions both the number- and the weight-average should be reported.

Similar to the molecular weight distribution, the number-average fiber length L_N is expressed as

$$L_N = \frac{\sum N_i l_i}{\sum N_i},\tag{1}$$

the weight-average fiber length L_W as

$$L_W = \frac{\sum N_i l_i^2}{\sum N_i},\tag{2}$$

and the total fiber length is described as

$$L_T = \sum N_i l_i\tag{3}$$

For the arbitrary LFT sample A shown in Figure 1, the averages are calculated from the complete population of fibers inside the sample's volume. Thus, it is valid to re-formulate Equations (1) and (2) by grouping the addends into sub-volumes B (shells) and C (core). The number average of the entire sample $L_{N(A)}$ can then be expressed as

$$L_{N(A)} = \frac{(\sum N_i l_i)_B + (\sum N_i l_i)_C}{(\sum_i N_i)_A}\tag{4}$$

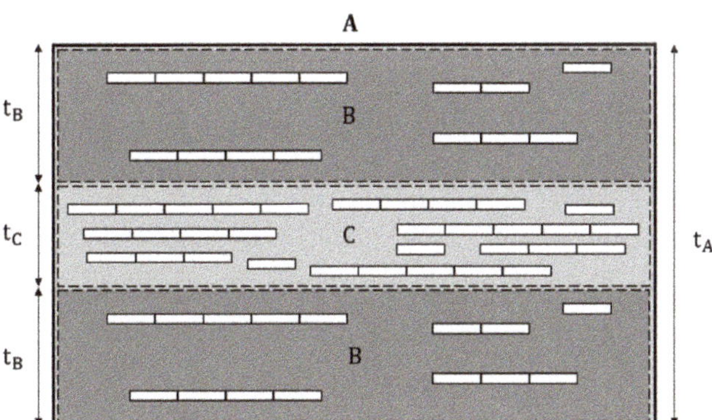

Figure 1. Schematic of a core–shell structure.

Assuming the sample's width and length are constant, Equation (4) can be formulated in terms of the local number-average fiber length

$$L_{N(A)} = \frac{L_{N(B)} t_B \phi_B + L_{N(C)} t_C \phi_C + L_{N(B)} t_B \phi_B}{t_B \phi_B + t_C \phi_C + t_B \phi_B} \quad \text{or} \quad \frac{\sum L_{N(K)} t_K \phi_K}{\sum t_K \phi_K},\tag{5}$$

where the index K represents individual layers along the thickness of the sample. The changes in fiber content (ϕ_K) have to be accounted for in order to satisfy mass conservation; that is, L_T should remain unchanged. Since the objective is determining the length in the core ($L_{N(C)}$), and both the global sample length ($L_{N(A)}$) and the shell sample length ($L_{N(B)}$) can be measured experimentally, Equation (5) can be solved for $L_{N(C)}$

$$L_{N(C)} = \frac{L_{N(A)}\left(2t_B\phi_B + t_C\phi_C\right) - 2L_{N(B)}t_B\phi_B}{t_C\phi_C} \tag{6}$$

The weight-average fiber length in the core ($L_{W(C)}$) can be calculated in the same way. This approach requires knowledge of the thickness of each layer and the through-thickness fiber content. This information can be obtained from µCT analysis.

3. Materials and Methods

3.1. Material

The material employed in this study was a 20 wt.% long glass fiber reinforced polypropylene (PPGF20, SABIC® STAMAX™). The constituent's main properties are listed in Table 2. The initial fiber length is uniform and equal to the nominal length of the pellets.

Table 2. SABIC® STAMAX™ long fiber-reinforced thermoplastic (LFT) material properties according to the material supplier.

Material Property	Value
Nominal fiber length (mm)	15.0
Fiber diameter (µm)	19 ± 1
Density of fibers (g/cm³)	2.550
Density of PP (g/cm³)	0.905
Secant modulus at 1% elongation of PP (MPa)	1800
Yield stress of PP (MPa)	37
Modulus of fibers (GPa)	73
Ultimate strength (MPa)	2600

3.2. Microstructure Measurement Techniques

To calculate the FLD information in the core region of an injection molded sample, the fiber microstructure needs to be fully characterized, starting with the global fiber length. Through-thickness FOD data are needed to identify the thickness of the core region. Additionally, through-thickness FC is required to solve for the core FLD in Equation (6). Various methods exist to quantify each of these properties. The following sections describe the techniques used in the present study.

3.2.1. µCT Analysis

Until recently, the determination of FOD involved physically sectioning the sample and analyzing the cross section via optical microscopy [9]. In the analyzed cross section, fibers are seen as ellipses and fiber orientation is quantified by measuring the aspect ratio and inclination of the ellipse's major axis. Similarly, FC has been obtained by quantifying the area fraction of the cross section covered by fibers. Alternatively, through-thickness FC can also be determined by milling thin layers and quantifying the fiber weight fraction via pyrolysis [24].

µCT technology has gained traction as a method to obtain FOD and FC in a fast and accurate way; it is a non-destructive testing method based on X-ray imaging to inspect the internal structure of a sample. For this study, FOD and FC were determined using an industrial µCT system (Metrotom 800, Carl Zeiss AG, Oberkochen, Germany). Since the fiber diameter is 19 µm, the µCT scan resolution needed to be high. Previous studies with the used material have shown that a voxel size of 5 µm adequately captures the fiber geometry [24]. Table 3 summarizes the acquisition parameters for the µCT scan.

Table 3. Micro computed tomography settings.

Parameter	Value
Voltage (V)	80
Current (A)	100
Integration Time (ms)	1000
Gain (-)	8
Voxel Size (µm)	4.5
Number of projections (-)	2200

The X-ray projections were used to reconstruct the scanned sample in 3D, after which an analysis was performed using VG StudioMAX (Version 2.2, Volume Graphics GmbH, Heidelberg, Germany) to obtain through thickness values of fiber volume fraction and second-order orientation tensor components.

3.2.2. Fiber Length Measurement

Measuring the fiber length for discontinuous fiber composites is a time-consuming task since even small samples contain millions of fibers [26]. The Polymer Engineering Center, UW-Madison, has developed a fiber length measurement technique adapting features from various measurement methods, aiming to reduce the manual input [26]. The main steps of the technique are depicted in Figure 2. A 30-mm diameter disk is cut out from the composite part and the matrix is removed via pyrolysis at 500 °C for 2.0 h. A representative subsample is extracted employing a variation of the epoxy-plug method described by Kunc [14], where UV curable resin is used instead of an epoxy. The subsample is carefully removed with tweezers and a second pyrolysis is performed to remove the resin. The loose fibers are dispersed inside a chamber using an ionized air stream and fall onto an optical glass sheet. The sheet with the fibers is scanned using a flatbed scanner (Epson Perfection V750 PRO; Seiko Epson Corporation, Nagano, Japan). The obtained digital image is optimized in Photoshop and analyzed using a Marching Ball algorithm based on the work of Wang [35]. The result is a FLD and its average values L_N and L_W. It is known that the down-sampling step skews the FLD since it preferentially captures longer fibers; thus, the Kunc correction is applied to the FLD data [14].

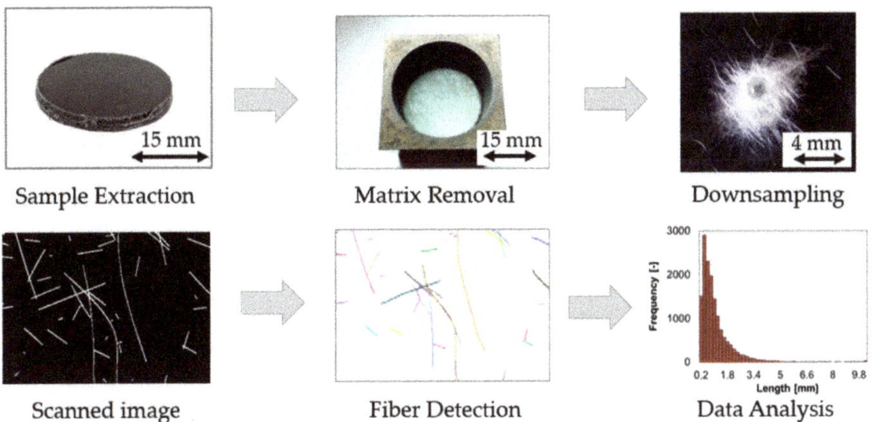

Figure 2. Overview of the steps of the employed fiber length measurement technique.

3.3. Shell Extraction and Experimental Validation

The mathematical approach to determine the fiber length in the core is described above. However, experimental validation was required before the method could be used to measure the FLD in an actual injection molded part. Since the approach is based on extracting a single shell layer and measuring its FLD, it needs to be assured that the extraction method does not damage the fibers. For this purpose, plates with an artificial core–shell structure were fabricated via compression molding, the core FLD was determined with the mathematical approach, and the result was compared with reference samples.

First, PPGF20 pellets were extruded using a single screw extruder (Extrudex Kunstoffmaschinen, Mühlacker, Germany) and a circular 3-mm diameter die, as depicted in Figure 3 (1). The extrudate was cut into 50-mm strands and placed on a rectangular mold with dimensions 50 mm × 75 mm × 1.1 mm. The strands were aligned parallel to the shorter side of the mold. The extrudate was compression molded using a hydraulic press (Carver 3889.1NE0, Carver Inc, Wabash, IN, USA) with heated platen at a temperature of 210 °C. The resulting thin plates correspond to the core layer of the artificial core–shell sample (Figure 3 (2)).

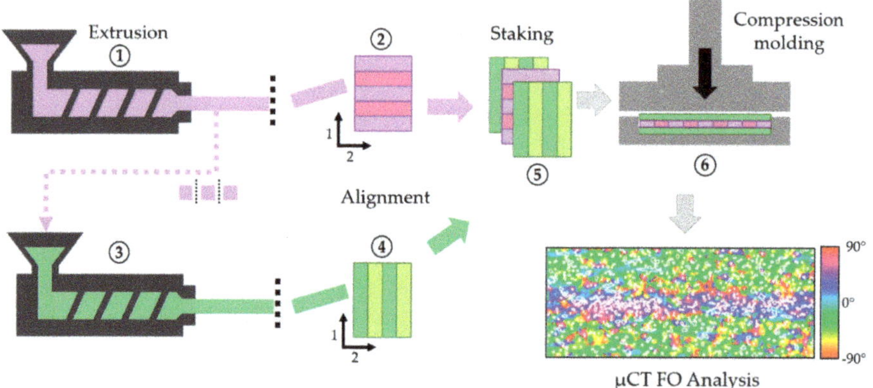

Figure 3. Sample preparation method for artificial core–shell plates.

To have different lengths for the core and shell, a fraction of the extrudate was pelletized to a length of 3.2 mm and re-extruded (3). The new extrudate was cut into 75 mm long strands, and compression molded using the same mold; in this case, the strands were aligned perpendicular to the shorter side of the mold. The resulting thin plates correspond to the shell layers of the artificial core–shell sample (4). Each core plate was stacked in between two shell plates (5), and compression molded into a 3-mm-thick plate (6). This small compression step aimed to fuse the layers together. For each molded plate variation, four specimens were manufactured in the hydraulic press. From each specimen, two samples were extracted and measured. The average fiber length of the core and shell plates and full stack was recorded to be used as a reference for the later validation (Table 4). µCT orientation analysis was performed in four specimens, which showed that distinct core and shell layers in the full stack sample were obtained (Figure 3).

Table 4. Average fiber length of compression molded plates.

Region	L_N (mm)	L_W (mm)
Shell	0.74	1.55
Core	1.49	5.03
Full stack	0.97	2.90

Using the A_{11} tensor component as guide, the thickness of the shell layer that is to remain after the material removal can be determined (highlighted in red in Figure 4a).

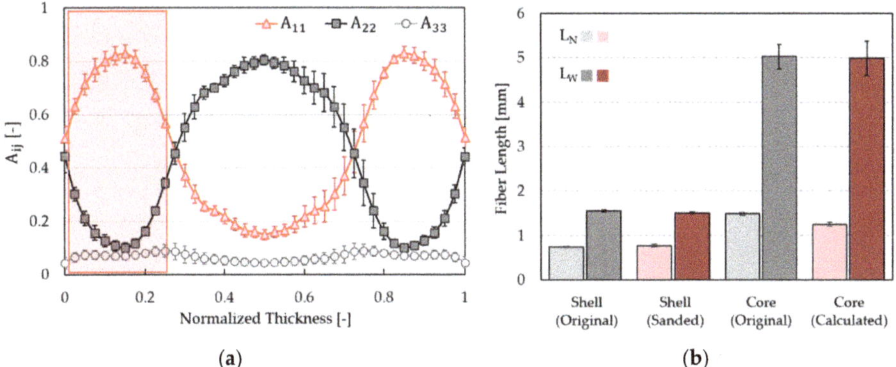

Figure 4. Artificial core–shell plate microstructure: (**a**) through-thickness fiber orientation distribution (FOD); and (**b**) fiber length values for individual core and shell layers.

The material removal is a critical step since damage to the fibers in the remaining layer must be avoided. For this step, 30-mm disks were cut out of the full stack sample and mounted in resin, in similar fashion to metallographic samples. The mounted samples were carefully grinded in two stages using a polishing disk (Autopolisher Metprep 3 PH-3, Allied High Tech Products Inc., Compton, CA, USA). In the first stage, an aggressive grinding cycle with a 180-grit sanding paper disk was used to remove around 90% of the material. In the second stage, a 600-grit sanding paper disk was used to remove the remaining material until the desired shell layer had been isolated. After the sanding process, the sample was removed from the resin and underwent the fiber length measurement procedure.

Results from this experimental validation are shown in Figure 4b. From these length values, it can be concluded that the material removal step does not affect the fiber length in a significant way. The reason the fibers are not excessively damaged is the highly planar fiber orientation in the sample (low A_{33} values) [36,37].

The main objective of this validation was to establish if the core length can be accurately determined with the approach described in Section 2. With this approach, the orientation data are used to find the thickness of each layer. The fiber length of the full stack sample and the sanded shell layer are used in Equation (6) to calculate the length in the core layer. Figure 4b shows the comparison between the fiber length of the original compression molded core layer and the calculated fiber length in the core layer. Based on these results, it can be concluded that the proposed approach can be used to measure fiber length in the shell and indirectly determine fiber length in the core, provided the off-plane orientation tensor component has a low value.

3.4. Injection Molding Plaques

A 130-ton IM machine (SM-130, Supermac Machinery, Gujarat, India) was used to mold a PPGF20 plaque with dimensions 102 mm × 305 mm × 2.85 mm (Figure 5). The processing parameters followed the suggested processing guidelines by SABIC® and are listed in Table 5.

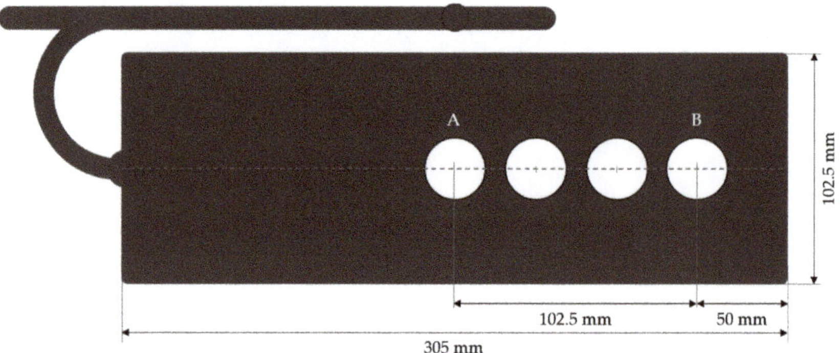

Figure 5. Sketch of the plaque geometry and illustration of the sample locations for microstructure analysis.

Table 5. Processing conditions for injection molding (IM) trials.

Molding Parameter	Value
Melt temperature (°C)	250
Mold temperature (°C)	50
Back pressure (bar)	5
Injection time (s)	2
Holding pressure (bar)	300
Holding time (s)	22

Preliminary analysis of microstructure showed well-defined core–shell layers away from the gate, between locations A and B [38]. The FOD profile, FC profile, and global FLD remained unchanged between these two locations.

In total, 16 samples were extracted for length analysis (four samples per plaque, as shown in Figure 5). Half of the samples were sanded to extract the shell layer. Additionally, μCT analysis of fiber orientation and fiber concentration was performed in locations A and B for each plaque.

4. Results and Discussion

4.1. Microstructural Analysis

The simple geometry of the injection cavity leads to a well-defined and predictable microstructure away from the gate region, where the material initially moves following a radial flow [7]. The fiber orientation analysis shows a clear transition between the core and shell regions (Figure 6a). For these particular injection trials, the core region covers about 15% of the sample thickness, which is expected of the PPGF20 material, as it has the lowest fiber content available commercially, and previous work has shown the thickness of the core region decreases with decreasing fiber content [24]. Unlike the artificial core–shell sample, there is a gradual transition in the orientation of the fibers between the central and outer layers. This can be observed in Figure 6b, which shows the 1-2 plane fiber orientation of section A-A. This section is slightly below the start of the core region, and yet it shows a wide range of colors associated to the fiber orientation.

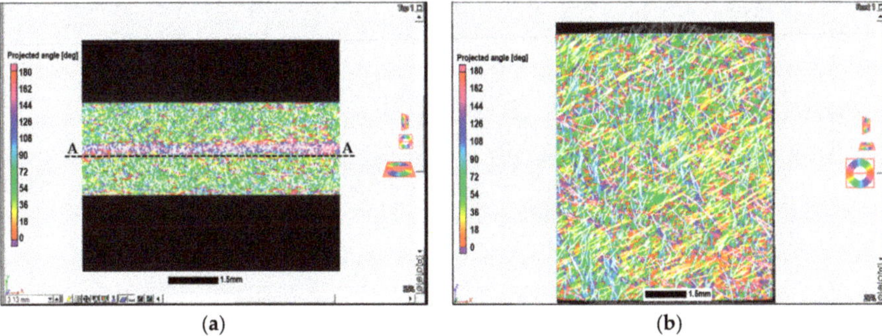

(a)   (b)

Figure 6. Reconstruction of fiber structure from micro-computed tomography (μCT) analysis using VG StudioMAX: (**a**) 2-3 plane cross section; and (**b**) 1-2 plane section A-A.

Figure 7a shows the diagonal orientation tensor components through the thickness of the plaque. Again, the characteristic core–shell structure is visible, as well as the low values of the A_{33} tensor component. The A_{33} value averaged over the sample thickness of the injection molded sample is 40% lower than the one measured in the artificial core–shell sample. These low values of the off-plane orientation tensor component are required for the length measurement approach to work.

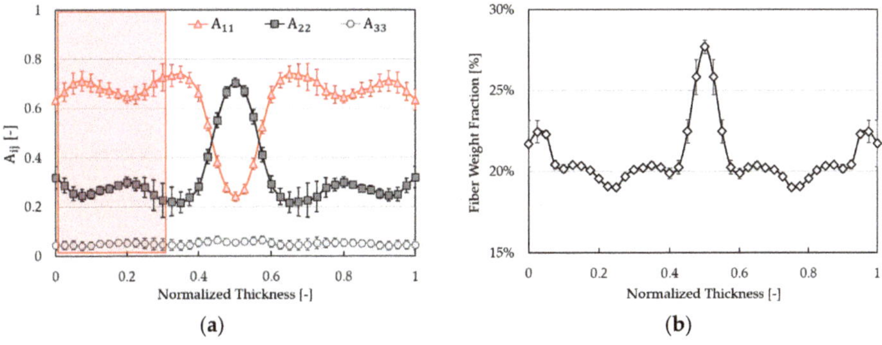

(a)   (b)

Figure 7. Injection molded plaque microstructure: (**a**) through-thickness FOD; and (**b**) through-thickness fiber content (FC).

Figure 7b shows the through-thickness fiber weight fraction. This microstructural variable also varies between the central and the outer regions. As it goes through the core region, there is a significant increase in fiber content, which is linked to the high level of alignment and little motion of the fibers in the low shear core region [39].

It has been suggested that uneven temperatures in the mold walls result in an unsymmetrical through-thickness microstructure [26,40]. The injection trials in the present work, however, showed good symmetry and are therefore considered symmetrical with respect to their mid-plane; one side of the plaque is thus a mirror image of the other.

Employing the orientation data, the shell layer to be extracted was identified (highlighted in red in Figure 7a). In this case, the extracted layer was slightly thinner than the shell region. This is to avoid measuring fibers shared by both regions. To account for the gradual transition in the microstructure, rather than calculating an average value for the core region, a distribution was used to recreate the fiber length (Equation (7)). The base line for the distribution is the fiber length measured in the shell. The spread of the distribution (σ) was adjusted to match the core thickness determined from the information in Figure 7. A factor (f) was included to scale the height of the distribution's peak until the global fiber length calculated through Equation (5) matched the experimental measurement.

$$L_N = L_{N(shell)} + \left[f \times e^{-\left(\frac{x-0.5}{\sigma}\right)^2} \right] \qquad (7)$$

As this fiber length calculation requires the through thickness fiber content values, the fiber content in the shell region was averaged, since small variations of fiber content in the shell do not imply a change in fiber length. The resulting fiber length profiles are presented in Figure 8.

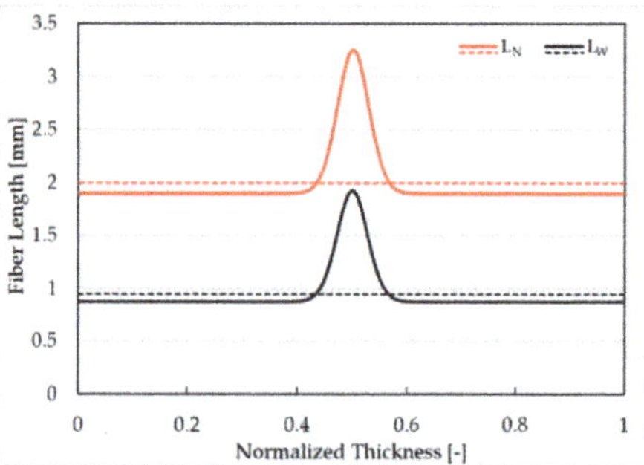

Figure 8. Experimentally determined through-thickness fiber length distribution (FLD) for the injection molded plaque. Dashed lines show global fiber length measured over the whole part's thickness.

4.2. Comparison with Length Prediction Model

While various studies develop empirical correlations to describe fiber attrition [41,42], very few mechanically based models have been used in mold filling simulations. Currently, the Phelps–Tucker model is the only one implemented in commercial software [19]. This model for fiber attrition is based on buckling failure as the driving mechanism for fiber breakage. The model uses three fitting parameters: ζ is the fiber drag coefficient which impacts the unbreakable length or steady state of the breakage process; C_B is the fiber breakage coefficient, which is a scale factor for the rate of deformation and impacts the transient portion of the breakage process; and S defines the shape of the final FLD.

Moldex3D^{TM} (Version R17, Moldex3D, Zhubei City, Taiwan) was used to run a mold filling simulation of the injection molded plaque. Process parameters were set to match the processing settings listed in Table 4. The Phelps–Tucker model parameters were manually adjusted to find a good agreement with the global fiber length measured experimentally. Additional to the three parameters, the initial fiber aspect ratio needs to be defined. The approximate nominal fiber length of 15 mm was used as the initial fiber length, and screw-induced fiber breakage was considered. Model parameters and initial aspect ratio are listed in Table 6.

Table 6. Input values for Phelps–Tucker model parameters.

Parameter	Value
Aspect ratio	700
ζ	1.1
CB	0.015
S	0.25

Figure 9 shows the through thickness L_W for both, experimental and predicted data. While the experimental length data were determined based on the thickness of the core region, the predicted length comes from a hydrodynamic stress-based failure criterion. Therefore, the predicted length profile follows the changes of the shear rate (Figure 9). Averaging the predicted L_W over the shell gives a length value just 10% lower than the experimental measurement.

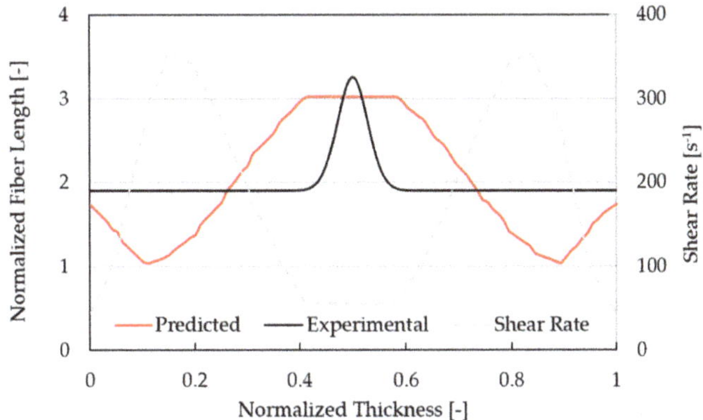

Figure 9. Comparison of experimental and predicted through-thickness FLD.

If the flat, low shear region in Figure 9 can be interpreted as the core, its thickness closely resembles the experimental core thickness. However, this variable is not dependent on the fiber attrition model. Instead, the coefficients of the Cross-WLF viscosity model used for this material are what determines the thickness of the core [43]. The predicted fiber length in the core is constant and its value is greater than the experimental L_W averaged over the core region.

By using optimized fitting parameters, we are comparing the through-thickness variation of the FLD, rather than validating the accuracy of the model. The average length obtained with Moldex3D^{TM} default parameters, underpredicted the fiber length in the region. One reason the simulation overpredicts damage with its default parameters, is due to the model's negligence of the fiber concentration effect on the rate and level of fiber damage [19]. Recent experimental studies have shown that damage increases as the nominal fiber content is increased [42,44].

4.3. Impact on Stiffness

The three microstructural variables considered until now have independent impact on the mechanical properties of the bulk material. Translating the microstructural data obtained through mold filling simulations into mechanical properties, is a critical step in the design process when using discontinuous fiber composites. To achieve this, a complex two-phase microstructure is homogenized through different micromechanical approaches to generate effective mechanical constants, that can then be used in traditional FEA simulations [45]. Many micromechanical models for non-dilute composite materials have evolved from a model originally proposed by Mori and Tanaka [46]. Tandon and Weng [47], for example, used the Mori–Tanaka approach to develop equations for the complete set of elastic constants of a short-fiber composite. Their equations describe the change of the elastic constants as function of aspect ratio and volume fraction.

The through-thickness FOD, FLD and FC were discretized into layers and used to create a representative volume element (RVE) using Digimat–MF, a mean field homogenization tool. For each layer, the tool uses as input the full orientation tensor, the fiber volume fraction, the aspect ratio distribution, and the mechanical properties of each of the phases. The Mori–Tanaka homogenization model was used to determine the stiffness constants for each individual layer and the RVE. The Mori–Tanaka model is accurate in predicting the effective properties of two-phase composites for moderate volume fractions of inclusions (around 25%). Since the maximum volume fraction measured in the sample was below 12%, it is appropriate to use the Mori–Tanaka model for the stiffness analysis.

To evaluate the impact of having through-thickness length data, a reference RVE was created. It had identical fiber orientation and fiber volume fraction, but with constant fiber length over the thickness. Figure 10 shows the aspect ratio distribution recreated from the experimental measurements for the global sample and the core and shell, independently. Since the core layer in the injection trials is thin, there is little change in the shape of the distribution between the global and shell data. The core layer in contrast has a considerable shift to the right and a wider spread compared to the global data.

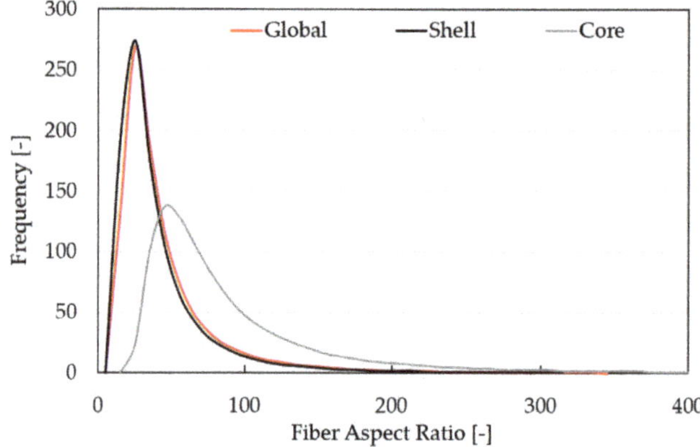

Figure 10. Aspect ratio distribution for the global sample, and the core and shell regions.

Table 7 lists the relevant longitudinal and transverse Young's modulus for the core and shell regions. In the regions marked as "Global", the global FLD was used (red line in Figure 10), and the regions marked as "Varying" used the local FLD. As expected, there is negligible change in the longitudinal stiffness in the shell, since the change in the FLD is small. Nguyen et al. performed a sensibility analysis introducing small variations to the shape of the FLD, and they concluded this had little to no impact in the mechanical properties [36]. A comparatively larger change of about 5% is observed in the transverse stiffness in the core region.

Table 7. Longitudinal and transverse Young's modulus for the core and shell layers.

Region	E_{11}	E_{22}
Shell (Global)	6243	-
Shell (Varying)	6218	-
Core (Global)	-	7418
Core (Varying)	-	7857

Stress–strain curves in the longitudinal and transverse directions for the RVE are plotted in Figure 11. The small local change of E_{22} in the core layer is effectively dissolved in the RVE, which shows no significant change of stiffness in either of the directions.

Since the global average aspect ratio for these injection trials is higher than 50, little change can be expected in the stiffness constants with increasing fiber length. Tandon and Weng theoretical equations show that these stiffness constants (E_{11}, E_{22}) had little variation at aspect ratios above 50. Schemme collected and summarized experimental data for various mechanical properties as function of the aspect ratio [2]. His work also suggests that the tensile modulus of the composite plateaus when the aspect ratio approaches 50, while other properties such as tensile strength and impact strength can still grow with fiber length and plateau much later when the aspect ratio reaches values of 400 and 1000, respectively.

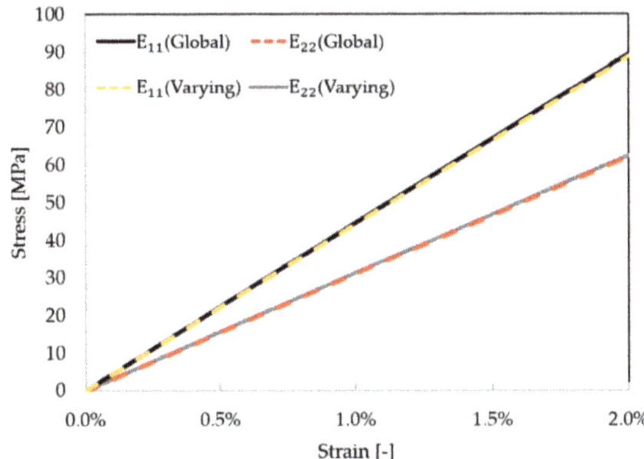

Figure 11. Stress–strain curves for representative volume element (RVE) with global fiber length and with varying fiber length.

5. Summary

The FLD in the core and shell regions of PP20GF injection molded plaques was measured with a new approach expanding on a currently used fiber length measurement technique. The approach involves extracting a single shell layer by grinding material away in a controlled manner and measuring its FLD. The FLD in the core can then be determined indirectly. Comparison with the through-thickness FLD from mold filling simulations does not show a good quantitative agreement. Stiffness predictions show that the longitudinal and transverse Young's modulus suffer little change when considering the through-thickness changes in FLD rather than a constant value over the whole thickness. This agrees with stiffness predictions from Tandon and Weng and experimental results collected by Schemme, as in both cases the tensile modulus levels-off when the fiber's aspect ratio approaches 50.

Author Contributions: A.B.S. conceived and designed the experiments, analyzed the data, and wrote the paper. T.O. supervised the project and was involved in all stages of the research. All authors have read and agreed to the published version of the manuscript.

Funding: This research was supported by the National Science Foundation under Grant No.1633967.

Acknowledgments: The authors thank SABIC® for their ongoing support and collaboration with our research group and their material supply.

Conflicts of Interest: The authors declare no conflict of interest.

References

1. Ning, H.; Lu, N.; Hassen, A.A.; Chawla, K.; Selim, M.; Pillay, S. A review of Long fibre thermoplastic (LFT) composites. *Int. Mater. Rev.* **2020**, *65*, 164–188. [CrossRef]
2. Osswald, T.A.; Ghandi, U.; Goris, S. *Discontinous Fiber Reinforced Composites*, 1st ed.; Carl-Hanser Verlag: Munich, Germany, 2020.
3. Thomason, J.L.; Vlug, M.A. Influence of fibre length and concentration on the properties of glass fibre-reinforced polypropylene: 4. Impact properties. *Compos. Part A Appl. Sci. Manuf.* **1997**, *28*, 277–288. [CrossRef]
4. Thomason, J.L. The influence of fibre length, diameter and concentration on the strength and strain to failure of glass fibre-reinforced polyamide 6,6. *Compos. Part A Appl. Sci. Manuf.* **2008**, *39*, 1618–1624. [CrossRef]
5. Von Bradsky, G.J.; Bailey, R.S.; Cervenka, A.J.; Zachmann, H.G.; Allan, P.S. Characterization of finite length composites: Part IV—Structural studies on injection moulded composites (Technical Report). *Pure Appl. Chem.* **1997**, *69*, 2523–2540. [CrossRef]
6. Toll, S.; Andersson, P.-O. Microstructure of long- and short-fiber reinforced injection molded polyamide. *Polym. Compos.* **1993**, *14*, 116–125. [CrossRef]

7. Osswald, T.A.; Menges, G. Material Science of Polymers for Engineers. In *Materials Science of Polymers for Engineers*, 3rd ed.; Carl Hanser Verlag GmbH & Co. KG: München, Germany, 2012; pp. I–XIX.

8. Advani, S.G.; Hsiao, K.-T. Introduction to composites and manufacturing processes. In *Manufacturing Techniques for Polymer Matrix Composites (PMCs)*; Elsevier: Amsterdam, The Netherlands, 2012; pp. 1–12.

9. Bay, R.S.; Tucker, C.L. Stereological measurement and error estimates for three-dimensional fiber orientation. *Polym. Eng. Sci.* **1992**, *32*, 240–253. [CrossRef]

10. Bernasconi, A.; Cosmi, F.; Hine, P.J. Analysis of fibre orientation distribution in short fibre reinforced polymers: A comparison between optical and tomographic methods. *Compos. Sci. Technol.* **2012**, *72*, 2002–2008. [CrossRef]

11. Hegler, R.P.; Mennig, G. Phase separation effects in processing of glass-bead- and glass-fiber-filled thermoplastics by injection molding. *Polym. Eng. Sci.* **1985**, *25*, 395–405. [CrossRef]

12. Lafranche, E.; Krawczak, P.; Ciolczyk, J.-P.; Maugey, J. Injection moulding of long glass fiber reinforced polyamide 66: Processing conditions/microstructure/flexural properties relationship. *Adv. Polym. Technol.* **2005**, *24*, 114–131. [CrossRef]

13. Bailey, R.; Kraft, H. A Study of Fibre Attrition in the Processing of Long Fibre Reinforced Thermoplastics. *Int. Polym. Process.* **1987**, *2*, 94–101. [CrossRef]

14. Kunc, V.; Frame, B.; Nguyen, B.N.; Tucker, C.L.; Velez-Garcia, G. Fiber length distribution measurement for long glass and carbon fiber reinforced injection molded thermoplastics. *Res. Gate* **2007**, *2*, 866–876.

15. Jeffery, G.B. The motion of ellipsoidal particles immersed in a viscous fluid. *Proc. R. Soc. Lond. Ser. A Contain. Pap. Math. Phys. Character* **1922**, *102*, 161–179.

16. Folgar, F.; Tucker, C.L. Orientation Behavior of Fibers in Concentrated Suspensions. *J. Reinf. Plast. Compos.* **1984**, *3*, 98–119. [CrossRef]

17. Wang, J.; O'Gara, J.F.; Tucker, C.L. An objective model for slow orientation kinetics in concentrated fiber suspensions: Theory and rheological evidence. *J. Rheol.* **2008**, *52*, 1179–1200. [CrossRef]

18. Phelps, J.H.; Tucker, C.L. An anisotropic rotary diffusion model for fiber orientation in short- and long-fiber thermoplastics. *J. Nonnewton. Fluid Mech.* **2009**, *156*, 165–176. [CrossRef]

19. Phelps, J.H.; Abd El-Rahman, A.I.; Kunc, V.; Tucker, C.L. A model for fiber length attrition in injection-molded long-fiber composites. *Compos. Part A Appl. Sci. Manuf.* **2013**, *51*, 11–21. [CrossRef]

20. Morris, J.F.; Boulay, F. Curvilinear flows of noncolloidal suspensions: The role of normal stresses. *J. Rheol.* **1999**, *43*, 1213–1237. [CrossRef]

21. Wan, X.F.; Pan, Y.; Liu, X.D.; Shan, Y.C. Influence of Material Anisotropy on Long Glass Fiber Reinforced Thermoplastics Composite Wheel: Dynamic Impact Simulation. In *Volume 14: Emerging Technologies; Safety Engineering and Risk Analysis; Materials: Genetics to Structures*; ASME: New York, NY, USA, 2015; Volume 14, pp. 1–8.

22. Desplentere, F.; Soete, K.; Bonte, H.; Debrabandere, E. Local mechanical properties of LFT injection molded parts: Numerical simulations versus experiments. In *AIP Conference Proceedings*; American Institute of Physics: College Park, MD, USA, 2014; Volume 1593, pp. 641–645.

23. Thi, T.B.N.; Morioka, M.; Yokoyama, A.; Hamanaka, S.; Yamashita, K.; Nonomura, C. Measurement of fiber orientation distribution in injection-molded short-glass-fiber composites using X-ray computed tomography. *J. Mater. Process. Technol.* **2015**, *219*, 1–9. [CrossRef]

24. Goris, S.; Osswald, T.A. Process-induced fiber matrix separation in long fiber-reinforced thermoplastics. *Compos. Part A Appl. Sci. Manuf.* **2018**, *105*, 321–333. [CrossRef]

25. Baranowski, T.; Dobrovolskij, D.; Dremel, K.; Hölzing, A.; Lohfink, G.; Schladitz, K.; Zabler, S. Local fiber orientation from X-ray region-of-interest computed tomography of large fiber reinforced composite components. *Compos. Sci. Technol.* **2019**, *183*, 107786. [CrossRef]

26. Goris, S.; Back, T.; Yanev, A.; Brands, D.; Drummer, D.; Osswald, T.A. A novel fiber length measurement technique for discontinuous fiber-reinforced composites: A comparative study with existing methods. *Polym. Compos.* **2018**, *39*, 4058–4070. [CrossRef]

27. Teßmann, M.; Mohr, S.; Gayetskyy, S.; Haßler, U.; Hanke, R.; Greiner, G. Automatic Determination of Fiber-Length Distribution in Composite Material Using 3D CT Data. *EURASIP J. Adv. Signal Process.* **2010**, *2010*, 545030. [CrossRef]

28. Pinter, P.; Bertram, B.; Weidenmann, K.A. A Novel Method for the Determination of Fibre Length Distributions from μCT-data. In Proceedings of the 6th Conference on Industrial Computed Tomography (iCT 2016), Wels, Austria, 9–12 February 2016; pp. 1–8.

29. Salaberger, D.; Kannappan, K.A.; Kastner, J.; Reussner, J.; Auinger, T. Evaluation of Computed Tomography Data from Fibre Reinforced Polymers to Determine Fibre Length Distribution. *Int. Polym. Process.* **2011**, *26*, 283–291. [CrossRef]
30. Miettinen, A.; Ojala, A.; Wikström, L.; Joffe, R.; Madsen, B.; Nättinen, K.; Kataja, M. Non-destructive automatic determination of aspect ratio and cross-sectional properties of fibres. *Compos. Part A Appl. Sci. Manuf.* **2015**, *77*, 188–194. [CrossRef]
31. Köpplmayr, T.; Milosavljevic, I.; Aigner, M.; Hasslacher, R.; Plank, B.; Salaberger, D.; Miethlinger, J. Influence of fiber orientation and length distribution on the rheological characterization of glass-fiber-filled polypropylene. *Polym. Test.* **2013**, *32*, 535–544. [CrossRef]
32. Hessman, P.A.; Riedel, T.; Welschinger, F.; Hornberger, K.; Böhlke, T. Microstructural analysis of short glass fiber reinforced thermoplastics based on x-ray micro-computed tomography. *Compos. Sci. Technol.* **2019**, *183*, 107752. [CrossRef]
33. Kastner, J.; Plank, B.; Salaberger, D. High resolution X-ray computed tomography of fibre- and particle-filled polymers. In Proceedings of the 18th World Conference on Nondestructuve Testing, Durban, South Africa, 16–20 April 2012; pp. 16–20.
34. O'Regan, D.; Akay, M. The distribution of fibre lengths in injection moulded polyamide composite components. *J. Mater. Process. Technol.* **1996**, *56*, 282–291. [CrossRef]
35. Wang, H. Accurate length measurement of multiple cotton fibers. *J. Electron. Imaging* **2008**, *17*, 031110. [CrossRef]
36. Nguyen, B.N.; Bapanapalli, S.K.; Holbery, J.D.; Smith, M.T.; Kunc, V.; Frame, B.J.; Phelps, J.H.; Tucker, C.L., III. Fiber Length and Orientation in Long-Fiber Injection-Molded Thermoplastics—Part I: Modeling of Microstructure and Elastic Properties. *J. Compos. Mater.* **2008**, *42*, 1003–1029. [CrossRef]
37. Bay, R.S.; Tucker, C.L., III. Fiber Orientation in Simple Injection Moldings. Part II: Experimental Results. *Polym. Compos.* **1992**, *13*, 332–341. [CrossRef]
38. Goris, S. Characterization of the Process-Induced Fiber Configuration of Long Glass Fiber-Reinforced Thermoplastics. Ph.D. Thesis, University of Wisconsin, Madison, WI, USA, 2017.
39. Phelps, J.H. *Processing-Microstructure Models for Short- and Long-Fiber Thermoplastic Composites*; University of Illinois at Urbana-Champaign: Champaign, IL, USA, 2009.
40. Rohde, M.; Ebel, A.; Wolff-Fabris, F.; Altstädt, V. Influence of processing parameters on the fiber length and impact properties of injection molded long glass fiber reinforced polypropylene. *Int. Polym. Process.* **2011**, *26*, 292–303. [CrossRef]
41. Shon, K.; Liu, D.; White, J.L. Experimental Studies and Modeling of Development of Dispersion and Fiber Damage in Continuous Compounding. *Int. Polym. Process.* **2005**, *20*, 322–331. [CrossRef]
42. Moritzer, E.; Gilmar, H. *Fiber Length Degradation of Glass Fiber Reinforced Polypropylene During Shearing*; ANTEC: Fremont, CA, USA, 2016; pp. 647–651.
43. Kantz, M.R.; Newman, H.D.; Stigale, F.H. The Skin-Core Morphology and Structureproperty Relationships in Injection-Molded Polypropylene. *J. Appl. Polym. Sci.* **1972**, *16*, 1249–1260. [CrossRef]
44. Goris, S. Experimental study on fiber attrition of long glass fiber-reinforced thermoplastics under controlled conditions in a couette flow. In Proceedings of the Annual Technical Conference—ANTEC, Anaheim, CA, USA, 8–10 May 2017; Volume 2017.
45. Tucker, C.L., III; Liang, E. Stiffness predictions for unidirectional short-fiber composites: Review and evaluation. *Compos. Sci. Technol.* **1999**, *59*, 655–671. [CrossRef]
46. Lomov, S.V.; Abdin, Y.; Jain, A. Mori-tanaka methods for micromechanics of random fibre composites. In Proceedings of the ICCM International Conferences on Composite Materials, Copenhagen, Denmark, 19–24 July 2015; Volume 2015.
47. Tandon, G.P.; Weng, G.J. The effect of aspect ratio of inclusions on the elastic properties of unidirectionally aligned composites. *Polym. Compos.* **1984**, *5*, 327–333. [CrossRef]

Journal of
Composites Science

Article

Validation of Fiber Breakage in Simple Shear Flow with Direct Fiber Simulation

Tzu-Chuan Chang [1,*], Abrahán Bechara Senior [1], Hakan Celik [2], Dave Brands [3], Angel Yanev [3] and Tim Osswald [1]

[1] Polymer Engineering Center, University of Wisconsin-Madison, Madison, WI 53706, USA; bechara@wisc.edu (A.B.S.); tosswald@wisc.edu (T.O.)
[2] IKV Institute for Plastics Processing in Industry and Craft, RWTH Aachen University, 52074 Aachen, Germany; Hakan.Celik@ikv.rwth-aachen.de
[3] Global Application Technology, SABIC, 6160 AH Geleen, The Netherlands; Dave.Brands@SABIC.com (D.B.); Angel.Yanev@SABIC.com (A.Y.)
* Correspondence: tchang68@wisc.edu

Received: 20 August 2020; Accepted: 7 September 2020; Published: 10 September 2020

Abstract: This study aims to use particle level simulation to simulate the breakage behavior of glass fibers subjected to simple shear flow. Each fiber is represented as a chain of rods that experience hydrodynamic, interaction, and elastic effects. In order to validate the approach of the model, the simulation results were compared to simple shear flow experiments conducted in a Couette Rheometer. The excluded volume force constants and critical fiber breakage curvature were tuned in the simulation to gain a better understanding of the system. Relaxation of the fiber clusters and a failure probability theory were introduced into the model to solve the fiber entanglement and thus, better fit the experimental behavior. The model showed agreement with the prediction on fiber length reduction in both number average length and weight average length. In addition, the simulation had a similar trend of breakage distribution compared to a loop test using glass fibers.

Keywords: long fiber reinforced plastics; fiber breakage; fiber length; mechanistic model

1. Introduction

Discontinuous fiber-reinforced polymers play a significant role in industrial applications due to their lightweight performance and lower manufacturing costs [1–3]. Currently in the industry, injection molding and compression molding of long fiber-reinforced thermoplastics are widely used to produce parts with outstanding mechanical properties [4]. For example, during injection molding, fiber breakage already starts in the plasticizing section of the machine and continues as the material is forced through the gate and mold filling. There are various modes of fiber breakage—from the melting process to the high shear during mold filling. It has been shown that a big contributor to fiber attrition is the melting zone of the plasticating unit, leading to a significantly shorter fiber length when polymer enters the metering zone from the compression zone [5]. As fiber length is crucial in improving the mechanical performance of a molded product, an increase in fiber length correlates in the increased strength of the product. Researchers have also found that parts are stronger in the direction of fiber alignment if fiber length and volume fraction are increased [6]. During polymer processing, however, fibers in the polymer melt often break because they are subjected to intense viscous forces during flow and deformation. Therefore, it is beneficial to understand the mechanism of fiber breakage during flow, in particular shear flow, which is dominant. This will allow the engineer to optimize the process to reduce the length degradation and, thus, gain better mechanical properties. However, there are still some aspects of processing that are not well understood, which will be addressed further in the paper.

To estimate the mechanical behavior of a product, it is necessary to predict the process-induced fiber breakage. Forgacs et al. [7] proposed in 1959 that the critical shear stress that provokes fiber buckling is described by the fibers' Young modulus and their geometrical properties. From this, it becomes clear that with high aspect ratios, the fibers tend to break under low loadings. Later, Hinch [8] used the Slender Body theory to obtain the deformation of an ideally elastic particle in shear flow. Using the Couette rheometer, Goris et al. [9] developed an experimental setup in combination with fiber length measurement to obtain repeatable length degradation of glass fibers at different fiber concentration, initial fiber length, residence time, melt temperature, and processing speed. This setup provides good insight into measuring fiber length. However, there is no simulation tool that can accurately predict final fiber length in a molded part and the phenomena of fiber breakage is not fully understood. Therefore, numerical simulation is applied to investigate the phenomena of fiber breakage. Some continuum models which are very different from particle level simulations have been developed to obtain the macroscopic picture of breakage by solving the balance equation of fiber length distribution [2,5]. Phelps et al. [2] presented a quantitative model to describe fiber attrition, which is based on buckling as the driving mechanism for fiber breakage during processing. However, Phelps did not explicitly state that the drag coefficient is independent on fiber concentration, which was shown from the Couette rheometer experiments [5]. To fully understand the micromechanical picture of fiber breakage, particle simulation is necessary to develop to understand the details of concentrated fiber suspension dynamics [10–14]. Single particle models are not accurate enough and well developed to investigate the degradation mechanisms at the fiber level due to expensive computation. Therefore, a particle level model developed at the Polymer Engineering Center (PEC) at the University of Wisconsin-Madison will be extended to gain a better understanding of fiber damage.

In this work, a simulation is conducted via a mechanistic model, to understand the effect of fiber breaking curvature and the magnitude of penalty forces that prevent fiber from overlapping during simulation to the fiber damage under shear flow condition, neglecting the effect of melting. In addition, probability breakage is introduced to the system to increase the reliability of the model and better describe the uncertainty of breakage in numerical terms. Moreover, to relieve the entanglement between fibers when generating an initial fiber cluster used for the simulation, a relaxation step is applied to reduce the initial breakage caused by an unsteady system. Then, the mechanistic model is used to predict fiber breakage in a Couette flow rheometer. By comparing the model's predictions to the experimental curve, the influence of adjusting parameters in the model to different conditions is assessed.

2. Direct Fiber Model

In the mechanistic model, which is based on a work done by Schmid et al. [11], models fibers are chains of rigid cylindrical rods, as shown Figure 1. At each segment node in a fiber, the position x_i, the velocity u_i and the angular velocity ω_i are calculated. Additionally, the segments experience hydrodynamic effects, fiber-fiber interaction, excluded volume effects, and elastic deformation within the flow field.

The fibers are immersed in a homogeneous shear flow, which is at a low Reynolds number; therefore, inertial effects can be neglected based on the experiments by Hoffman [15] and Barnes [16]. In addition, long-range hydrodynamic interactions are neglected due to high viscosity for polymers [17]. Furthermore, as fluid forces are not high enough to cause fibers to stretch or have torsional deformation, only bending deformation is included in the model. Buoyant effects are neglected as well [18,19].

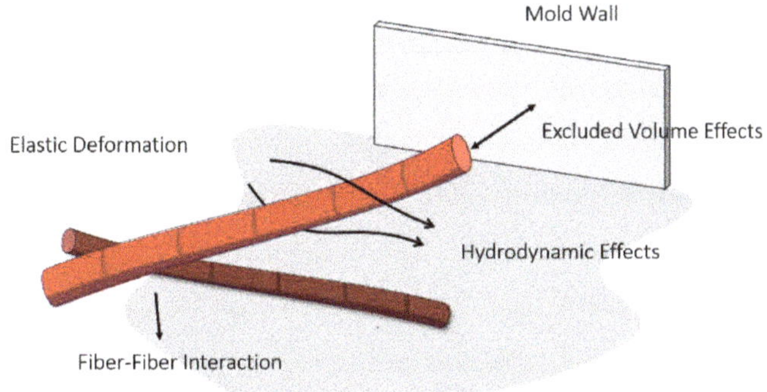

Figure 1. Particle level simulation: modeling single fiber and macroscale interactions.

The excluded volume force stops fibers from overlapping and is used to model inter-fiber interaction; it is implemented as a discrete penalty method. Discretizing the fibers into more than two nodes or one beam allows fiber bending to occur, where the back coupling from fiber motion to fluid is not considered due to expensive computational cost [18]; the force equilibrium are written as:

$$0 = \underline{F}_i^H + \sum_j \underline{F}_{ij}^C + \underline{X}_i - \underline{X}_{i+1} \tag{1}$$

where $\underline{F}_i^H$ is the drag forces from the surrounding fluid, $\underline{F}_{ij}^C$ the inter-fiber interaction force with rod j, and $\underline{X}_i$, $\underline{X}_{i+1}$ the intra-fiber forces exerted by adjacent rods.

Likewise, the moment equilibrium is analogous but includes an elastic recovery term $\underline{M}^b$ and a hydrodynamic torque $\underline{T}^H$, where $\underline{r}_{ij}$ describes the shortest distance vector between two segments:

$$0 = \underline{T}_i^H - \underline{r}_i \times \underline{X}_{i+1} + \sum_j \underline{r}_{ij} \times \underline{F}_{ij}^C + \underline{M}_i^b - \underline{M}_{i+1}^b \tag{2}$$

Additionally, if a fiber has more than one segment, an extra constraint that enforces connectivity between the different segments is used with $\underline{u}_i$, the speed of rod i, and angular speed $\underline{\omega}_i$:

$$0 = \underline{u}_i + \underline{\omega}_i \times \underline{r}_i - \underline{u}_{i+1} \tag{3}$$

Using a chain of beads to represent the rod-like geometry of a fiber for hydrodynamic effects reduces the complex solution as compared with an ellipsoid geometry [20]. The hydrodynamic force $\underline{F}_i^H$ is calculated as the summation of forces experienced by the beads $\underline{F}_k^H$ and is given as:

$$\underline{F}_i^H = \sum_{k=1}^m \underline{F}_k^H \tag{4}$$

where $\underline{F}_k^H$ is the hydrodynamic force and k describes the number of beads given by Stokes law:

$$\underline{F}_k^H = 6\pi\mu a\left(\underline{u}_k^\infty - \underline{u}_k\right) \tag{5}$$

where $\underline{u}_k^\infty$ is the surrounding fluid velocity, a is the radius of the bead, and $\underline{u}_k$ is the velocity bead k, which can also be represented as $(\underline{u}_k = \underline{u}_i + \underline{\omega}_i \times \underline{r}_k)$. This allows the final $\underline{F}_i^H$ equation to be written as shown below:

$$\underline{F}_i^H = 6\pi\mu a\left(\sum_{k=1}^m \underline{u}_k^\infty - m\underline{u}_i - \underline{w}_i \times \sum_{k=1}^m \underline{r}_k\right) \tag{6}$$

Similarly, the torque exerted on a rod is:

$$\underline{T}^H = \sum_{k=1}^{m} \underline{T}_k^H \tag{7}$$

where the hydrodynamic contribution $\underline{F}_k^H$ of bead k, where $\underline{\Omega}_k^\infty$ is the vorticity of the surrounding fluid and $\underline{\omega}_k$ is the angular velocity of the bead k.

$$\underline{T}_k^H = 8\pi\mu a^3\left(\underline{\Omega}_k^\infty - \underline{w}_k\right) + 6\pi\mu a \underline{r}_k \times \left(\underline{u}_k^\infty - \underline{u}_i - \underline{w}_i \times \underline{r}_k\right) \tag{8}$$

By substituting the expression of $\underline{T}_k^H$, the expression of $\underline{u}_k$ into the fluid's vorticity can be written as:

$$\underline{T}^H = 8\pi\mu a^3\left(\sum_{k=1}^{m}\underline{\Omega}_k^\infty - m\underline{w}_i\right) + 6\pi\mu a \underline{r}_k \times \left(\sum_{k=1}^{m}\left(\underline{r}_k \times \underline{u}_k^\infty\right) + \underline{u}_i \times \sum_{k=1}^{m}\underline{r}_k - \sum_{k=1}^{m}\underline{r}_k \times \left(\underline{w}_i \times \underline{r}_k\right)\right) \tag{9}$$

Using the elastic beam theory, the approach below is similar to Schmid [11]. The bending moment of a fiber where the radius of curvature of a beam subjected to pure bending is given by:

$$\frac{1}{\rho} = \frac{M}{EI} \tag{10}$$

where M is the bending moment, E the Young's modulus, ρ the radius of curvature of the beam, and I is the inertial moment of the beam's cross section.

On approximation by linear segments, which is connected with elastic joints, the bending moment will be related to the length of the segment ℓ and angle α; then:

$$\underline{M}_i^b = \frac{\alpha IE}{\ell}\underline{e} \tag{11}$$

$$\underline{e} = \frac{\underline{r}_i \times \underline{r}_{i-1}}{\left|\underline{r}_i \times \underline{r}_{i-1}\right|} \tag{12}$$

The model also includes mechanical interaction between fibers. The fiber-fiber interaction force is the sum of a normal force and a tangential force, as seen below, where $\underline{F}^N$ is the normal force, and $\underline{F}^T$ is the tangential force representing the friction between rods.

$$\underline{F}_{ij}^C = \underline{F}_{ij}^N + \underline{F}_{ij}^T \tag{13}$$

The collision response between fibers is represented as a discrete penalty method [21,22]. The penalty method implemented in the model starts with selecting a force dependent on the penetration distance. The equation below of the excluded volume force is often used in a particle level simulation for fiber suspension [11,23,24], where A and B are parameters [25,26], d_{ij} is the shortest distance between rod i and rod j, b is the fiber radius, and $\underline{n}_{ij}$ is the vector along the closest distance between the rods.

$$\underline{F}_{ij}^N = A\,\exp\left[-B\left(\frac{2d_{ij}}{D} - 2\right)\right]\underline{n}_{ij} \tag{14}$$

The force increases exponentially as fibers get closer. In this research, B is chosen as 2 [18], and A is tuned for each system which represent as excluded volume force constant later.

The friction force between segment i and j is calculated as a force in the direction of the relative velocity of the rods and is computed using the equation below, where μ_f is the coulomb coefficient between fibers and Δu_{ij} is the relative velocity between segments i and j.

$$\underline{F}_{ij}^{T} = \mu_f \left| \underline{F}_{ij}^{N} \right| \frac{\Delta u_{ij}}{\left| \Delta u_{ij} \right|} \tag{15}$$

Due to the non-linear behavior, the model uses the previous time step to calculate future steps. For this reason, the initial time step is zero to avoid the absence of data points.

3. Fiber Deformation and Breakage

In the mechanistic model, bending of a fiber is the only mechanism attributed to fiber deformation. Elongation and shear-deformation due to tensile, compression, and shear loads are neglected. Thus, the bending behavior is implemented with the elastic beam theory. In the model, the forces experienced by the fibers within the polymer matrix, which lead to bending and breaking, are approximated within the linear segments interconnected with flexible joints. To determine the breaking point, the local degree of bending is characterized by the radius of curvature at the connection points of two rods, as shown in Figure 2. Furthermore, the critical curvature is used as an input parameter for the model to initiate fiber breakage. Once a fiber's segment curvature is below the assigned input parameter, the fiber will break at the joint of two connecting rods.

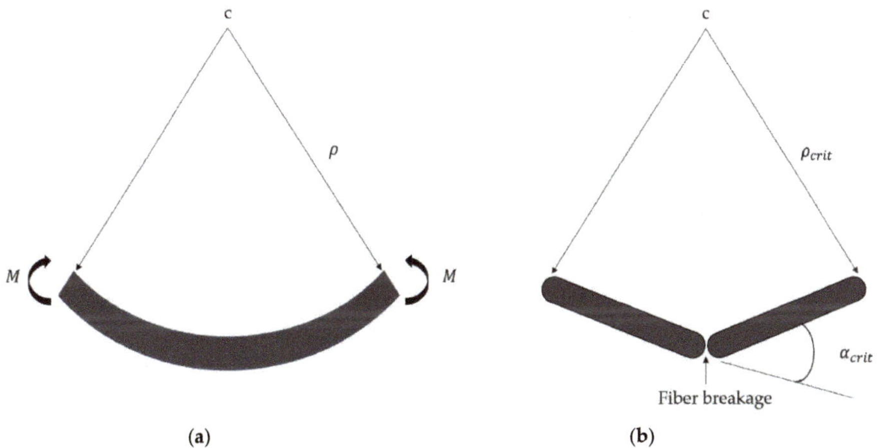

Figure 2. Fiber deformation and approximation with bending theory: (**a**) Beam deformation with pure bending; (**b**) Approximation with rods and the expression of critical curvature during fiber breakage.

To describe the fiber behavior during breakage under realistic conditions, a bending method was presented by Sinclair [27] using glass fibers. The tensile strength and Young's modulus were measured by twisting a loop in a fiber and pulling the ends until the loop breaks. To understand the phenomena in a microscopic level of fiber breakage, a test was further developed and improved to obtain the critical radius of curvature for individual fibers [28]. The tested fibers exhibit an elastic modulus of 73,000 N/mm^2 and a tensile strength of 2600 N/mm^2. To measure the critical radius of curvature at the break point, the glass fiber was bent into a loop and placed between two glass slides; a drop of glycerin was added to prevent the loop from unfolding and to create a film between the slides. This ensured the fiber would not break due to friction. One end of the loop was fixed to the bottom slide and the other was attached to an actuator. The setup was placed under a microscope and a video was recorded while the actuator pulled the end of the fiber until breakage occurred, as seen in Figure 3. The last frame when the loop was still intact was captured and the radius of curvature was measured. A total of 48 experiments were performed and the non-Gaussian distribution resulted in an average value of 204.6 µm for the critical radius of curvature. The values varied from 119 to 371 µm, as seen in Figure 4.

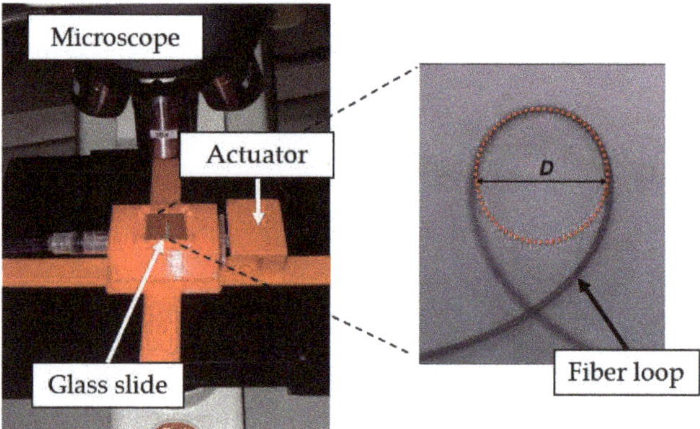

Figure 3. An example of observing glass fiber breakage under microscope level in a loop test.

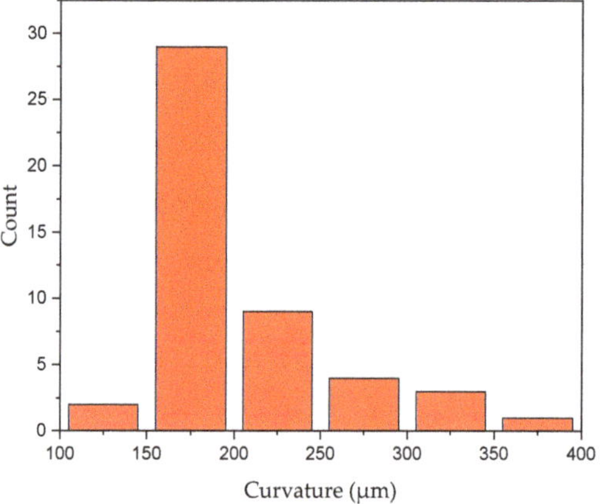

Figure 4. Breakage Distribution of the loop test using glass fibers.

4. Direct Fiber Simulation

The mechanistic model was first implemented to examine the effect of the excluded volume force constant. To validate the model, the result was compared with the Couette flow experiments using glass fiber-reinforced polypropylene (PP). To set up a simulation, 1000 fibers with equal length of 2.5 mm were placed in a shear cell (Figure 5). The segment length in each fiber is a fixed value of 0.1 mm with aspect ratio of 5.26 to ensure flexible rotation in the flow field [18]. In addition, Lee-Edwards periodic boundaries [29] were applied to all the cell walls to represent periodic conditions during the simulation. A simple shear field with a shear rate of 16.65 s^{-1} in x–y plane was applied to the polymer matrix, which corresponds to the hydrodynamic forces discussed previously. The material used in this work was SABIC® STAMAX (Saudi Basic Industries Corporation (SABIC), Riyadh, Saudi Arabia), a reinforced polypropylene material. Table 1 shows the physical properties of fibers and the matrix.

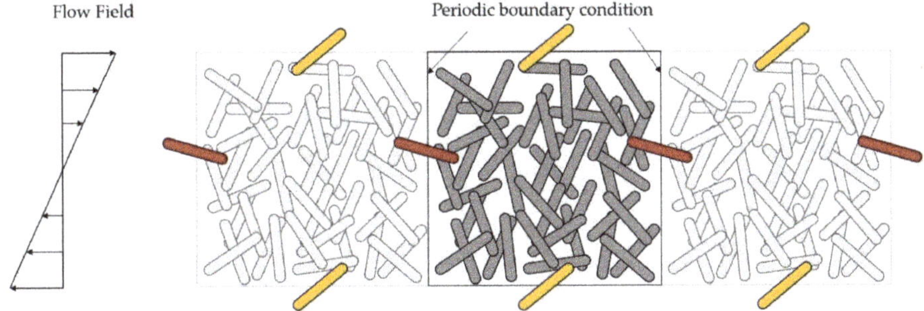

Figure 5. Shear cell with periodic boundary conditions.

Table 1. Constant values of shear cell properties in simulation.

Parameter	Value
Viscosity (Matrix) [Pa s]	200
Shear Rate [s^{-1}]	16.65
Fiber Young's Modulus [GPa]	73
Fiber diameter [μm]	19
Fiber Weight Fraction [%]	30
Time step [s]	10^{-6}–10^{-8}

Stated in ref. [18], the values of A and B in Equation (14) for a hydrodynamic effect still remain unknown and need to be adjusted for different processing conditions. Values A and B have been chosen empirically, which are usually set in such a way that no fiber intersections are perceived, nor high repulsive forces are created. As suggested by [18], B was chosen to be 2 and A was tuned for each system with the relationship of *shear rate* × *viscosity*/2 for the mechanistic model simulation, which is 1665 in this case. Figure 6 shows the number average length (L_n) and the weight average length (L_w) for the simulation. It is clear that when using the empirical method, breakage occurs too fast, when compared to experimental data. This shows that the traditional algorithm used for the simulation, where $A = 1665$, results in much higher excluded forces than the actual values. This causes fibers to break as they move closer to each other within the cluster. Thus, it is necessary and critical to find a proper repeatable method to determine the value for simulating fiber breakage.

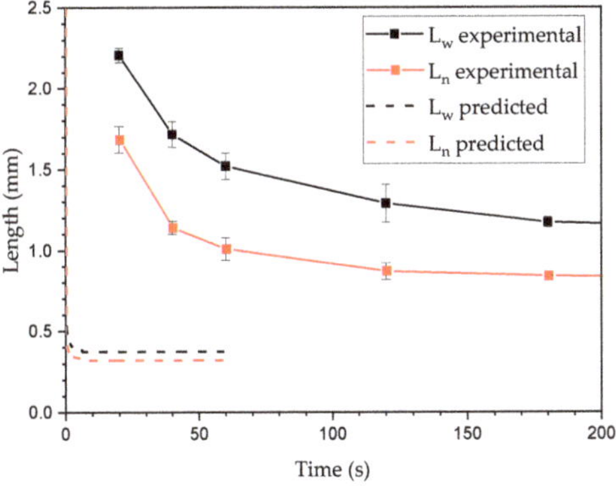

Figure 6. Length evolution result by a mechanistic model using an empirical excluded volume force constant, compared to the Couette flow experiments.

To further validate the model, the critical fiber breaking curvature and excluded volume force constant *A* were tuned to examine the variation of fiber length reduction. Table 2 shows variables used in the simulation.

Table 2. Values varied in simulation.

Parameter	Value
Fiber Critical Curvature [μm]	200, 250, and 300
Excluded Volume Force Constant [-]	250, 500, and 750

Figure 7 shows the 5 s time evolution of L_n at varying pre-demand breaking curvature from 200 μm, 250 μm, and 300 μm while keeping the same excluded volume force constant. As fibers start to bend, they first reach a larger critical curvature and result in a faster rate of breakage. On increasing the force constant, the fiber experiences higher repulsive force as they approach surrounding fibers. As two fibers approach each other, the force increases until it reaches the maximum excluded volume force, which is determined by the value of constant *A* in Equation (14). Thus, the influence of excluded volume force on fiber length reduction becomes more significant with a higher critical breaking curvature. This trend can be seen in Figure 8, where the excluded volume force constant varies from 250 to 750, while breaking curvature is the same.

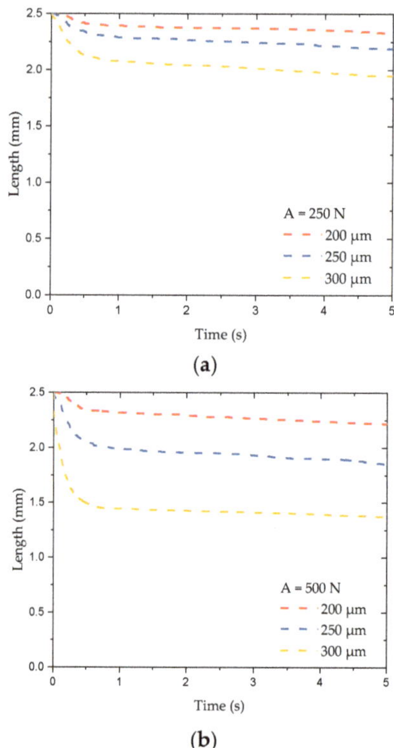

(a)

(b)

Figure 7. *Cont.*

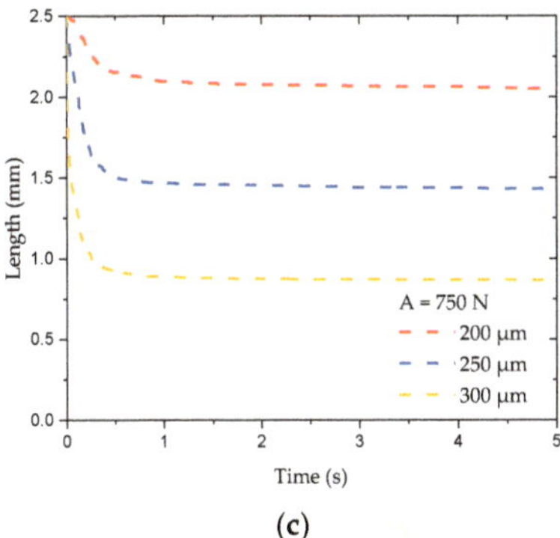

(c)

Figure 7. Predicted L_n of simple shear flow simulation at varying fiber critical breaking curvature with three excluded volume force constants A from: (**a**) 250 N, (**b**) 500 N, (**c**) 750 N.

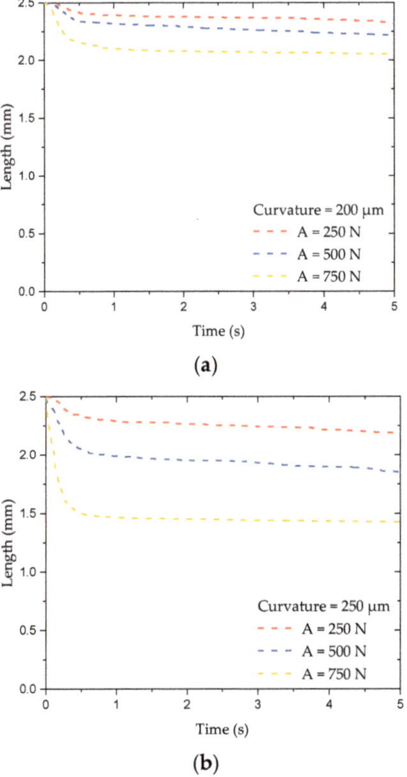

(a)

(b)

Figure 8. Cont.

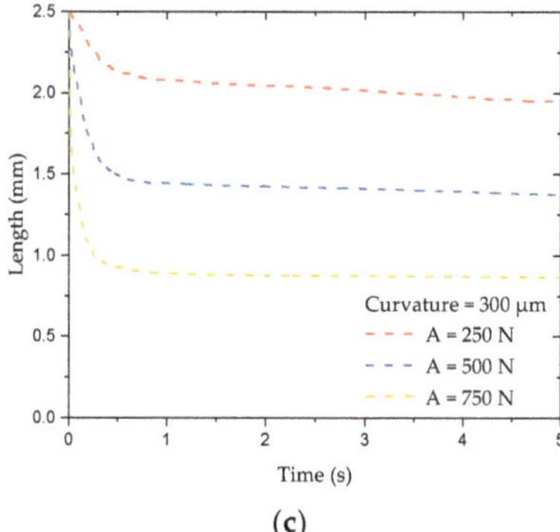

(**c**)

Figure 8. Predicted L_n of simple shear flow simulation at varying excluded volume force constant A with three critical curvature from: (**a**) 200 μm, (**b**) 250 μm, (**c**) 300 μm.

5. Fiber Relaxation

As discussed above, fibers experience a higher breaking rate than those in the experimental results. In order to place thousands of fibers within a small cell, fibers are forced into position with a critical angle, which leads to entanglement and bending of the fibers within the cluster. This entanglement was a big issue during breakage prediction, and can cause overestimation of breakage in the early stage of the simulation. To achieve relaxation of the entanglements, the shear rate was not stepped up instantaneously, but rather, increased in a stepwise fashion from 0 to 16.65 s^{-1}, which is same as the experiment within the first second of simulation time. This allows relaxation of the bent fibers inside the cluster. The simulation shear rate remained at 16.65 s^{-1} for the remaining 149 s (Figure 9). This technique allowed fibers to straighten out and thus reduce the number of critical angles between the connecting joints, without leading to excessive and unrealistic fiber attrition at the beginning of the simulation.

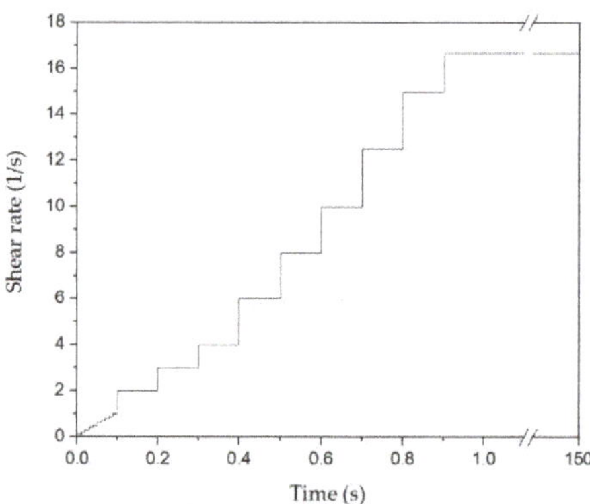

Figure 9. Stepwise increase profile of shear rate for fiber relaxation.

To assess the effect of this relaxation at acceptable computational costs, a relaxation test with a 1 s increase profile was done by tuning critical curvature from 200 μm to 300 μm while keeping the force constant *A* equal to 500 (see Figure 10). Relaxation in the first second of the simulation significantly reduced the initial fiber breakage caused by entanglement during cluster generation, compared to Figure 7b. After the first second, the interaction with neighboring fibers caused by the flow field leads to decreases in fiber curvature, which results in breakage when the assigned critical curvature is reached.

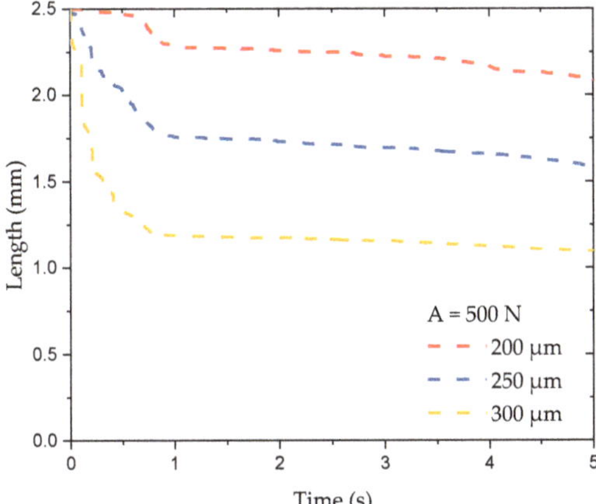

Figure 10. Predicted L_n of simple shear flow simulation at varying fiber critical breaking curvature with fiber relaxation applied to the first second in the simulation.

Figure 11 presents a comparison between the Couette experiment and the simulation for a critical curvature of 200 μm and an excluded volume constant A of 500, with a shear rate increase from 0 to 16.65 s^{-1} within the first second and remaining at 16.65 throughout the rest of the simulation. As seen in Figure 11, the initial breakage caused by unsteady system is reduced significantly. The simulation matches the experimental results much better. However, there is a slightly faster reduction rate for both L_n and L_w, as well as a higher final unbreakable length.

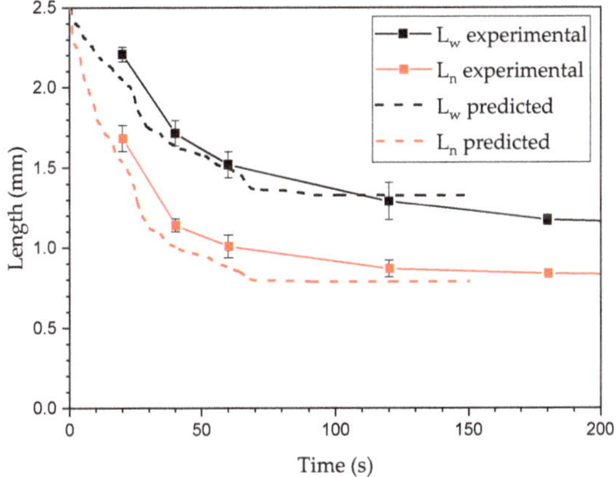

Figure 11. Fiber length evolution compared to the Couette flow experiment with fiber relaxation applied to the first second in simulation.

6. Fiber Probability Breakage

The final part of the fiber attrition model is to introduce the fiber probability measurements. Using the Weibull probability plot (WPP), the mechanistic model was first extended so that the empirically determined fracture behavior of the fibers can be simulated and validated [28]. In Figure 12, the corrected experimental data, calibration curve, and data generated with the mechanistic model are presented in a WPP.

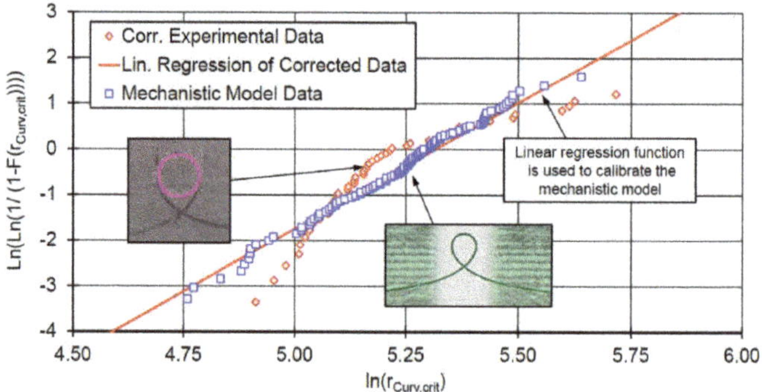

Figure 12. Calibration and validation of fiber breakage model and experiment [28].

First, it should be noted that linear regression cannot accurately represent experimental data. Consequently, the parameter determination leads to a loss of accuracy and the experimental data cannot be replicated precisely. Thus, a probability function was further introduced into the breakage simulation. Figure 13 presents the breakage criteria with (b) and without (a) an experimental probability of failure distribution. Fibers begin to have a probability to break when its curvature is smaller than the maximum curvature, and will break when curvature reaches the minimum curvature (Figure 13b). Comparing to the fixed curvature, probability theory provides powerful tools to explain the breakage behavior as seen below, where m is chosen as 15 for this research (Figure 13b).

$$x = \frac{\mathrm{max}curvature - curvature}{\mathrm{max}curvature - \mathrm{min}curvature} \tag{16}$$

$$Probability = \frac{exp(m * x) - 1}{exp(m) - 1} \tag{17}$$

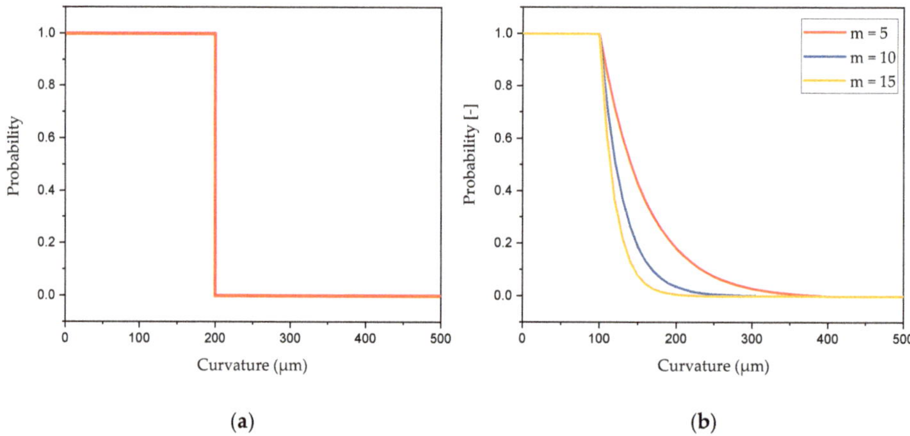

(a)

(b)

Figure 13. Different breakage models implemented in the mechanistic model: (a) Original breakage model; (b) probability breakage model when $m = 15$.

In Figure 14, the model showed a lower initial breakage rate and achieved nearly the experimental steady state fiber length. Unlike the fixed critical breaking curvature, the relaxation allowed fibers to relieve the entanglement and achieve a steady state in the system, which slowed the rate of breakage. As long fibers break into shorter fibers, the chance for fibers to contact each other is reduced, which also results in a higher alignment of fibers in the flow direction. Thus, fibers would gradually align in the flow direction where the rate of breakage significantly reduces after about 60 s. However, due to the Jeffrey orbits, the fibers experience shear flow; if allowing a sufficient amount of time, the fibers eventually rotate and break, which can be seen at around 150 s.

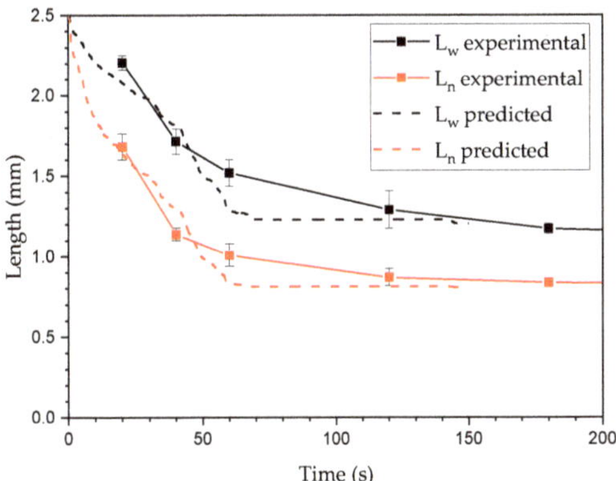

Figure 14. Fiber length evolution compared to the Couette flow experiment with fiber relaxation and probability breakage applied in simulation.

The curvature at break point for Figure 14 was recorded throughout the simulation and shown in Figure 15. Here, only the first 60 s are represented in the distribution, as the length in the simulation remained relatively constant after this period. Compared to Figure 4, which shows the loop experimental result from [28], the predicted distribution showed a similar trend to the experimental data. There was a higher deviation in the range of 200 to 250 μm and 350 to 400 μm. This may be due to the value of m in Equation (17), selected for this research, being too high for the system. Lowering the value of m will shift the curve to the right (Figure 13b), which increases the probability for fibers to break at a larger curvature. In addition, fiber critical curvature is related to fiber mechanical properties, such as carbon or glass fiber, fiber length, fiber diameter, and Young's modulus. For example, the minimum critical curvature will be even smaller than the value discussed here for fibers with a lower Young's modulus.

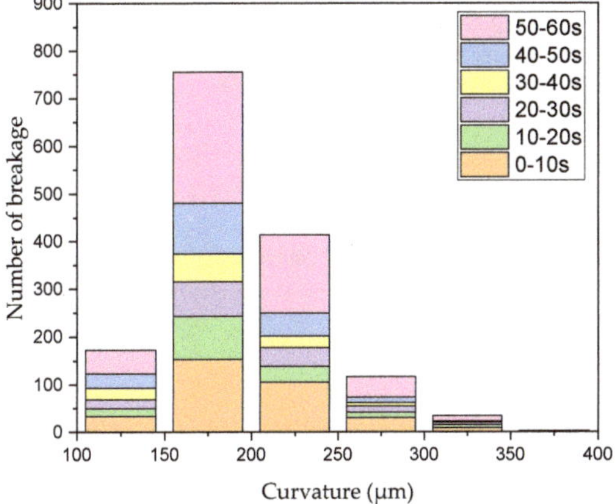

Figure 15. Breakage distribution over time of simulation results with fiber relaxation and probability breakage model applied.

7. Conclusions and Outlook

A particle level model for computing the breakage behavior for glass fiber in a polypropylene matrix under simple shear flow was studied. The simulation first showed how the variables can be tuned to obtain detailed information about breakage during the processing and develop an understanding of fiber length reduction. Based on the results obtained from the simulation, the length validation on fiber breakage was performed and compared with a Couette flow experiment. However, unsteady initial breakage was observed due to the high fiber volume fraction material used for the experiment. A relaxation of artificial fiber entanglement was introduced to the system. Moreover, a loop test with glass fibers showed a breakage distribution. To better describe breakage behavior, this probability theory was introduced in the simulation. The results had good agreement with the experimental data.

The variations in accuracy of the breakage prediction shows that there is a lack of understanding of the mechanisms involved in the fiber breakage and the respective translation to numerical models. Furthermore, there is not enough experimental data to evaluate the relationship between the excluded volume force and process condition. Thus, future work should conduct relevant experiments and verify the result with mechanistic model simulation to further understand and better describe the effect of excluded volume forces on different systems. Additionally, melting and fiber dispersion effects should be part of future research.

Author Contributions: Conceptualization, T.-C.C., A.B.S., H.C., D.B., A.Y., and T.O.; resources, D.B. and A.Y.; investigation, A.B.S. and H.C.; software, T.-C.C. and H.C.; validation, T.O.; visualization, T.-C.C., formal analysis, T.-C.C.; writing—original draft preparation, T.-C.C.; writing—review and editing, A.B.S., D.B., and T.O.; supervision, T.O. All authors have read and agreed to the published version of the manuscript.

Funding: This research received no external funding.

Acknowledgments: The authors would like to thank SABIC® for the material supply and collaboration with the Polymer Engineering Center. Additionally, the authors would like to thank their colleagues at the Polymer Engineering Center, who provided assistance to the research.

Conflicts of Interest: The authors declare no conflict of interest.

References

1. Osswald, T.; Menges, G. *Materials Science of Polymers for Engineers*, 3rd ed.; Hanser Publisher: Munich, Germany, 2012.
2. Phelps, J.H.; Abd EI-Rahman, A.I.; Kunc, V.; Tucker, C.L. A model for fiber length attrition in injection molded long-fiber composites. *Compos. Part Appl. Sci. Manuf.* **2013**, *51*, 11–21. [CrossRef]
3. Thomason, J.L. The influence of fibre length and concentration on the properties of glass fibre reinforced polypropylene. 7. Interface strength and fibre strain in injection moulded long fibre PP at high fibre content. *Compos. Part Appl. Sci. Manuf.* **2007**, *38*, 210–216. [CrossRef]
4. Ning, H.; Lu, N.; Hassen, A.A.; Chawla, K.; Selim, M.; Pillay, S. A review of long fibre thermoplastic (LFT) composites. *Int. Mater. Rev.* **2019**, *65*, 164–188. [CrossRef]
5. Huang, C.T.; Tseng, H.C. Simulation prediction of the fiber breakage history in regular and barrier structure screws in injection molding. *Polym. Eng. Sci.* **2018**, *58*, 452–459. [CrossRef]
6. Fu, S.Y.; Hu, X.; Yue, C.Y. Effects of fiber length and orientation distributions on the mechanical properties of short-fiber-reinforced polymers: A Review. *Int. J. Mater. Res.* **1999**, *5*, 74–83. [CrossRef]
7. Forgacs, O.L.; Mason, S.G. Particle motions in sheared suspensions. IX. Spin and deformation of threadlike particles. *J. Colloid Sci.* **1959**, *14*, 457–472. [CrossRef]
8. Hinch, E.J. The distortion of a flexible inextensible thread in a shearing flow. *J. Fluid Mech.* **1976**, *74*, 317–333. [CrossRef]
9. Goris, S.; Simon, S.; Montoya, C.; Bechara, A.; Candal, M.V.; Brands, D.; Yanev, A.; Osswald, T. Experimental study on fiber attrition of long glass fiber-reinforced thermoplastics under controlled conditions in a Couette flow. In Proceedings of the Annual Technical Conference of the Society of Plastics Engineers, Anaheim, CA, USA, 8–10 May 2017.

10. Ross, R.F.; Klingenberg, D.J. Dynamic simulation of flexible fibers composed of linked rigid bodies. *J. Chem. Phys.* **1997**, *106*, 2949–2960. [CrossRef]

11. Schmid, C.F.; Switzer, L.H.; Klingenberg, D.J. Simulations of fiber flocculation: Effects of fiber properties and interfiber friction. *J. Rheol.* **2000**, *44*, 781–809. [CrossRef]

12. Joung, C.G.; Phan-Thien, N.; Fan, X.J. Direct simulation of flexible fibers. *J. Non-Newton. Fluid Mech.* **2001**, *99*, 1–36. [CrossRef]

13. Londoño-Hurtado, A.; Osswald, T. Fiber jamming and fiber matrix separation during compression molding. *J. Plast. Tech.* **2006**, *2*, 1–17.

14. Sasayama, T.; Inagaki, M. Simplified bead-chain model for direct fiber simulation in viscous flow. *J. Non-Newton. Fluid Mech.* **2017**, *250*, 52–58. [CrossRef]

15. Hoffman, R.L. Discontinuous and dilatant viscosity behavior in concentrated suspensions. 1. Observation of a flow instability. *Trans. Soc. Rheol.* **1972**, *16*, 155–173. [CrossRef]

16. Barnes, H.A. Shear-thickening (dilatancy) in suspension of nonaggregating solid particles dispersed in Newtonian liquids. *J. Rheol.* **1989**, *33*, 329–366. [CrossRef]

17. Sundararajakumar, R.R.; Koch, D.L. Structure and properties of sheared fiber suspensions with mechanical contacts. *J. Non-Newton. Fluid Mech.* **1997**, *73*, 205–239. [CrossRef]

18. Pérez, C. The Use of a Direct Particle Simulation to Predict Fiber Motion in Polymer Processing. Ph.D. Thesis, University of Wisconsin-Madison, Madison, WI, USA, 2016.

19. Osswald, T.; Ghandi, U.; Goris, S. *Discontinous Fiber Reinforced Composites*, 1st ed.; Carl-Hanser: Munich, Germany, 2020.

20. Wang, G.; Yu, W.; Zhou, C. Optimization of the rod chain model to simulate the motions of a long flexible fiber in simple shear flows. *Eur. J. Mech. B Fluids* **2006**, *25*, 337–347. [CrossRef]

21. Anczurowski, E.; Mason, S.G. The kinetics of owing dispersions: II. Equilibrium orientations of rods and discs (theoretical). *J. Colloid Sci.* **1967**, *23*, 522–532. [CrossRef]

22. Anczurowski, E.; Mason, S.G. The kinetics of owing dispersions: III. Equilibrium orientations of rods and discs (experimental). *J. Colloid Sci.* **1967**, *23*, 533–546. [CrossRef]

23. Switzer, L.H.; Klingenberg, D.J. Rheology of sheared flexible fiber suspensions via fiber-level simulations. *J. Rheol.* **2003**, *47*, 759–778. [CrossRef]

24. Lindstrm, S.B.; Uesaka, T. Simulation of the motion of flexible fibers in viscous fluid flow. *Phys. Fluids* **2007**, *19*, 113307. [CrossRef]

25. Mason, S.G.; Manley, R.S.J. Particle Motions in Sheared Suspensions: Orientations and Interactions of Rigid Rods. *Proc. R. Soc. Lond. A Math. Phys. Eng. Sci.* **1956**, *238*, 117–131.

26. Goldsmith, H.L.; Mason, S.G. The flow of suspensions through tubes. I. Single spheres, rods, and discs. *J. Colloid Sci.* **1962**, *17*, 448–476. [CrossRef]

27. Sinclair, D. A Bending Method for Measurement of the Tensile Strength and Young's Modulus of Glass Fibers. *J. Appl. Phys.* **1950**, *21*, 380–386. [CrossRef]

28. Çelik, H. Numerical Study to Predict Fiber Motion and Fiber Attrition of Long-Fiber-Reinforced Thermoplastics during the Injection Molding Process using a Single Particle Model. Master's Thesis, IKV, RWTH Aachen University, Aachen, Germany, 2017.

29. Evans, D.J.; Morris, O.P. Non-Newtonian molecular dynamics. *Comput. Phys. Rep.* **1984**, *1*, 297–343. [CrossRef]

Journal of
Composites Science

Article

Parameter Identification of Fiber Orientation Models Based on Direct Fiber Simulation with Smoothed Particle Hydrodynamics

Nils Meyer [1,*], Oleg Saburow [1], Martin Hohberg [1], Andrew N. Hrymak [2], Frank Henning [1,3] and Luise Kärger [1]

[1] Karlsruhe Institute of Technology (KIT), Institute of Vehicle System Technology, 76131 Karlsruhe, Germany; oleg.saburow@kit.edu (O.S.); martin.hohberg@kit.edu (M.H.); frank.henning@ict.fraunhofer.de (F.H.); luise.kaerger@kit.edu (L.K.)

[2] Department of Chemical & Biochemical Engineering, Western University, London, ON N6A 5B9, Canada; ahrymak@uwo.ca

[3] Fraunhofer Institute for Chemical Technology (ICT), 76327 Pfinztal, Germany

* Correspondence: nils.meyer@kit.edu

Received: 29 May 2020; Accepted: 14 June 2020; Published: 22 June 2020

Abstract: The behavior of fiber suspensions during flow is of fundamental importance to the process simulation of discontinuous fiber reinforced plastics. However, the direct simulation of flexible fibers and fluid poses a challenging two-way coupled fluid-structure interaction problem. Smoothed Particle Hydrodynamics (SPH) offers a natural way to treat such interactions. Hence, this work utilizes SPH and a bead chain model to compute a shear flow of fiber suspensions. The introduction of a novel viscous surface traction term is key to achieve full agreement with Jeffery's equation. Careful modelling of contact interactions between fibers is introduced to model suspensions in the non-dilute regime. Finally, parameters of the Reduced-Strain Closure (RSC) orientation model are identified using ensemble averages of multiple SPH simulations implemented in PySPH and show good agreement with literature data.

Keywords: short fiber reinforcement; process simulation; smoothed particle hydrodynamics

---

1. Introduction

Molding of discontinuous reinforced polymer fiber suspensions, e.g., glass fibers in polymer melt, leads to fiber reorientation. Understanding and predicting the behavior of such fiber suspensions is crucial for achieving high quality products in common composite manufacturing processes such as injection molding and compression molding. Due to the high cost of production facilities and molds, it is desirable to simulate the flow in a computationally efficient and reliable way before running experiments. The fiber reorientation is also of high interest for subsequent anisotropic structural simulations in the framework of a continuous CAE chain [1,2].

Jeffery [3] was the first to analytically derive the motion of a single rigid, ellipsoidal shaped body in a viscous Newtonian flow without buoyancy or inertia. Jeffery's work was extended by Folgar and Tucker [4] to account for fiber interactions by introducing a fiber interaction coefficient that models isotropic diffusivity of fiber orientation. Other phenomenological parameters were introduced to capture experimentally observed orientation delays in the SRF and RSC model [5] and to account for anisotropic diffusion in the Anisotropic Rotary Diffusion (ARD) model [6] or the improved Anisotropic Rotary Diffusion (iARD) model [7]. These phenomenological parameters account for interactions of multiple fibers in non-dilute suspensions and are fitted to experimental observations, but do not describe a two-phase suspension. Instead, they model the fiber orientation state with second and fourth order moments of the fiber orientation distribution function $\Psi(\mathbf{p})$ introduced by Advani and Tucker [8] as fiber orientation tensors

$$\mathbf{A} = \int_S \mathbf{p} \otimes \mathbf{p} \, \Psi(\mathbf{p}) \, \mathrm{d}p \tag{1}$$

and

$$\mathbb{A} = \int_S \mathbf{p} \otimes \mathbf{p} \otimes \mathbf{p} \otimes \mathbf{p} \, \Psi(\mathbf{p}) \, \mathrm{d}p. \tag{2}$$

Here, $\mathbf{p}$ describes a fiber direction and $\mathrm{d}p$ is the surface element on a unit sphere $S := \{\mathbf{p} \in \mathbb{R}^3 : \|\mathbf{p}\| = 1\}$. Typically, a closure approach for the fourth order fiber orientation tensor is employed to describe the evolution of the second order fiber orientation tensor. Whenever this work utilizes a closure approximation, the invariant-based optimal fitting (IBOF) closure is chosen [9]. Determining the parameters in these macroscopic models from experiments can be time consuming. Thus, a direct fiber simulation may be utilized to identify these parameters.

1.1. Point-Wise Interaction Methods

A common approach for the simulation of fiber suspensions is the treatment of fibers as slender bodies that interact with the fluid at discrete points. Exact solutions from Stokesian dynamics [10] or slender-body theory [11] are utilized to describe long-range hydrodynamic interaction between fluid and particles for creeping flows. Several authors proposed models for single flexible fibers suspended in a fluid. Hinch [12] started by modeling inextensible threads. Yamamoto et al. [13,14] developed a fiber model consisting of individual beads that experience Stokesian drag forces from the fluid. These bead chain models are computationally expensive and authors have combined multiple beads to rods leading to more efficient rod chains [15–17]. Alternatively, spheres [18,19] and spheroids [20] connected with hinges were suggested. Lindström and Uesaka [21] use a discrete field of point forces to ensure that the fluid is experiencing forces opposed to the fiber (two-way coupling). Several authors investigated rheological properties of multiple rigid fibers suspended in a fluid based on these models for single fibers. Yamane et al. [22] described the motion of multiple rigid rods with hydrodynamic drag force and torque on the individual rods. They added a lubrication force for rods that are in close proximity to each other in order to capture short range hydrodynamic effects. However, they did not account for long range hydrodynamic interaction between particles, which was addressed later by Fan et al. [23] using slender-body theory. Sundararajakumar and Koch [24] showed that lubrication forces alone do not prevent penetration of fibers and included contact forces. Suspensions of multiple flexible fibers were investigated with rod-chain models [25] and simplified bead chain models [26]. In addition to these rheological investigations, few direct simulations have been applied on component scale with flexible fibers or bundles to investigate effects such as fiber-matrix separation [27–30].

1.2. Resolved Methods

In contrast to models based on discrete point-wise interaction forces, the suspension flow might be also described directly by a two-phase flow, in which one phase represents fibers and the other one the suspending fluid. Solving this fully coupled two-phase system with mesh-based approaches such as Finite Elements or Finite Volumes raises the problem of large mesh deformations. A simulation using an immersed boundary method resolves the remeshing problem but requires a mesh resolution significantly smaller than the fiber diameter to track the interface and is therefore computationally expensive. An alternative approach for the simulation of two-phase fiber suspensions is the use of particle methods. Meshless particle methods can offer a natural way to represent fluid-structure interaction at the fiber surface and have been studied less than point-wise interaction methods. Bian and Ellero propose a splitting scheme for separate integration of long-range hydrodynamic interactions and short-range lubrication [31], which was applied in an SPH simulation of a concentrated spherical particle suspension [32]. The fine resolution of suspended particles leads to a high accuracy, but comes at high computational costs. Yashiro et al. [33,34] connected particles to rigid bodies for modeling the injection molding of dilute short-fiber-reinforced composites using a moving particle semi-implicit method. Recently, a very similar approach using SPH was reported by He et al. [35,36]. However, both showed only short time periods when comparing the simulation to Jeffery's equation and did not investigate the interaction between fibers. Fiber suspensions were also investigated in the framework of dissipative particle dynamics with a focus on Boger fluids [37]. This investigation used a Langrange multiplier to constrain fiber extension and multiple parameters had to be calibrated to analytical solutions in order to achieve correct hydrodynamic forces on the fiber. Yang et al. [38] employed the SPH method in order to simulate a single flexible fiber in a viscous fluid with a focus on determining its drag coefficient. However, they limited their work to a single fiber without interactions.

The present work extends Yang's model with a viscous surface traction term necessary to meet Jeffery's equation precisely and allows for the simulation of non-dilute fiber suspensions through the introduction of fiber contact forces. First, rotation and bending of a single fiber is compared to Jeffery's equation to validate the approach. Then, parameters of the RSC fiber orientation model are determined using SPH simulations.

2. Theory

SPH was developed independently by Lucy [39] and Gingold & Monaghan [40] for astrophysical problems in 1977. Since then, multiple formulations were developed and were applied to various fields. The core idea of SPH is the conversion of a partial differential equation (PDE) for a continuum into a set of ordinary differential equations (ODE) for multiple Lagrangian particles. The ODEs are then integrated for each particle to determine their properties (such as mass density or velocity) and position for the next time step. Since particles carry the mass and are moved according to the velocity field, this method conserves mass exactly and treats pure advection correctly. Any arbitrary property $\alpha(x)$ in an n-dimensional domain $\Omega \subset \mathbb{R}^n$ at a position $\mathbf{x}, \mathbf{x}' \in \Omega$ can be described using an integral of the form

$$\alpha(\mathbf{x}) = \int_\Omega \alpha(\mathbf{x}')\delta(\mathbf{x} - \mathbf{x}')d\mathbf{x}' \tag{3}$$

employing the Dirac distribution $\delta(\mathbf{x})$. The two fundamental approximations of SPH are the description of the continuous integral over the domain using a sum over interpolation points, that can be interpreted as particles, and the replacement of the Dirac distribution with a smooth kernel $W(\mathbf{x} - \mathbf{x}', h)$. Here, h describes the smoothing length that controls the kernel size and is usually chosen similar to the average initial distance between particles. A very common smooth kernel is the cubic spline kernel defined as

$$W = \beta_n \begin{cases} (2-q)^3 - 4(1-q)^3 & 0 \leq q < 1 \\ (2-q)^3 & 1 \leq q < 2 \\ 0 & 2 \leq q \end{cases} \tag{4}$$

with $q = \frac{\|\mathbf{x}-\mathbf{x}'\|}{h}$ and a dimension-dependent normalization factor β_n. Employing the kernel $W(\mathbf{x}_i - \mathbf{x}_j, h) = W_{ij}$, Equation (3) can be approximated as

$$\alpha(\mathbf{x}) \approx \check{\alpha}(\mathbf{x}_i) = \check{\alpha}_i = \sum_{j \in \Omega} \alpha_j W_{ij} \frac{m_j}{\rho_j} \tag{5}$$

denoting a particle's property α at position x_j as α_j and by expressing the associated volume as ratio between mass m_j and density ρ_j of the particle. The key is the usage of an analytically differentiable kernel such that

$$\frac{\partial \check{\alpha}_i}{\partial \mathbf{x}} = \sum_{j \in \Omega} \alpha_j \frac{m_j}{\rho_j} \nabla_i W_{ij} \tag{6}$$

can be computed analytically with the kernel gradient $\frac{\partial W}{\partial \mathbf{x}}(\mathbf{x} - \mathbf{x}_j, h)|_i = \nabla_i W_{ij}$, hence turning the PDE to an ODE. The domain $\Omega = \Omega_m \cup \Omega_s$ is the set of all particles in the model with subsets for the fluid particles Ω_m and solid particles $\Omega_s = \Omega_w \cup \Omega_f$ consisting of wall particles Ω_w and fiber particles Ω_f. There is a wide range of variants and actual implementations for this concept and Monaghan gives a comprehensive review in one of his later publications [41].

The modelling of the fluid phase and its interaction with solid particles follows the work of Adami et al. [42,43]. The basic fiber model is adapted from Yang et al. [38] and extended to account for viscous surface traction as well as fiber interactions.

2.1. Fluid Model

In this work, a Transport Velocity Formulation [43] is used to model the suspending fluid, since it is fairly robust against stability issues such as the tensile instability [44]. Adami et al. [43] separate momentum velocity $\mathbf{v}$ and advection velocity $\tilde{\mathbf{v}}$ leading to a Navier-Stokes equation of the form

$$\frac{\tilde{d}(\rho \mathbf{v})}{dt} = -\nabla p + \eta \nabla^2 \mathbf{v} + \rho \mathbf{g} + \nabla \cdot (\rho \mathbf{v} \otimes (\tilde{\mathbf{v}} - \mathbf{v})) \tag{7}$$

with density ρ, pressure p, viscosity η and body accelerations $\mathbf{g}$. The last term is a momentum convection that compensates the deviation between advection velocity and momentum velocity. The difference is based on a virtual background pressure p_b that effectively suppresses tensile instability, but does not influence the actual momentum. Hence, the term $\sigma^A = \rho \mathbf{v} \otimes (\tilde{\mathbf{v}} - \mathbf{v})$ is called an artificial stress. The SPH-discretized version of (7) used in this work models the acceleration of particle i in the fluid domain by

$$\begin{aligned}
\tilde{\mathbf{a}}_i &= \frac{1}{m_i} \sum_{j \in \Omega} \left(V_i^2 + V_j^2 \right) \left(-\frac{\rho_j p_j + \rho_i p_i}{\rho_i + \rho_j} \nabla_i W_{ij} \right) \\
&+ \frac{1}{m_i} \sum_{j \in \Omega_s} \left(V_i^2 + V_j^2 \right) \left(\frac{2\eta_i \eta_j}{\eta_i + \eta_j} \frac{\nabla_i W_{ij} \cdot (\mathbf{x}_i - \mathbf{x}_j)}{\|\mathbf{x}_i - \mathbf{x}_j\| + \varepsilon} (\mathbf{v}_i - \hat{\mathbf{v}}_j) \right) \\
&+ \frac{1}{m_i} \sum_{j \in \Omega_f} \left(V_i^2 + V_j^2 \right) \left(\frac{2\eta_i \eta_j}{\eta_i + \eta_j} \frac{\nabla_i W_{ij} \cdot (\mathbf{x}_i - \mathbf{x}_j)}{\|\mathbf{x}_i - \mathbf{x}_j\| + \varepsilon} (\mathbf{v}_i - \mathbf{v}_j) + \frac{1}{2}(\sigma_i^A + \sigma_j^A) \nabla_i W_{ij} \right), \quad i \in \Omega_m
\end{aligned} \tag{8}$$

with the volume attributed to each particle $V_i = m_i/\rho_i$ and the velocity of an adjacent solid $\hat{\mathbf{v}}_j$, which is explained in the next section. Indices are used to refer to the particle based density ρ_i, pressure p_i, viscosity η_i, mass m_i, position $\mathbf{x}_i$ and velocity $\mathbf{v}_i$. The parameter ε is a small value to avoid singularities in the formulation. The last term is a momentum convection that compensates the deviation between advection velocity and momentum velocity.

Finally, the discrete conservation of mass is written as

$$\rho_i = m_i \sum_{j \in \Omega} W_{ij}, \quad i \in \Omega_m \tag{9}$$

and an equation of state is used to relate mass density to pressure. The equation of state has the form

$$p_i = p_0 \left(\left(\frac{\rho_i}{\rho_0} \right)^\gamma - 1 \right), \quad i \in \Omega_m, \tag{10}$$

where ρ_0 describes the nominal density and p_0 is a reference pressure chosen large enough to keep the density variation small. The latter is achieved by setting $p_0 = \frac{\rho_0 c^2}{\gamma}$ with the speed of sound c set to the tenfold of the maximum speed in the flow, thus limiting density variation to approximately 1%. Following Adami et al. [43], the parameter γ is set as $\gamma = 1$.

2.2. Interaction between Fluid and Solid Particles

The summation of the first term in Equation (8) includes fiber particles and solid wall particles. Adami et al. [42] determined the pressure of a solid particle from a force balance along the centerline of a solid-fluid particle pair as

$$p_i = \frac{\sum_{j \in \Omega_m} p_j W_{ij} + (\mathbf{g} - \mathbf{a}_i) \cdot \sum_{j \in \Omega_m} \rho_j (\mathbf{x}_i - \mathbf{x}_j) W_{ij}}{\sum_{j \in \Omega_m} W_{ij}}, \quad i \in \Omega_s \tag{11}$$

with a prescribed acceleration of the solid $\mathbf{a}_i$. The corresponding density may be computed by inverting (10). To ensure a no-slip condition, the velocity of solid particles is modified before insertion in (8) to

$$\hat{\mathbf{v}}_i = 2v_i - \frac{\sum_{j \in \Omega_m} v_j W_{ij}}{\sum_{j \in \Omega_m} W_{ij}}, \quad i \in \Omega_s, \tag{12}$$

where the fluid velocity field is extrapolated and subtracted to ensure zero velocity at the interface between solid particles and fluid particles [42].

In this work, wall particles are represented by three particle layers to ensure full kernel support and move with a constant prescribed velocity

$$\mathbf{v}_i = \text{const}, \quad \mathbf{a}_i = 0, \quad i \in \Omega_w. \tag{13}$$

Fibers are represented by a single chain of particles that experience hydrodynamic forces from the fluid, elastic forces from neighbors within the fiber and contact forces from other fibers. The first contribution to acceleration is a hydrodynamic interaction

$$\mathbf{a}_i^{\text{hydro}} = \frac{1}{m_i} \sum_{j \in \Omega_f} \left(V_i^2 + V_j^2 \right) \left(-\frac{\rho_j p_j + \rho_i p_i}{\rho_i + \rho_j} \nabla_i W_{ij} + \frac{2\eta_i \eta_j}{\eta_i + \eta_j} \frac{\nabla_i W_{ij} \cdot (\mathbf{x}_i - \mathbf{x}_j)}{\|\mathbf{x}_i - \mathbf{x}_j\| + \varepsilon} (\hat{\mathbf{v}}_i - \mathbf{v}_j) \right), \quad i \in \Omega_f \tag{14}$$

that balances the momentum together with (8). Modelling the fiber as a single layered chain of SPH particles has also been applied by other authors [38,45], but it has some implications, which are discussed further in Section 2.4. One implication is that the particle spacing Δx is directly related to the fiber diameter by

$$d = \frac{2\Delta x}{\sqrt{\pi}} \tag{15}$$

to ensure that fiber particles and fluid particles describe equal volumes in the beginning of a simulation.

2.3. Basic Model for Flexible Fibers

Besides the description of the fluid phase with SPH, a suitable model for the elastic fiber is needed. Thus, the straight-forward linear elastic bead chain formulation of Yang et al. [38] for tensile forces

$$\mathbf{F}_{ij}^t = EA \left(\frac{\| \mathbf{x}_{ij} \|}{x_{ij}^0} - 1 \right) \frac{\mathbf{x}_{ij}}{\| \mathbf{x}_{ij} \|} \tag{16}$$

and bending moment

$$\mathbf{M}_i^b = \frac{EI}{2} \frac{\boldsymbol{\theta}_i - \boldsymbol{\theta}_i^0}{\| \mathbf{x}_{i(i-1)} \| + \| \mathbf{x}_{i(i+1)} \|} \tag{17}$$

can be used as a basis for further model development. Here, E describes Young's modulus, while A and I are the fiber's cross sectional area and second moment of area respectively. The vector between two neighboring particles with indices i and j is $\mathbf{x}_{ij}$ and the enclosed angle is denoted as $\boldsymbol{\theta}_i$. Here, the vector notation of $\boldsymbol{\theta}_i$ indicates that its direction resembles the axis of rotation. The undeformed reference configuration is denoted with a superscript $(\cdot)^0$ in both, Equations (16) and (17). The bending moment can be converted to pairs of forces

$$\mathbf{F}^{b}_{ij} = \frac{1}{2} \frac{\mathbf{M}^{b}_{i} \times \mathbf{x}_{ij}}{\| \mathbf{x}_{ij} \|^2} \tag{18}$$

that act on the particle and its neighbors. Figure 1 illustrates these forces and it can been seen, that generally $\mathbf{F}^{b}_{ij} \neq \mathbf{F}^{b}_{ji}$. It is assumed that torsional torque of the fiber is of minor importance to the orientation evolution investigated in this work. Finally, the acceleration on a particle i due to elastic forces can be summarized as

$$\mathbf{a}^{elastic}_{i} = \frac{1}{m_i} \Big(\mathbf{F}^{t}_{i(i+1)} + \mathbf{F}^{t}_{i(i-1)}$$
$$+ \mathbf{F}^{b}_{i(i-1)} + \mathbf{F}^{b}_{i(i+1)} - \mathbf{F}^{b}_{(i-1)i} - \mathbf{F}^{b}_{(i+1)i} \Big). \tag{19}$$

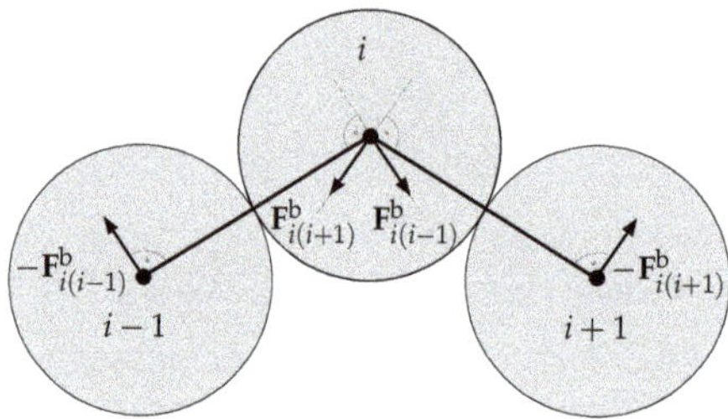

Figure 1. Force pairs representing bending moment on segments between fiber particles.

If the angle between two adjacent particle pairs exceeds a certain critical value θ_c, the neighborhood relation between these particles may be revoked permanently to model fracture of the fiber. Such a criterion is motivated by brittle fiber fracture, as it is typical for glass fibers.

2.4. Viscous Traction at Fiber Surface

A fiber modeled as a particle chain cannot capture a variation of a property in thickness direction of the fiber, as it stores properties at the center line only. Hence, a fiber placed horizontally in a shear flow with periodic boundary conditions, as depicted in Figure 2, would not experience any moment caused by friction forces on its surface. In order to model a physical thickness of the fiber, this work uses an analytical derivation to apply appropriate moment from surface friction to the fiber segment. Figure 3 illustrates a cylindrical fiber segment of length ΔL and diameter d with the orientation of its centerline $\mathbf{p}$. One exemplary surface normal $\mathbf{n}$ is shown with its parametrization angle ψ.

The fiber direction $\mathbf{p}$ and any arbitrary surface normal $\mathbf{n}^0$ are perpendicular unit-vectors. Any other surface normal can be constructed from this arbitrary normal by rotating it around $\mathbf{p}$. The surface normal can be parameterized using $\mathbf{n}^0$ and ψ employing a rotational tensor $\mathbf{R}$ around axis $\mathbf{p}$ [46] . The parameterized normal becomes

$$\mathbf{n}(\psi) = \mathbf{R}(\psi)\mathbf{n}_0. \tag{20}$$

Using this normal, the viscous traction on the surface can be expressed as

$$\mathbf{t}(\psi) = (-p\mathbf{I} + \eta \nabla \mathbf{v}) \, \mathbf{n}(\psi) = -p\mathbf{n}(\psi) + \eta \nabla \mathbf{v}\mathbf{n}(\psi) \tag{21}$$

if Newtonian viscosity is assumed. This term represents the force acting on each infinitesimal area of the fiber surface. The resulting moment can be computed by integrating $\mathbf{t}$ with its corresponding leverage $\frac{d}{2}\mathbf{n}(\psi)$ as

$$\mathbf{M}^{\mathrm{v}} = \Delta L \int_0^{2\pi} \frac{d}{2}\mathbf{n}(\psi) \times \mathbf{t}(\psi)\frac{d}{2}\mathrm{d}\psi \tag{22}$$

with a constant fiber diameter d. The term $\frac{d}{2}\mathrm{d}\psi$ represents an infinitesimal circumferential line segment on the cylinder surface. As the cross product of a vector with itself vanishes, this can be simplified to

$$\mathbf{M}^{\mathrm{v}} = \Delta L \int_0^{2\pi} \frac{d^2}{4}\mathbf{n}(\psi) \times \eta \nabla \mathbf{v}\mathbf{n}(\psi)\mathrm{d}\psi. \tag{23}$$

The diameter is finite and thus provides some leverage for the traction to generate a moment. The discrete evaluation of (23) is explained in Appendix A. The equation for angular momentum is then multiplied with the distance to its neighbors as a cross product leading to the acceleration of individual particles due to viscous friction

$$\mathbf{a}_i^{\mathrm{traction}} = \frac{1}{2}\frac{\mathbf{M}_i^{\mathrm{v}} \times \mathbf{x}_{ij}}{J}, \quad j \in [i-1, i+1] \tag{24}$$

Here, J denotes the moment of inertia for a cylindrical body around its first principle axis of inertia and the factor $\frac{1}{2}$ is chosen to represent the moment by two equal forces at both neighboring particles. This acceleration is then applied to the two neighboring particles and implies hereby a rotational acceleration of a fiber segment, consisting of three particles $\Delta L = 3\Delta x$. The central particle is used to evaluate the velocity gradient of Equation (21) as

$$\nabla \mathbf{v}_i = -\frac{1}{\rho_i}\sum_j m_j(\mathbf{v}_i - \mathbf{v}_j)\nabla_i W_{ij}. \tag{25}$$

It is assumed that the velocity gradient is approximately constant within each segment. In theory, the velocity gradient could be determined at each point of the cylindrical surface from kernel functions, but this comes at much higher computational costs and the difference in the resulting moment is expected to be small.

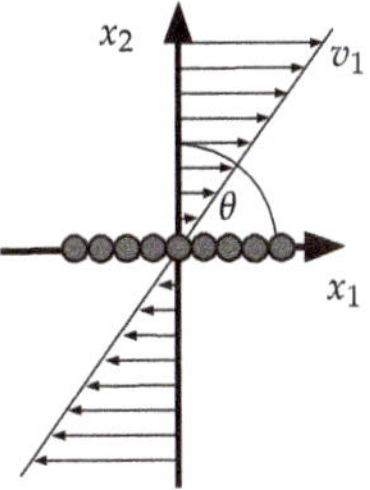

Figure 2. A fiber is placed horizontally with $\phi = \pi/2$ in a shear flow. The top and bottom walls have a prescribed velocity and a no-slip condition, the boundary conditions in x_1 direction are periodical.

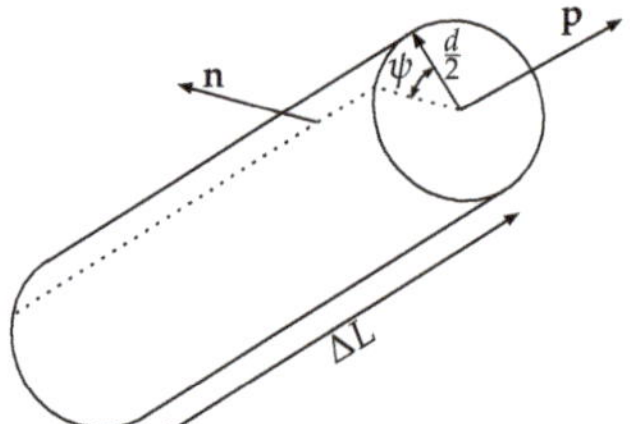

Figure 3. Cylindrical fiber segment with length ΔL, orientation **p** and an arbitrary surface normal **n**.

2.5. Fiber Interactions

If suspended objects come in close contact (10–50% of the radius [26,47]), lubrication forces oppose the relative velocity between fibers. Sundararajakumar and Koch [24] showed that lubrication forces alone do not necessarily prevent penetration at higher fiber volume fractions and added contact forces. The pressure gradient computed from SPH is generally not sufficient to counteract the accumulated forces on the fiber bead chain and does not necessarily prevent penetration of fibers.

It is too simple to apply contact forces directly bead to bead, because this would lead to entangled fibers that interlock at two bead gaps. For the determination of the contact properties at each potential contact pair (i, j) within the kernel radius, different cases need to be considered:

Surface-Surface: If both interacting particles are located at the center of the fiber (e.g., they have two neighbor particles each, compare Figure 4 at t_0), the normal direction of contact pair (i, j) can be computed using the cross product

$$\mathbf{n}_{ij} = [\![\mathbf{p}_i \times \mathbf{p}_j]\!] \tag{26}$$

of the involved fiber direction vectors $\mathbf{p}_i$ and $\mathbf{p}_j$. The operator $[\![\cdot]\!] = \frac{(\cdot)}{\|\cdot\|}$ is used to conveniently denote the normalization of a vector. Solving the small linear system of equations

$$[\mathbf{p}_i, \mathbf{n}_{ij}, -\mathbf{p}_j][\mathcal{P}_{ij}, D_{ij}, \mathcal{P}_{ji}]^\top = \mathbf{x}_{ij} \tag{27}$$

with its adjugate matrix leads to the solution for the distance between the fibers D_{ij} and the projections to source and destination vectors $\mathcal{P}_{ij}$ and $\mathcal{P}_{ji}$, respectively.

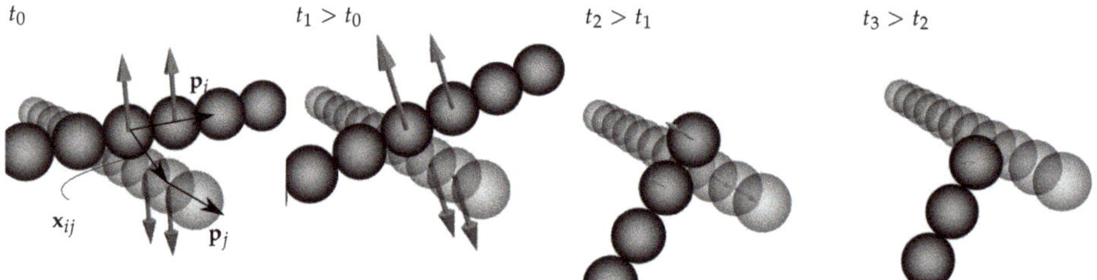

Figure 4. Snapshots of two fibers in contact. At contact initiation t_0, $\mathbf{p}_i$ and $\mathbf{p}_j$ denote unit vectors for the fiber directions and $\mathbf{x}_{ij}$ is the vector between particles of one active contact pair (i, j). The contact forces are indicated by arrows which scale with contact force magnitude and rotate in the subsequent time steps (t_1, t_2, t_3).

Surface-End and End-Surface: If a particle of a fiber end interacts with a central particle of another fiber, the vector between these two particles $\mathbf{x}_{ij}$ can be used to obtain the normal direction by projection. It is assumed that $\mathbf{p}_i$ describes a unit vector in fiber direction at one fiber particle at position $\mathcal{A}$. Let $\mathbf{x}_{ij}$ be the vector from another fibers' end particle $\mathcal{B}$ to the point $\mathcal{A}$. The closest point to $\mathcal{B}$ on a line with direction $\mathbf{p}_i$ is denoted as $\mathcal{C}$ and can be used to define the normal direction as

$$\mathbf{n}_{ij} = [\![\overrightarrow{\mathcal{BC}}]\!] = [\![\overrightarrow{\mathcal{AC}} - \overrightarrow{\mathcal{AB}}]\!]. \tag{28}$$

Using the definitions above and the fact that $\mathcal{C}$ is the projection of $\mathcal{B}$ to the line with direction $\mathbf{p}_i$, Equation (28) can be rewritten as

$$\mathbf{n}_{ij} = [\![\mathcal{P}_{ij}\mathbf{p}_i - (-\mathbf{x}_{ij})]\!] = [\![\mathbf{x}_{ij} + \mathcal{P}_{ij}\mathbf{p}_i]\!]. \tag{29}$$

The projection on the destination fiber is given as $\mathcal{P}_{ij} = -\mathbf{p}_i \cdot \mathbf{x}_{ij}$ and the contact distance is computed as $D_{ij} = \|\mathbf{x}_{ij} + \mathcal{P}_{ij}\mathbf{p}_i\|$.

End-End: The simplest case is the interaction of two fiber ends. Here, the vector between those two particles can be simply determined by

$$\mathbf{n}_{ij} = [\![\mathbf{x}_{ij}]\!] \tag{30}$$

with the corresponding distance $D_{ij} = \|\mathbf{x}_{ij}\|$.

A penalty approach is proposed to prevent fiber penetration, if the distance between fibers falls below the fiber diameter.

The penalty force is formulated as a Hertzian contact force between two cylinders [48]

$$F_{ij}^c = \begin{cases} \frac{4}{3} E^* R^* (d - D_{ij})^{1.5} & D_{ij} < d \\ 0 & \text{else} \end{cases} \tag{31}$$

with

$$E^* = \frac{E}{2(1 - v^2)} \quad \text{and} \quad R^* = \sqrt{\frac{d}{4}} \tag{32}$$

where E is Young's modulus and v denotes the Poisson ratio. Strictly, the contact force varies slightly at fiber ends due to different contact areas. However, the exact pressure distribution in the contact area is not the focus of this work and for high fiber aspect ratios, the portion of fiber end contact becomes relatively small. Thus, the contact force at fiber ends may be rather interpreted as a penetration penalty.

Finally, the particle acceleration due to contact forces is computed as

$$\mathbf{a}_i^{\text{contact}} = \sum_j \frac{w_{ij} F_{ij}^c}{m_i} \mathbf{n}_{ij} \tag{33}$$

for each proximity point j with the contact normal $\mathbf{n}_{ij}$ and a weighting factor w_{ij}. The weighting factor is necessary to distribute the force at a contact point between particles associated to this contact point and is defined as

$$w_{ij} = \begin{cases} \frac{d_{\text{p}} - \mathcal{P}_{ij}}{d_{\text{p}}} & 0 < \mathcal{P}_{ij} < d_{\text{p}} \\ \frac{d_{\text{n}} + \mathcal{P}_{ij}}{d_{\text{n}}} & -d_{\text{n}} < \mathcal{P}_{ij} \le 0 \\ 1 & \text{two fiber ends} \\ 0 & \text{else} \end{cases} \tag{34}$$

with the distance to the previous fiber particle $d_{\text{p}} = \|\mathbf{x}_i - \mathbf{x}_{i-1}\|$ and next particle $d_{\text{n}} = \|\mathbf{x}_i - \mathbf{x}_{i+1}\|$. Friction in tangential direction is neglected and fiber surfaces are assumed to be smooth. However, friction could be easily incorporated at this point, if the friction coefficient is available. The acceleration is computed for all particles that might possibly contact any other particle and this way, equal force magnitudes on destination and source fibers are ensured.

2.6. Time Integration and Implementation

The time integration is an extension of the kick-drift scheme used in the original transport velocity formulation [43]. The advection velocities are computed for a half step as

$$
\mathbf{v}_i^{n+\frac{1}{2}} = \mathbf{v}_i^n + \frac{\Delta t}{2}\left[\tilde{\mathbf{a}}_i + \mathbf{g}\right]_{n-\frac{1}{2}} \qquad , \quad i \in \Omega_\mathrm{m} \tag{35}
$$

$$
\tilde{\mathbf{v}}_i^{n+\frac{1}{2}} = \mathbf{v}_i^{n+\frac{1}{2}} + \frac{\Delta t}{2}\left[\frac{p_\mathrm{b}}{m_i}\sum_{j\in\Omega_\mathrm{m}}\left(V_i^2 + V_j^2\right)\nabla_i W_{ij}\right]_{n-\frac{1}{2}} \qquad , \quad i \in \Omega_\mathrm{m} \tag{36}
$$

$$
\tilde{\mathbf{v}}_i^{n+\frac{1}{2}} = \tilde{\mathbf{v}}_i^n + \frac{\Delta t}{2}\left[\mathbf{a}_i^\mathrm{hydro} + \mathbf{a}_i^\mathrm{elastic} + \mathbf{a}_i^\mathrm{traction} + \mathbf{a}_i^\mathrm{contact} + \mathbf{g}\right]_{n-\frac{1}{2}} \qquad , \quad i \in \Omega_\mathrm{f} \tag{37}
$$

based on the previous accelerations at step $n - \frac{1}{2}$. Equation (36) utilizes the background pressure p_b to move fluid particles such that no agglomerations or voids form. This is done by modifying the momentum velocity $\mathbf{v}_i^{n+\frac{1}{2}}$ to the advection velocity $\tilde{\mathbf{v}}_i^{n+\frac{1}{2}}$. The difference in momentum velocity and advection velocity is corrected by the artificial stress σ_i^A in the momentum balance. Fiber particles do not experience a background pressure, as they should not be used to fill voids etc. and thus, their advection velocity is equal to their physical velocity. Then, all particles are moved according to

$$
\mathbf{x}_i^{n+1} = \mathbf{x}_i^n + \Delta t\, \tilde{\mathbf{v}}_i^{n+\frac{1}{2}} \qquad , \quad i \in \Omega \tag{38}
$$

for the full step. Finally, the velocities are updated as

$$
\mathbf{v}_i^{n+1} = \mathbf{v}_i^{n+\frac{1}{2}} + \frac{\Delta t}{2}\left[\tilde{\mathbf{a}}_i + \mathbf{g}\right]_{n+\frac{1}{2}} \qquad , \quad i \in \Omega_\mathrm{m} \tag{39}
$$

$$
\tilde{\mathbf{v}}_i^{n+1} = \tilde{\mathbf{v}}_i^{n+\frac{1}{2}} + \frac{\Delta t}{2}\left[\mathbf{a}_i^\mathrm{hydro} + \mathbf{a}_i^\mathrm{elastic} + \mathbf{a}_i^\mathrm{traction} + \mathbf{a}_i^\mathrm{contact} + \mathbf{g}\right]_{n+\frac{1}{2}} \qquad , \quad i \in \Omega_\mathrm{f}. \tag{40}
$$

As this is an explicit time integration scheme, it is only conditionally stable. The maximum time step is computed by

$$
\Delta t = \min\left(0.4\frac{\Delta x}{1.1c_0}, 0.125\frac{\Delta x^2}{\eta}, 0.25\sqrt{\frac{\Delta x}{g}}, 0.5\Delta x\sqrt{\frac{\rho_0}{E}}, 0.5\Delta x^2\sqrt{\frac{\rho_0 A}{2EI}}\right) \tag{41}
$$

as the minimum of the CFL condition, a viscous condition, a body force condition, a tensile elastic condition and a elastic bending condition. The implementation was realized in PySPH [49] due to its flexibility in implementing the additional equations on top of the transport velocity scheme. The code is publicly available as fork of the original PySPH project (https://github.com/nilsmeyerkit/pysph/tree/fibers).

3. Results

3.1. Rotation and Bending in a Simple Shear Flow

This section presents simulation results for a 3D shear flow (cf. Figure 2) with periodic boundaries in x_1 and x_3. The lateral dimensions of the modelled fluid domain are $B = 1.2L_\mathrm{f}$ and the dimension in flow direction is $L = 2B$ with the length of a fiber L_f. The Reynolds number

$$Re = \frac{\rho_0 L_f \frac{GB}{2}}{\eta} \tag{42}$$

is set to $Re = 0.5$ to be small enough for a quasi-creeping flow, but also as large as possible to increase the time increment. A dimensionless measure for the fiber stiffness is

$$S = \frac{E\pi}{4\eta G r_e^4} \tag{43}$$

for a given shear rate G and ellipsoidal aspect ratio r_e [50]. Bending of fibers starts with an increased aspect ratio, higher shear rates and reduced bending stiffness [51]. The dimensionless stiffness S summarises these effects in one parameter.

First, a stiff fiber with $S = 100$ is considered. Such a fiber spins in a rigid manner without significant bending and can be compared to the solution of Jeffery's equation

$$\frac{D\mathbf{p}}{Dt} = \boldsymbol{\omega}\mathbf{p} + \zeta \left(\mathbf{D}\mathbf{p} - (\mathbf{p} \otimes \mathbf{p} \otimes \mathbf{p})\, \mathbf{D} \right) \tag{44}$$

with vorticity tensor $\boldsymbol{\omega}$ and symmetric strain rate tensor $\mathbf{D}$. The shape factor ζ is an alternative measure for the (equivalent) ellipsoidal aspect ratio r_e and is defined as

$$\zeta(r_e) = \frac{r_e^2 - 1}{r_e^2 + 1}. \tag{45}$$

The solution to Equation (44) is periodic with

$$T = \frac{2\pi}{G} \left(r_e + \frac{1}{r_e} \right) \tag{46}$$

being the time for a full rotation of a fiber. When solving Jeffery's equation for a cylinder with geometric aspect ratio r_p instead of an ellipsoid, Jeffery's equation can still be applied, but an equivalent aspect ratio has to be used. Such an equivalent aspect ratio r_e was derived by Cox [52] based on slender-body theory and Zhang et al. [53] utilizing Finite Element Analysis. The latter is applicable for small aspect ratios and therefore Zhang's cubic fit

$$r_e(r_p) = 0.000035 r_p^3 - 0.00467 r_p^2 + 0.764 r_p + 0.404 \tag{47}$$

is used here.

Figure 5 depicts the orientation ϕ of a single fiber in shear flow for fiber length $L_f = 5\Delta x$ and $L_f = 11\Delta x$ (e.g., the fiber is represented by 5 or 11 particles in a chain). The rotation period obtained with the present SPH approach is sligthly faster than the solution to Jeffery's equation. One possible reason for the difference is the finite Reynolds number and finite simulation domain with periodic boundaries, whereas Jeffery used an infinite domain with strictly no effect of inertia. Another reason might be the coarse resolution with only one layer of particles for the fiber. Due to the averaging nature of SPH, the exact flow field close to the suspended particles cannot be resolved exactly. The presented approach solves the entire fluid field, but the low resolution and smoothing makes it less accurate than e.g., Stokesian Dynamics simulations. However, the simplicity and computational efficiency make it attractive for engineering applications, such as the parameter fitting presented later.

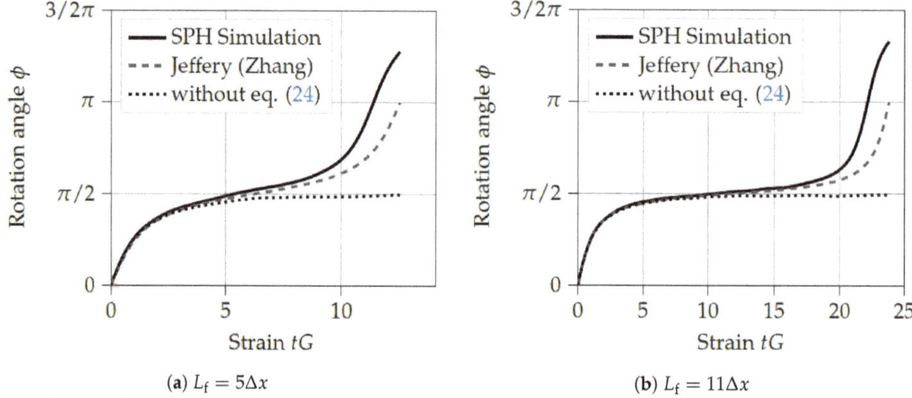

(**a**) $L_f = 5\Delta x$ (**b**) $L_f = 11\Delta x$

Figure 5. Orientation angle ϕ for fiber in a shear flow. The fiber length L_f is expressed in multiples of the particle spacing Δx. The dashed gray line represents the solution to Jeffery's equation with Zhang's [53] fit. The solid black line represents the solution obtained with this SPH implementation. If the viscous surface traction term (Equation (24)) is neglected, the fiber stops rotating after one half rotation, as shown with the black dotted line.

Bending modes were shown in Yamamoto and Matsuka's [13] numerical results and the corresponding SPH simulation in Table 1 agree well with their observations. Differences can be attributed to the fact that Yamamoto and Matsuka used an inextensible fiber, while the fibers simulated in this work experience stretching, because the Young's modulus is taken into account for tensile stiffness as well.

Table 1. Bending modes of a fiber with length $L_f = 11\Delta x$ for varied dimensionless stiffness. One example with $S = 10$ and critical fiber bending angle $\theta_c = \frac{\pi}{4}$ is shown to demonstrate fiber fracture.

Strain	0	$\frac{1}{4}TG$	$\frac{1}{2}TG$	$\frac{3}{4}TG$	TG
$S = 5, \theta_c = \infty$					
$S = 10, \theta_c = \infty$					
$S = 10, \theta_c = \frac{\pi}{4}$					
$S = 20, \theta_c = \infty$					
$S = 100, \theta_c = \infty$					

This section illustrated that the implementation based on SPH can reproduce the rotation periods quantitatively. Furthermore, numerically obtained bending modes can be described qualitatively. In addition, a fully coupled solution for the fluid field is computed. It can be noted that the fluid field obtained for these single fiber setups does not significantly differ from the ideal field. This is reasonable, since a single flexible fiber does not offer much resistance to a highly viscous flow. However, a significant difference can be expected as soon as multiple fibers interact in a concentrated suspension. In that scenario, the presence of suspended fibers and their interactions are expected to raise the macroscopically observed effective viscosity and fiber interactions affect the fiber orientation evolution.

3.2. Parameter Identification for the Orientation Evolution in a Non-Dilute Short Fiber Suspensions

A fully resolved computation of all fibers is often not feasible for full components made from composite material. It can be sufficient to give a reasonable description of the fiber orientation function in terms of a fiber orientation tensor, if these are accurate and scale-separation applies. A common two-parameter model is the RSC model [5] with fiber interaction coefficient C_I and a phenomenological factor κ that models a delay to compensate an over-prediction in the change of orientation observed in the classical Folgar-Tucker model. In tensor notation, the RSC model reads

$$\frac{D\mathbf{A}}{Dt} = \omega\mathbf{A} - \mathbf{A}\omega + \xi\Big(\mathbf{D}\mathbf{A} + \mathbf{A}\mathbf{D} - 2\big(\mathbb{A} + (1-\kappa)(\mathbb{L} - \mathbb{M}\mathbf{A})\big)[\mathbf{D}]\Big) + 2\kappa C_I G\Big(\mathbf{1} - 3\mathbf{A}\Big) \qquad (48)$$

with $\mathbb{L} = \sum_{i=1}^{3}\lambda_i \mathbf{e}_i \otimes \mathbf{e}_i \otimes \mathbf{e}_i \otimes \mathbf{e}_i$ and $\mathbb{M} = \sum_{i=1}^{3}\mathbf{e}_i \otimes \mathbf{e}_i \otimes \mathbf{e}_i \otimes \mathbf{e}_i$ using the eigenvalues λ_i and eigenvectors $\mathbf{e}_i$ of the second order fiber orientation tensor $\mathbf{A}$. Setting $\kappa = 1$ would reduce this model to the Folgar-Tucker model and setting $C_I = 0$ reduces it to Jeffery's Equation (44). The choice of feasible parameters $C_I \in [0, 0.1]$ and $\kappa \in [0, 1.0]$ remains. Hence, this section computes the orientation evolution in terms of the second order fiber orientation tensor for a small set of fibers in a 3D shear flow and compares the solution obtained with SPH to macroscopic models.

The investigated domain is a cube with edge length $L = 15\Delta x$ and Lees-Edwards boundary conditions [54,55] are employed to induce a shear rate G on the periodic fluid domain. Essentially, these boundary conditions shift dummy particles and particles leaving the domain in x_1-direction according to

$$x_2' = (x_2 \pm GLt) \bmod L \qquad (49)$$

with the sign depending on the direction of the shift and the modulo operator mod. Additionally, the velocity is adjusted

$$v_2' = (v_2 \pm GL) \qquad (50)$$

for the shift. Conventional periodic boundary conditions apply in x_2- and x_3-direction, as depicted in Figure 6. All fibers are initially oriented in x_1-direction and randomly positioned in the cube. This unidirectional arrangement is chosen because it can be easily achieved without a micro-structure generator. Instead of analyzing larger representative volumes [26], this work performs multiple realizations of the random process to create a statistical representative behavior. The advantage of this approach is that scatter and standard deviation between random realizations on the micro scale can be observed. The ensemble average of a property $\langle \cdot \rangle_{t_n}$ is defined as the mean across multiple realizations at the same time step t_n.

To obtain optimal parameters at different volume fractions, a least squares fit

$$\underset{C_i,\kappa}{\text{minimize}} \quad \sum_{n}\big\|\mathbf{A}(t_n) - \langle\tilde{\mathbf{A}}\rangle_{t_n}\big\|^2 \qquad (51)$$

is applied to minimize the squared difference of $\mathbf{A}$ obtained by Equation (48) and the ensemble average of the discrete second order fiber orientation tensor $\tilde{\mathbf{A}}$ of the SPH analysis. This tensor can be computed as

$$\tilde{\mathbf{A}} = \frac{1}{N} \sum_{i=1}^{N} \mathbf{p}_i \otimes \mathbf{p}_i \tag{52}$$

with each fiber's orientation $\mathbf{p}_i$. A flexible fiber has different tangential orientations and the tensor's definition becomes ambiguous then. Consequently, the following examples use fibers with a high stiffness $S = 100$ to ensure enough rigidity for an unambiguous interpretation of $\mathbf{A}$. However, the method is in no way limited to rigid fibers.

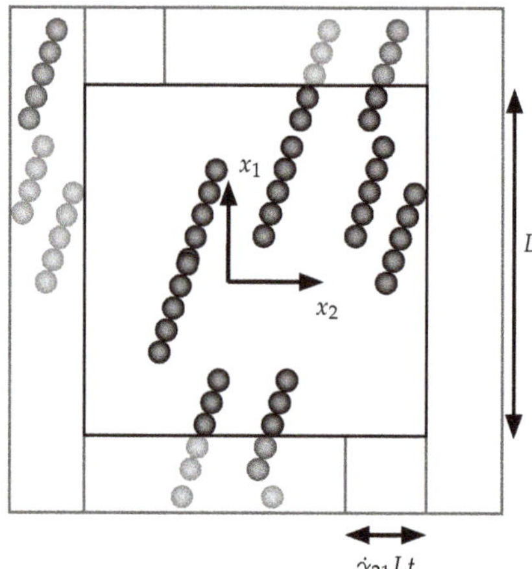

Figure 6. Setup for the 3D shear with fibers in a cube of edge length $L = 15\Delta x$. Lees-Edwards boundary conditions [54] are employed to induce a shear rate G. Therefore dummy particles (light gray) and particles leaving the domain in x_1-direction are shifted periodically in x_2 during each domain update. Conventional periodic boundary conditions apply to all other sides of the cube. The initial state is generated from fibers aligned in x_1-direction at unique random positions in the entire volume. (Some fibers appear longer in this figure due to other overlapping fibers behind it.)

Figure 7 shows the non-trivial components of the ensemble average $\langle \tilde{\mathbf{A}} \rangle_{t_n}$ computed from simulations with five different initial random realizations as a solid green line. The standard deviation at each time step is depicted as light filled area in the background. The gray solid line represents orientation tensor components $\mathbf{A}$ computed with optimal parameters according to the RSC model given in Equation (48). The simulation time is $10T$, which is the time of 10 full rotations of a single rigid fiber. The corresponding strain is approximately 150.

The simple parameter fitting approach proposed in (51) works well and generally shows a good agreement of macroscopic models with the SPH micro-model. All results show a decreasing orientation amplitude and a trend towards a stationary state with a significant non-zero component in x_3-direction. This trend arises from a combination of fiber contacts and long range pertubations of the flow field that push fibers out of their original trajectories. Figure 8 shows, how fibers are oriented after 30 strains in the case of 10 % fiber volume fraction. Several fibers have left the sheared $x_1 x_2$-plane due to interactions at this point. Eventually, fiber interactions and shear-induced reorientation balance each other and lead to a stationary orientation state. The stationary orientation state is reached faster for increased fiber volume fractions.

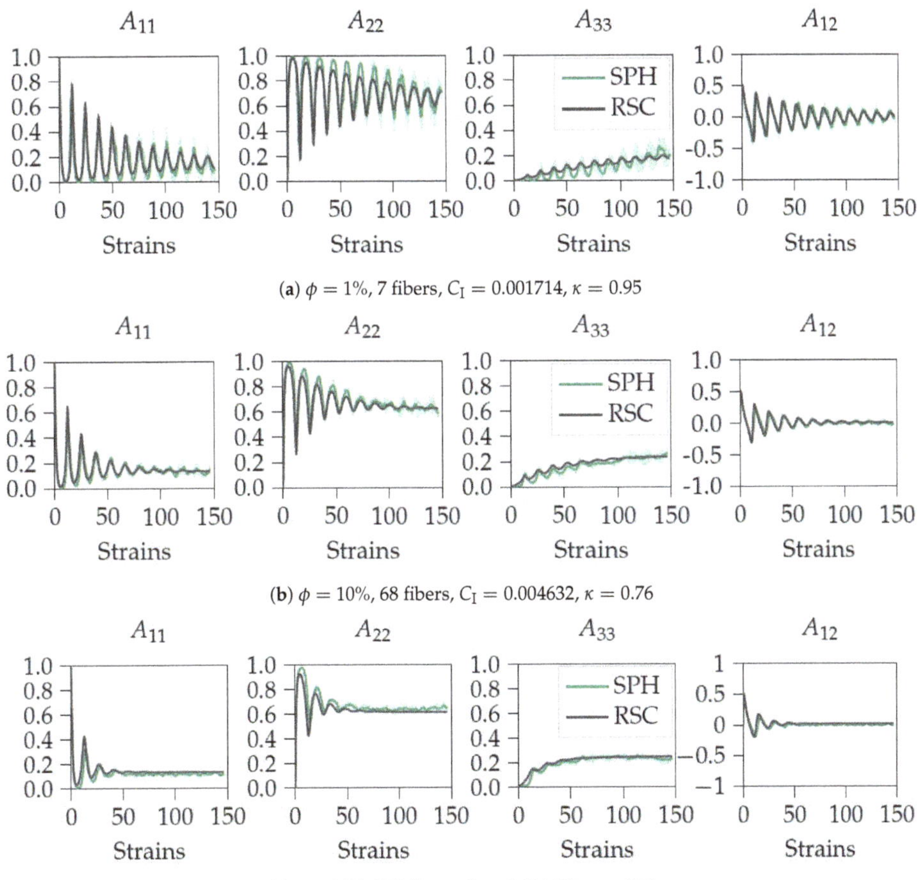

(**a**) $\phi = 1\%$, 7 fibers, $C_I = 0.001714$, $\kappa = 0.95$

(**b**) $\phi = 10\%$, 68 fibers, $C_I = 0.004632$, $\kappa = 0.76$

(**c**) $\phi = 30\%$, 202 fibers, $C_I = 0.011405$, $\kappa = 0.74$

Figure 7. Ensemble average of fiber orientation tensor components for fibers with aspect ratio $r_p = 4.43$ ($L_f = 5$) in a 3D shear flow and its comparison to the reduced strain model with optimal parameters. Each simulation result was obtained from five independently sampled initial configurations. The standard deviation is indicated by a light green filled area for each volume fraction.

The deviations between different initial configurations decrease for increasing fiber volume fraction, as the sample size increases. The deviation between individual realizations at 1% volume fraction is large and it might not be appropriate to describe such a system with a macroscopic fiber orientation model. This highlights that a macroscopic description requires a sufficient scale separation and a sufficient number of fibers to provide reliable results. The proposed SPH simulation can be used to quickly evaluate different configurations and may be used as a tool to not only determine parameters, but also quantify deviations from macromodels, if the underlying conditions of such models are in question for a specific application.

The obtained parameters of the interaction coefficient are compared to Folgar and Tucker's original work [4] and the values obtained by Phan-Thien et al. [47] in Figure 9. The results from SPH simulations show a good agreement with literature data and support the use of this approach for parameter identification.

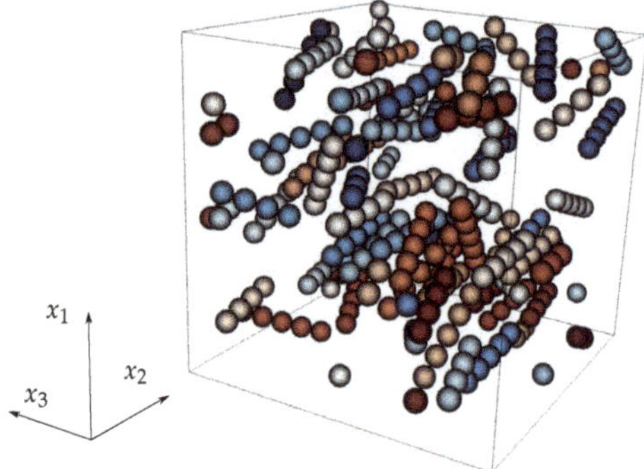

Figure 8. Snapshot of fibers at 10% fiber volume fraction after 30 strains. The colors are introduced to distinguish fiber particles and represent the particle ID.

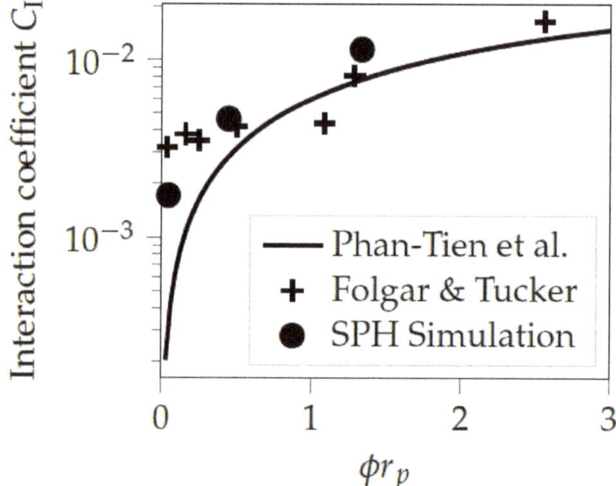

Figure 9. Comparison of C_I values to the original work of Folgar and Tucker [4] and a fit based on simulation results by Phan-Tien et al. [47]. The values obtained in the presented work are reasonably close to these literature results.

4. Conclusions

Fiber suspensions are treated as flexible bead chains of SPH particles surrounded by other particles representing the fluid domain. The bead chains are connected by elastic tension- and bending forces and interact with the fluid particles in a two-way coupled manner. A novel viscous surface traction term is introduced to compensate the missing fiber thickness that is introduced by the line representation of a fiber. In addition, contact forces are introduced to model fiber interactions in a non-dilute suspension.

The fiber orientation evolution of a single stiff fiber shows good agreement with the rotation periods based on Jeffery's equation thanks to the introduction of a new surface traction term. Bending modes of single fibers are consistent with results reported in literature.

A periodic domain with Lees-Edwards boundary conditions and suspended fibers is subjected to shear. The investigation of suspensions with different volume fractions of fibers can be used to directly compute the fiber orientation tensor. If several of these computations are evaluated in a statistical sense, the ensemble average can be used to fit optimal parameters of fiber orientation models. For relatively stiff fibers of length $L_f = 5\Delta x$, good parameters of the RSC fiber orientation model are found based on the SPH simulation.

In future, the authors plan to extend the investigation to arbitrary initial configurations, flexible and breaking fibers as well as other flow types beyond shear flows. Further, the assessment of standard deviations may enable modeling of uncertainties related to the fiber orientation models.

Author Contributions: N.M.: writing–original draft preparation, methodology, software, visualization; O.S. and M.H.: conceptualization, writing–review and editing; A.N.H.: supervision, conceptualization, writing–review and editing; F.H.: supervision, funding acquisition; L.K.: supervision, methodology, writing–review and editing. All authors have read and agreed to the published version of the manuscript.

Funding: The research documented in this manuscript has been funded by the German Research Foundation (DFG) within the International Research Training Group "Integrated engineering of continuous-discontinuous long fiber-reinforced polymer structures" (GRK 2078). The support by the German Research Foundation (DFG) is gratefully acknowledged.

Acknowledgments: We acknowledge support by the KIT-Publication Fund of the Karlsruhe Institute of Technology.

Conflicts of Interest: The authors declare no conflict of interest. The funders had no role in the design of the study; in the collection, analyses, or interpretation of data; in the writing of the manuscript, or in the decision to publish the results.

Appendix A. Evaluation of the Surface Traction Integral

An arbitrary normal to the fiber direction $\mathbf{p}_i$ named $\mathbf{n}_0$ must be generated first. This can be achieved by forming the cross product of $\mathbf{p}_i$ with an arbitrary other vector that is not parallel to $\mathbf{p}_i$. Chosing the arbitrary vector $[1, 0, 0]^\top$, the parametrized normal in (20) becomes

$$\mathbf{n}_i(\psi) = \begin{pmatrix} -\sin(\psi)\sqrt{p_{i2}^2 + p_{i3}^2} \\ \dfrac{p_{i1}\sin(\psi)p_{i2} + \cos(\psi)p_{i3}}{\sqrt{p_{i2}^2 + p_{i3}^2}} \\ -\dfrac{-p_{i1}\sin(\psi)p_{i3} + \cos(\psi)p_{i2}}{\sqrt{p_{i2}^2 + p_{i3}^2}} \end{pmatrix}. \tag{A1}$$

The integration in (23) leads to

$$\mathbf{M}_i^{\mathrm{v}} = \frac{3\Delta x \frac{d^2}{4}\eta\pi}{\sqrt{p_{i2}^2 + p_{i3}^2}} \begin{pmatrix} (p_{i1}p_{i2}^2 p_{i3} + p_{i1}p_{i3}^3)\frac{\partial v_2}{\partial x_1} + (-p_{i1}^2 p_{i2}p_{i3} + p_{i2}p_{i3})\frac{\partial v_2}{\partial x_2} + (-p_{i1}^2 p_{i3}^2 - p_{i2}^2)\frac{\partial v_2}{\partial x_3} + \\ (-p_{i1}p_{i2}^3 - p_{i1}p_{i2}p_{i3}^2)\frac{\partial v_3}{\partial x_1} + (p_{i1}^2 p_{i2}^2 + p_{i3}^2)\frac{\partial v_3}{\partial x_2} + (p_{i1}^2 p_{i2}p_{i3} - p_{i2}p_{i3})\frac{\partial v_3}{\partial x_3} \\[1em] (-p_{i1}p_{i2}^2 p_{i3} - p_{i1}p_{i3}^3)\frac{\partial v_1}{\partial x_1} + (p_{i1}^2 p_{i2}p_{i3} - p_{i2}p_{i3})\frac{\partial v_1}{\partial x_2} + (p_{i1}^2 p_{i3}^2 + p_{i2}^2)\frac{\partial v_1}{\partial x_3} + \\ (-p_{i2}^4 - 2p_{i2}^2 p_{i3}^2 - p_{i3}^4)\frac{\partial v_3}{\partial x_1} + (p_{i1}p_{i2}^3 + p_{i1}p_{i2}p_{i3}^2)\frac{\partial v_3}{\partial x_2} + (p_{i1}p_{i2}^2 p_{i3} + p_{i1}p_{i3}^3)\frac{\partial v_3}{\partial x_3} \\[1em] (p_{i1}p_{i3}^3 + p_{i1}p_{i2}^2 p_{i3})\frac{\partial v_1}{\partial x_1} + (-p_{i1}^2 p_{i2}^2 - p_{i3}^2)\frac{\partial v_1}{\partial x_2} + (-p_{i1}^2 p_{i2}p_{i3} + p_{i2}p_{i3})\frac{\partial v_1}{\partial x_3} + \\ (p_{i2}^4 + 2p_{i2}^2 p_{i3}^2 + p_{i3}^4)\frac{\partial v_2}{\partial x_1} + (-p_{i1}p_{i2}^3 - p_{i1}p_{i2}p_{i3}^2)\frac{\partial v_2}{\partial x_2} + (-p_{i1}p_{i2}^2 p_{i3} - p_{i1}p_{i3}^3)\frac{\partial v_2}{\partial x_3} \end{pmatrix} \tag{A2}$$

with the velocity gradient from Equation (25). For the special case that $\mathbf{p}_i$ is equivalent to $[1, 0, 0]^\top$, the moment is computed in the same way with a different arbitrary initial direction, e.g., $[0, 1, 0]^\top$.

References

1. Kärger, L.; Bernath, A.; Fritz, F.; Galkin, S.; Magagnato, D.; Oeckerath, A.; Schön, A.; Henning, F. Development and validation of a CAE chain for unidirectional fibre reinforced composite components. *Compos. Struct.* **2015**, *132*, 350–358. [CrossRef]

2. Görthofer, J.; Meyer, N.; Pallicity, T.D.; Schöttl, L.; Trauth, A.; Schemmann, M.; Hohberg, M.; Pinter, P.; Elsner, P.; Henning, F.; et al. Virtual process chain of sheet molding compound: Development, validation and perspectives. *Compos. Part B Eng.* **2019**, *169*, 133–147. [CrossRef]

3.   Jeffery, G.B. The motion of ellipsoidal particles immersed in a viscous fluid. *Proc. R. Soc. Lond. Ser. A Contain. Pap. Mathem. Phys. Character* **1922**, *102*, 161–179. [CrossRef]

4.   Folgar, F.; Tucker, C.L. Orientation Behavior of Fibers in Concentrated Suspensions. *J. Reinf. Plast. Compos.* **1984**, *3*, 98–119. [CrossRef]

5.   Wang, J.; O'Gara, J.F.; Tucker, C.L. An objective model for slow orientation kinetics in concentrated fiber suspensions: Theory and rheological evidence. *J. Rheol.* **2008**, *52*, 1179–1200. [CrossRef]

6.   Phelps, J.H.; Tucker, C.L. An anisotropic rotary diffusion model for fiber orientation in short- and long-fiber thermoplastics. *J. Non-Newt. Fluid Mech.* **2009**, *156*, 165–176. [CrossRef]

7.   Tseng, H.C.; Chang, R.Y.; Hsu, C.H. An objective tensor to predict anisotropic fiber orientation in concentrated suspensions. *J. Rheol.* **2016**, *60*, 215–224. [CrossRef]

8.   Advani, S.G.; Tucker, C.L. The Use of Tensors to Describe and Predict Fiber Orientation in Short Fiber Composites. *J. Rheol.* **1987**, *31*, 751–784. [CrossRef]

9.   Chung, D.H.; Kwon, T.H. Invariant-based optimal fitting closure approximation for the numerical prediction of flow-induced fiber orientation. *J. Rheol.* **2002**, *46*, 169–194. [CrossRef]

10.   Brady, J.F.; Bossis, G. Stokesian Dynamics. *Ann. Rev. Fluid Mech.* **1988**, *20*, 111–157. [CrossRef]

11.   Batchelor, G.K. Slender-body theory for particles of arbitrary cross-section in Stokes flow. *J. Fluid Mech.* **1970**, *44*, 419–440. [CrossRef]

12.   Hinch, E.J. The distortion of a flexible inextensible thread in a shearing flow. *J. Fluid Mech.* **1976**, *74*, 317–333. [CrossRef]

13.   Yamamoto, S.; Matsuoka, T. A method for dynamic simulation of rigid and flexible fibers in a flow field. *J. Chem. Phys.* **1993**, *98*, 644–650. [CrossRef]

14.   Yamamoto, S.; Matsuoka, T. Viscosity of dilute suspensions of rodlike particles: A numerical simulation method. *J. Chem. Phys.* **1994**, *100*, 3317–3324. [CrossRef]

15.   Wang, G.; Yu, W.; Zhou, C. Optimization of the rod chain model to simulate the motions of a long flexible fiber in simple shear flows. *Europ. J. Mech. B/Fluid.* **2006**, *25*, 337–347. [CrossRef]

16.   Meirson, G.; Hrymak, A.N. Two-dimensional long-flexible fiber simulation in simple shear flow. *Polym. Compos.* **2016**, *37*, 2425–2433. [CrossRef]

17.   Meirson, G.; Hrymak, A.N. Two dimensional long-flexible fiber orientation simulation in squeeze flow. *Polym. Compos.* **2018**, *39*, 4656–4665. [CrossRef]

18.   Skjetne, P.; Ross, R.F.; Klingenberg, D.J. Simulation of single fiber dynamics. *J. Chem. Phys.* **1997**, *107*, 2108–2121. [CrossRef]

19.   Joung, C.G.; Phan-Thien, N.; Fan, X.J. Direct simulations of flexible fibers. *J. Non-Newt. Fluid Mech.* **2001**, *99*, 1–36. [CrossRef]

20.   Ross, R.F.; Klingenberg, D.J. Dynamic simulation of flexible fibers composed of linked rigid bodies. *J. Chem. Phys.* **1997**, *106*, 2949–2960. [CrossRef]

21.   Lindström, S.B.; Uesaka, T. Simulation of the motion of flexible fibers in viscous fluid flow. *Phys. Fluids* **2007**, *19*, 113307. [CrossRef]

22.   Yamane, Y.; Kaneda, Y.; Dio, M. Numerical simulation of semi-dilute suspensions of rodlike particles in shear flow. *J. Non-Newt. Fluid Mech.* **1994**, *54*, 405–421. [CrossRef]

23.   Fan, X.; Phan-Thien, N.; Zheng, R. A direct simulation of fibre suspensions. *J. Non-Newt. Fluid Mech.* **1998**, *74*, 113–135. [CrossRef]

24.   Sundararajakumar, R.; Koch, D.L. Structure and properties of sheared fiber suspensions with mechanical contacts. *J. Non-Newt. Fluid Mech.* **1997**, *73*, 205–239. [CrossRef]

25.   Schmid, C.F.; Switzer, L.H.; Klingenberg, D.J. Simulations of fiber flocculation: Effects of fiber properties and interfiber friction. *J. Rheol.* **2000**, *44*, 781–809. [CrossRef]

26.   Sasayama, T.; Inagaki, M. Simplified bead-chain model for direct fiber simulation in viscous flow. *J. Non-Newt. Fluid Mech.* **2017**, *250*, 52–58. [CrossRef]

27.   Londoño-Hurtado, A.; Hernandez-Ortiz, J.P.; Osswald, T. Mechanism of fiber–matrix separation in ribbed compression molded parts. *Polym. Compos.* **2007**, *28*, 451–457. [CrossRef]

28.   Ramirez, D. Study of Fiber Motion in Molding Processes by Means of a Mechanistic Model. Ph.D. Thesis, University of Wisconsin Madison, Madison, WI, USA, 2014.

29. Kuhn, C.; Walter, I.; Täger, O.; Osswald, T. Simulative Prediction of Fiber-Matrix Separation in Rib Filling During Compression Molding Using a Direct Fiber Simulation. *J. Compos. Sci.* **2017**, *2*, 2. [CrossRef]

30. Meyer, N.; Schöttl, L.; Bretz, L.; Hrymak, A.; Kärger, L. Direct Bundle Simulation approach for the compression molding process of Sheet Molding Compound. *Compos. Part A Appl. Sci. Manuf.* **2020**, *132*, 105809. [CrossRef]

31. Bian, X.; Ellero, M. A splitting integration scheme for the SPH simulation of concentrated particle suspensions. *Comput. Phys. Commun.* **2014**, *185*, 53–62. [CrossRef]

32. Vázquez-Quesada, A.; Ellero, M. Rheology and microstructure of non-colloidal suspensions under shear studied with Smoothed Particle Hydrodynamics. *J. Non-Newt. Fluid Mech.* **2016**, *233*, 37–47. [CrossRef]

33. Yashiro, S.; Okabe, T.; Matsushima, K. A Numerical Approach for Injection Molding of Short-Fiber-Reinforced Plastics Using a Particle Method. *Adv. Compos. Mater.* **2011**, *20*, 503–517. [CrossRef]

34. Yashiro, S.; Sasaki, H.; Sakaida, Y. Particle simulation for predicting fiber motion in injection molding of short-fiber-reinforced composites. *Compos. Part A Appl. Sci. Manuf.* **2012**, *43*, 1754–1764. [CrossRef]

35. He, L.; Lu, G.; Chen, D.; Li, W.; Lu, C. Three-dimensional smoothed particle hydrodynamics simulation for injection molding flow of short fiber-reinforced polymer composites. *Model. Simul. Mater. Sci. Eng.* **2017**, *25*, 055007. [CrossRef]

36. He, L.; Lu, G.; Chen, D.; Li, W.; Chen, L.; Yuan, J.; Lu, C. Smoothed particle hydrodynamics simulation for injection molding flow of short fiber-reinforced polymer composites. *J. Compos. Mater.* **2018**, *52*, 1531–1539. [CrossRef]

37. Duong-Hong, D.; Phan-Thien, N.; Yeo, K.S.; Ausias, G. Dissipative particle dynamics simulations for fibre suspensions in newtonian and viscoelastic fluids. *Comput. Method. Appl. Mech. Eng.* **2010**, *199*, 1593–1602. [CrossRef]

38. Yang, X.; Liu, M.; Peng, S. Smoothed particle hydrodynamics and element bending group modeling of flexible fibers interacting with viscous fluids. *Phys. Rev. E* **2014**, *90*, 063011. [CrossRef] [PubMed]

39. Lucy, L.B. A numerical approach to the testing of the fission hypothesis. *Astronom. J.* **1977**, *82*, 1013. [CrossRef]

40. Gingold, R.A.; Monaghan, J.J. Smoothed particle hydrodynamics: theory and application to non-spherical stars. *Mon. Not. R. Astronom. Soc.* **1977**, *181*, 375–389. [CrossRef]

41. Monaghan, J.J. Smoothed particle hydrodynamics. *Rep. Prog. Phys.* **2005**, *68*, 1703–1759. [CrossRef]

42. Adami, S.; Hu, X.; Adams, N. A generalized wall boundary condition for smoothed particle hydrodynamics. *J. Comput. Phys.* **2012**, *231*, 7057–7075. [CrossRef]

43. Adami, S.; Hu, X.; Adams, N. A transport-velocity formulation for smoothed particle hydrodynamics. *J. Comput. Phys.* **2013**, *241*, 292–307. [CrossRef]

44. Meleán, Y.; Sigalotti, L.D.G.; Hasmy, A. On the SPH tensile instability in forming viscous liquid drops. *Comput. Phys. Commun.* **2004**, *157*, 191–200. [CrossRef]

45. Akinci, N.; Ihmsen, M.; Akinci, G.; Solenthaler, B.; Teschner, M. Versatile rigid-fluid coupling for incompressible SPH. *ACM Trans. Graph.* **2012**, *31*, 1–8. [CrossRef]

46. Cole, I.R. Modelling CPV. Ph.D. Thesis, Loughborough University, Loughborough, UK, 2015.

47. Phan-Thien, N.; Fan, X.J.; Tanner, R.; Zheng, R. Folgar–Tucker constant for a fibre suspension in a Newtonian fluid. *J. Non-Newt. Fluid Mech.* **2002**, *103*, 251–260. [CrossRef]

48. Johnson, K.L.; Johnson, K.L. *Contact Mechanics*; Cambridge University Press: Cambridge, UK, 1987.

49. Ramachandran, P. PySPH: A reproducible and high-performance framework for smoothed particle hydrodynamics. In Proceedings of the 15th Python in Science Conference, Austin, TX, USA, 11–17 July 2016; pp. 127–135.

50. Baird, D.G.; Collias, D.I. *Polymer Processing: Principles and Design*; John Wiley & Sons: Hoboken, NJ, USA, 2014.

51. Forgacs, O.; Mason, S. Particle motions in sheared suspensions. *J. Colloid Sci.* **1959**, *14*, 457–472. [CrossRef]

52. Cox, R.G. The motion of long slender bodies in a viscous fluid. Part 2. Shear flow. *J. Fluid Mech.* **1971**, *45*, 625–657. [CrossRef]

53. Zhang, D.; Smith, D.E.; Jack, D.A.; Montgomery-Smith, S. Numerical Evaluation of Single Fiber Motion for Short-Fiber-Reinforced Composite Materials Processing. *J. Manuf. Sci. Eng.* **2011**, *133*, 51002. [CrossRef]

54. Lees, A.W.; Edwards, S.F. The computer study of transport processes under extreme conditions. *J. Phys. C Solid State Phys.* **1972**, *5*, 1921–1928. [CrossRef]

55. Pan, D.; Hu, J.; Shao, X. Lees–Edwards boundary condition for simulation of polymer suspension with dissipative particle dynamics method. *Mol. Simul.* **2016**, *42*, 328–336. [CrossRef]

Journal of
Composites Science

Article

Calibration of Fiber Orientation Simulations for LFT—A New Approach

Fabian Willems *, Philip Reitinger * and Christian Bonten

Institut für Kunststofftechnik (IKT), University of Stuttgart, Pfaffenwaldring 32, 70569 Stuttgart, Germany;
christian.bonten@ikt.uni-stuttgart.de
* Correspondence: fabian.willems@ikt.uni-stuttgart.de (F.W.); philip.reitinger@ikt.uni-stuttgart.de (P.R.);
Tel.: +49-711-685-62883 (F.W.); +49-711-685-62870 (P.R.)

Received: 18 August 2020; Accepted: 26 October 2020; Published: 30 October 2020

Abstract: Short fiber reinforced thermoplastics (SFT) are extensively used due to their excellent mechanical properties and low processing costs. Long fiber reinforced thermoplastics (LFT) show an even more interesting property profile and are increasingly used for structural parts. However, their processing by injection molding is not as simple as for SFT, and their anisotropic properties resulting from the fiber microstructure (fiber orientation, length, and concentration) pose a challenge with regard to the engineering design process. To reliably predict the structural mechanical properties of fiber reinforced thermoplastics by means of micromechanical models, it is also necessary to reliable predict the fiber microstructure. Therefore, it is crucial to calibrate the underlying prediction models, such as the fiber orientation model, within the process simulation. In general, these models may be adjusted manually, but this is usually ineffective and time-consuming. To overcome this challenge, a new calibration method was developed to automatically calibrate the fiber orientation model parameters of the injection molding simulation by means of optimization methods. This optimization routine is based on experimentally determined fiber orientation distributions and leads to optimized parameters for the fiber orientation prediction model within a few minutes. To better understand the influence of the model parameters, different versions of the fiber orientation model, as well as process and material influences on the resulting fiber orientation distribution, were investigated. Finally, the developed approach to calibrate the fiber orientation model was compared with a classical approach, a direct optimization of the whole process simulation. Thereby, the new optimization approach shows a calculation time reduced by the factor 15 with comparable error variance.

Keywords: lightweight design; long fiber reinforced thermoplastics; process simulation; fiber microstructure; parameter-optimization; fiber orientation models; calibration

---

1. Introduction

Due to the ever-decreasing development times and the steady progress of digitization, numerical methods for predicting unknown target and design variables are gaining in importance. This also applies to the processing of plastics and the resulting part properties in general. Especially in the case of fiber-reinforced plastics, the strongly varying part properties directly depend on the fiber microstructure and must be taken into account during the design process [1–4]. In the following, the fiber microstructure is used as a generic term for fiber orientation, fiber length, and fiber concentration.

A reliable prediction of the process-induced fiber microstructure, especially for long fiber reinforced thermoplastics (LFT), still poses a considerable challenge today [5–7]. In contrast to short fiber reinforced thermoplastics (SFT), not only the fiber orientation must be taken into account in the prediction of the mechanical properties of LFTs, but effects such as fiber breakage and the migration of fibers during processing also play a significant role. In order to predict these properties correctly, suitable calculation models are required on the one hand, and correspondingly well-calibrated model parameters on the other hand.

However, the reliable prediction of fiber orientation in general, but especially for LFTs, is still the subject of current research and requires further improvement in the achievable prediction quality. Current commercial software environments that calculate the injection molding process do not allow a purposeful and fast optimization of the resulting fiber orientation distribution. Only a manual adaptation of model parameters with a subsequent recalculation of the entire flow field is possible. Due to the lack of automation and the necessary recalculations, the calibration of the model parameters becomes remarkably inefficient and time-consuming. Therefore, a new calibration method was developed to automatically adjust the resulting fiber orientation distribution.

1.1. State of the Art

During the processing of discontinuous fiber reinforced plastics, a specific fiber microstructure (fiber orientation, fiber length and fiber concentration) is formed within the part cavity that depends on the material, the geometry of the molded part, and the process settings. This microstructure determines the parts main properties. With suitable accuracy of the predicted fiber microstructure the resulting anisotropic mechanical properties can be reliably calculated [8–12]. Thus, the part's behavior can also be determined in structural simulations.

1.1.1. Fiber Microstructure

The resulting part properties of fiber reinforced materials are mainly determined by process induced fiber microstructure and its properties (Figure 1), such as:

- fiber orientation [13–22],
- fiber length and diameter [10,20,23–28],
- fiber concentration [10,23,29], and
- fiber matrix adhesion [29,30].

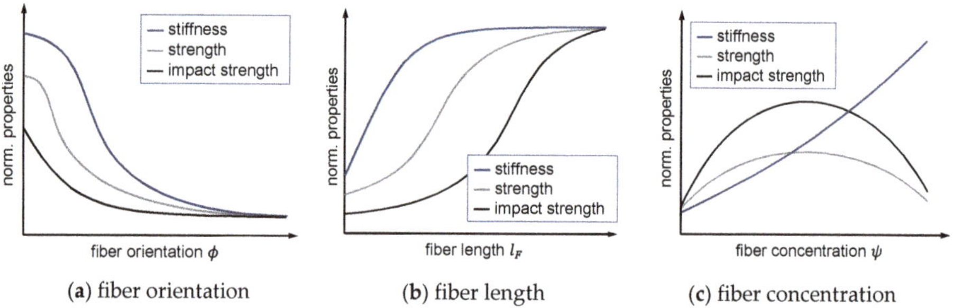

(**a**) fiber orientation   (**b**) fiber length   (**c**) fiber concentration

Figure 1. Influence of fiber orientation, length, and concentration on the resulting mechanical properties (schematically) according to [16,26,29].

The increase in the respective fiber microstructure properties (orientation, length, and concentration) also leads to an increase in mechanical properties of the composite material, such as stiffness, strength, and impact strength [23]. Only at high fiber concentrations can a decrease in strength and impact strength be observed. With regard to an entire fiber reinforced part, the local distribution of these properties along the flow path as well as along the cavity height (part thickness) must also be taken into account [31]. Even if in some cases a classification cannot be clearly distinguished, fiber reinforced materials are categorized into three classes according to their initial or resulting fiber length [32]: short fiber reinforced plastics (0.1–1 mm), long fiber reinforced plastics (1–50 mm), and continuous fiber reinforced plastics (>50 mm).

1.1.2. Process Induced Microstructure

The properties of the fiber microstructure (orientation, length, and concentration), especially in the case of discontinuously reinforced plastics such as SFT or LFT, can usually only be influenced to a very limited extent by its constituent properties and compound constitution (initial fiber orientation, initial fiber length and initial fiber concentration). However, these properties are rather a result of the processing [13,26,33–37]. One of the most widely used processing methods is injection molding, in which even complex parts can be produced in large quantities [32,37].

The materials used are subjected to special stresses during the injection molding process. This can be seen in the orientation of the fibers as well as in the migration and fracture phenomena. During the filling of a part with molten plastic by injection molding, a flow field is formed inside the cavity that depends on the material, the part geometry, and the process settings. This varies not only along the flow direction, for example through cooling effects, but also along the part's thickness or the cavity height, as shown in Figure 2. Due to this velocity profile and the resulting shear stresses, characteristic layers are formed along the part thickness (Figure 3) in which the fibers

- are aligned differently (fiber orientation) [31,38–45],
- break more frequently (fiber length) [43,46–48], or
- migrate unevenly (fiber concentration) [5,14,45,49,50].

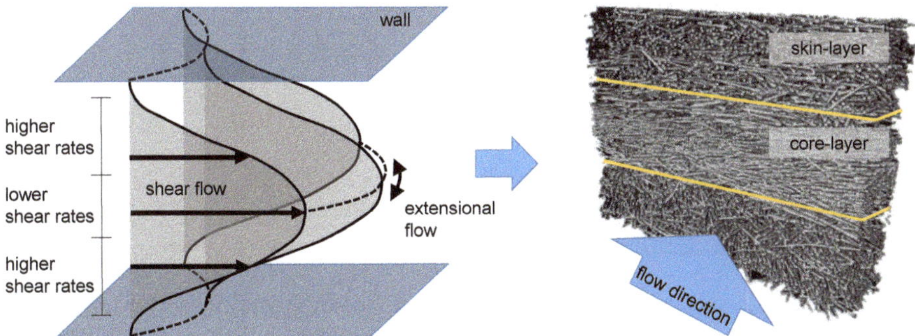

Figure 2. Velocity profile and the resulting fiber orientation distribution according to [51].

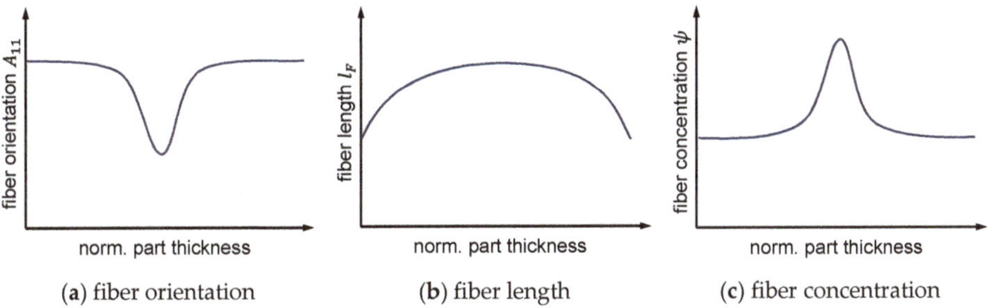

Figure 3. Locally varying fiber microstructure along the normalized part thickness (schematically).

These individual microstructure properties also affect each other. Due to the interaction of the fibers, the fiber orientation depends on the fiber length and concentration [5], which also applies to interchanged dependencies. The fiber length depends on the fiber orientation and concentration while the distribution of fibers depends inversely on the fiber orientation and length [36]. Due to wall adhesion and mold cooling, high shear rates are present in the skin or shear layers, and lower shear rates are found in the core layer in the middle of the cavity. A characteristic profile for each property is created inside a part, which can be subdivided into different layers [38,41,52] as shown in Figure 3.

According to the crystalline structure [34,53] or the fiber orientation distribution [46,54], the flow profile can be divided into at least a core layer and two shear or skin layers. This three-layer model is shown schematically in Figure 2 (right), and the individual microstructure properties are shown in Figure 3. The specific characteristic ratio of the layers is determined by the quotient of the individual layer thicknesses and is generally referred to as the skin core layer ratio.

The influence of the velocity field on the resulting fiber orientation distribution is a complex interrelation. The flow field changes in response to many different influencing factors. Important parameters are, for example, the injection speed and the processing temperature, as well as the material's viscosity [14]. Figure 4 shows the correlation of the shear rate and injection speed [55] along the cavity height.

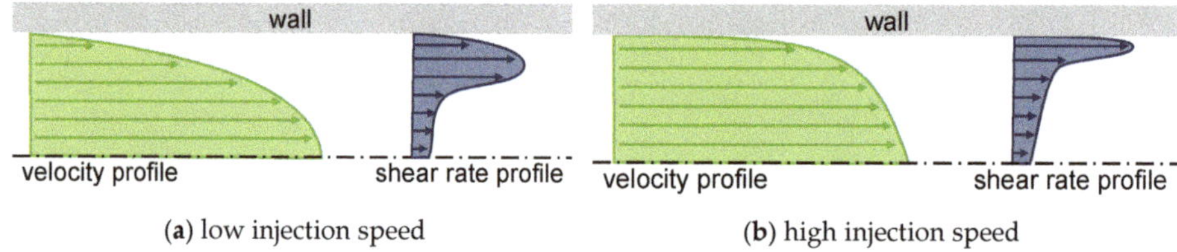

(**a**) low injection speed   (**b**) high injection speed

Figure 4. Influence of the injection speed on the velocity profile according to [55].

At high injection speeds (Figure 4b), the maximum shear rate shifts towards the mold wall, and the core layer width increases [39]. Whereas, for lower injection speeds (Figure 4a) the maximum of the shear rate profile shifts towards the cavity core, the range of the high shear rates generally increases, and the core layer width decreases. Depending on the shear and extensional flow inside the cavity, individual layers are formed with their associated fiber orientation distribution [19].

Furthermore, the viscosity of the melt changes with the type of matrix polymer and the reinforcing filler material (material, geometry, concentration, length, diameter, orientation, adhesion etc.) [42] as shown in Figure 5a. For example, a higher fiber concentration results in a higher viscosity, which leads to a different velocity and shear rate profile, as shown in Figure 5b,c [32,56]. In areas of low shear rates in particular, the viscosity of filled polymers differs strongly from that of unfilled polymers. Higher fiber concentrations mostly result in smaller skin layers due to their shear thinning behavior [19,39,57,58]. For this reason, different viscosity models are required to describe specific material behavior at low shear rates for reinforced or filled polymers (Figure 5b). For injection molding simulation, a frequently used model was developed by Herschel and Bulkley. A detailed description of the Herschel–Bulkley model is given in Section 1.1.7.

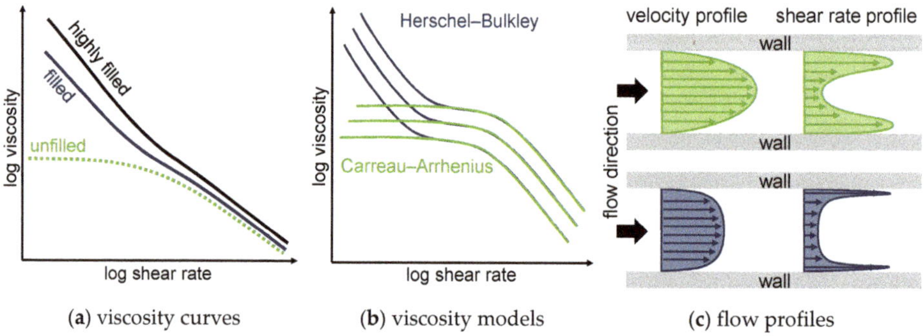

(**a**) viscosity curves   (**b**) viscosity models   (**c**) flow profiles

Figure 5. Influence of the viscosity to the velocity profile according to [59].

1.1.3. Fiber Orientation Distribution of Short and Long Fiber Reinforced Thermoplastics

As mentioned before, the properties of the reinforcement itself and the fiber microstructure properties also affect each other. In this context, the resulting fiber length inside a part plays a crucial role in the resulting fiber orientation [1]. The fiber length in particular causes a varying layer ratio, which can be seen in the fiber orientation distribution of short and long fiber reinforced plastic parts [39,44]. Even if the fiber breakage and the resulting fiber length are also determined by its orientation inside the flow, the resulting fiber orientation is more dominated by the fiber length. As Figure 6 schematically demonstrates, the degree of orientation in the skin layer decreases and the core layer width increases with increased fiber length [39,44,47,54]. Furthermore, in some cases, the degree of fiber orientation perpendicular to the flow direction (A_{22}) in the core layer can also exceed the degree of fiber orientation parallel to the flow direction (A_{11}), as shown in Figure 6b, for a long fiber reinforced plastic. In that special case, the mechanical properties perpendicular to the flow can also exceed the mechanical properties in the direction of flow.

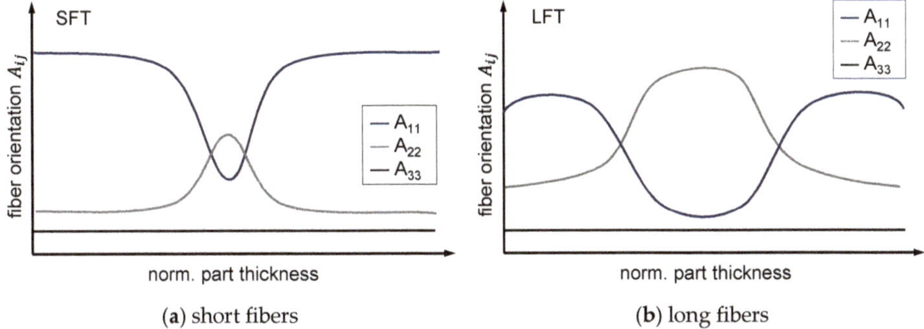

(**a**) short fibers (**b**) long fibers

Figure 6. Characteristic fiber orientation distributions along the parts thickness in discontinuous fiber reinforced plastics.

1.1.4. Tensorial Description of the Fiber Orientation

The fiber orientation is usually described as a 2nd order orientation tensor $A = a_{ij}$ according to Advani and Tucker [60]. The tensor describes the orientation of all fibers in a defined volume or element where $\Psi(p)$ is a probability density function, which is a statistical description of orientation states. The orientation tensor A is calculated as an integral over all possible fiber orientations from the distribution function of the fibers $\Psi(p)$ and the dyadic product of the unit vectors p. The first entry a_{11} describes the orientation of the fibers in the flow direction and is represented in Figure 7a. The eigenvectors Λ_i of the tensor indicate the principal direction of fiber orientation, while the eigenvalues e_i indicate the degree of fiber orientation, i.e., the statistical proportion of fibers in the respective principal direction.

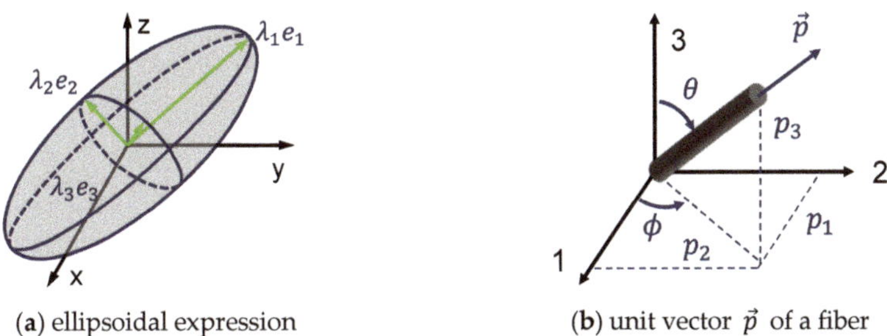

(**a**) ellipsoidal expression (**b**) unit vector $\vec{p}$ of a fiber

Figure 7. Tensorial Description of the Fiber Orientation.

The calculation of fiber orientation is based on a previously computed flow field. The velocity gradient tensor L and the shear rate $\dot{\gamma}$ are required as input variables. The result is a tensor A that describes the orientation of the fibers in the melt or flow channel in three dimensions [60].

$$A = a_{ij} = \begin{bmatrix} a_{11} & a_{11} & a_{11} \\ a_{11} & a_{22} & a_{11} \\ a_{11} & a_{11} & a_{33} \end{bmatrix} \tag{1}$$

The diagonal entries a_{11}, a_{22}, and a_{33} describe the orientation in the three spatial directions.

1.1.5. Experimental Determination of the Fiber Orientation

Several methods have been developed to analyze the fiber microstructure, such as ultrasound methods [61,62], optical methods like grinding pattern analysis [35,40,60,63–70], electron microscopy [14], Radiography [45,71,72] and X-ray computed tomography (CT) [73–78]. However, all methods are usually associated with high effort and are therefore used infrequently. To determine the spatial fiber orientation by means of cross section analysis, each fiber ellipsis, as shown in Figure 8a, has to be measured by its major and minor axis according to Figure 8b. With the trigonometric functions depicted in Figure 8c, the fiber orientation tensor components can be calculated for each fiber [40,60]. In addition, the reduced cutting probabilities of the inclined fibers can be considered by applying a weighting function according to [40,79].

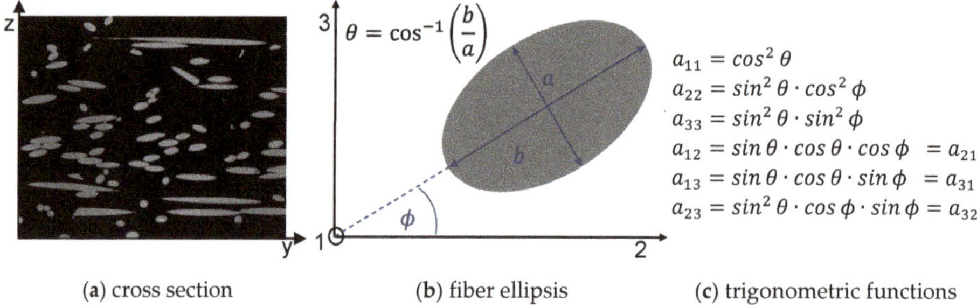

(a) cross section	(b) fiber ellipsis	(c) trigonometric functions

Figure 8. Cross section analysis to experimentally determine fiber orientation.

1.1.6. Injection Molding Simulation

For the simulation of the unsteady, non-Newtonian and non-isothermal injection molding process, the three-dimensional basic Equations for fluid mechanics are usually solved using the finite volume method (FVM). The basic equations for conservation of mass (2), conservation of linear momentum (3), and energy (4), are solved by FVM numerically [16,80].

conservation of mass:

$$\frac{\partial \rho}{\partial t} + \nabla \cdot \rho \vec{v} = 0 \tag{2}$$

conservation of momentum:

$$\rho \left(\frac{D \vec{v}}{Dt} \right) = \rho \vec{g} + \nabla \cdot \underline{\tau} - \nabla \cdot p \tag{3}$$

conservation of energy:

$$\frac{\partial \rho T}{\partial t} = -\nabla \cdot (\rho \vec{v} T) + \nabla \cdot \left(\frac{\lambda}{c_p} \cdot \nabla T \right) + \underline{\tau} : \nabla \vec{v} \tag{4}$$

Based on these three equations, flow velocities ∇ and their gradients $\nabla \vec{v}$ are calculated, which represent the major determining variables for the subsequent fiber orientation calculation. The velocity gradient tensor $\nabla \vec{v}$ describes the change of velocity in the three spatial directions of a stationary flow and can be split into a symmetrical $\underline{D}$ and a rotational $\underline{W}$ component (Equation (5)). From the deformation velocity tensor $\underline{D}$, the shear rate $\dot{\gamma}$ can be derived (Equation (6)). However, for a detailed description of the FVM for injection molding simulation, refer to [80].

$$\nabla \vec{v} = L = \frac{1}{2}\left(\nabla \vec{v} + \left(\nabla \vec{v}\right)^T\right) + \frac{1}{2}\left(\nabla \vec{v} - \left(\nabla \vec{v}\right)^T\right) = \underline{D} + \underline{W} \tag{5}$$

$$\dot{\gamma} = |\underline{D}| = \sqrt{\frac{1}{2}\left(\underline{D} : \underline{D}^T\right)} \tag{6}$$

1.1.7. Viscosity Models

To describe the flow behavior of plastic melts [81], several different viscosity models were developed. The Herschel–Bulkley model is frequently used to accurately represent the viscosity of long fiber reinforced thermoplastics especially in the range of low shear rates. In this model, on the one hand, the yield stress viscosity at low shear rates is considered, which can be determined experimentally by rotational rheometry [82,83]. On the other hand, the range of high shear rates is precisely covered where so-called shear thinning [84] is present. This range can experimentally be determined by means of capillary rheometry. The model is composed of the Herschel–Bulkley yield stress (first term) and the Cross-WLF viscosity (second term), as shown in Equation (7) with power-law index n, temperature T, pressure p, and zero shear viscosity η_0.

$$\eta\left(\dot{\gamma}, T, P\right) = \frac{\tau_y}{\dot{\gamma}} + \frac{\eta_0(T, p)}{1 + \left(\frac{\eta_0}{\tau^*}\dot{\gamma}\right)^{1-n}} \tag{7}$$

The model consists of seven parameters N, τ^*, $D1$, $D2$, $D3$, $A1$, and $A2$ in order to approximate the shear rate distribution of the materials viscosity.

1.1.8. Fiber Orientation Models

To calculate fiber orientation distributions, the tensorial description developed by Advani and Tucker [60] is usually used. The corresponding calculation models are mostly based on the extension of the Jeffery model [85] made by Folgar and Tucker [86]. Current commercial injection molding solvers also use combined model extensions, such as the RPR (retarding principal rate) and ARD or iARD (improved anisotropy rotary diffusion) models. The iARD-RPR model is a combination of Jeffery's hydrodynamic model (HD), the extension of the IRD model (isotropic rotary diffusion) from Folgar–Tucker to the iARD model and the RPR model (Equation (12)), according to Tseng et al. [87–89]. With C_i (fiber–fiber interaction coefficient), C_m (fiber–matrix interaction coefficient), and α (slow-down parameter), a changing of the fiber orientation based on the existing flow field can be calculated. The influencing factors of the calculated fiber orientation are in this case the gradient tensor L and the shear rate $\dot{\gamma}$ of the flow field.

$$\dot{A} = \dot{A}_{HD} + \dot{A}_{iARD}(C_i, C_m) + \dot{A}_{RPR}(\alpha) \tag{8}$$

This leaves three model parameters C_i, C_m, and α, which can be used to adjust or calibrate the fiber orientation distribution. The first part is defined as iARD (Equation (9)) [87].

$$\dot{A}^{iARD} = \dot{\gamma}[2D_r - 2tr(D_r)A - 5D_r\,A - 5A\,D_r - 10A_4] \tag{9}$$

In addition to the second order orientation tensor, the fourth order orientation tensor is required for the calculation, and is solved with invariant-based closure approximation (IBOF5) according to Chung and Kwon [90], which is a model that offers a good compromise of a precise approximation and computation effort.

The rotary diffusion tensor D_r (Equation (11)) is determined by the scalar $\| D^2 \|$ (Equation (10)) of the square of the symmetrical part D from the velocity gradient tensor L.

$$\| D^2 \| = \sqrt{\frac{1}{2} D^2 : D^2} \tag{10}$$

$$D_r = C_I \left(I - C_m \frac{D^2}{\| D^2 \|} \right) \tag{11}$$

The second part describes the retarded principal rate (Equation (12))

$$\dot{A}^{RPR} = -R\, \Lambda^{IOK}\, R^T \tag{12}$$

with Λ^{IOK} (Intrinsic Orientation Kinetics), as the substantial derivative of a certain diagonal tensor. In general, an eigenvalue calculation of both the orientation tensor A and its change $\dot{A}$ is coupled here. This is realized as follows (Equations (13) and (14)):

$$\Lambda^{IOK}_{ii} = \alpha \dot{\Lambda}_i,\ \text{with } i = 1,2,3 \tag{13}$$

$$R = [e_1, e_2, e_3] \tag{14}$$

where $\dot{\Lambda}_i$ represents the eigenvalues of the orientation tensor and R the rotation matrix, which is composed of the eigenvectors of the change in the orientation tensor. This relationship is extended as follows (Equation (15))

$$\Lambda^{IOK}_{ii} = \alpha \left[\dot{\Lambda}_i - \beta \left(\dot{\Lambda}_i^2 + 2\dot{\Lambda}_j \dot{\Lambda}_k \right) \right],\ \text{with } i,j,k = 1,2,3 \tag{15}$$

with an additional parameter β, which is defined as a time constant.

In order to determine the different influences of the velocity field more precisely, the parameters C_i and α were calculated depending on the shear rate according to Tseng et al. [91]. Figure 9 shows that α was adjusted in the core layers, which damps this parameter at lower shear rates, while the opposite is true for C_i, which is damped at higher shear rates.

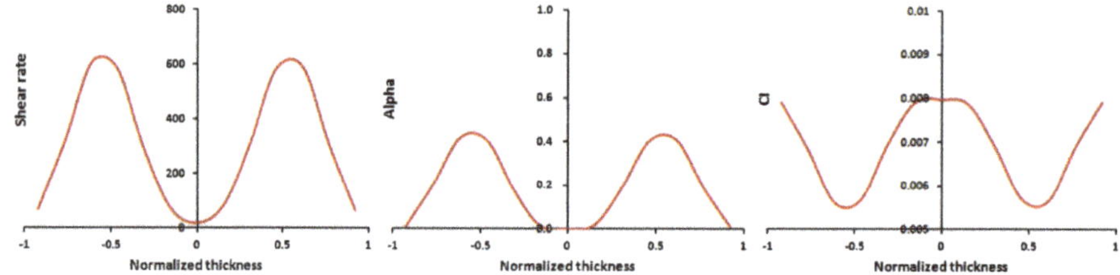

Figure 9. Shear rate-dependent adaption of the parameters C_i and α [91].

For this purpose, the parameters are changed as follows (Equations (16) and (17)) [91].

$$\frac{C_i(\dot{\gamma})}{C_{i0}} = \frac{1}{1 + \left(\frac{\dot{\gamma}}{\dot{\gamma}_c} \right)^2} \tag{16}$$

$$\frac{\alpha(\dot{\gamma})}{\alpha_0} = 1 - \frac{1}{1 + \left(\frac{\dot{\gamma}}{\dot{\gamma}_c} \right)^2} \tag{17}$$

The α_0 and C_{i0} input values are given by the user and then changed to a specific value for each element of the simulation model. With $\dot{\gamma}_c$, a critical shear rate is defined, which has to be determined empirically. In commercial simulation tools like Moldex3D® R17 this variable is implemented as a constant determined by experience of the developers.

1.1.9. Calibration of Fiber Orientation Models

The model parameters play a crucial role in the use of fiber orientation models, because only with an appropriately calibrated model the fiber orientation can be reliably predicted [7,40,88,92–94]. The main challenge here is that the model parameters are not exclusively material-dependent, but may even depend on the process and part geometry [94,95]. Therefore, standard parameters can only be used rarely or only for a few specific cases. Especially for LFTs, a reliable prediction with current models is not possible when using standard parameters.

Most publications concerning the prediction of fiber orientation either use standard parameters [2] or simply adjust or fine tune them manually by trial and error [1,96]. Sometimes there is a so-called fit, but no algorithm for the parameter identification is described [8,88,92,97,98]. Independently of the geometry, objective fitting methods has been described for simple shear flows of short fiber [99] and also long fiber [100] reinforced plastics. Some of the necessary model parameters like the fiber-fiber interaction coefficient C_i were extensively analyzed and associated approximation functions [95,101,102] were developed, respectively. The influence of the parameters of recent fiber orientation models [87,92,103,104] or viscosity [6,105] has been investigated in some publications, but basic correlations to the material or approximated functions to evaluate the model parameters are still missing.

Several methods are available to optimize the necessary parameters of the prediction models for the microstructure of fiber reinforced plastics. In a first step, however, the properties of the microstructure must be determined, or methods as described in Section 3.2 must be developed in order to compare certain properties with each other directly [106]. Scalar values, such as the spatial fiber concentration, are less problematic than the fiber orientation and length, which are subject to a certain distribution even in local areas. If these requirements are satisfied, the parameters of the respective prediction models can be optimized by means of a reverse engineering procedure, as shown in Figure 10. However, even though it is possible to automatically optimize the parameters of the fiber orientation model, only a few algorithms are described in the literature [94,106–109].

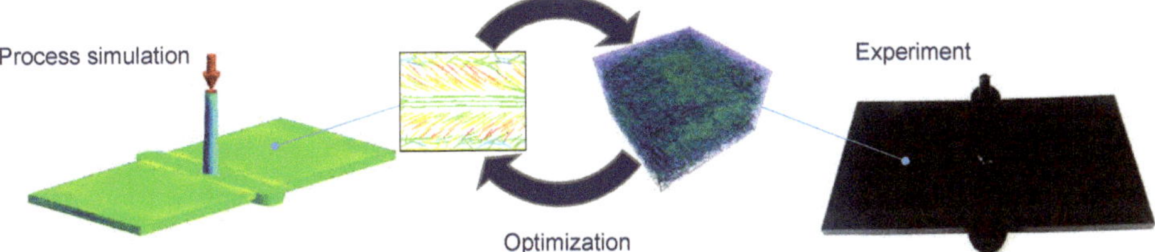

Figure 10. Reverse engineering of fiber microstructure properties at a specific position.

1.1.10. Optimization Methods

The aim of this study is to predict the experimentally determined fiber orientation with minimum error variance by optimizing the parameters of the fiber orientation model (C_i, C_m, and α for example for the iARD model, see Section 1.1.8). According to the previous Section 1.1.9, the most suitable calibration of the fiber orientation model should be identified. Since the intention of this work is not the invention of an improved optimization algorithm, the basics of optimization methods will only be described in principle. A detailed description and a comparison of different approaches to optimization will be omitted and reference will be made to further literature [110,111].

In general, parameter optimization deals with the problem of finding a parameter vector x^* for which a given target function is minimized. Non-linear variation methods of the form $min f(x)$ are solved. This function can be approximated by quadratic Taylor series approximation Equation (18).

$$min f(x) \approx h(x) := f(\bar{x}) + \nabla f(\bar{x})^T (x - \bar{x}) + \frac{1}{2}(x - \bar{x})^t H(\bar{x})(x - \bar{x})f(x) \tag{18}$$

where $\nabla f(x)$ represents the gradient of $f(x)$, and $H(x)$ represents the Hessian matrix of $f(x)$. The gradient of $h(x)$ is calculated as follows by finding the minimum of the gradient (Equation (19))

$$0 = \nabla f(\bar{x}) + H(\bar{x})(x - \bar{x}) \tag{19}$$

For the use of the Newton method for problems with constraints, an additional state variable α_C is added to the original Equation, which leads to the following variation Equation (20).

$$(x - \bar{x}) = -\alpha_C H(\bar{x})^{-1} \nabla f(\bar{x}) \tag{20}$$

In order to determine α_C, various line search techniques or methods can be used to directly satisfy the decisive variable constraints. In addition, the Hessian matrix can be approximated by a diagonal matrix D_H This simplifies the optimization to a quasi-Newton method [112,113].

1.2. Objectives and Hypotheses

According to the major challenge to reliably predict the behavior of discontinuously fiber reinforced parts produced by injection molding, a holistic approach to predict all considerable properties of the fiber microstructure is desirable. In a first step, the following hypotheses are drawn and investigated to improve the predicted fiber orientation. In the future, these investigations can be extended to the remaining microstructural properties such as fiber length and fiber concentration.

On the one hand, the prediction accuracy of the fiber orientation distribution can be significantly improved by using optimization methods. On the other hand, the required optimization time can be significantly reduced with a new calibration approach.

In order to optimize the parameters of the fiber orientation model to predict the experimentally determined fiber orientation with a minimum error variance, two different calibration approaches are investigated. However, since there have been several further developments of the iARD fiber orientation model in the last few years; the most suitable version is initially identified. In the first calibration approach, a commercial simulation software is used to calculate the fiber orientation model, and only the parameters of the fiber orientation model are externally optimized in another tool to obtain the most accurate prediction for the fiber orientation distribution. As second approach, a new method is developed to calibrate the fiber orientation model. The aim of the new approach is to achieve the optimization goal as quickly and efficiently as possible. Therefore, in contrast to the first approach, which requires a time-consuming recalculation of the entire flow field with each optimization step, the new approach is based on only one initial calculation of the flow field. The subsequent parameter optimization of the fiber orientation model is performed without any further calculation of the flow field. With this approach, a suitable calibration of the fiber orientation model should be achieved within a few minutes.

Even though the new approach focuses on the calibration of the fiber orientation model, the method should be designed in a way that it can be easily transferred to other models to predict microstructural properties. For example, in the future, this approach should also be used to calibrate the models to predict fiber length and fiber concentration.

2. Experimental Setup

2.1. Materials, Parts, and Processing

The fiber orientation distribution was experimentally determined for injection molded plates according to ISO 294-5 with dimensions of 80 × 80 × 4 mm, as shown in Figure 11a.

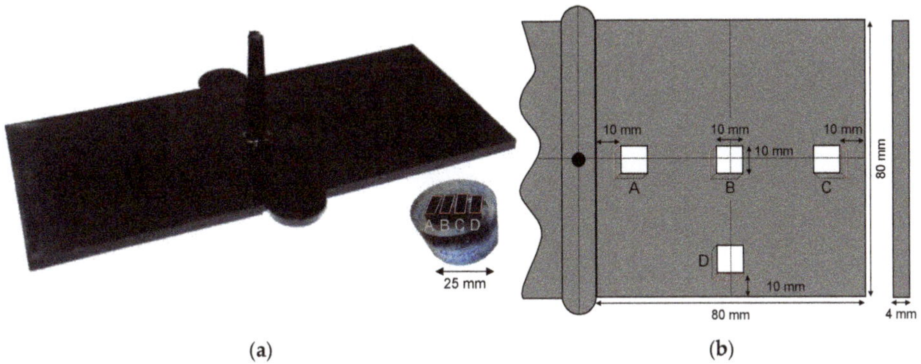

(a) (b)

Figure 11. Injection molded plates (**a**) and scheme of evaluation points along one side of the plate (**b**).

A long fiber reinforced polypropylene (PP-LGF) from TechnoCompound GmbH, Bad Sobernheim, Germany was chosen, because it is the most frequently used LFT material [114]. The granules are 10 mm long, rod-shaped pellets with fiber mass fractions of 20%, 40%, and 60% of the product type TechnoFiber. The plates were injection molded with two constant injection speeds of 30 cm^3/s (low) and 100 cm^3/s (high), a melt temperature of 230 °C and a mold temperature of 80 °C on an injection molding machine, Allrounder 520S 1600–400 from Arburg, Lossburg, Germany.

For the experimental determination of the fiber orientation distribution, small samples (10 mm × 10 mm) were taken at the beginning (A), middle (B), end (C) and aside (D) of the flow path according to Figure 11b. These samples were analyzed by inspecting the cross sections highlighted with red in Figure 11, using optical reflection microscopy as described in the following section.

2.2. Experimental Determination of the Fiber Orientation Distribution

With the image processing program FIJI (ImageJ), developed by Wayne Rasband, in combination with MATLAB® by MathWorks®, Natick, USA, spatial fiber orientation for the individual cross section were determined using the fiber orientation tensor A as explained in Section 1.1.5. First of all the high-resolution subsection images (1.920 × 1.200 px) of each cross section (ca. 150) were stitched together and converted to a binary image by means of different image processing algorithms (image stitching, background subtraction, local threshold, watershed) as shown in Figure 12a before and b after digital image processing. Based on these binary images a particle analysis is performed in ImageJ. By means of the second order central moment to fit the best ellipse, the major and minor axis of each particle is determined. Afterwards, the fiber orientation tensor of each ellipsis was calculated in a MATLAB® tool by using the equations given in Figure 8c, developed by [60], and corrected with the weighting function according to [40]. In order to analyze the fiber orientation distribution along the thickness with sufficient accuracy, the whole cross section was discretized along the z-direction into a structured grid with at least twenty individual fibers per cell. Subsequently, the fiber orientation tensors of all fibers within a discretized cell were averaged according to the cells marked with red in Figure 12.

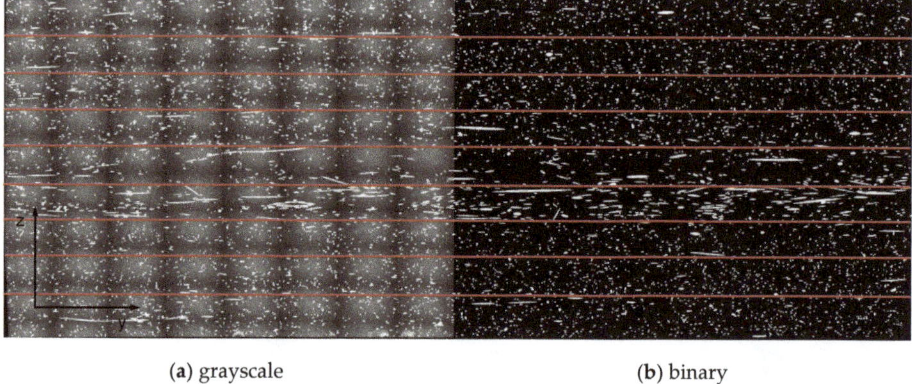

(a) grayscale   **(b)** binary

Figure 12. Different states of image processing to determine the fiber orientation.

The fiber orientation distribution along the sample thickness (z-direction) of each individual sample is obtained from overall 15 grid points by the calculated cell averages. In Addition, two cross sections of each material and process setting were analyzed and averaged to eliminate measurement and processing deviations. For example, Figure 13a shows the fiber orientation results for two samples and the averaged curve.

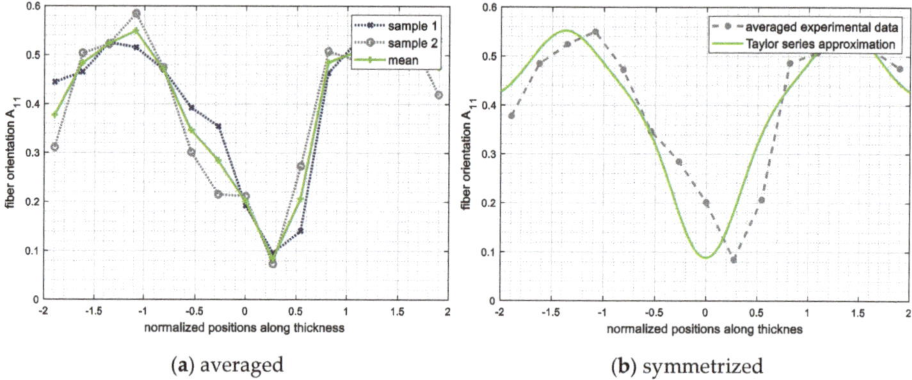

(a) averaged   **(b)** symmetrized

Figure 13. Processing of the fiber orientation results for objective comparison (PP-LGF20 v30).

Even though two samples were averaged, an objective calibration of the fiber orientation model on the basis of only few measuring points is usually not reasonable. Furthermore, the experimentally determined fiber orientation shows a slightly off-center position of the core layer, which is caused by an uneven mold cooling as already observed in [101]. To obtain comparable results for experiment and simulation, the measuring points are symmetrized towards the center and smoothed by approximation. However, the characteristic properties of the curve progression regarding the fiber properties in the core and shear layer should be retained. For this purpose, the measuring points are represented by a fifth-order Taylor series approximation, as shown in Figure 13b for the orientation tensor entry A_{11}. This method allows an objective comparison between simulation and experiment, independently of the number of measured data points. In addition, the curves are symmetrized without a significant change in their characteristic shape.

2.3. Process Simulation

The process simulation was performed with Moldex3D® of CoreTech System, Chupei City, Taiwan, which numerically solves the governing Equations (2)–(4). The calculated geometry is a complete model of the injection molded plate as shown in Figure 11a. The mesh was made by Rhinozeros 5 from Robert McNeel & Associates, Seattle, USA, and is shown in Figure 14a. The plates were meshed with a structured grid consisting of 21 hexahedral elements along the thickness of 4 mm, whereas tetrahedral elements were used for the runner system and sprue. A constant velocity according to the injection speed was set at the Inlet. The viscosities of the individual materials PP-LGF20, PP-LGF40, and PP-LGF60 were determined by means of a high-pressure capillary rheometer in the range of high shear rates, and a rotational rheometry for low shear rates. The experimental data was fitted by the Herschel–Bulkley viscosity model as shown in Figure 14b.

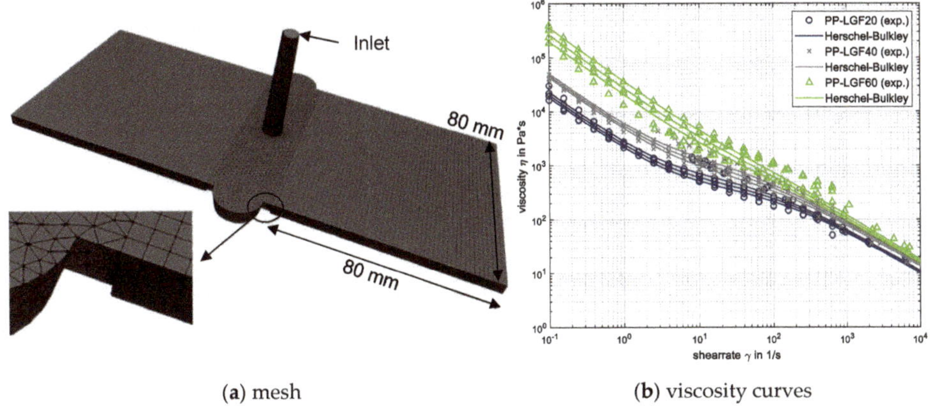

(a) mesh	(b) viscosity curves

Figure 14. Geometry and material setup of the simulation model.

The results of the viscosity determination show the expected rise of the viscosity with higher fiber content, especially for relatively low shear rates. The respective model parameters for the prediction of the fiber orientation model are determined by different iterative optimization methods described in the following sections, which compare the simulation results to the experimental measured data. The pvT-model for the thermodynamic state relations according to [115] was used to describe the density of the fluid as a function of temperature and pressure. All other necessary boundary conditions, such as injection speed as well as melt and mold temperatures, were defined according to the injection molding processing conditions described in Section 2.1.

In order to investigate the influence of the different versions of the iARD fiber orientation model as explained in Section 1.1.8 on the resulting fiber orientation distribution the following software versions of Moldex3D® with the different fiber orientation model versions

- basic iARD model in version R13,
- enhanced iARD model with shear rate dependency in version R16,
- and enhanced iARD model with shear rate dependency and fiber coupled viscosity in R17 were used.

Because fiber length is significantly reduced within the plastification unit, and this is not part of the simulation, the number averaged fiber length (see Table 1) was experimentally measured as described in [36] at the sprue and specified as a boundary condition in the respective process simulations. As expected, the results show that, with an increasing fiber content, the probability of fiber–fiber interaction increases significantly, and fiber breakage occur more frequently.

Table 1. Experimentally determined number average fiber length at the sprue.

Material	PP-LGF20	PP-LGF40	PP-LGF60
Average fiber length in mm	1.701	1.08	0.795
Aspect ratio (L/D)	104.36	66.26	55.71

2.4. Adjustment of the Fiber Orientation Model within Process Simulation

2.4.1. Comparison of Simulation and Experimental Data

In order to objectively compare the predicted fiber orientation distribution along the thickness z with the experimentally determined data, the least square error S is calculated by Equation (21) at all positions of a sample according to Figure 11b.

$$S = \sum_{z=0}^{d} \left(A_{exp}(z) - A_{sim}(z) \right)^2 \tag{21}$$

To reduce the optimization problem, only the three orientation tensor components A_{11}, A_{22}, and A_{33} were added as least square functions. Furthermore, the minimum square error functions can be weighted differently if the adaptation to a certain orientation direction takes priority.

2.4.2. Influence of the Fiber Orientation Model Parameters

Another considerable factor is the accurate calibration of the model parameters C_i, C_m, and α, which can influence the core-layer thickness (mainly α) as well as the degree of orientation in the skin layers (mainly C_i and C_m). In comparison of different fiber orientation model versions implemented in Moldex3D® as described in Section 2.3 the influence of the three parameters changes significantly. As shown in Figure 15a change of C_i and C_m within the basic iARD model mainly result in a different degree of orientation inside the skin layers, whereas α leads to a change in the core layer thickness. With more recent enhancements of the orientation model iARD with shear rate dependency in R16 and also with fiber coupled viscosity in R17, the parameter α shows hardly any change in the core layer thickness Figure 15b,c. This also leads to a very small transition zone between the core and skin layers.

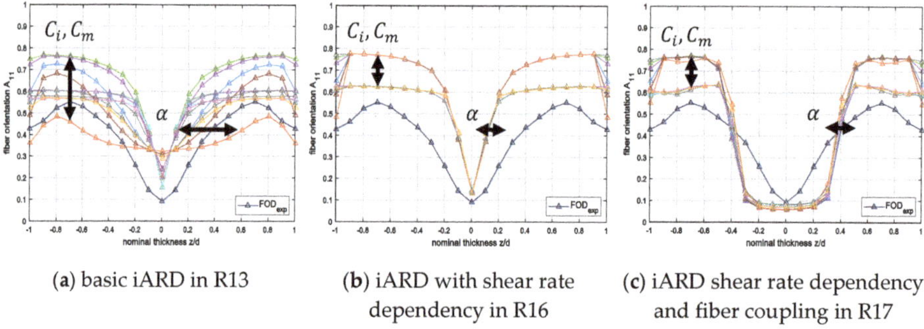

(**a**) basic iARD in R13   (**b**) iARD with shear rate dependency in R16   (**c**) iARD shear rate dependency and fiber coupling in R17

Figure 15. Influence of fiber orientation model parameters on different software versions.

Even though the basic iARD fiber orientation model implemented in Moldex3D® version R13 is considerably older than the other versions, only this version allows to adjust the core layer thickness by means of the corresponding fiber orientation model parameters. According to this study, the basic iARD model has the most individual configuration parameters to represent the experimentally determined fiber orientation as accurately as possible.

2.5. Calibration of Fiber Orientation Model

Based on the calibration and optimization methods explained in Sections 1.1.9 and 1.1.10, a detailed description of the methods used in this article is given below.

2.5.1. Direct Optimization

In order to directly optimize the results of the fiber orientation model iARD, the respective model parameters C_i, C_m and α have to be adjusted until the defined optimization objective is reached. However, the optimization algorithms require input variables as well as output variables, which can be described by the functions or directly adjusted. Therefore, a specific routine is needed, which changes the fiber orientation model parameters inside the simulation environment, starts the simulation run with the defined parameters, waits until the simulation is successfully calculated, reads out the defined results (i.e., fiber orientation tensor component A_{11} along the thickness at a specified location of the part), and transfers these results as input parameters to the optimization algorithm, and compares the results with a defined optimization objective. Finally, depending on the optimization objective and the associated accuracy, the optimization procedure must be terminated or the routine must be restarted. In this work, the described optimization approach was implemented in a MATLAB® script, which directly accesses the input data of Moldex3D®, and thus creates new runs and triggers their calculation.

2.5.2. A New Calibration Method

In a new method, the fiber orientation calculation and its optimization were performed with a MATLAB® script according to the process scheme shown in Figure 16.

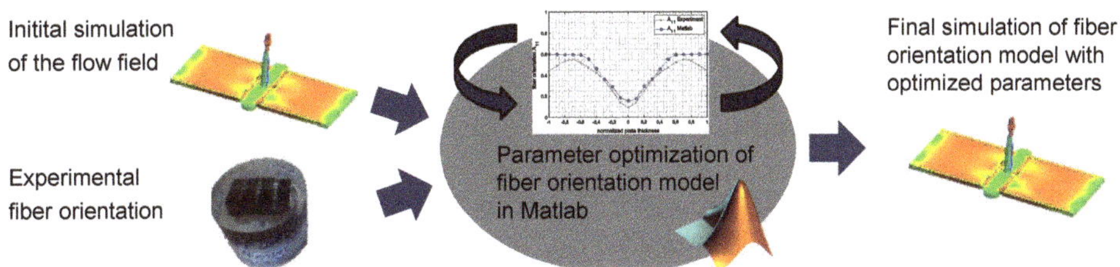

Figure 16. Process scheme of the fiber orientation optimization tool.

To obtain the necessary flow field for the injection molding process, the results (velocity gradient tensor L) of a single initial injection molding simulation in Moldex3D® were exported and used in MATLAB® to calculate the corresponding fiber orientation model by means of Equations (5)–(15) (basic iARD model). Based on the initial calculation of the flow field, the whole parameter optimization of the fiber orientation model was carried out within a few minutes.

The main intention of this method was not the development a new optimization algorithm, but the removal of the time-consuming recalculation of the entire flow field for each optimization step (in contrast to the direct optimization described in Section 2.5.1). By using the flow field of an initial injection molding simulation, the computation is reduced to the calculation of the fiber orientation model.

The goal of the optimization was to minimize the objective function, consisting of the error sum of squares between simulative calculated and experimental determined fiber orientations at pre-defined positions. A gradient-based algorithm according to Section 1.1.10 was used for this purpose. This means that at the current point of the optimization, the first derivative for the direction of the optimization step and second derivatives for the length of the iteration step for each variable are used.

Because this case is a multi-parameter optimization, a system of Equations (22) is required to solve a variation step. This is as follows for the three parameters of the fiber orientation model.

$$
\begin{bmatrix} C_i \\ C_m \\ \propto \end{bmatrix} = \begin{bmatrix} \partial^2 S\,(C_i) & 0 & 0 \\ 0 & \partial^2 S\,(C_m) & 0 \\ 0 & 0 & \partial^2 S\,(\propto) \end{bmatrix}^{-1} \begin{bmatrix} \partial S(C_i) \\ \partial S(C_m) \\ \partial S(\propto) \end{bmatrix}
\tag{22}
$$

In this form, only the main diagonals of the matrix are filled, which changes the optimization to a quasi-Newtonian method. In this form, fewer derivatives are needed to calculate the variation vector, which makes the calculation much more timesaving. Furthermore, the addition of scaling is also simplified. The Hessian matrix is positive definite, and thus the correct direction of the variation step is determined. Additionally, all three parameters are assigned to limits.

- $C_i \in [0-0.1]$
- $C_m \in [0-1]$
- $\alpha \in [0-1]$

For this reason, an additional line search method is introduced to guarantee that these limits are observed. To prevent the optimization from getting stuck at these limits, a damping factor $\propto_D$ is introduced in Equation (23) in addition to the line search factor $\propto_k$.

$$
\begin{bmatrix} C_i^* \\ C_m^* \\ \propto^* \end{bmatrix} = \begin{bmatrix} C_i \\ C_m \\ \propto \end{bmatrix} \propto_k \propto_D \;.\; \text{with } C_i^* = C_{i,i-1} + \propto_k \; C_i \geq 0.1 \text{ and } \propto_k = \frac{0.1 - C_{i,i-1}}{C_i}
\tag{23}
$$

From all restrictions, the value $\propto_k$ is determined. Furthermore, a line search factor is used to prevent oscillation. For this purpose, it is verified whether the target function value increases contrary to the expectation during the optimization. In this case, the optimization step is shortened greatly with $\propto_k$ to allow only a small deterioration. For this purpose, the factor $\propto_k = 0.05$ is specified if $S_i > S_{i-1}$. Because the optimization problem can be dominated by the restrictions, the following is suggested for the damping factor:

- $\propto_D = 0.8$ if $\propto_k$ is active
- $\propto_D = 0.95$ without active constraint

This is to prevent the optimizer from getting stuck at the given variable boundaries. This problem is further improved by the introduction of scaling for the Hessian matrix (Equation (24)).

$$
\begin{bmatrix} C_i \\ C_m \\ \propto \end{bmatrix} = \begin{bmatrix} f_1\,\partial^2 S\,(C_i) & 0 & 0 \\ 0 & f_2\,\partial^2 S\,(C_m) & 0 \\ 0 & 0 & f_3\,\partial^2 S\,(\propto) \end{bmatrix}^{-1} \begin{bmatrix} \partial S(C_i) \\ \partial S(C_m) \\ \partial S(\propto) \end{bmatrix} \Bigg| f_2 = f_3 = (0.1 - C_i^*)^{-m}
\tag{24}
$$

With the damping factor $\propto_D$, the barrier is approached more slowly, and the other two parameters are weighted more heavily.

3. Results

In this Section, the influences of different material and process settings are evaluated. Based on the experimental data, the optimization results for the simulation are shown, and the two different optimization methods are compared.

3.1. Experimental Results

3.1.1. Fiber Orientation at Different Positions

The following Figure 17a shows the fiber orientation distribution along the plate thickness at three different positions (according to Figure 13, Pos. A, B, and C) along the flow path. In addition, the experimental results were smoothed and symmetrized according to Section 3.2 in order to be compared with the simulated results, Figure 17b.

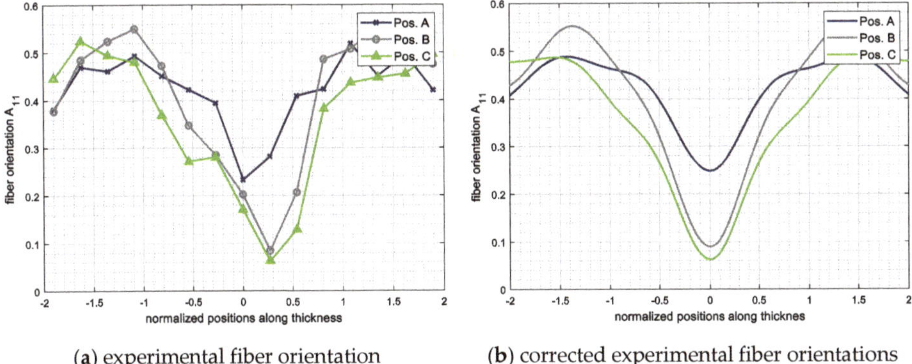

(**a**) experimental fiber orientation   (**b**) corrected experimental fiber orientations

Figure 17. Experimentally determined fiber orientations along the flow path for PP-LGF20 v30.

The analysis of the fiber orientation distribution along the flow path (position A–C) shows an increasing core layer, which also means that more fibers are oriented perpendicular to the flow direction. This also leads to enhancement of the mechanical properties in this direction, while the properties in the flow direction are diminished. However, the increase in shear layer thickness along the flow path has often been observed and is not a particular phenomenon of long fiber reinforced plastics.

3.1.2. Fiber Orientation of Different Materials and Process Conditions

In the following section, the influence of fiber concentration and injection speed is analyzed. In general, it can be observed (Figure 18) that all fiber concentrations and processing conditions of the injection-molded long glass fiber reinforced polypropylene lead to a low degree of fiber orientation in the shear layers, and a wide core layer with a high degree of orientation.

All the analyzed specimens show a higher degree of orientation perpendicular to the flow (A_{22}) than parallel to the flow direction (A_{11}) in the core layer region, whereas in the shear layers, the degree of orientation is nearly the same in both directions (parallel A_{11} and perpendicular to the flow A_{22}). This effect changes only imperceptibly, even with a variation in fiber concentration and injection speed. In contrast to short fiber reinforced plastics, the wide core layer with its high degree of fiber orientation in the long fiber reinforced material also leads to higher mechanical properties (e.g., stiffness, strengthm and impact resistance as shown in Figure 1) perpendicular to the flow. A unique detail of the analyzed parts can be observed in the shifted core layers, which are not completely centered. This effect is caused by a different cooling performance (comparable to [101]) of the mold halves and has no influence on the interpretation of the results.

The analysis of the three different fiber concentrations (fiber mass fractions 20%, 40%, and 60%) of the long glass fiber reinforced polypropylene also show that an increase in fiber concentration results in a more distinct core layer Figure 18a,c,e. This means that a more concentrated suspension leads to an increased number of fibers that are oriented perpendicular to the flow direction A_{22}. The same effect can also be seen in the variation of the injection speed. With increasing flow velocity, a wider core layer can be observed for all analyzed fiber concentrations.

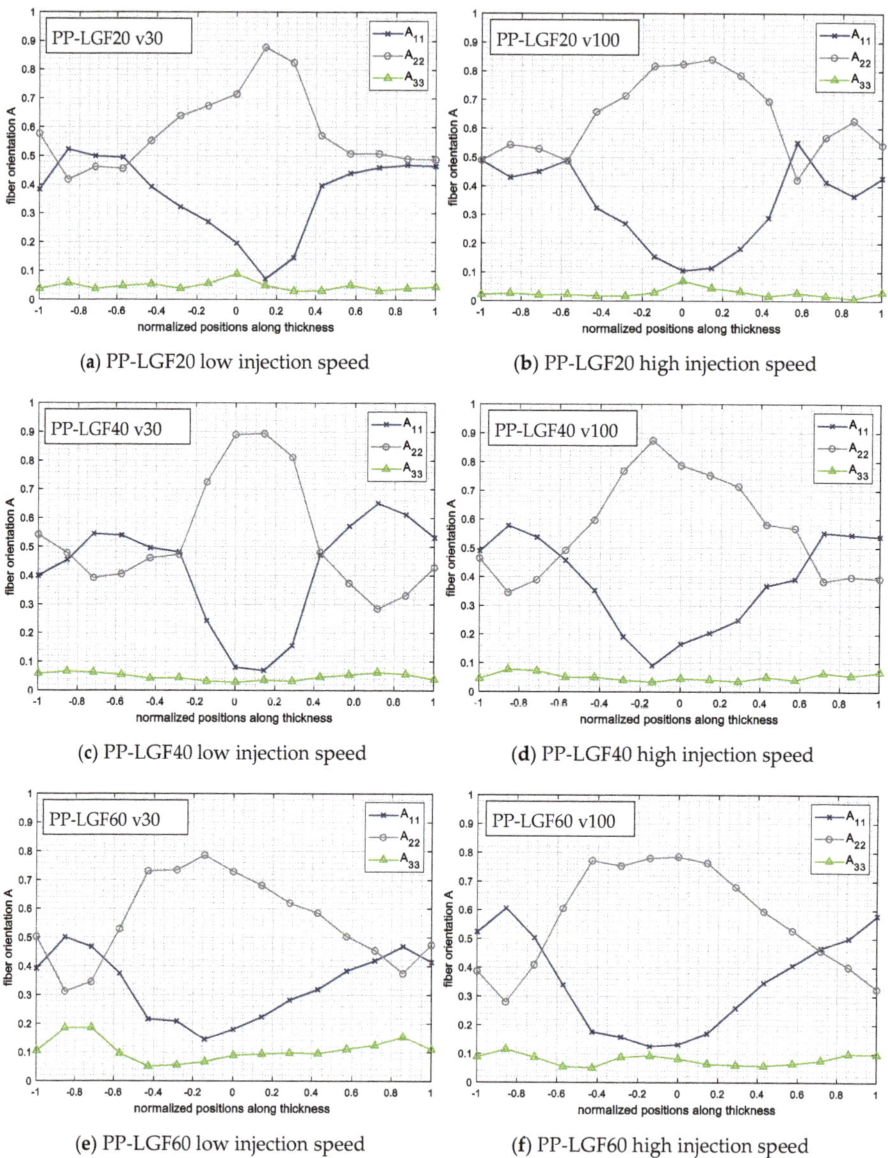

(**a**) PP-LGF20 low injection speed

(**b**) PP-LGF20 high injection speed

(**c**) PP-LGF40 low injection speed

(**d**) PP-LGF40 high injection speed

(**e**) PP-LGF60 low injection speed

(**f**) PP-LGF60 high injection speed

Figure 18. Resulting fiber orientation along the thickness of different fiber concentrations (20, 40 and 60% mass fraction long glass fiber) and different injection speeds (low = 30 cm^3/s, high = 100 cm^3/s).

In order to compare the skin core layer ratio systematically, two different methods were developed and used in this study. As a first method, the A_{11} entry of the fiber orientation tensor is approximated by a function, and the transition from skin to core layer (green line) is determined by the mean value of the two inflection points of its derivative as shown in Figure 19a. The determined threshold lines are also depicted in the grinded pattern of the cross section as shown in Figure 19b. Subsequently, the skin layer ratio can be calculated by the ratio of the associated thickness values. As a second method, the two intersections between the fiber orientation curves A_{11} and A_{22} are determined and their mean value is defined as the transition from skin to core.

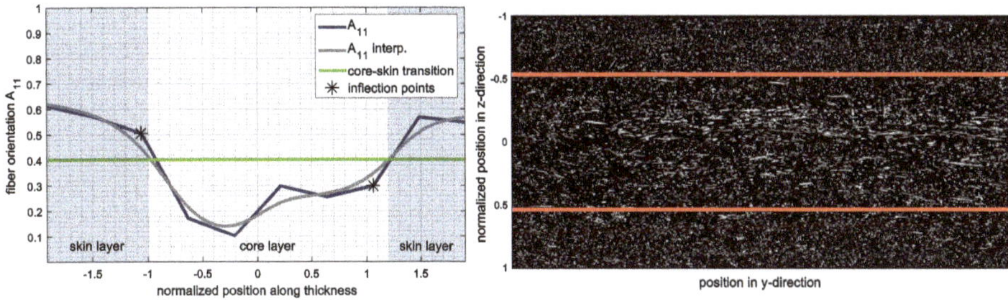

(**a**) determination of skin core layer   (**b**) cross section with skin core layer

Figure 19. Method to determine the skin core layer ratio of a reinforced material by objective criteria.

For the different materials and injection speeds the resulting skin core layer ratios showed only slight differences between the two described methods. Therefore, only the results of the intersection method of A_{11} and A_{22} were chosen and shown in Table 2.

Table 2. Skin core layer ratio for different fiber concentrations and injection speeds of PP-LGF.

Injection Speed	Low (v = 30 cm³/s)			High (v = 100 cm³/s)		
Fiber mass fraction	20	40	60	20	40	60
Skin core layer ratio	55%	36.25%	70%	56.25%	62.5%	70%

The results for the calculated skin-to-core layer ratios show an increase in the core-layer thickness with increasing fiber content. Thus, the previously recognized tendency of increase in core layer thickness can be proven objectively. The effect can be seen in general for low as well as for high injection speeds, except the outlying value for a very thin core layer thickness with 36% for PP-LGF40 at a low injection speed. No reason has been found to explain this effect.

3.2. Results of the Parameter Optimization

As shown in Figure 15, the resulting fiber orientation distribution varies very strongly, depending on the chosen fiber orientation model parameters. Figure 20 shows the results of direct parameter optimization, as explained in Section 2.5.1, by coupling MATLAB® and Moldex3D® with different fiber orientation models, (a) basic iARD and (b) iARD with shear rate dependency.

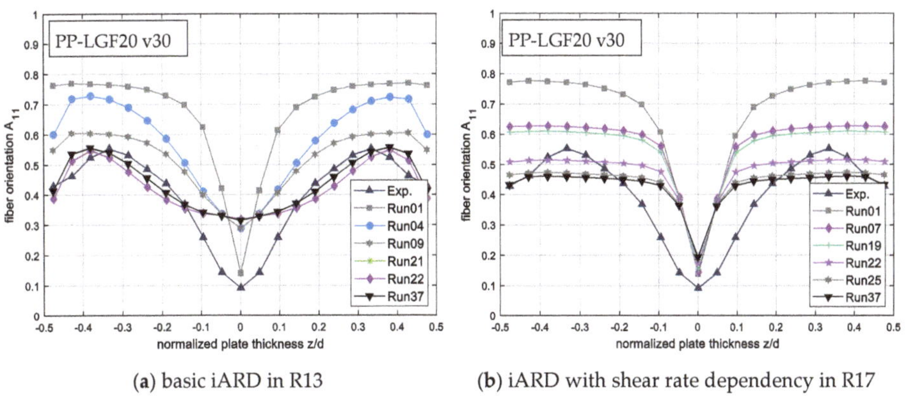

(**a**) basic iARD in R13   (**b**) iARD with shear rate dependency in R17

Figure 20. Results of the direct parameter optimization by coupling MATLAB® and Moldex3D®.

To prove the functionality of the optimization routine, deviant values were chosen as initial parameters for the corresponding fiber orientation model (Run 1), which did not match the experiments. The experimentally determined fiber orientation is depicted as blue line with triangular markers. For a clearer presentation of the optimization progress, only some selected optimization steps are shown. However, the run numbers still correspond to the respective iteration step. The results obviously show a better agreement to the experiment with increasing iterations. In the skin layer, the optimization result of the basic iARD model shows very good accordance with the experimental data, whereas in the core layer, the simulation result does not represent the low degree of orientation determined by the experiment. However, this cannot be further improved by optimizing the model parameters, as there is no parameter set that can represent the entire orientation distribution from the experiment. This means that the best possible parameter set was found after 37 optimization steps for the basic iARD fiber orientation model.

For the optimization of the iARD model with shear rate dependency in version R17, the adjustment of the parameters has only a very one-dimensional influence on the change of the fiber orientation. Because the change from core to shear layer is very abrupt, no real transition zone is formed. Therefore, the degree of orientation in the skin layers can only be adjusted by varying the parameters C_i and C_m. All the other areas remain largely constant despite the parameter variation.

3.3. Influence of Parameters and Viscosity

In addition to the model parameters, the defined shear rate-dependent viscosity also has a significant influence on the resulting velocity profile as already explained in Section 1.1.2 as well as the results of the fiber orientation model. For this reason, the parameters of the defined viscosity model (Herschel–Bulkley) were varied and the influence on the fiber orientation distribution was investigated. Figure 21 shows the analyzed viscosity curves, which are moved slightly towards higher and lower viscosities (deviating from experimental data).

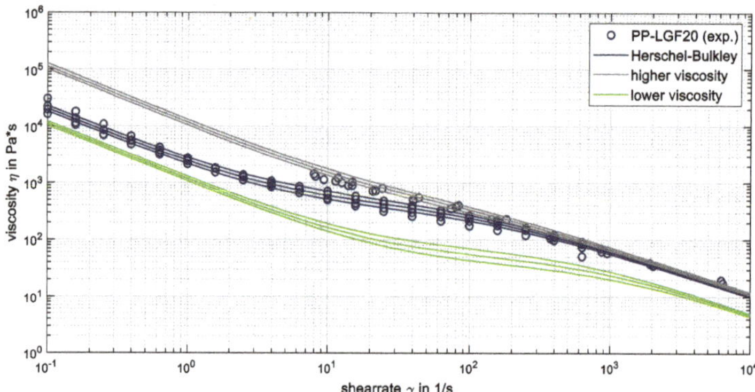

Figure 21. Shear rate and temperature dependent viscosity and analyzed viscosity curves.

In Figure 22 the influence of the viscosity on the resulting fiber orientation distribution is investigated for different fiber orientation model versions and parameters (comparable to Figure 15). On the left side of Figure 22a,c,e, the fiber orientation was calculated with the experimentally determined viscosity for PP-LGF20 (blue curve in Figure 21), and on the right side of the Figure 22b,d,f, a higher viscosity, especially in the area of low shear rates (grey curve in Figure 21), is used. In the first row (a, b) the basic iARD model in Moldex3D® version R13, and in the second row (c, d) the iARD model with shear rate dependency in version R16, and the last row (e, f) the iARD model with shear rate dependency and fiber coupling in version R17 is used. For the comparison of the three fiber orientation model versions, the same variation of fiber orientation model parameters was used. The parameter α was varied in the range of 0–0.99 for high and also low C_i and C_m values.

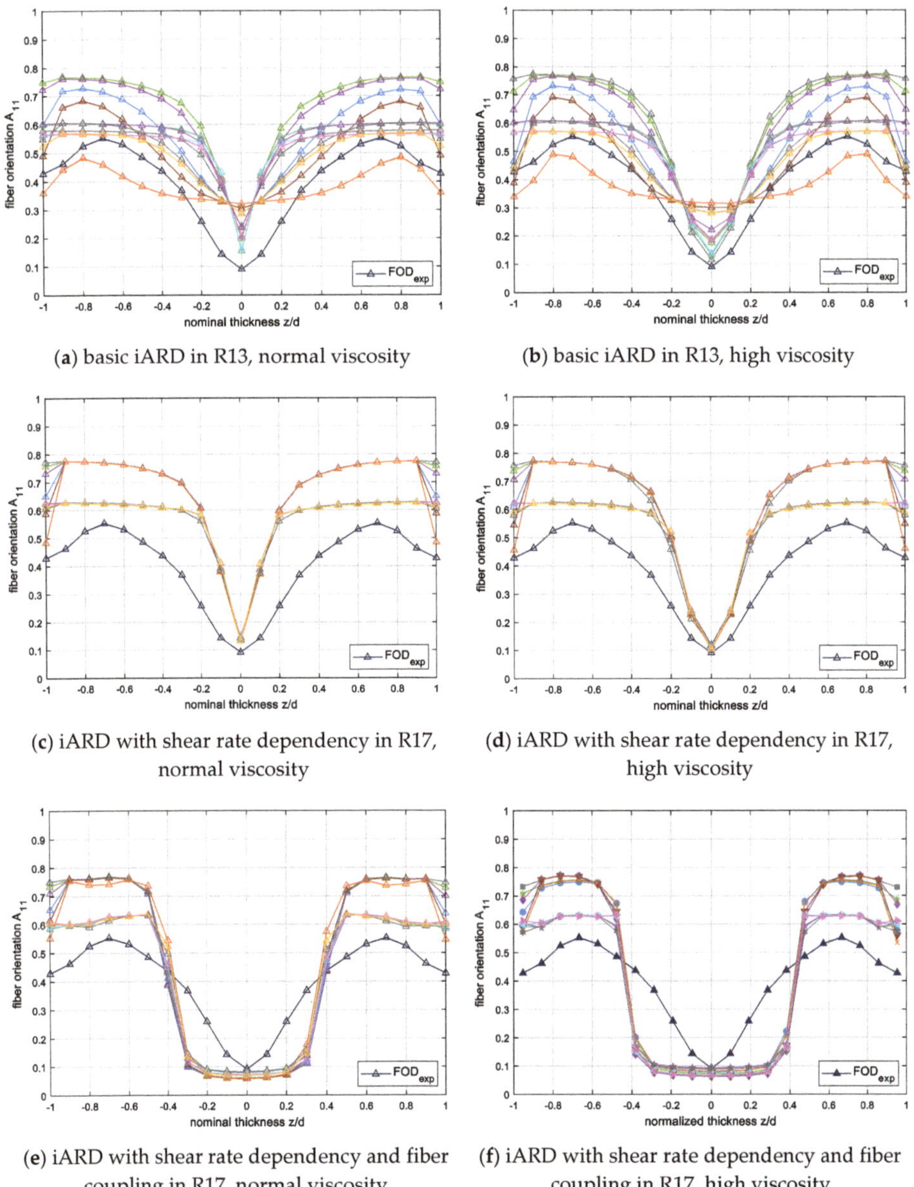

(**a**) basic iARD in R13, normal viscosity

(**b**) basic iARD in R13, high viscosity

(**c**) iARD with shear rate dependency in R17, normal viscosity

(**d**) iARD with shear rate dependency in R17, high viscosity

(**e**) iARD with shear rate dependency and fiber coupling in R17, normal viscosity

(**f**) iARD with shear rate dependency and fiber coupling in R17, high viscosity

Figure 22. Influence of the viscosity for different fiber orientation model versions for PP-LGF20v30.

First of all, all results show a change of the fiber orientation distribution within the core layer due to the change in viscosity. However, the fiber orientation distribution in the shear layers remains almost unchanged for all settings (viscosity models, fiber orientation models and parameters) due to the same viscosity in the range of high shear rates (blue and grey curve in Figure 21). The viscosity change mainly only affects the thickness of the core layer. The core layer is widened with a higher viscosity curve, which corresponds more closely to the experimental determined fiber orientation distribution. However, even if the adjustment of the viscosity curve improves the thickness of the core layer, a deviation between the predicted and measured fiber orientation still remains for all analyzed fiber orientation models, as well as for the viscosity models and their respective parameters.

3.4. Validation and Results of the New Calibration Approach

In order to ensure the correct implementation of the new optimization routine, a test case was chosen from literature [87]. Thus, the different parts of the implemented orientation model were tested for performance and verified by reference values [87]. The investigated test case is a simple shear flow within a single element. Two different versions of the fiber orientation models explained in Section 1.1.8, the basic iARD model (Equation (9)) and the iARD-RPR model (Equation (12)), were tested. The parameters of the fiber orientation model are chosen to make the iARD model ($C_m = 0, C_i = 0.01$ and $\alpha = 0.9$) correspond to the standard model of Folgar-Tucker (FT) [86]. In Figure 23 the reference values are shown as points and the calculated fiber orientation as lines.

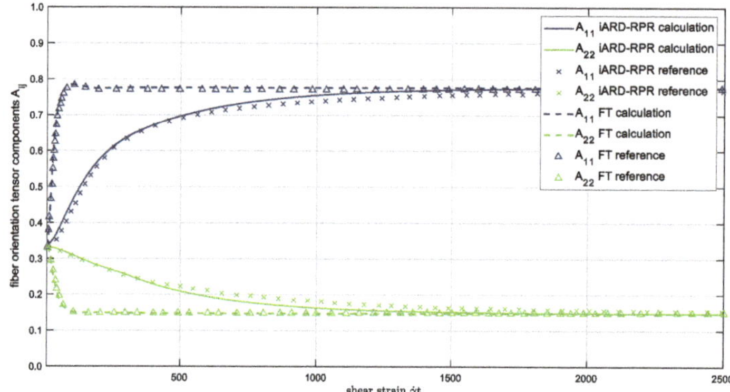

Figure 23. Validation of the implemented fiber orientation calculation within the calibration tool.

The results of the fiber orientation calculation with the new calibration tool in Figure 23 show quite good agreement with the results from literature. Thus, a correct implementation of the fiber orientation model is assumed. A further validation can be obtained by comparing the calculation of the fiber orientation by Moldex3D® with the fiber orientation calculated by the new calibration tool in MATLAB® on the basis of the same flow field exported from Moldex3D®. The results of the calculated fiber orientation distribution are almost identical, as shown in Figure 24a.

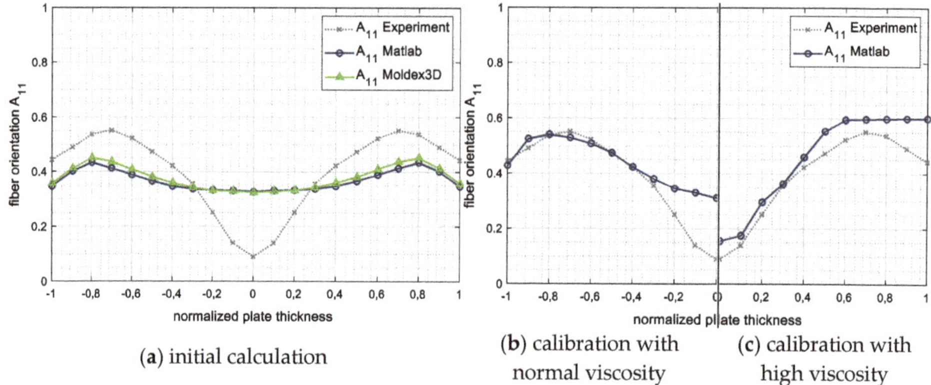

(**a**) initial calculation

(**b**) calibration with normal viscosity

(**c**) calibration with high viscosity

Figure 24. Calculation and calibration results of the fiber orientation calibration tool.

Based on this initial calculation, a parameter optimization was performed with the calibration tool in MATLAB®. The calibration results of the parameter optimization are shown in Figure 24b for the experimentally measured viscosity, and in (c) for the increased viscosity according to the gray curve in Figure 21. Even if the fiber orientation model does not represent the entire orientation distribution, the adjusted viscosity curve (high viscosity) leads to a significant improvement in the resulting fiber orientation distribution in the core layer.

Compared with direct optimization by Moldex3D® (Section 2.5.1), the new calibration approach (Section 2.5.2) fully implemented in MATLAB® is significantly faster, as the entire calculation is based on only one initial calculation of the flow field without any necessary recalculation. The following Figure 25 clearly demonstrates that a significant amount (approximately factor 15) of the required calculation time can be saved by the new approach. Because the calculation of the fiber orientation is performed as an independent post-processing step, the calculation of the flow field is independent of the fiber orientation calculation and remains the same for each optimization step.

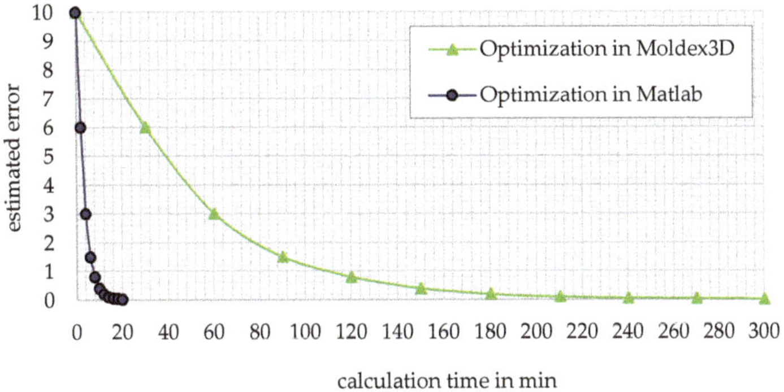

Figure 25. Comparison of calculation time of direct optimization method and the new optimization approach implemented in the fiber orientation optimization tool.

4. Conclusions and Outlook

A novel method for the objective comparison of experimentally determined and by process simulation predicted fiber orientation was developed. As a first step, a new standardized method for the objective comparison of experimentally determined and simulation predicted fiber orientation based on the error deviation was defined. Furthermore, a corresponding validation of the method showed that the fiber orientation model has successfully been implemented in the novel calibration tool, and that the developed problem-specific optimization algorithm can adjust the model parameters simultaneously in order to minimize the objective function.

After only a few iteration loops and within a few minutes, the automated method shows an optimized fiber orientation distribution for long fiber reinforced materials based on the calculated flow field. Thus, the model parameters for the fiber orientation model are determined in the most accurate way and can be used for further predictions. This also allows for calculating the structural mechanics of fiber reinforced parts with high prediction accuracy. These results clearly confirm the scientific hypotheses stated at the beginning of this study.

The investigation also showed that there is further optimization potential which should be investigated and considered in the future. The parameter optimization is currently only implemented for the fiber orientation model, but can easily be extended to the prediction of fiber length distribution and fiber concentration. Further development of this method also allows a transfer of the calculated properties to structural mechanics with high quality predictions.

J. Compos. Sci. **2020**, *4*, 163

Author Contributions: Conceptualization, F.W. and P.R.; methodology, F.W. and P.R.; software, F.W. and P.R.; validation, F.W. and P.R.; formal analysis, F.W. and P.R.; investigation, F.W. and P.R.; resources, F.W. and P.R.; data curation, F.W. and P.R.; writing—original draft preparation, F.W. and P.R.; writing—review and editing, F.W., P.R. and C.B.; visualization, F.W. and P.R.; supervision, C.B.; project administration, F.W.; funding acquisition, C.B. All authors have read and agreed to the published version of the manuscript.

Funding: This research was funded by the German Federal Ministry of Economics and Energy (BMWi) within the Central Innovation Program for SMEs (ZIM). Grant number [ZF4041122BL7].

Conflicts of Interest: The authors declare no conflict of interest.

References

1. Tseng, H.-C.; Chang, R.-Y.; Hsu, C.-H. Numerical prediction of fiber orientation and mechanical performance for short/long glass and carbon fiber-reinforced composites. *Compos. Sci. Technol.* **2017**, *144*, 51–56. [CrossRef]

2. Mu, Y.; Chen, A.; Zhao, G.; Cui, Y.; Feng, J.; Ren, F. Prediction for the mechanical property of short fiber-reinforced polymer composites through process modeling method. *J. Thermoplast. Compos. Mater.* **2019**, *32*, 1525–1546. [CrossRef]

3. Fu, S.-Y.; Lauke, B.; Mai, Y.W. Science and engineering of short fibre reinforced polymers composites. In *Woodhead Publishing in Materials*; Woodhead Publishing: Cambridge, UK; Boca Raton, FL, USA, 2009.

4. Folkes, M.J. Short fibre reinforced thermoplastics. In *Polymer Engineering Research Studies Series*; Research Studies Press: Chichester, UK, 1985; Volume 1.

5. Goris, S.; Osswald, T.A. Process-induced fiber matrix separation in long fiber-reinforced thermoplastics. *Compos. Part A Appl. Sci. Manuf.* **2018**, *105*, 321–333. [CrossRef]

6. Tseng, H.-C.; Chang, R.-Y.; Hsu, C.-H. Improved fiber orientation predictions for injection molded fiber composites. *Compos. Part A Appl. Sci. Manuf.* **2017**, *99*, 65–75. [CrossRef]

7. Willems, F.; Bonten, C. The use of micromechanical models to predict fiber reinforced plastics. In Proceedings of the Europe/Africa Conference Dresden 2017—Polymer Processing Society PPS, Dresden, Germany, 27–29 June 2017. [CrossRef]

8. Nguyen, B.N.; Bapanapalli, S.K.; Holbery, J.D.; Smith, M.T.; Kunc, V.; Frame, B.J.; Phelps, J.H.; Tucker, C.L. Fiber Length and Orientation in Long-Fiber Injection-Molded Thermoplastics—Part I: Modeling of Microstructure and Elastic Properties. *J. Compos. Mater.* **2008**, *42*, 1003–1029. [CrossRef]

9. Tseng, H.-C.; Goto, M.; Chang, R.-Y.; Hsu, C.-H. Accurate predictions of fiber orientation and mechanical properties in long-fiber-reinforced composite with experimental validation. *Polym. Compos.* **2018**, *39*, 3434–3445. [CrossRef]

10. Willems, F.; Bonten, C. Prediction of Fiber-Reinforced Plastics Considering Local Fiber Length and Orientation. W05-495. In *Society of Plastics Engineers, (SPE), Proceedings of the SPE ANTEC 2018, Orlando, FL, USA, 7–10 May 2018*; Curran Associates, Inc.: Red Hook, NY, USA, 2018; pp. 2649–2656.

11. Garesci, F.; Fliegener, S. Young's modulus prediction of long fiber reinforced thermoplastics. *Compos. Sci. Technol.* **2013**, *85*, 142–147. [CrossRef]

12. van Haag, J.; Hopmann, C. Simulate "Longer". A New Calculation Model Allows the Prediction of Strength of Long Fiber-Reinforced Thermoplastics by means of Integrative Simulation. *Kunstst. Int.* **2016**, *2*, 42–45.

13. Kardos, J.L. Critical issues in achieving desirable mechanical properties for short fiber composites. *Pure Appl. Chem.* **1985**, *57*, 1651–1657. [CrossRef]

14. Akay, M.; Barkley, D. Fibre orientation and mechanical behaviour in reinforced thermoplastic injection mouldings. *J. Mater. Sci.* **1991**, *26*, 2731–2742. [CrossRef]

15. Mortazavian, S.; Fatemi, A. Effects of fiber orientation and anisotropy on tensile strength and elastic modulus of short fiber reinforced polymer composites. *Compos. Part B Eng.* **2015**, *72*, 116–129. [CrossRef]

16. Osswald, T.A. Understanding polymer processing. In *Processes and Governing Equations*, 2nd ed.; Hanser Publishers: Cincinnati, OH, USA; Munich, Germany, 2017.

17. Lees, J.K. A study of the tensile modulus of short fiber reinforced plastics. *Polym. Eng. Sci.* **1968**, *8*, 186–194. [CrossRef]

18. Lees, J.K. A study of the tensile strength of short fiber reinforced plastics. *Polym. Eng. Sci.* **1968**, *8*, 195–201. [CrossRef]

19. De Monte, M.; Moosbrugger, E.; Quaresimin, M. Influence of temperature and thickness on the off-axis behaviour of short glass fibre reinforced polyamide 6.6—Quasi-static loading. *Compos. Part A Appl. Sci. Manuf.* **2010**, *41*, 859–871. [CrossRef]

20. Fu, S.-Y.; Lauke, B. Effects of fiber length and fiber orientation distributions on the tensile strength of short-fiber-reinforced polymers. *Compos. Sci. Technol.* **1996**, *56*, 1179–1190. [CrossRef]

21. Fu, S.-Y.; Lauke, B. The elastic modulus of misaligned short-fiber-reinforced polymers. *Compos. Sci. Technol.* **1998**, *58*, 389–400. [CrossRef]

22. Lauke, B.; Fu, S.-Y. Strength anisotropy of misaligned short-fibre-reinforced polymers. *Compos. Sci. Technol.* **1999**, *59*, 699–708. [CrossRef]

23. Thomason, J.L.; Vlug, M.A. Influence of fibre length and concentration on the properties of glass fibre-reinforced polypropylene: 1. Tensile and flexural modulus. *Compos. Part A Appl. Sci. Manuf.* **1996**, *27*, 477–484. [CrossRef]

24. Thomason, J.L. Structure–property relationships in glass reinforced polyamide, part 2: The effects of average fiber diameter and diameter distribution. *Polym. Compos.* **2007**, *28*, 331–343. [CrossRef]

25. Thomason, J.L.; Vlug, M.A. Influence of fibre length and concentration on the properties of glass fibre-reinforced polypropylene: 4. Impact properties. *Compos. Part A Appl. Sci. Manuf.* **1997**, *28*, 277–288. [CrossRef]

26. Thomason, J.L. The influence of fibre length and concentration on the properties of glass fibre reinforced polypropylene: 5. Injection moulded long and short fibre PP. *Compos. Part A Appl. Sci. Manuf.* **2002**, *33*, 1641–1652. [CrossRef]

27. Toll, S.; Aronsson, C.-G. Notched strength of long- and short-fibre reinforced polyamide. *Compos. Sci. Technol.* **1992**, *45*, 43–54. [CrossRef]

28. Vu-Khanh, T.; Denault, J. Toughness of Reinforced Ductile Thermoplastics. *J. Compos. Mater.* **1992**, *26*, 2262–2277. [CrossRef]

29. Thomason, J.L. The influence of fibre length and concentration on the properties of glass fibre reinforced polypropylene: 7. Interface strength and fibre strain in injection moulded long fibre PP at high fibre content. *Compos. Part A Appl. Sci. Manuf.* **2007**, *38*, 210–216. [CrossRef]

30. Kumar, K.S.; Bhatnagar, N.; Ghosh, A.K. Development of Long Glass Fiber Reinforced Polypropylene Composites: Mechanical and Morphological Characteristics. *J. Reinf. Plast. Compos.* **2007**, *26*, 239–249. [CrossRef]

31. Gupta, M.; Wang, K.K. Fiber orientation and mechanical properties of short-fiber-reinforced injection-molded composites: Simulated and experimental results. *Polym. Compos.* **1993**, *14*, 367–382. [CrossRef]

32. Bonten, C. Plastics Technology. In *Introduction and Fundamentals*; Carl Hanser Verlag GmbH & Co. KG: München, Germany, 2019.

33. Barbosa, S.E.; Kenny, J.M. Analysis of the Relationship between Processing Conditions-Fiber Orientation-Final Properties in Short Fiber Reinforced Polypropylene. *J. Reinf. Plast. Compos.* **1999**, *18*, 413–420. [CrossRef]

34. Singh, P.; Kamal, M.R. The effect of processing variables on microstructure of injection molded short fiber reinforced polypropylene composites. *Polym. Compos.* **1989**, *10*, 344–351. [CrossRef]

35. Sharma, B.N.; Naragani, D.; Nguyen, B.N.; Tucker, C.L.; Sangid, M.D. Uncertainty quantification of fiber orientation distribution measurements for long-fiber-reinforced thermoplastic composites. *J. Compos. Mater.* **2018**, *52*, 1781–1797. [CrossRef]

36. Willems, F.; Bonten, C. Influence of processing on the fiber length degradation in fiber reinforced plastic parts. In Proceedings of the Regional Conference Graz 2015—Polymer Processing Society PPS: Conference Papers, Graz, Austria, 21–25 September 2015; p. 20003. [CrossRef]

37. Gandhi, U.N.; Goris, S.; Osswald, T.A.; Song, Y.-Y. Discontinuous fiber-reinforced composites. In *Fundamentals and Applications*; Hanser Publishers: Munich, Germany; Cincinnati, OH, USA, 2020.

38. Hegler, R.P. Faserorientierung beim Verarbeiten kurzfaserverstärkter Thermoplaste. *Kunststoffe* **1984**, *74*, 271–277.

39. Bailey, R.; Rzepka, B. Fibre Orientation Mechanisms for Injection Molding of Long Fibre Composites. *Int. Polym. Process.* **1991**, *6*, 35–41. [CrossRef]

40. Bay, R.S.; Tucker, C.L. Fiber orientation in simple injection moldings. Part II: Experimental results. *Polym. Compos.* **1992**, *13*, 332–341. [CrossRef]

41. Folkes, M.J.; Russell, D.A.M. Orientation effects during the flow of short-fibre reinforced thermoplastics. *Polymer* **1980**, *21*, 1252–1258. [CrossRef]

42. Karger-Kocsis, J. Microstructure and Fracture Mechanical Performance of Short-Fibre Reinforced Thermoplastics. In *Application of Fracture Mechanics to Composite Materials*; Composite Materials Series; Elsevier: Amsterdam, The Netherlands, 1989; Volume 6, pp. 189–247.

43. Toll, S.; Andersson, P.-O. Microstructure of long- and short-fiber reinforced injection molded polyamide. *Polym. Compos.* **1993**, *14*, 116–125. [CrossRef]

44. Spahr, D.E.; Friedrich, K.; Schultz, J.M.; Bailey, R.S. Microstructure and fracture behaviour of short and long fibre-reinforced polypropylene composites. *J. Mater. Sci.* **1990**, *25*, 4427–4439. [CrossRef]

45. Bright, P.F.; Crowson, R.J.; Folkes, M.J. A study of the effect of injection speed on fibre orientation in simple mouldings of short glass fibre-filled polypropylene. *J. Mater. Sci.* **1978**, *13*, 2497–2506. [CrossRef]

46. Bailey, R.; Kraft, H. A Study of Fibre Attrition in the Processing of Long Fibre Reinforced Thermoplastics. *Int. Polym. Process.* **1987**, *2*, 94–101. [CrossRef]

47. Karger-Kocsis, J.; Friedrich, K. Fracture behavior of injection-molded short and long glass fiber—Polyamide 6.6 composites. *Compos. Sci. Technol.* **1988**, *32*, 293–325. [CrossRef]

48. O'Regan, D.; Akay, M. The distribution of fibre lengths in injection moulded polyamide composite components. *J. Mater. Process. Technol.* **1996**, *56*, 282–291. [CrossRef]

49. Mondy, L.A.; Brenner, H.; Altobelli, S.A.; Abbott, J.R.; Graham, A.L. Shear-induced particle migration in suspensions of rods. *J. Rheol.* **1994**, *38*, 444–452. [CrossRef]

50. Lafranche, E.; Krawczak, P.; Ciolczyk, J.P.; Maugey, J. Injection moulding of long glass fibre reinforced polyamide 6-6: Guidelines to improve flexural properties. *Express Polym. Lett.* **2007**, *1*, 456–466. [CrossRef]

51. Gruber, G.; Wartzack, S. Evaluierung der orientierungsbezogenen Leichtbaugüte. *Lightweight Des.* **2013**, *6*, 18–23. [CrossRef]

52. Kenig, S. Fiber orientation development in molding of polymer composites. *Polym. Compos.* **1986**, *7*, 50–55. [CrossRef]

53. Katti, S.S.; Schultz, M. The microstructure of injection-molded semicrystalline polymers: A review. *Polym. Eng. Sci.* **1982**, *22*, 1001–1017. [CrossRef]

54. Friedrich, K. Microstructural efficiency and fracture toughness of short fiber/thermoplastic matrix composites. *Compos. Sci. Technol.* **1985**, *22*, 43–74. [CrossRef]

55. Jiang, J.; Wang, S.; Hou, J.; Zhang, K.; Wang, X.; Li, Q.; Liu, G. Effect of injection velocity on the structure and mechanical properties of micro injection molded polycarbonate/poly(ethylene terephthalate) blends. *Mater. Des.* **2018**, *141*, 132–141. [CrossRef]

56. Osswald, T.A.; Rudolph, N. Polymer rheology. In *Fundamentals and Applications*; Hanser Publications: Cincinnati, OH, USA, 2014.

57. Guell, D.C.; Papathanasiou, T.D. *Flow-Induced Alignment in Composite Materials*; Woodhead: Cambridge, UK, 1997.

58. Stommel, M.; Stojek, M.; Korte, W. *FEM Zur Berechnung von Kunststoff- und Elastomerbauteilen*, 2nd ed.; Hanser: München, Germany, 2018.

59. Sigmasoft: 3D Moulding Simulation—The 'Whole Process' Approach. *Met. Powder Rep.* **2013**, *68*, 30–32. Available online: https://www.sciencedirect.com/science/article/abs/pii/S0026065713700948?via%3Dihub (accessed on 29 October 2020). [CrossRef]

60. Advani, S.G.; Tucker, C.L. The Use of Tensors to Describe and Predict Fiber Orientation in Short Fiber Composites. *J. Rheol.* **1987**, *31*, 751–784. [CrossRef]

61. Predak, S.; Solodov, I.Y.; Busse, G.; Bister, V.H.; Vöhringer, M.C.; Haberstroh, E.; Ehbing, H. Faserorientierungsmessung an kurzfaserverstärkten PUR-RIM-Bauteilen: Kombination zerstörungsfreier Prüfmethoden zur Optimierung von Simulation und Herstellungsprozess, (Fiber Orientation Measurement on Short Fiber Reinforced PUR-RIM Components: Combination of Nondestructive Testing Methods for Optimization of Simulation and Production Process). *TM Tech. Mess.* **2006**, *73*, 3. [CrossRef]

62. Predak, S. *Mikrowellen-Orientierungsmessungen zur Zerstörungsfreien Charakterisierung Kurzfaserverstärkter Kunststoffe*; Publication Server of the University of Stuttgart: Stuttgart, Germany, 2007.

63. Toll, S.; Andersson, P.-O. Microstructural characterization of injection moulded composites using image analysis. *Composites* **1991**, *22*, 298–306. [CrossRef]

64. Hine, P.J.; Davidson, N.; Duckett, R.A.; Ward, I.M. Measuring the fibre orientation and modelling the elastic properties of injection-moulded long-glass-fibre-reinforced nylon. *Compos. Sci. Technol.* **1995**, *53*, 125–131. [CrossRef]

65. McGee, S.H.; McCullough, R.L. An Optical Technique for Measuring Fiber Orientation in Short Fiber Composites. In *The Role of the Polymeric Matrix in the Processing and Structural Properties Composite Materials*; Seferis, J.C., Nicolais, L., Eds.; Springer: Boston, MA, USA, 1983; Volume 64, pp. 425–436.

66. Mlekusch, B. Fibre orientation in short-fibre-reinforced thermoplastics II. Quantitative measurements by image analysis. *Compos. Sci. Technol.* **1999**, *59*, 547–560. [CrossRef]

67. Fakirov, S.; Fakirova, C. Direct determination of the orientation of short glass fibers in an injection-molded poly(ethylene terephthalate) system. *Polym. Compos.* **1985**, *6*, 41–46. [CrossRef]

68. Eberhardt, C.; Clarke, A. Fibre-orientation measurements in short-glass-fibre composites. Part I: Automated, high-angular-resolution measurement by confocal microscopy. *Compos. Sci. Technol.* **2001**, *61*, 1389–1400. [CrossRef]

69. Vélez-García, G.M.; Wapperom, P.; Baird, D.G.; Aning, A.O.; Kunc, V. Unambiguous orientation in short fiber composites over small sampling area in a center-gated disk. *Compos. Part A Appl. Sci. Manuf.* **2012**, *43*, 104–113. [CrossRef]

70. Willems, F.; Beerlink, A.; Metayer, J.-F.; Kreutzbruck, M.; Bonten, C. Bestimmung der Faserorientierung langglasfaserverstärkter Thermoplaste mittels Bildoptischer Analyse und Computertomografie; Deutsche Gesellschaft für Zerstörungsfreie Prüfung e.V, Ed.; DGZfP Jahrestagung: Leipzig, Germany, 7–9 May 2018.

71. Kim, E.G.; Park, J.K.; Jo, S.H. A study on fiber orientation during the injection molding of fiber-reinforced polymeric composites. *J. Mater. Process. Technol.* **2001**, *111*, 225–232. [CrossRef]

72. Darlington, M.W.; McGinley, P.L.; Smith, G.R. Structure and anisotropy of stiffness in glass fibre-reinforced thermoplastics. *J. Mater. Sci.* **1976**, *11*, 877–886. [CrossRef]

73. Grote, F. Schaumstoffe mit CT charakterisieren. *Kunstst. Plast Eur.* **1999**, *89*, 110–111.

74. Bernasconi, A.; Cosmi, F.; Hine, P.J. Analysis of fibre orientation distribution in short fibre reinforced polymers: A comparison between optical and tomographic methods. *Compos. Sci. Technol.* **2012**, *72*, 2002–2008. [CrossRef]

75. Shen, H.; Nutt, S.; Hull, D. Direct observation and measurement of fiber architecture in short fiber-polymer composite foam through micro-CT imaging. *Compos. Sci. Technol.* **2004**, *64*, 2113–2120. [CrossRef]

76. Maisl, M.; Scherer, T.; Reiter, H.; Hirsekorn, S. Nondestructive Investigation of Fibre Reinforced Composites by X-Ray Computed Tomography. In *Nondestructive Characterization of Materials*; Höller, P., Hauk, V., Dobmann, G., Ruud, C.O., Green, R.E., Eds.; Springer: Berlin/Heidelberg, Germany, 1989; pp. 147–154.

77. Robb, K.; Wirjadi, O.; Schladitz, K. Fiber Orientation Estimation from 3D Image Data: Practical Algorithms, Visualization, and Interpretation. In Proceedings of the 7th International Conference on Hybrid Intelligent Systems, (HIS 2007), Kaiserslautern, Germany, 17–19 September 2007; pp. 320–325. [CrossRef]

78. Sun, X.; Lasecki, J.; Zeng, D.; Gan, Y.; Su, X.; Tao, J. Measurement and quantitative analysis of fiber orientation distribution in long fiber reinforced part by injection molding. *Polym. Test.* **2015**, *42*, 168–174. [CrossRef]

79. Fischer, G.; Eyerer, P. Measuring spatial orientation of short fiber reinforced thermoplastics by image analysis. *Polym. Compos.* **1988**, *9*, 297–304. [CrossRef]

80. Kennedy, P.; Zheng, R. *Flow Analysis of Injection Molds*, 2nd ed.; Hanser Publishers: Munich, Germany; Cincinnati, OH, USA, 2013.

81. Advani, S.G. (Ed.) Flow and rheology in polymer composites manufacturing. In *Composite Materials Series, Vol. 10*; Elsevier: Amsterdam, The Netherlands, 1994.

82. Papanastasiou, T.C. Flows of Materials with Yield. *J. Rheol.* **1987**, *31*, 385–404. [CrossRef]

83. Han, S.; Wang, K.K.; Hieber, C.A.; Cohen, C. Characterization of the rheological properties of a fast-curing epoxy-molding compound. *J. Rheol.* **1997**, *41*, 177–195. [CrossRef]

84. Cross, M.M. Relation between viscoelasticity and shear-thinning behaviour in liquids. *Rheol. Acta* **1979**, *18*, 609–614. [CrossRef]

85. Jeffery, G.B. The motion of ellipsoidal particles immersed in a viscous fluid. *Proc. R. Soc. Lond. A* **1922**, *102*, 161–179. [CrossRef]

86. Phelps, J.H.; Tucker, C.L. An anisotropic rotary diffusion model for fiber orientation in short- and long-fiber thermoplastics. *J. Non-Newton. Fluid Mech.* **2009**, *156*, 165–176. [CrossRef]

87. Tseng, H.-C.; Chang, R.-Y.; Hsu, C.-H. Method and computer readable media for determining orientation of fibers in a fluid. U.S. Patent 8,571,828, 29 October 2013.

88. Tseng, H.-C.; Chang, R.-Y.; Hsu, C.-H. Phenomenological improvements to predictive models of fiber orientation in concentrated suspensions. *J. Rheol.* **2013**, *57*, 1597–1631. [CrossRef]

89. Tseng, H.-C.; Chang, R.-Y.; Hsu, C.-H. An objective tensor to predict anisotropic fiber orientation in concentrated suspensions. *J. Rheol.* **2016**, *60*, 215–224. [CrossRef]

90. Du Chung, H.; Kwon, T.H. Invariant-based optimal fitting closure approximation for the numerical prediction of flow-induced fiber orientation. *J. Rheol.* **2002**, *46*, 169–194. [CrossRef]

91. Tseng, H.-C.; Chang, R.-Y.; Hsu, C.-H. The use of shear-rate-dependent parameters to improve fiber orientation predictions for injection molded fiber composites. *Compos. Part A Appl. Sci. Manuf.* **2018**, *104*, 81–88. [CrossRef]

92. Tseng, H.-C.; Chang, R.-Y.; Hsu, C.-H. Comparison of recent fiber orientation models in injection molding simulation of fiber-reinforced composites. *J. Thermoplast. Compos. Mater.* **2020**, *33*, 35–52. [CrossRef]

93. Willems, F.; Bonten, C. Structural prediction of injection molded long fiber reinforced plastics based on process induced fiber microstructure. In Proceedings of the International Conference on Composite Materials, 22nd, Melbourne, Australia, 11–16 August 2019; ICCM22. ICCM 22 2019. Engineers Australia, Ed.; Engineers Australia: Barton, Australia, 2019; pp. 4822–4831.

94. Morak, M.; Tscharnuter, D.; Lucyshyn, T.; Hahn, W.; Göttlinger, M.; Kummer, M.; Steinberger, R.; Gross, T. Optimization of fiber prediction model coefficients in injection molding simulation based on micro computed tomography. *Polym. Eng. Sci.* **2019**, *59*, E152–E160. [CrossRef]

95. Meyer, K.J.; Hofmann, J.T.; Baird, D.G. Prediction of short glass fiber orientation in the filling of an end-gated plaque. *Compos. Part A Appl. Sci. Manuf.* **2014**, *62*, 77–86. [CrossRef]

96. Nguyen, B.N.; Jin, X.; Wang, J.; Phelps, J.; Tucker III, C.L.; Kunc, V.; Bapanapalli, S.K.; Smith, M.T. *Implementation of New Process Models for Tailored Polymer Composite Structures into Processing Software Packages*; Pacific Northwest National Lab. (PNNL): Richland, WA, USA, 2010.

97. Foss, P.H.; Tseng, H.-C.; Snawerdt, J.; Chang, Y.-J.; Yang, W.-H.; Hsu, C.-H. Prediction of fiber orientation distribution in injection molded parts using Moldex3D simulation. *Polym. Compos.* **2014**, *35*, 671–680. [CrossRef]

98. Phelps, J.H. Processing-Microstructure Models for Short- and Long-Fiber Thermoplastic Composites. Ph.D. Thesis, University of Illinois at Urbana-Champaign, Champaign, IL, USA. Available online: https://search.proquest.com/docview/304896821?accountid=14133 (accessed on 16 September 2017).

99. Eberle, A.P.R.; Vélez-García, G.M.; Baird, D.G.; Wapperom, P. Fiber orientation kinetics of a concentrated short glass fiber suspension in startup of simple shear flow. *J. Non-Newton. Fluid Mech.* **2010**, *165*, 110–119. [CrossRef]

100. Ortman, K.; Baird, D.; Wapperom, P.; Whittington, A. Using startup of steady shear flow in a sliding plate rheometer to determine material parameters for the purpose of predicting long fiber orientation. *J. Rheol.* **2012**, *56*, 955–981. [CrossRef]

101. Bay, R.S. Fiber Orientation in Injection-Molded Composites: A Comparison of Theory and Experiment. Ph.D. Thesis, University of Illinois at Urbana-Champaign, Champaign, IL, USA, 1991.

102. Phan-Thien, N.; Fan, X.-J.; Tanner, R.I.; Zheng, R. Folgar–Tucker constant for a fibre suspension in a Newtonian fluid. *J. Non Newton. Fluid Mech.* **2002**, *103*, 251–260. [CrossRef]

103. Favaloro, A.J.; Tucker, C.L. Analysis of anisotropic rotary diffusion models for fiber orientation. *Compos. Part A Appl. Sci. Manuf.* **2019**, *126*, 105605. [CrossRef]

104. Kugler, S.K.; Kech, A.; Cruz, C.; Osswald, T. Fiber Orientation Predictions—A Review of Existing Models. *J. Compos. Sci.* **2020**, *4*, 69. [CrossRef]

105. Willems, F.; Bonten, C. Prediction of the Mechanical Properties of Long Fiber Reinforced Thermoplastics. In Proceedings of the PPS2019—PPS Europe-Africa Regional Conference, Pretoria, South Africa, 18–21 November 2019.

106. Wonisch, A.; Wüst, A. More Precise Part Design. Accurate Simulation of Fiber Orientation of Glass Fiber-Reinforced Plastics. *Kunstst. Int.* **2014**, *9*, 80–83.

107. Li, T.; Luyé, J.-F. Optimization of Fiber Orientation Model Parameters in the Presence of Flow-Fiber Coupling. *J. Compos. Sci.* **2018**, *2*, 73. [CrossRef]

108. Kugler, S.K.; Lambert, G.M.; Cruz, C.; Kech, A.; Osswald, T.A.; Baird, D.G. Efficient parameter identification for macroscopic fiber orientation models with experimental data and a mechanistic fiber simulation. In Proceedings of the 35th International Conference of the Polymer Processing Society, (PPS-35), Cesme-Izmir, Turkey, 26–30 May 2019; AIP Publishing: Melville, NY, USA, 2020; p. 20050. [CrossRef]

109. Reitinger, P.; Willems, F.; Bonten, C. Calibration of Models to Predict the Fiber Microstructure of LFRTs. In Proceedings of the PPS2019—PPS Europe-Africa 2019 Regional Conference, Pretoria, South Africa, 18–21 November 2019.

110. Nocedal, J.; Wright, S. *Numerical Optimization. Springer Series in Operations Research and Financial Engineering*; Springer: New York, NY, USA, 2006.

111. Poljak, B.T.; Poljak, B.T. Introduction to optimization. Translations Series in mathematics and engineering. In *Optimization Software*; Publications Division: New York, NY, USA, 1987.

112. Feilmeier, M. Parameteroptimierung. In *Hybridrechnen. International Series of Numerical Mathematics/ Internationale Schriftenreihe zur Numerischen Mathematik/Série Internationale D'Analyse Numérique*; Birkhäuser: Basel, Switzerland, 1974; Volume 2, pp. 243–259.

113. Bertsekas, D.P. Projected Newton Methods for Optimization Problems with Simple Constraints. *SIAM J. Control Optim.* **1982**, *20*, 221–246. [CrossRef]

114. Ning, H.; Lu, N.; Hassen, A.A.; Chawla, K.; Selim, M.; Pillay, S. A review of Long fibre thermoplastic, (LFT) composites. *Int. Mater. Rev.* **2020**, *65*, 164–188. [CrossRef]

115. van Krevelen, D.W. *Properties of Polymers: Their Correlation with Chemical Structure; Their Numerial Estimation and Prediction from Additive Group Contributions*, 3rd ed.; Elsevier Science: Oxford, UK, 1997.

Publisher's Note: MDPI stays neutral with regard to jurisdictional claims in published maps and institutional affiliations.

 Journal of
Composites Science

Article

A Flow-Dependent Fiber Orientation Model

Susanne Katrin Kugler [1,2,*,†], **Argha Protim Dey** [1,3,†], **Sandra Saad** [1,4], **Camilo Cruz** [1], **Armin Kech** [1] **and Tim Osswald** [2]

[1]   Corporate Sector Research and Advance Engineering, Robert Bosch GmbH, 71272 Renningen, Germany; argha.pro@gmail.com (A.P.D.); Sandra.Saad@de.bosch.com (S.S.); Camilo.Cruz@de.bosch.com (C.C.); Armin.Kech@de.bosch.com (A.K.)

[2]   Polymer Engineering Center, University of Wisconsin-Madison, Madison, WI 53706, USA; tosswald@wisc.edu

[3]   Institute of Applied Mechanics, Chair of Material Theory, Universität Stuttgart, 70174 Stuttgart, Germany

[4]   LAMPA, ENSAM ParisTech, 49035 Angers, France

*   Correspondence: Susanne.Kugler2@de.bosch.com

†   These authors contributed equally to this work.

Received: 24 June 2020; Accepted: 17 July 2020; Published: 22 July 2020

Abstract: The mechanical performance of fiber reinforced polymers is dependent on the process-induced fiber orientation. In this work, we focus on the prediction of the fiber orientation in an injection-molded short fiber reinforced thermoplastic part using an original multi-scale modeling approach. A particle-based model developed for shear flows is extended to elongational flows. This mechanistic model for elongational flows is validated using an experiment, which was conducted for a long fiber reinforced polymer. The influence of several fiber descriptors and fluid viscosity on fiber orientation under elongational flow is studied at the micro-scale. Based on this sensitivity analysis, a common parameter set for a continuum-based fiber orientation macroscopic model is defined under elongational flow. We then develop a novel flow-dependent macroscopic fiber orientation, which takes into consideration the effect of both elongational and shear flow on the fiber orientation evolution during the filling of a mold cavity. The model is objective and shows better performance in comparison to state-of-the-art fiber orientation models when compared to μCT-based fiber orientation measurements for several industrial parts. The model is implemented using the simulation software Autodesk Moldflow Insight Scandium® 2019.

Keywords: fiber orientation; modeling; polymer composites

---

1. Introduction

The current needs for weight reduction and eco-friendly products (e.g., natural fiber composites) have fueled the use of polymers in various industries. They are of particular importance in the automotive and aerospace industries due to their low weight and durability. The polymers in most cases are further reinforced with fibers to enhance their mechanical properties. Those are strongly influenced by fiber volume fraction, fiber length distribution, and fiber orientation.

Fiber reinforced polymers show typically anisotropic material properties. The local orientation of fibers depends on the manufacturing process and the complexity of the part being produced. The prediction of the failure and lifetime of a part using simulative methods instead of experimental methods can significantly help to reduce the costs and speed up the development of new parts. The final fiber orientation induced by the manufacturing process is an important parameter for the structural simulation of the part. The current state-of-the-art fiber orientation models based on Jeffrey's equation [2] have been developed and implemented in different commercial software. These seek to emulate several effects like strain reduction [3–5] and anisotropic rotary diffusion [6–8] on fiber orientation. All these models require phenomenological parameters derived from experimental or microsimulation results to obtain a suitable prediction accuracy.

The experimental determination of the phenomenological parameter to describe the steady state fiber orientation by diffusion was for example conducted by Folgar and Tucker or Bay and Tucker [9,10]. In order to determine phenomenological parameters, which describe the fiber orientation evolution by the retarding rate, it is necessary to measure the transient fiber orientation development. Consequently, the methods are more demanding. Stover et al. [11] measured transient fiber orientations in semi-dilute solutions using a Couette device, a tracer fiber, and video cameras. For highly concentrated solutions, in situ methods with optical methods are not applicable yet. To overcome this shortcoming, experiments in homogeneous flows with a repeatable starting condition have been developed. Eberle et al. [12] measured fiber orientation evolution for a 30% wt. short fiber reinforced polybutylene terephthalate (PBT) in a cone and plate rheometer with specially shaped donut-like samples. Ortman et al. [13] used a sliding plate rheometer, where it is possible to control the initial conditions [14], to measure fiber orientation evolution. Kugler et al. [15] used the same setup to determine phenomenological parameters for a short fiber reinforced PBT-GF30. Recently, Perumal et al. [16] measured transient fiber orientation evolution for a 30% wt. glass fiber reinforced Nylon-6 in a parallel plate rheometer. All stated approaches determine fiber orientation parameters in shear flow. Lambert et al. [17,18] determined fiber orientation parameters for the first time in elongational flow.

Additionally to experimental determination, microscopic fiber simulation can be used to evaluate macroscopic fiber orientation parameters. Modeling approaches on the microscopic scale have the aim to approximate the physical behavior of the composite more accurately. Many authors determined the parameters based on microscopic simulation, for example Mezher et al. [19] and Perez [1]. For a detailed review, refer for example to [20].

In this work, we use a multi-scale simulation chain to enhance the fiber orientation prediction for complex industrial parts. The simulation workflow consists of three steps. Firstly, a virtual flow test on the particle scale (micro) is conducted. Secondly, macroscopic fiber orientation model parameters are determined based on the resulting fiber orientation evolution from the virtual flow test. Thirdly, the final fiber orientation in a part is obtained using continuum-based macro models with the material-dependent optimal parameters. The multi-scale process simulation is sketched in Figure 1.

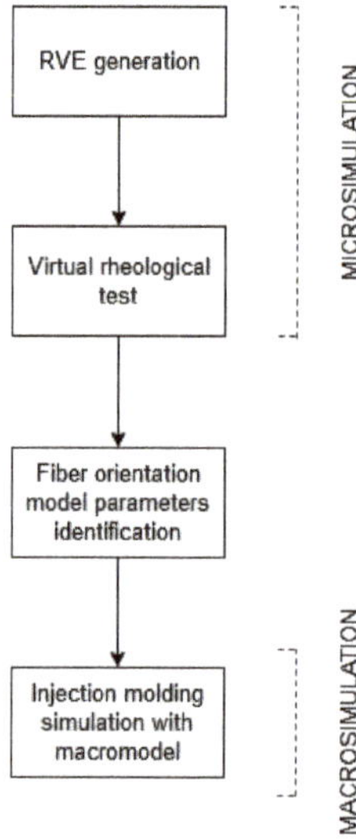

Figure 1. Multi-scale simulation process for short fiber reinforced thermoplastics (SFRT). RVE, representative volume element.

Current fiber orientation models, on both scales, are mostly focused on the effects of shear flow. However, experiments in elongation flow [17,18] showed that fiber orientation develops differently under elongational flows. On the other hand, during injection molding, complex flow fields combining shear and elongational flow usually take place. Considering only shear-fitted fiber orientation models may lead to discrepancies in the final fiber orientation predictions in a part. Recently, Chen et al. [21,22] introduced a flow-dependent strain reduction factor for the different state-of-the-art macroscopic fiber orientation models. This last model approach is able to differentiate the orientation evolution speed between shear, elongational, and rotational flow.

In this work, we focus primarily on the influence of elongational flow on the fiber orientation phenomenon using a simulation at the particle level. Finally, a novel flow-dependent fiber orientation model, scaling between shear and elongational flows, is proposed and implemented in Autodesk Moldflow Insight Scandium® 2019. In comparison to the approach proposed by Chen et al. [21,22], we propose a more general model since it not only considers the change in strain reduction, but also in diffusion.

2. Theory

We briefly describe the different modeling techniques for fiber orientation estimation of short fiber reinforced thermoplastics (SFRT) in this section. Firstly, we give an overview of the fiber orientation tensor. Then, the common macroscopic models in the continuum scale are depicted starting from the basic models to the current state-of-the-art models. Finally, a mechanistic model based on the discrete element method (DEM) for the simulation on the particle level is discussed.

2.1. Fiber Orientation Tensor

A fiber can be represented as a unit vector under the assumption that the fibers are rigid, cylindrical, and have a uniform length and diameter [23]. Considering the above assumptions, we can define the fiber in space by a unit vector **p** and two angles (θ, ϕ) as shown in Figure 2.

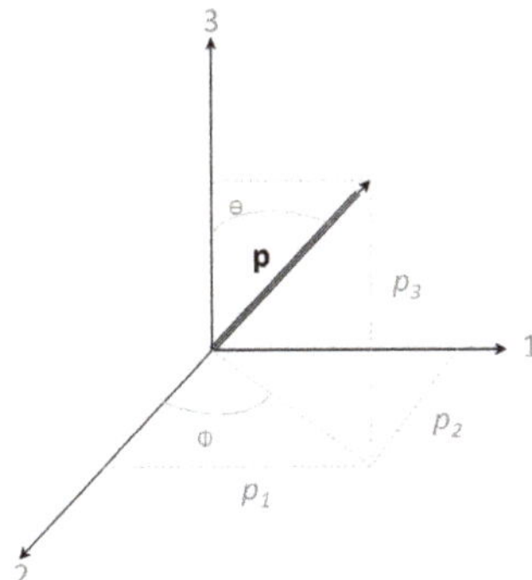

Figure 2. Unit vector representing the orientation of a single rigid fiber.

The components of **p** can be represented in terms of θ and ϕ as:

$$p_1 = \sin\theta\cos\phi \tag{1}$$

$$p_2 = \sin\theta\sin\phi \tag{2}$$

$$p_3 = \cos\theta. \tag{3}$$

A probability distribution function (PDF) $\psi(\mathbf{p}, t)$ can be used to represent the orientation of a population of fibers in space and time [9]. The PDF $\psi(\mathbf{p}, t)d\mathbf{p}$ gives the probability of a fiber being directed between **p** and **p**+d**p** at a time t. The properties of the PDF are:

$$\mathcal{B}(\psi) = [0, 1] \tag{4}$$

$$\oint \psi(\mathbf{p}, t) d\mathbf{p} = 1 \tag{5}$$

$$\psi(\mathbf{p}) = \psi(-\mathbf{p}) \tag{6}$$

$$\frac{D\psi}{Dt} = -\nabla_S \cdot (\psi \dot{\mathbf{p}}), \tag{7}$$

where $\mathcal{B}$ indicates the image of the PDF. Equation (6) is only valid when the fibers are axis symmetrical and when neither of the ends of the fiber have a preferred head. Equation (7) is the continuity condition where ∇_S is the gradient on the surface of the sphere formed by all possible orientations of an axis-symmetrical filler.

The representation of fiber orientation using the PDF is complex and inefficient for numerical simulation. A reduced representation was proposed for better computational efficiency [23]. This representation considers the moments of the PDF, which are calculated in the following way:

$$a_{ij} = \int p_i p_j \psi(\mathbf{p}) d\mathbf{p} \tag{8}$$

$$a_{ijkl} = \int p_i p_j p_k p_l \psi(\mathbf{p}) d\mathbf{p} \tag{9}$$

$$a_{i...n} = \int p_i \dots p_n \psi(\mathbf{p}) d\mathbf{p}. \tag{10}$$

They are called fiber orientation tensors. Since the PDF is of even order, as depicted in Equation (6), the odd ordered tensor integrals are equal to zero. Hence, we only consider the tensors of even order. There can be an infinite number of even ordered tensors. Among these even ordered tensors, mostly the second and fourth order tensors are used [23]. We introduce a compact notation $\mathbf{A} = a_{ij}$ and $\mathbb{A} = a_{ijkl}$ for further discussion.

The important properties of the orientation tensors are discussed only for the second and fourth order tensors in this work. The second order tensor is symmetric, and its trace is one [23],

$$\mathbf{A}_{ij} = \mathbf{A}_{ji} \tag{11}$$

$$\mathrm{tr}\mathbf{A} = 1. \tag{12}$$

The fourth order tensor is symmetric for any pair of indices:

$$\mathbb{A}_{ijkl} = \mathbb{A}_{jikl} = \mathbb{A}_{ijlk} = \mathbb{A}_{kjil} = \mathbb{A}_{ljki} = \mathbb{A}_{ikjl} = \mathbb{A}_{ilkj} \tag{13}$$

and it includes the entire information contained in the second order tensor [23]:

$$\mathbb{A}_{ijkk} = \mathbf{A}_{ij}. \tag{14}$$

The major advantages of using orientation tensors for computational purposes are that the unit sphere around a fiber does not need discretization and the three-dimensional calculations can be feasibly done independently of the basis system [23].

2.2. Continuum-Based Models for Fiber Orientation

A substantial amount of research work has been done to model the fiber orientation for fiber reinforced polymers since they play an important role in determining the structural strength of final parts. One of the very first models depicting the motion of a single fiber in a non-turbulent Newtonian fluid was given by Jeffrey [2]. The fiber is modeled as a rigid ellipsoid, which does not exhibit bending, nor attrition.

$$\dot{\mathbf{p}} = \mathbf{W} \cdot \mathbf{p} + \zeta(\mathbf{D} \cdot \mathbf{p} - \mathbf{D} : \mathbf{ppp}) \tag{15}$$

$$\mathbf{W} = \frac{1}{2}(\nabla\mathbf{u} - \nabla\mathbf{u}^T) \tag{16}$$

$$\mathbf{D} = \frac{1}{2}(\nabla\mathbf{u} + \nabla\mathbf{u}^T), \tag{17}$$

where $\mathbf{W}$ is the vorticity tensor and $\mathbf{D}$ is the rate of strain tensor with $\nabla\mathbf{u}$ being the flow gradient. The term ζ represents the particle shape parameter where $\zeta = \frac{\lambda^2 - 1}{1 + \lambda^2}$ and λ is the fiber aspect ratio.

The Jeffrey model is valid only in dilute solutions, where the dominant mechanism is fluid-fiber interaction since it was developed as a model for single fiber motion [2]. Industrial thermoplastics are typically highly charged in fiber (concentrated regime), where fiber-fiber interaction is dominant. Hence, newer models taking into account the fiber-fiber interactions were developed. All models for the prediction of fiber orientation, except the Jeffrey model, are phenomenological.

The Folgar and Tucker (FT) model [9] uses an additional diffusion term to take into account the fiber-fiber interaction in semi-dilute and concentrated regimes. This model considers the fibers as rigid and uniform cylinders, but large enough to prevent Brownian motion. The fluid is incompressible and viscous enough in order to neglect the inertia of particle and its buoyancy. This model was developed in terms of the rate of change of the second order oriented tensor by Advani and Tucker [23]:

$$\frac{D\mathbf{A}}{Dt} = \dot{\mathbf{A}} = \dot{\mathbf{A}}^h + \dot{\mathbf{A}}^d \tag{18}$$

$$\dot{\mathbf{A}}^h = (\mathbf{W} \cdot \mathbf{A} - \mathbf{A} \cdot \mathbf{W}) + \zeta(\mathbf{D} \cdot \mathbf{A} + \mathbf{A} \cdot \mathbf{D} - 2\mathbb{A} : \mathbf{D}) \tag{19}$$

$$\dot{\mathbf{A}}^d = 2C_I\dot{\gamma}(\mathbf{I} - 3\mathbf{A}), \tag{20}$$

where $\dot{\mathbf{A}}^h$ corresponds to the hydrodynamic effects and $\dot{\mathbf{A}}^d$ is the diffusion term, which takes care of the fiber-fiber interactions. The term C_I is the interaction coefficient, and $\dot{\gamma} = \sqrt{2\mathbf{D} : \mathbf{D}}$ is the magnitude of the rate of the strain tensor. We see that in Equation (19), the fourth order orientation tensor is used. Hence, a closure approximation is necessary. There are several types of closure approximations proposed in the literature as for example fitted closures such as orthotropic closures [24–26], exact closures [27–29], or simple closures [23,30]. In our work, we use a fitted closure, the invariant-based optimal fitted (IBOF) closure approximation [25] based on the findings of Kugler et al. [15].

The Folgar–Tucker model predicts a faster evolution of fiber orientation when compared to experimental results [3]. Therefore, newer models introducing a retardation of the fiber orientation evolution have been developed. All the aforementioned assumptions are valid for the following models as they are based on [9].

The reduced strain closure (RSC) model introduced by Wang et al. [3] is based on the spectral decomposition of $\mathbf{A}$ where $\mathbf{A} = \sum_{i=1}^{3} \lambda_i \mathbf{e}_i \mathbf{e}_i$. The modified growth rate is given as:

$$\dot{\lambda}_i^{RSC} = \kappa\dot{\lambda}_i \tag{21}$$

$$\dot{\mathbf{e}}_i^{RSC} = \dot{\mathbf{e}}_i, \tag{22}$$

where κ is a constant such that $\kappa \in [0,1]$. The Folgar–Tucker Equation (18) along with the modified growth rate Equation (21) can be expressed as:

$$\dot{\mathbf{A}} = \dot{\mathbf{A}}^{RSC} + \kappa\dot{\mathbf{A}}^d \tag{23}$$

$$\dot{\mathbf{A}}^{RSC} = \mathbf{W} \cdot \mathbf{A} - \mathbf{A} \cdot \mathbf{W} + \zeta[\mathbf{D} \cdot \mathbf{A} + \mathbf{A} \cdot \mathbf{D} - 2(\mathbb{A} + (1 - \kappa)(\mathbb{L} - \mathbb{M} : \mathbb{A})) : \mathbf{D}] \tag{24}$$

$$\mathbb{L} = \sum_{i=1}^{3} \lambda_i \mathbf{e}_i \mathbf{e}_i \mathbf{e}_i \mathbf{e}_i \tag{25}$$

$$\mathbb{M} = \sum_{i=1}^{3} \mathbf{e}_i \mathbf{e}_i \mathbf{e}_i \mathbf{e}_i. \tag{26}$$

The output of the Folgar–Tucker model could not be properly fitted with the experimental data [15]. Hence, Phelps and Tucker [6] formulated an anisotropic rotary diffusion model applicable to both short and long fiber thermoplastics. In this model, the interaction coefficient C_I is substituted by a rotary diffusion tensor $\mathbf{C}$. It can be additionally noted that in this model, the state of fiber orientation has an influence on the rotary diffusion effect.

$$\dot{\mathbf{A}} = \dot{\mathbf{A}}^h + \dot{\mathbf{A}}^{ARD} \tag{27}$$

$$\dot{\mathbf{A}}^{ARD} = \dot{\gamma}[2\mathbf{C} - 2\mathrm{tr}(\mathbf{C})\mathbf{A} - 5(\mathbf{C} \cdot \mathbf{A} + \mathbf{A} \cdot \mathbf{C}) + 10\mathbb{A} : \mathbf{C}]. \tag{28}$$

To reduce the amount of phenomenological parameters, the pARD model was developed by Tseng et al. [7]. In their model, the rotary diffusion tensor $\mathbf{C}$ is represented as:

$$\mathbf{C} = C_I \mathbf{R}_A \begin{pmatrix} D_1 & 0 & 0 \\ 0 & D_2 & 0 \\ 0 & 0 & D_3 \end{pmatrix} \mathbf{R}_A^\top, \tag{29}$$

where $\mathbf{R}_A$ is the eigen matrix. The work of [7] used $D_1 = 1$, $D_2 = c$ and $D_3 = 1 - c$ to reduce the number of parameters. This implies that C_I determines the rotary diffusion factor in the first principal direction for fiber orientation. The second and the third principal fiber orientation directions are scaled with the factors $c \cdot C_I$ and $(1 - c) \cdot C_I$, respectively.

An ARD-RSC model is also introduced based on the work of [6] to further retard the kinetics of the ARD model:

$$\begin{aligned}
\dot{\mathbf{A}}^{ARD,RSC} =& \mathbf{W} \cdot \mathbf{A} - \mathbf{A} \cdot \mathbf{W} + \zeta[\mathbf{D} \cdot \mathbf{A} + \mathbf{A} \cdot \mathbf{D} - 2(\mathbb{A} + (1 - \kappa)(\mathbb{L} - \mathbb{M} : \mathbb{A})) : \mathbf{D}] \\
&+ \dot{\gamma}[2(\mathbf{C} - (1 - \kappa)\mathbb{M} : \mathbf{C}) - 2\kappa(\mathrm{tr}(\mathbf{C}))\mathbf{A} - 5(\mathbf{C} \cdot \mathbf{A} + \mathbf{A} \cdot \mathbf{C}) \\
&+ 10[\mathbb{A} + (1 - \kappa)(\mathbb{L} - \mathbb{M} : \mathbb{A})] : \mathbf{C}],
\end{aligned} \tag{30}$$

which reduces to the ARD model if $\kappa = 1$. Another novel approach is the MRD (Moldflow rotational diffusion) model from the work of Bakharev et al. [8],

$$\dot{\mathbf{A}} = \dot{\mathbf{A}}^h + \dot{\mathbf{A}}^{MRD} \tag{31}$$

$$\dot{\mathbf{A}}^{MRD} = 2\dot{\gamma}(\mathbf{C} - \mathrm{tr}(\mathbf{C})\mathbf{A}). \tag{32}$$

2.3. Particle-Based Mechanistic Model

We use a mechanistic model for the determination of fiber positions under the influence of different flow types and varying boundary conditions. The fiber model is based on the discrete element method (DEM) and is derived from the works of Perez [1] and Lindström [31].

A fiber is represented by a chain of cylindrical rods. The cylindrical rods are rigid and interconnected via ball and socket joints as shown in Figure 3. The cylindrical rods are known as segments, and the ends of each segment are called nodes. In each segment, the positions $\mathbf{x}_i$, velocities $\mathbf{u}_i$, and angular velocities ω_i are stored at every node i. The fibers are suspended in a constant viscosity fluid, which is submitted to an external flow field $\mathbf{U}^\infty$. The system is partially coupled such that the hydrodynamic forces on fibers due to the flow are taken into account, but the force exerted on the fluid by the fibers is neglected. The different forces acting on each segment i are: hydrodynamic forces $\mathbf{F}_i^H$ from the fluid, the interaction force with neighboring segment j $\mathbf{F}_{ij}^C$, and intra-fiber force from the adjacent segment i $\mathbf{X}_i$. $\mathbf{M}_i^b$ and $\mathbf{M}_{i+1}^b$ are the bending moments. These forces and moments are used to formulate the translational equation of motion (33) and the rotational equation of motion (34).

$$\mathbf{F}_i^H + \sum_j \mathbf{F}_{ij}^C + \mathbf{X}_i - \mathbf{X}_{i+1} = 0 \tag{33}$$

$$\mathbf{T}_i^H - \mathbf{r}_i \times \mathbf{X}_{i+1} + \sum_j d_{ij} \times \mathbf{F}_{ij}^C + \mathbf{M}_i^b - \mathbf{M}_{i+1}^b = 0 \tag{34}$$

$$\mathbf{u}_i + \omega_i \times \mathbf{r}_i - \mathbf{u}_{i+1} = 0, \tag{35}$$

where $\mathbf{r}_i$ is the segment vector and d_{ij} is the minimum distance between two neighboring segments, as shown in Figure 3. Equation (35) is necessary to enforce connectivity if the fibers are divided into multiple segments.

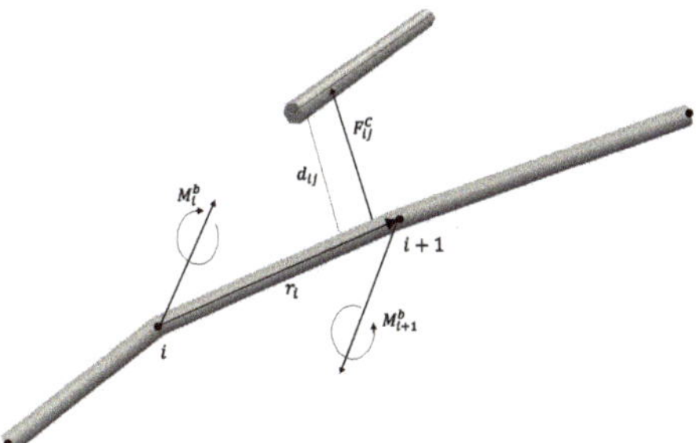

Figure 3. Fiber represented as a chain of segments. A neighboring fiber is also represented (adapted from [31]).

The definitions used in this model for the hydrodynamic and bending forces are similar to [1]. The interaction force between fibers is modeled based on the work of [31]. The interaction between fibers is divided into three regimes: Coulomb friction and normal contact forces $\mathbf{F}_{ij}^{mc}$ when a mechanical contact exists between fiber segments, transitional forces $\mathbf{F}_{ij}^{trans}$ at distances lower than the fiber surface roughness δ_{sur}, and lubrication forces $\mathbf{F}_{ij}^{lub}$, which takes place at distances larger than δ_{sur}, but shorter than the threshold δ_{lub} (Equation (36)).

$$\mathbf{F}_{ij}^C = \begin{cases} \mathbf{F}_{ij}^{lub} & \delta_{sur} < d_{ij} < \delta_{lub} \\ \mathbf{F}_{ij}^{trans} & |d_{ij}| < \delta_{sur} \\ \mathbf{F}_{ij}^{mc} & d_{ij} < -\delta_{sur}. \end{cases} \tag{36}$$

In the case of the inclusion of walls in the system, the interaction among fibers and walls is computed in the same manner as the computation of the interaction between fibers. The solution of Equations (33)–(35) gives the position of each fiber at every time step. In post-processing, from fiber positions at every time step, the second order orientation tensor can be computed as follows [32]:

$$\mathbf{A}_{ij}(t) = \frac{\sum_{n=1}^N p_{n,j}(t)\, p_{n,i}(t)}{N}, \tag{37}$$

where N is the total number of fibers in the system, t is time, and $p_n(t)$ is the position of the fiber n at time t. The evolution of the fiber orientation in the directions of the global system can be obtained from the diagonal components of the orientation tensor, $\mathbf{A}_{11}$, $\mathbf{A}_{22}$, and $\mathbf{A}_{33}$, which correspond to the fraction of total fibers oriented preferentially in the X-, Y-, and Z-directions, respectively.

3. Fiber Orientation in Elongational Flows

The mechanistic model described in the previous section is now used to simulate the effect of elongational flow on fiber orientation.

3.1. Simulation Method and Comparison with the Experiment

Kugler et al. [32] used the mechanistic model to simulate the effect of shear flow using periodic boundary conditions. In our work, this mechanistic model is extended for simulating also fiber orientation under elongational flow. The simulation is performed with the help of a representative volume element (RVE) created using the method developed by Schneider [33]. The elongational flow on the RVE is applied by emulating walls in two directions and imposing the following velocity field:

$$u_x = \dot{\epsilon}x, \quad u_y = -\dot{\epsilon}y, \quad u_z = 0. \tag{38}$$

where $\dot{\epsilon}$ is the rate of elongation. This flow field can be interpreted as a fluid flowing from the y-direction towards the x-direction, but constrained in the z-direction. Walls are defined on the top and bottom of the RVE, which are moving with velocities v_w^{upper} and v_w^{lower} towards the negative y-direction and the positive y-direction, respectively. They are given according to:

$$v_{w|y}^{upper} = -\frac{1}{2}\dot{\epsilon}\,h(t), \quad v_{w|y}^{lower} = \frac{1}{2}\dot{\epsilon}\,h(t), \tag{39}$$

where $h(t)$ is the distance between the walls as they approach each other during the simulation. It should be noted that the wall velocities $v_{w|y}$ reduce as the height of the cell decreases. The wall velocities after a certain time become too small for the numerical procedure to continue. Hence, a lower limit is added to the wall velocities beyond which they are kept constant. This lower limit depends on the elongational rate and time. We expect this to have a negligible influence on the estimated fiber orientation, since only a small amount of fibers close to the walls are directly affected. The walls are also positioned 2×10^{-4} m away from the edge fibers in the initial RVE configuration. In that way, we avoid any initial instability arising out of a fiber with ends outside the RVE, where the velocities cannot be numerically resolved.

Additionally, walls in the xy-plane are included. Although there is no flow defined in the z-direction, it prevents fibers from leaving the RVE. The RVE along with the boundary walls is shown in Figure 4.

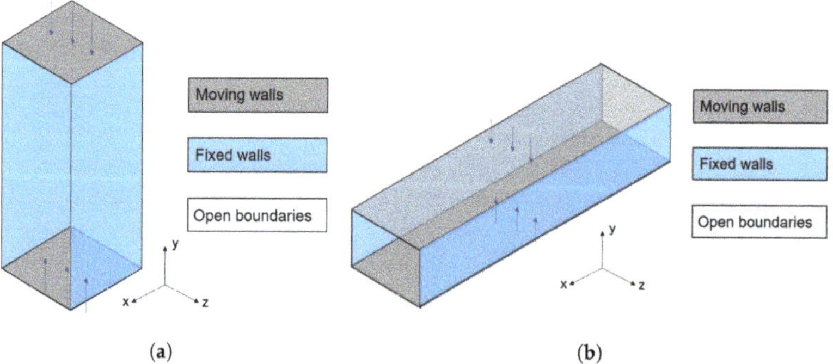

(**a**)   (**b**)

Figure 4. Visualization of the boundary walls of the RVE for elongational flow. (**a**) Before the start of simulation. (**b**) After the end of simulation.

There have been no experimental works so far conducted for elongational flow using short fiber reinforced thermoplastics. However, there were two experiments conducted by Lambert et al. [17] and Lambert and Baird [18] using long fibers, under the influence of elongational flow. In the following sections, we use the mechanistic model for the simulation of elongational flow with long fibers to compare the results with the experimental work.

3.1.1. Validation with Experimental Non-Lubricated Squeeze Flow

Lambert et al. [17] used a commercial composite Verton MV006S (30% wt. glass fiber reinforced polypropylene (PP)) for the experiment. Samples were fabricated by compression molding; for a detailed description, refer to [17]. The squeeze flow experiment was done using a custom-built device described in [17]. There was no lubricant used at the interaction of the mold walls and the samples. The process settings for the experiment are given in Table 1.

Table 1. Process settings of the non-lubricated squeeze flow experiment for a 30% wt. glass fiber reinforced polypropylene (PP) in [17].

Material	Sample Thickness	Temperature	Strain Rate	Max Strain	Total Time
PP -GF30	8×10^{-3} m	$200\,^{\circ}\text{C}$	$0.5\,\text{s}^{-1}$	1	$2\,\text{s}$

The mechanistic model defined in the previous section is now used to simulate the elongational flow with long fibers mimicking the settings in Table 2. The initial fiber orientation obtained from [17] was $A_{11} = A_{33} = 0.45$ and $A_{22} \cong 0.1$ (averaged over three samples). This initial fiber orientation was used to generate an RVE with a squared cross-section of $4\,\text{mm}^2$ using the method given in [33]. The RVE cross-section was smaller than the experimental sample due to computational constraints. The viscosity was approximated as three times the zero shear viscosity of the pure PP matrix to account for elongational viscosity.

Table 2. Simulation parameters for comparison with the non-lubricated squeeze flow experiment.

Height of RVE	Number of Fibers	Fiber Volume	Fiber Radius	Fiber Length	Elongational Rate	Viscosity
8×10^{-3} m	51725	13%	8×10^{-6} m	3.09×10^{-4} m	$0.5\,\text{s}^{-1}$	$900\,\text{Pa s}$

The flow field and boundary conditions were similar to the ones we described above. It is important to note that during the simulation, we considered no friction between the walls of the RVE and the fibers, as opposed to the experimental setup. The RVE was divided into five layers over the thickness direction (y-direction) for post-processing. We plot in Figure 5 the diagonal components of the orientation tensor A_{11}, A_{22}, and A_{33} over the normalized thickness position obtained from the layers with the experimental results found in [17].

The simulation results showed good agreement in the middle of the thickness. However, they slightly under predicted A_{33} at the core. The overall shape of the fiber orientation profile from the experiment was not well captured by the simulation. This was evident from the mismatch of the simulation and experimental results at the boundaries. This can be attributed to the fact that we did not account for friction at walls in the simulation.

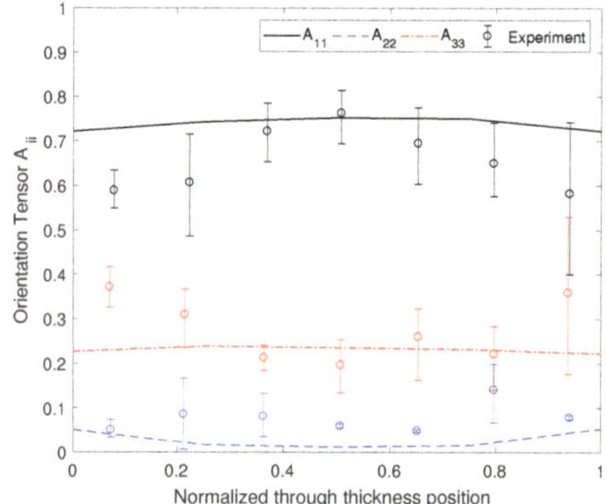

Figure 5. Final fiber orientation through thickness after a elongation time of 2 s at a constant strain rate of 0.5 s^{-1} with experimental data from [17] in points and simulation data in continuous lines.

3.1.2. Validation with Experimental Constant Planar Extension

Lambert and Baird [18] used a 10% wt. glass fiber reinforced PP for the constant planar extension experiment. All samples were made using compression molding. During the course of the experiment, a silicone lubricant (DCF 203, Dow Corning, Midland, MI, USA) with low viscosity was applied to prevent friction and reduce the shear between the mold wall and the melt [18]. The samples were squeezed in a custom-built device at a constant strain rate. The process settings are described in Table 3, and the experiment is explained in detail in [18].

Table 3. Process settings of the constant planar extensional flow experiment in [18].

Material	Sample Thickness	Temperature	Strain Rate	Total Time
PP-GF10	8×10^{-3} m	200 °C	0.05 s^{-1}	70 s

The introduced mechanistic model was then used for comparison with the experimental results from [18]. Due to computational constraints, the RVE used had a reduced squared cross-section of 9 mm^2. The simulation parameters are listed in Table 4. The RVE was compressed in the y-direction at a constant strain rate during 40 s.

Table 4. Simulation parameters for comparison with the planar extensional flow experiment.

Height of RVE	Number of Fibers	Fiber Volume	Fiber Radius	Fiber Length	Elongational Rate	Viscosity
8×10^{-3} m	12160	3.9%	7×10^{-6} m	1.5×10^{-3} m	0.05 s^{-1}	900 Pa s

The second order fiber orientation tensor **A** was computed at each time step. The evolution of the fiber orientation in the principal directions A_{11}, A_{22}, and A_{33} is plotted against time along with the experimental results from [18] in Figure 6.

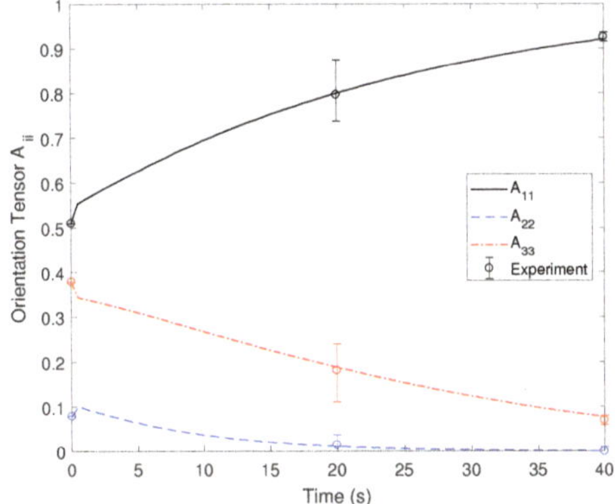

Figure 6. Fiber orientation evolution under constant planar extension at a strain rate of $0.05\ \text{s}^{-1}$ with experimental results from [18] in points and simulation results in continuous lines.

The good agreement of the simulation with the experimental results for constant planar extensional flow can be seen. The evolution of fiber orientation occurs at a very fast rate for this type of flow, which was confirmed by both the experimental and the mechanistic model results in our case. Since there was no parameter fitting involved for the output from our mechanistic model, it can be said that it gave a good estimation of fiber orientation under elongational flow. This model can now be used to predict fiber orientation evolution in elongational flows for various different materials.

3.2. Numerical Analysis: Factors Influencing Fiber Orientation in Elongational Flow

In this section, we study the effects of flow rate, several fiber descriptors, and polymer viscosity on the fiber orientation evolution in elongational flow.

To study the effect of the elongational rate, the simulation was performed with the following elongational rates: $1.0\ \text{s}^{-1}$ and $100.0\ \text{s}^{-1}$. Plotted over strain, there was almost no difference in the final fiber orientation for the analyzed flow rates, as seen from Figure 7. Thus, it seemed that the elongational rate scaled linearly with orientation evolution.

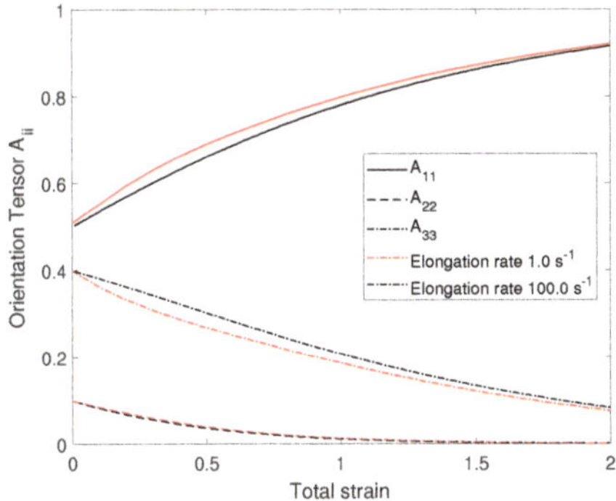

Figure 7. Fiber orientation evolution under elongational flow with different elongation rates.

We applied a similar procedure for the evaluation of the influence of fiber length. The fiber radius, the fiber volume fraction, and fluid viscosity were kept constant, and only the fiber length was varied. The different fiber lengths ranging from short fibers to long fibers were 2.5×10^{-4} m, 8×10^{-4} m, and 1.5×10^{-3} m. It can be observed from Figure 8 that the fiber length had a negligible effect on the fiber orientation.

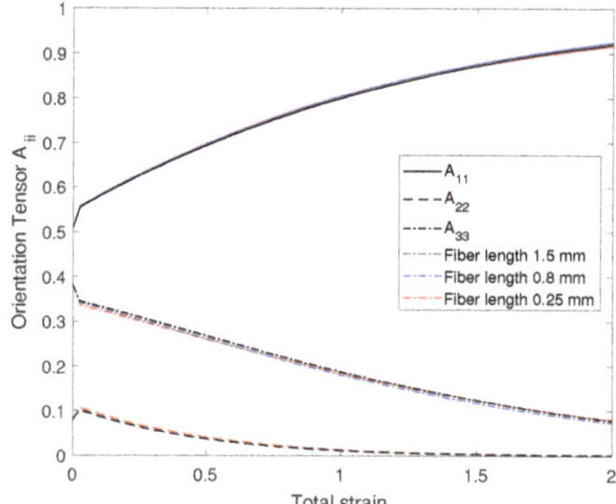

Figure 8. Fiber orientation evolution under elongational flow with different fiber lengths.

The influence of the volume fraction on fiber orientation was investigated using a fixed fiber length, fiber radius, and fluid viscosity. We chose short fibers having a length of 2.5×10^{-4} m and a radius of 5×10^{-6} m, since they are used extensively in industrial parts. The fluid viscosity was fixed at $900\,\mathrm{Pa\,s}$. The RVE had a squared cross-section of $2.5\,\mathrm{mm}^2$ and a height of 3×10^{-3} m. The volume fraction of each RVE is listed in Table 5. An elongational rate of $0.05\,\mathrm{s}^{-1}$ was applied until reaching a strain value of two. The fiber orientation evolution over strain is depicted in Figure 9.

Table 5. Volume fractions and number of fibers in the respective RVEs for analyzing the influence of fiber content on fiber orientation under elongational flow.

Fiber Volume	Number of Fibers
3.9%	1490
8%	3056
13%	4966
18%	6875
25%	9549
30%	11459

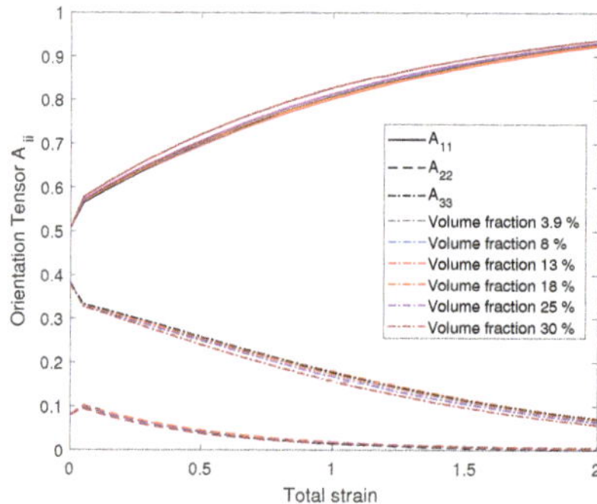

Figure 9. Fiber orientation evolution under elongational flow with different RVEs having different volume fractions.

There was no remarkable difference in fiber orientation evolution with different volume fractions, as seen from Figure 9. In the case of close examination of the evolution curves, we saw that at higher volume fractions of 30% and 25%, the fiber orientation evolution was slightly faster in the principal directions $\mathbf{A}_{11}$ and $\mathbf{A}_{33}$.

As the last factor, the effect of the polymer matrix viscosity on the fiber orientation evolution was studied. A flow rate of $0.5\,\mathrm{s}^{-1}$, a volume fraction of 18%, a fiber radius of 5×10^{-6} m, and a fiber length of 2.5×10^{-4} m were used. The RVE height was 3×10^{-3} m. The viscosity was varied from 300 Pa s to 1200 Pa s. The fiber orientation evolution with respect to strain is plotted in Figure 10.

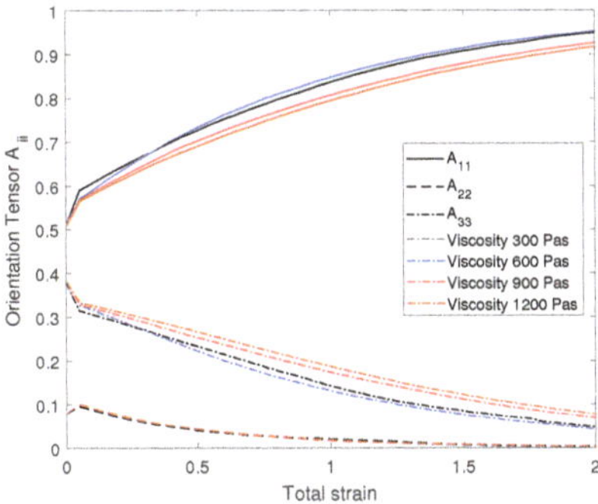

Figure 10. Fiber orientation evolution under elongational flow with different matrix viscosities.

We can observe from Figure 10 that there was a difference in fiber orientation evolution between the lower viscosities (300 Pa s and 600 Pa s) and the higher viscosities (900 Pa s and 1200 Pa s). A possible explanation is that the entire system became less stiff as the fluid viscosity reduced. This was evident given that a lower viscosity reduced the lubrication forces arising in the system, and hence, the fibers could orient more easily under the influence of flow.

In summary, we observed that there was a negligible effect of elongational rate and fiber length. Fiber volume fraction and matrix viscosity had a slight effect on the fiber orientation evolution under elongational flow, but this variation was minor when compared to the fiber orientation evolution under other types of flow (e.g., simple shear). Therefore, it seems that the fiber orientation evolution under elongational flow is universal and independent of strain rate intensity and material descriptors.

It has to be clarified that all effects were studied separately. Additional research is necessary to determine the effect of volume fraction variations with long fibers as well as the effect of fiber length at high volume fractions. The combination of high volume fraction and high fiber length has not been studied, since fibers have been modeled as rigid. The stated combination seems critical under this assumption. Overall, the assumption of rigid fibers seems less severe in elongational flow than in shear flow, since fibers do not perform orbits. In shear flow, the critical length where fiber flexibility has to be incorporated for single fibers movement is around an aspect ratio of 100 [34]. In elongational flow, we assume that the movement of single fibers can be modeled as rigid. Nevertheless, we do not assume that this assumption holds true with increased fiber contact at high volume fractions for long fibers.

3.3. Evaluation of Macroscopic Fiber Orientation Models in Elongational Flows

We conducted a calibration of macroscopic fiber orientation models in elongational flows. Since the mechanistic model is able to reproduce the fiber orientation phenomenon under elongational flow, a virtual elongational test for a 30 wt.% glass fiber reinforced PP was conducted. The simulation settings are stated in Table 6.

Table 6. Simulation settings for the virtual elongational test for PP-GF30.

RVE			Fiber			Process	
Height	Width	Number	Volume	Length	Radius	Rate	Viscosity
6000 μm	500 μm	13904	18.2%	250 μm	5 μm	$0.05\,\mathrm{s}^{-1}$	900 Pa s

A generic algorithm was used to determine the best fit of the diverse macroscopic fiber orientation model parameters. It can be observed that a scalar diffusion rate was sufficient to fit the evolution data in elongational flows. Figure 11 displays a comparison of the best fits of the FT, RSC, and pARD-RSC model. The other models are omitted since they show similar results. Consequently, only one parameter was fitted $C_I = 0.0019$. In all models, the parameters were set such that the FT model was retrieved, e.g., $\kappa = 1$ or $D_2 = D_3 = 1$.

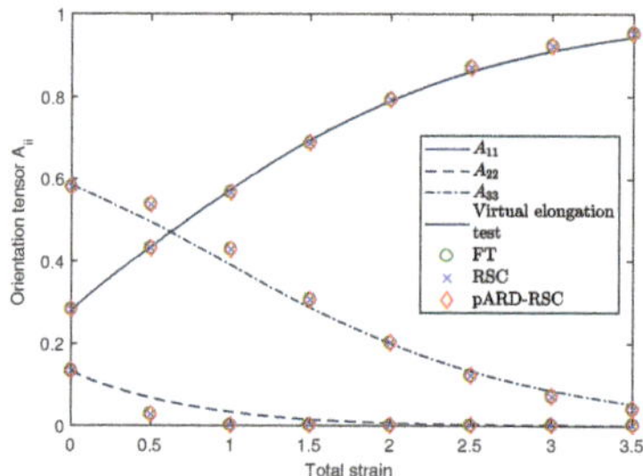

Figure 11. Comparison of the best fit of the FT, RSC, and pARD-RSC model with the virtual elongational test for PP-GF30. Simulation settings are stated in Table 6.

The fit is repeated for the different volume fractions and viscosity data in Section 3.2 to evaluate whether these changes at the micro scales had an impact on the macroscopic fiber orientation model parameters.

For low viscosities (300 Pa s to 600 Pa s), the fitted FT model slightly underpredicted the evolution speed of the orientation. The interaction coefficient for all viscosities varied between 0.001 and 0.003. Thus, we assumed that one constant fiber orientation model parameter set under elongational flow was valid for a broad variety of materials.

4. Influence of Flow Type on Fiber Orientation

In this section, we propose a macroscopic model for incorporating the flow-type effect on fiber orientation and implement it in the injection molding software Autodesk Moldflow Insight Scandium® 2019. The existing and newly implemented models from the work of Kugler et al. [15] took into account the effect of shear flow on fiber orientation. In this work, we introduce a novel model approach to consider the effect of both shear and elongational flows for the estimation of fiber orientation in an injection-molded part.

4.1. Flow-Dependent Fiber Orientation Model Definition

During injection molding, shear and elongational flows occur. To determine the main flow regime of a velocity field at a given position and time, an objective measure is necessary. We used the Manas-Zloczower number M_z [35]. Instead of the Manas-Zloczower number, the objective measure β by Astarita [36] can also be used.

The Manas-Zloczower number is defined by:

$$M_z = \frac{\|\mathbf{D}\|}{\|\mathbf{D}\| + \|\mathbf{W}\|} \tag{40}$$

where $\|\mathbf{D}\|$ is the magnitude of the rate of the strain tensor and $\|\mathbf{W}\|$ is the magnitude of the vorticity tensor. The magnitudes are calculated by using the square roots of the second invariants of the rate of strain and vorticity tensor. Consequently, the Manas-Zloczower number M_z is objective. The second invariant of the rate of strain tensor is defined by $\operatorname{tr} \mathbf{D}^2$ [37], so:

$$\|\mathbf{D}\| = \sqrt{\operatorname{tr} \mathbf{D}^2}. \tag{41}$$

The second invariant of the vorticity tensor is defined as $\operatorname{tr} \mathbf{W}^2 = \frac{1}{2}\|\mathbf{\Omega}\|^2$, where $\mathbf{\Omega} = \nabla \mathbf{u}$ is the vorticity vector [37]. Therefore, the magnitude of the vorticity tensor is defined by:

$$\|\mathbf{W}\| = \sqrt{\frac{1}{2}\|\mathbf{\Omega}\|^2}. \tag{42}$$

The Manas-Zloczower number characterizes the flow type in the following way:

$$M_z = \begin{cases} 1 & elongational\ flow \\ 0.5 & shear\ flow \\ 0 & rotational\ flow. \end{cases} \tag{43}$$

In our work, we consider the effect of shear and elongational flow. Hence, we applied a lower limit of 0.5 to the flow type descriptor M_z. If at a given point during the flow, $M_z < 0.5$ holds true, then it is set to 0.5, assuming that the flow is controlled by shear. We believe that this hypothesis is weak given the low frequency of rotational flow in front of shear and elongational flows in injection molding.

We assumed that fiber orientation for mixed shear and elongational flow is a linear combination of the shear- and elongational-driven fiber orientation models. This leads to the following flow-dependent fiber orientation model:

$$\dot{\mathbf{A}} = 2(M_z - 0.5)\dot{\mathbf{A}}_e + 2(1 - M_z)\dot{\mathbf{A}}_s. \tag{44}$$

In pure shear, this equals $\dot{\mathbf{A}} = \dot{\mathbf{A}}_s$, and in pure elongational flow, $\dot{\mathbf{A}} = \dot{\mathbf{A}}_e$. The derivatives $\dot{\mathbf{A}}_s$ and $\dot{\mathbf{A}}_e$ can be calculated using a macroscopic fiber orientation model with the respective parameters for shear and elongation. Based on the findings of Kugler et al. [15], we used the pARD-RSC model to model fiber orientation in shear flow. From the mechanistic model simulation and the respective macroscopic fiber orientation model fits in Section 3.3, it was shown that the FT model is sufficient in elongational flows. For the sake of consistency, we also used the pARD-RSC model in elongation. The number of parameters for the pARD-RSC model is then reduced in the case of elongational flow by setting $D_1 = D_2 = D_3 = 1$ in Equation (29). Additionally, no retarding rate is necessary; therefore, we set $\kappa = 1$. Consequently, only the interaction coefficient C_I has to be fitted.

This leads to the following definition of the flow-dependent fiber orientation model:

$$\dot{\mathbf{A}}^{\mathrm{pARD-RSC}} = 2(M_z - 0.5)\dot{\mathbf{A}}_e^{\mathrm{pARD-RSC}} + 2(1 - M_z)\dot{\mathbf{A}}_s^{\mathrm{pARD-RSC}}. \tag{45}$$

with:

$$\dot{\mathbf{A}}_e^{\mathrm{pARD-RSC}} = \dot{\mathbf{A}}^{\mathrm{pARD-RSC}}(C_I^e, 1, 1, 1) \tag{46}$$

$$\dot{\mathbf{A}}_s^{\mathrm{pARD-RSC}} = \dot{\mathbf{A}}^{\mathrm{pARD-RSC}}(C_I^s, D_2^s, D_3^s, \kappa^s). \tag{47}$$

This linearly interpolated fiber orientation model was implemented in Autodesk Moldflow Insight Scandium® 2019 using the Solver API feature. The flow-dependent model can estimate the fiber orientation of any fiber reinforced plastic part using user-defined model parameters.

4.2. Workflow for Fiber Orientation Estimation Using the Flow-Dependent Model

In this section, we propose a workflow for the identification of the shear and elongational parameters for the flow-dependent fiber orientation model.

The parameters for shear flow can either be experimentally fitted like in [15] or determined by a virtual shearing test with a mechanistic fiber simulation [38].

The parameters for elongational flow can be determined by the virtual elongational test proposed in Section 3. Additionally, the experimental setup of [17] can be used. This workflow is displayed in Figure 12.

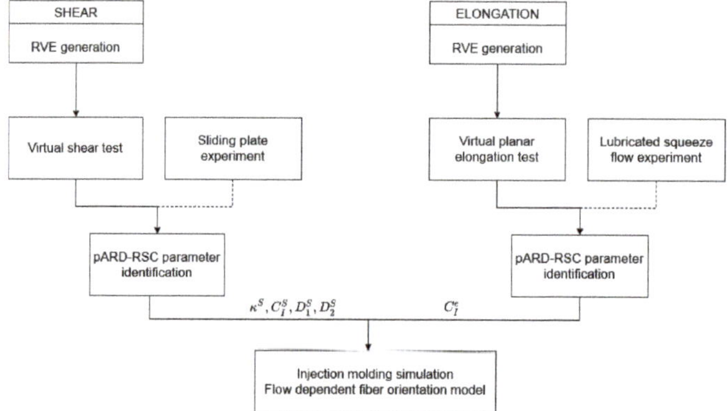

Figure 12. Workflow for parameter identification of the flow-dependent fiber orientation model.

4.3. Flow-Dependent Model Parameters for a PBT-GF30

In this section, the workflow is applied to define the parameters for a PBT-GF30. In shear, we used the experimental parameters determined in [15]. In elongation, we used the parameters from the virtual elongational test with the mechanistic model from Section 3.3. The test was conducted for a PP-GF30, but since the variation in the polymer matrix viscosity was small, we supposed that the same parameter set under elongational flow was also valid for our PBT-GF30. The parameters of the flow-dependent fiber orientation model PBT-GF30 are given in Table 7.

Table 7. Parameters of the flow-dependent pARD-RSC model for a PBT -GF30.

Parameters	Fitted Values
C_i^e	0.0019
C_i^s	0.012
κ^s	0.28
$D_2{}^s$	0.65

5. Results and Discussion

In this section, we evaluate the accuracy of diverse fiber orientation models by comparing with experimental fiber orientation determined by microcomputed tomography (μCT) in different parts. We use the following macroscopic fiber orientation models:

- RSC model
- MRD model

- pARD-RSC model with parameters from shear flow only
- pARD-RSC model with parameters from elongational flow only
- Flow-dependent pARD-RSC model.

We simulated a total of three geometries with the help of these models. All parts were injection-molded with a PBT-GF30, the same material referred to in Section 4.3. The parts included a 2 mm plate, a window gear lift housing, and a sensor cover. µCT scans were conducted at different positions in each geometry. The experimental fiber orientations were calculated from the analysis of the µCT data. The geometries and the positions of the µCT scans are depicted in Figure 13. The sensor cover cannot be fully displayed due to company restrictions. Position 3 at the sensor cover was located close to the gate. The locations were chosen in order to cover different flow lengths, different wall thicknesses, flow singularities like weld lines, and areas with specific mechanical requirements under assembling and/or operation.

The pARD-RSC model parameters for shear and elongational flow are given in Table 7. All geometries were simulated using a fill-and-pack analysis in Autodesk Moldflow Insight® 2019. The RSC and MRD models are pre-defined in Autodesk Moldflow®. The different versions of the pARD-RSC model were implemented in an extended technology preview of the software (Project Scandium) using the Solver API feature. For a validation of the implementation, refer to [15].

The simulation of the different geometries with the flow-dependent model requires the calculation of the Manas-Zloczower number during flow. We calculated the Manas-Zloczower number for each node at every time step during the filling of the mold. The coherence of the computation of the Manas-Zloczower number was evaluated with the help of the plate.

The Manas-Zloczower number was written as an output during the analysis using the Solver API. Figure 14a shows the cross-section of the plate with the different values of M_z after 0.41 s of filling. For information, the cavity was completely filled after 0.63 s. We could observe that M_z exhibited higher values at the core of the sprue and plate. The value of M_z was close to 0.5 near the boundaries of the plate. This agreed with the fact that shear flow was dominant at the boundaries due to the friction with the mold walls. The flow gradually changed to elongational flow towards the core of the plate. This can also be observed in Figure 14b where the Manas-Zloczower number is plotted against the thickness of the plate at Position 1 (Figure 13) once the node is fully filled for the first time.

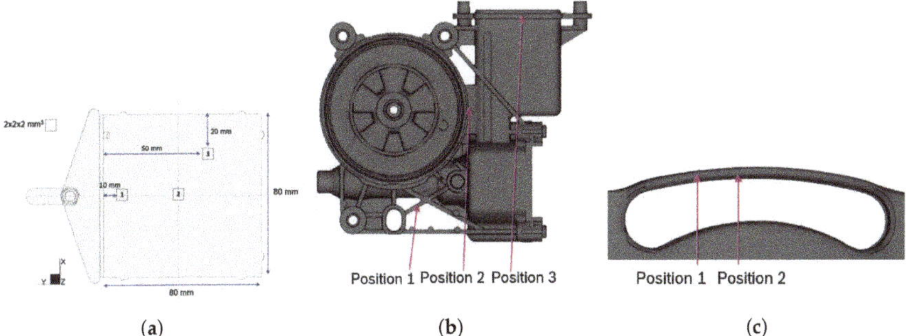

Figure 13. Different geometries with the position of the µCT scans. (**a**) Plate. (**b**) Window lift drive housing. (**c**) Sensor cover (partly displayed).

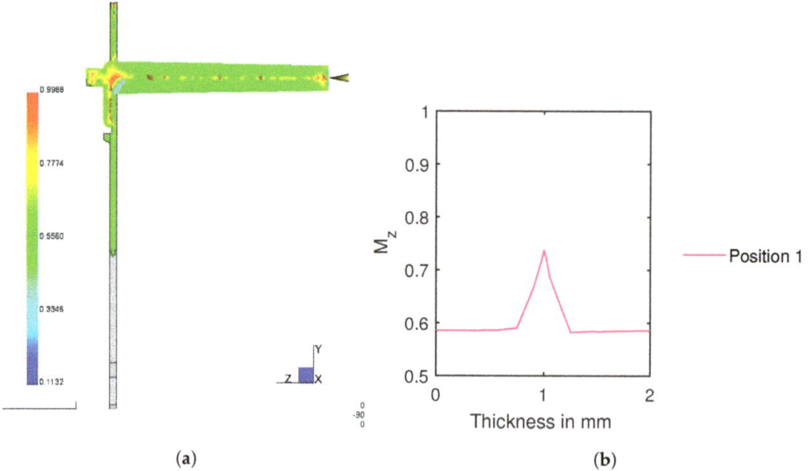

(a)   (b)

Figure 14. Evaluation of the Manas-Zloczower number calculation. (**a**) Cross-section of plate after 0.41 s of filling time. (**b**) Graph of M_z against the thickness of the plate at Position 1 once the node is fully filled for the first time.

We evaluated the performance of the different models with the help of the root mean squared deviation (RMSD) between the eigenvalues of the fiber orientation tensor obtained from the µCT scans and the eigenvalues of the final fiber orientation tensor from simulation. The eigenvalues of **A** were used in the comparison instead of the diagonal entries $\mathbf{A}_{ii}$ for $i = 1, 2, 3$ due to the non-coinciding coordinate axes of the µCT analyses and those of the simulation. We depict the RMSD values of each model at all µCT scan positions and the average over all the positions in Table 8. The columns Shear and Elongation represent the pARD-RSC model with parameters from shear flow and elongational flow, respectively.

We can observe from the RMSD values for the different parts that the flow-dependent model gives similar or better results in comparison to the standard fiber orientation models in Moldflow. In the plate, models without a retarding rate yield better results (MRD and elongational pARD-RSC model). On the contrary, for parts that are more intricate, models using a retarding rate yield better results (RSC model and shear pARD-RSC model). In the sensor cover, it can be observed that different models yield the best results at different positions of the part. Close to the sprue, elongational flows are dominant, and the MRD and elongation models are the best. At Positions 1 and 2, the models using a retarding rate are superior. The only model that has a high prediction accuracy with an RMSD smaller than 0.1 in all points is the novel flow-dependent model since it is able to scale between flow regimes, as it applies a retarding rate in shear flows and neglects it in elongational flows.

Table 8. RMSD values between the experimental and simulated fiber orientation tensor in different geometries.

Geometry	Position	RSC	MRD	Shear	Elongation	Flow-Dependent
Plate	1	0.07	0.06	0.11	0.07	0.06
	2	0.10	0.07	0.13	0.09	0.08
	3	0.09	0.08	0.09	0.06	0.06
	averaged	0.09	0.07	0.11	0.07	0.07
Window gear lift housing	1	0.07	0.14	0.03	0.06	0.03
	2	0.05	0.06	0.07	0.13	0.09
	3	0.05	0.04	0.06	0.03	0.03
	averaged	0.06	0.08	0.05	0.07	0.05
Sensor cover	1	0.03	0.09	0.04	0.09	0.05
	2	0.04	0.08	0.05	0.07	0.05
	3	0.08	0.07	0.09	0.08	0.08
	averaged	0.05	0.08	0.06	0.08	0.06
averaged		0.07	0.08	0.07	0.07	0.06

6. Conclusions

We propose a user-friendly multi-scale virtual workflow for estimating the fiber orientation of an injection-molded fiber reinforced thermoplastic part. To do this, we use a particle-based mechanistic model, which is able to evaluate the fiber orientation under shear flow and elongational flow. We validate the mechanistic model under elongational flow in front of experimental data using long fibers [17,18]. The estimations of the mechanistic model are in agreement with the experimental results. By exploiting the mechanistic model, we find that the fiber orientation evolution under elongational flow is independent of the fiber length, the fiber volume content, and the elongational rate for short fibers. There is however a slight sensitivity to variations of the matrix polymer viscosity. In future work, it would be interesting to study the influence of fiber flexibility on the fiber orientation phenomenon, in particular for long fibers.

At the macro-scale, we introduce a novel flow-dependent fiber orientation model that is able to adjust the fiber orientation evolution as a function of the local flow type by using an objective scaling parameter: the Manas-Zloczower number. This is reached by defining flow specific retarding and anisotropic diffusion parameters. We implement the 3D flow-dependent fiber orientation model in Moldflow using a Solver API feature. The fiber orientation results from our flow-dependent fiber orientation model are compared to μCT scans in three different injection-molded parts. The novel model gives not only comparable, but in some cases, more accurate results than current existing models. Additionally, it is the only model, that provides good results in a plate and a complex part. This shows that the flow dependency provides a more general modeling approach in front of the existing macroscopic fiber orientation models. We believe that the global performance of the model can still increase by incorporating fiber orientation-viscosity coupling in the macroscopic simulation. In our flow-dependent model, the fiber orientation evolution speed scales linearly between shear and elongation flows. A future enhancement could be the use of a different scaling approach and the incorporation of rotational flows. This novel multi-scale simulation approach consisting of the mechanistic model and the flow-dependent macro-model is significantly faster when compared to workflows with experimental parameter identification.

7. Patents

This work is partially covered by Patent No. 102020204586.0.

Author Contributions: Conceptualization, S.K.K., A.P.D., T.O., C.C., and A.K.; methodology, S.K.K., A.P.D., and T.O.; software, S.S., S.K.K., and A.P.D.; validation, S.K.K., A.P.D., and S.S.; writing, original draft preparation,

A.P.D. and S.K.K.; writing, review and editing, S.S., T.O., C.C., and A.K.; visualization, A.P.D. and S.K.K. All authors read and agreed to the published version of the manuscript.

Funding: This research received no external funding.

Acknowledgments: The authors would like to express their thanks to Thomas Riedel (Robert Bosch GmbH) for performing the µCT scans for all parts.

Conflicts of Interest: The authors declare no conflict of interest

Abbreviations

The following abbreviations are used in this manuscript:

µCT	Computed tomography
RVE	Representative volume element
SFRT	Short fiber reinforced thermoplastic
DEM	Discrete element method
PDF	Probability distribution function
FT	Folgar–Tucker
RSC	Reduced strain closure
ARD	Anisotropic rotary diffusion
pARD	Principal anisotropic rotary diffusion
MRD	Moldflow rotary diffusion
RMSD	Root mean squared deviation

References

1. Pérez, C. *The Use of a Direct Particle Simulation to Predict Fiber Motion in Polymer Processing*; The University of Wisconsin-Madison: Madison, WI, USA, 2016.
2. Jeffery, G.B. The motion of ellipsoidal particles immersed in a viscous fluid. *Proc. R. Soc. London. Ser. A Contain. Pap. A Math. Phys. Character* **1922**, *102*, 161–179.
3. Wang, J.; O Gara, J.F.; Tucker, C.L., III. An objective model for slow orientation kinetics in concentrated fiber suspensions: Theory and rheological evidence. *J. Rheol.* **2008**, *52*, 1179–1200. [CrossRef]
4. Huynh, H. *Improved Fiber Orientation Predictions for Injection-molded Composites*; University of Illinois at Urbana-Champaign: Urbana, IL, USA, 2001.
5. Tseng, H.C.; Chang, R.Y.; Hsu, C.H. Phenomenological improvements to predictive models of fiber orientation in concentrated suspensions. *J. Rheol.* **2013**, *57*, 1597–1631. [CrossRef]
6. Phelps, J.H.; Tucker III, C.L. An anisotropic rotary diffusion model for fiber orientation in short-and long-fiber thermoplastics. *J. Non-Newton. Fluid Mech.* **2009**, *156*, 165–176. [CrossRef]
7. Tseng, H.C.; Chang, R.Y.; Hsu, C.H. The use of principal spatial tensor to predict anisotropic fiber orientation in concentrated fiber suspensions. *J. Rheol.* **2018**, *62*, 313–320. [CrossRef]
8. Bakharev, A.; Yu, H.; Ray, S.; Speight, R.; Wang, J. Using new anisotropic rotational diffusion model to improve prediction of short fibers in thermoplastic injection molding. In Proceedings of the SPE ANTEC Conference, Technical Papers, Orlando, FL, USA, 7 May 2018; pp. 7–19.
9. Folgar, F.; Tucker, C.L. Orientation Behavior of Fibers in Concentrated Suspensions. *J. Reinf. Plast. Compos.* **1984**, *3*, 98–119. [CrossRef]
10. Bay, R.; Tucker, C. Fiber orientation in simple injection moldings. Part II: Experimental results. *Polym. Compos.* **1992**, *13*, 332–341. [CrossRef]
11. Stover, C.A.; Koch, D.L.; Cohen, C. Observations of fibre orientation in simple shear flow of semi-dilute suspensions. *J. Fluid Mech.* **1992**, *238*, 277–296. [CrossRef]
12. Eberle, A.P.; Vélez-García, G.M.; Baird, D.G.; Wapperom, P. Fiber orientation kinetics of a concentrated short glass fiber suspension in startup of simple shear flow. *J. Non-Newton. Fluid Mech.* **2010**, *165*, 110–119. [CrossRef]
13. Ortman, K.C.; Agarwal, N.; Eberle, A.P.R.; Baird, D.G.; Wapperom, P.; Jeffrey Giacomin, A. Transient shear flow behavior of concentrated long glass fiber suspensions in a sliding plate rheometer. *J. Non-Newton. Fluid Mech.* **2011**, *166*, 884–895. [CrossRef]

14. Ortman, K.; Baird, D.; Wapperom, P.; Whittington, A. Using startup of steady shear flow in a sliding plate rheometer to determine material parameters for the purpose of predicting long fiber orientation. *J. Rheol.* **2012**, *56*, 955–981. [CrossRef]

15. Kugler, S.K.; Lambert, G.M.; Cruz, C.; Kech, A.; Osswald, T.A.; Baird, D.G. Macroscopic fiber orientation model evaluation for concentrated short fiber reinforced polymers in comparison to experimental data. *Polym. Compos.* **2020**. [CrossRef]

16. Perumal, V.; Gupta, R.K.; Bhattacharya, S.N.; Costa, F.S. Fiber orientation prediction in nylon-6 glass fiber composites using transient rheology and 3-dimensional x-ray computed tomography. *Polym. Compos.* **2019**, *40*, E392–E398. [CrossRef]

17. Lambert, G.; Wapperom, P.; Baird, D. Obtaining short-fiber orientation model parameters using non-lubricated squeeze flow. *Phys. Fluids* **2017**, *29*, 121608. [CrossRef]

18. Lambert, G.M.; Baird, D.G. Evaluating rigid and semiflexible fiber orientation evolution models in simple flows. *J. Manuf. Sci. Eng.* **2017**, *139*, 031012. [CrossRef]

19. Mezher, R.; Perez, M.; Scheuer, A.; Abisset-Chavanne, E.; Chinesta, F.; Keunings, R. Analysis of the Folgar & Tucker model for concentrated fibre suspensions in unconfined and confined shear flows via direct numerical simulation. *Compos. Part A Appl. Sci. Manuf.* **2016**, *91*, 388–397. [CrossRef]

20. Kugler, S.K.; Kech, A.; Cruz, C.; Osswald, T. Fiber Orientation Predictions—A Review of Existing Models. *J. Compos. Sci.* **2020**, *4*, 69. [CrossRef]

21. Chen, H.; Wapperom, P.; Baird, D.G. The use of flow type dependent strain reduction factor to improve fiber orientation predictions for an injection molded center-gated disk. *Phys. Fluids* **2019**, *31*, 123105. [CrossRef]

22. Chen, H.; Wapperom, P.; Baird, D.G. A model incorporating the effects of flow type on fiber orientation for flows with mixed flow kinematics. *J. Rheol.* **2019**, *63*, 455–464. [CrossRef]

23. Advani, S.G.; Tucker, C.L., III. The use of tensors to describe and predict fiber orientation in short fiber composites. *J. Rheol.* **1987**, *31*, 751–784. [CrossRef]

24. Cintra, J.S.; Tucker, C.L. Orthotropic closure approximations for flow-induced fiber orientation. *J. Rheol.* **1995**, *39*, 1095–1122. [CrossRef]

25. Chung, D.H.; Kwon, T.H. Improved model of orthotropic closure approximation for flow induced fiber orientation. *Polym. Compos.* **2001**, *22*, 636–649. [CrossRef]

26. Kuzmin, D. Planar and orthotropic closures for orientation tensors in fiber suspension flow models. *SIAM J. Appl. Math.* **2016**, *78*, 3040–3059. [CrossRef]

27. Verley, V.; Dupret, F. Numerical prediction of the fiber orientation in complex injection molded parts. *Trans. Eng. Sci.* **1994**. [CrossRef]

28. Montgomery-Smith, S.; He, W.; Jack, D.; Smith, D. Exact tensor closures for the three-dimensional Jeffery's equation. *J. Fluid Mech.* **2011**, *680*, 321–335, [CrossRef]

29. Montgomery-Smith, S.; Jack, D.; Smith, D.E. The Fast Exact Closure for Jeffery's equation with diffusion. *J. Non-Newton. Fluid Mech.* **2011**, *166*, 343–353. [CrossRef]

30. Advani, S.G.; Tucker, C.L. Closure approximations for three-dimensional structure tensors. *J. Rheol.* **1990**, *34*, 367–386. [CrossRef]

31. Lindström, S.B.; Uesaka, T. Simulation of the motion of flexible fibers in viscous fluid flow. *Phys. Fluids* **2007**, *19*, 113307. [CrossRef]

32. Kugler, S.K.; Cruz, C.; Kech, A.; Chang, T.C.; Osswald, T.A. Workflow for Enhanced Fiber Orientation Prediction of Short Fiber-reinforced Thermoplastics. In Proceedings of the SPE ANTEC Conference, 2020. (Not published yet).

33. Schneider, M. The sequential addition and migration method to generate representative volume elements for the homogenization of short fiber reinforced plastics. *Comput. Mech.* **2017**, *59*, 247–263. [CrossRef]

34. Forgacs, O.; Mason, S. Particle motions in sheared suspensions. X. Orbits of flexible threadlike particles. *J. Colloid Sci.* **1959**, *14*, 473–491. [CrossRef]

35. Cheng, J.; Manas-Zloczower, I. Flow field characterization in a banbury mixer. *Int. Polym. Process.* **1990**, *5*, 178–183. [CrossRef]

36. Astarita, G. Objective and generally applicable criteria for flow classification. *J. Non-Newton. Fluid Mech.* **1979**, *6*, 69–76. [CrossRef]

37. Tesch, K. On invariants of fluid mechanics tensors. *Task Q.* **2013**, *17*, 1000–1008.
38. Kugler, S.K.; Lambert, G.M.; Cruz, C.; Kech, A.; Osswald, T.A.; Baird, D.G. Efficient parameter identification for macroscopic fiber orientation models with experimental data and a mechanistic fiber simulation. In *AIP Conference Proceedings*; AIP Publishing LLC: New York, NY, USA, 2020; Volume 2205, p. 020050.

Journal of
Composites Science

Article

Metamodelling of the Correlations of Preform and Part Performance for Preform Optimisation in Sheet Moulding Compound Processing

Christian Hopmann, Jonas Neuhaus *, Kai Fischer, Daniel Schneider and René Laschak Pinto Gonçalves

Institute for Plastics Processing (IKV), Seffenter Weg 201, 52074 Aachen, Germany;
office@ikv.rwth-aachen.de (C.H.); kai.fischer@ikv.rwth-aachen.de (K.F.);
daniel.schneider@ikv.rwth-aachen.de (D.S.); rene.laschak@rwth-aachen.de (R.L.P.G.)
* Correspondence: jonas.neuhaus@ikv.rwth-aachen.de

Received: 29 June 2020; Accepted: 19 August 2020; Published: 21 August 2020

Abstract: In the design of parts consisting of long-fibre-reinforced Sheet Moulding Compounds (SMC), the potential for the optimisation of processing parameters and geometrical design is limited due to the high number of interdependent variables. One of the influences on fibre orientations and therefore mechanical part performance is the initial filling state of the compression moulding tool, which is defined by the geometry and positioning of the SMC preform. In the past, response surface methodology and linear regression analysis were successfully used for a simulation-based optimisation of rectangular preform size and position in regard to a part performance parameter. However, the computational demand of these increase exponentially with an increase in the number of design variables, such as in the case of more complex preform geometries. In this paper, these restrictions are addressed with a novel approach for metamodelling the correlation of preform and the resulting mechanical part performance. The approach is applied to predicting the maximum absolute deflection of a plate geometry under bending load. For metamodelling, multiple neural networks (NN) are trained on a dataset obtained by process and structural simulation. Based on the discretisation of the plate geometry used in these simulation procedures, the binary initial filling states (completely filled/empty) of each element are used as inputs of the NNs. Outputs of the NNs are combined by ensemble modelling to form the metamodel. The metamodel allows an accurate prediction of maximum deflection; subsequent validation of the metamodel shows differences in predicted and simulated maximum deflection ranging from 0.26% to 2.67%. Subsequently, the metamodel is evaluated using a mutation algorithm for finding a preform that reduces the maximum deflection.

Keywords: computational modelling; compression moulding; moulding compounds; optimisation

---

1. Introduction

Sheet Moulding Compounds (SMC) compression moulding is the largest market segment in the processing of Glass Fibre-Reinforced Plastics (GFRP) in terms of production volume [1]. The benefits of compression moulding include the economical production of near-net-shape components, minimising the need for subsequent assembly steps. The deformation of the SMC preform during compression moulding causes the formation of inhomogeneous and transient flow fields, which in turn cause a change in the orientation of the contained fibres [2–8]. Mechanical properties of the resulting SMC part are dominated by these fibre orientations, which therefore play an important role in the design of SMC parts and setup of the compression moulding process such as definition preform position and size [4,9–11].

Martulli et al. reported a property difference between specimens cut from carbon fibre-reinforced (53 wt %) vinylester-based SMC with a preferential orientation of 0° and 90° of 150% for tensile stiffness, 260% for tensile strength, 120% for compressive stiffness, and 32% for compressive strength, respectively [11]. For a polyester-based SMC with 30 wt % glass fibre content, Oldenbo et al. determined a difference in tensile stiffness for preferential orientation of 0° and 90° upwards of 25% [10].

Fibre orientation can be predicted by the application of process simulation procedures, on which extensive work has been conducted since the early 1980s [12–16]. Initially, these were based on 2D and 2.5D modelling approaches (the latter considering through-thickness variations in flow and material properties). Lee, Folgar, and Tucker applied the generalised Hele–Shaw model for calculation of the filling of thin-walled structures during the SMC compression moulding process; however, they did initially not consider the influence of temperature on the SMC viscosity [15,17,18]. Barone and Caulk performed experimental analysis on SMC flow and reported boundary effects such as slippage of the SMC on the mould surface due to temperature influences, for which a model was subsequently proposed [19,20].

In recent years, these methods were expanded by the development of simulation procedures capable of 3D calculation, of which certain functionalities have been implemented in programs such as Moldex3D and Moldflow [21,22]. Hohberg employed the Coupled Eulerian Lagrangian framework within Abaqus for the calculation of SMC flow and flow-induced deformation of local reinforcements [23]. A research group led by Osswald developed a direct fibre simulation procedure, with which fibre bending and fibre–matrix separation can be calculated in long fibre-reinforced polymers [21]. This was expanded on by Meyer et al. with a direct bundle simulation approach [24].

However, process simulation has remained computationally demanding and time-consuming [4,9,25]. This limits the potential in part or process optimisation, since finding the "optimal" solution usually necessitates simulating a high number of variable variations, which is especially prevalent when varying only one at a time [25–30]. In the field of SMC compression moulding, advanced optimisation procedures have been presented, which are based on alternative approaches making use of approximations of the complex interactions of influencing parameters (also called metamodels or surrogate models) [25,30,31]. Metamodels provide a "Model of the Model", which may be used to replace computationally expensive simulation models in a wide number of engineering disciplines and have also been used in SMC modelling and process optimisation [25,30,32–34]. Huang et al. used a mesoscale metamodelling approach to accurately predict the stiffness matrices of chopped carbon fibre SMC from the fibre orientation tensor using individual Kriging models for each element of the stiffness matrix [34]. Sabiston et al. presented a neural networks (NN)-based procedure for the prediction of preform position and geometry-dependent fibre orientations in a SMC seat back component. With this approach, near instantaneous prediction of fibre orientation is achieved, the caveat being that a high number of data points (3000) is required for training the NN [8].

For determination of the optimal SMC peform placement for a hood scoop part, Ankenman, Bisgaard, and Osswald proposed an iterative optimisation approach based on evaluating simulation results with the response surface methodology [25]. The optimisation had two goals, being to minimise the "fill time tolerance" (standard deviation of the time necessary for filling all nodes of the discretised part geometry, thus describing the filling uniformity) and the uniformness of fibre orientation. However, this procedure necessitated significant human interaction in iteratively defining and evaluating different experimental setups. This was later expanded on by Twu and Lee, who automated the procedure by applying linear regression analysis. However, they noted that this methodology (as well as statistical regression methods or various mathematical approximation theories in general) had major drawbacks, as the number of necessary simulations would increase exponentially with the number of design variables [35]. Alternatively, they proposed employing neural networks (NN), which they subsequently applied in increasing the curing homogeneity of an SMC part with varying thickness by optimising the heating channel location inside the mould [30]. Initially, a start-up search is employed (necessitating 19 curing simulations). The parameter variations and simulation results are used for initial training of a feed-forward NN (FF-NN). Subsequently, the FF-NN is iteratively evaluated and retrained by supplementary curing simulation, finding the optimal design in less than 60 simulations (in comparison with the statistical approach necessitating 729 simulations when using a 3-level quadratic model without domain search) [30].

Alternatively, Kim, Lee, Han, and Vautrin used a genetic algorithm for optimising the preform size and placement. The goal of the optimisation was to minimise the maximum deflection of a symmetric car hood and an arbitrary non-symmetric geometry that resembles a fender [31].

Most of the optimisation procedures mentioned apply initial evaluations of the significance of design factors, excluding non-significant factors, to make the design cases more manageable [25,30]. Additionally, problem-dependent constraint handling techniques are used to rule out non-feasible solutions (e.g., limiting the ranges of the design factors to physical limits such as the mould dimensions). For curing homogeneity optimisation, Twu and Lee applied metamodelling to four design factors [30]. Ankenman, Bisgaard, and Osswald limited the number of design factors to three (charge size, length-to-width ratio, and position relative to one mould edge), while Kim, Lee, Han, and Vautrin employed the penalty function method and a repair algorithm in the handling of four design variables (size and position of the preform in x and y direction) [25,31].

Thus, promising approaches have been shown addressing the presented challenges, in which a wide range of metamodelling procedures are successfully used. However, to the best of the authors' knowledge, no metamodelling procedures that expand on the geometrical freedom of the preform (e.g., non-rectangular preforms) have been presented in the field of SMC processing.

Therefore, the focus of the presented study is the implementation of a metamodelling approach without inherent geometry restrictions, for which an ensemble metamodel comprising multiple FF-NN is proposed. The procedure directly makes use of the spatial geometry discretisation used in 2D and 2.5D process simulation for the description of the part and preform geometry and approximation of the correlation of the preform and resulting mechanical properties. The metamodel is subsequently used in a procedure for the optimisation of the SMC preform. As the design goal, minimisation of the maximum absolute deflection of a plate geometry under bending load is pursued.

Sampling for FF-NN training is conducted by evaluating rectangular preform geometries (which in sum span the totality of the part surface) and the resulting maximum absolute deflection using a coupled process and structural simulation. In the following subsection, methods used for the metamodelling of preform and part behaviour and subsequent metamodel-based preform optimisation are presented.

The presented procedures were implemented in MATLAB R2020a, Mathworks, Natick, MA, USA, when not stated otherwise.

2. Materials and Methods

The procedures were applied to a plate geometry exhibiting a cantilever load case, which is shown in Figure 1a. The goal of the optimisation of the preform was the minimisation of plate deflection. As this paper concentrates on procedure development, this fairly simple geometry is chosen.

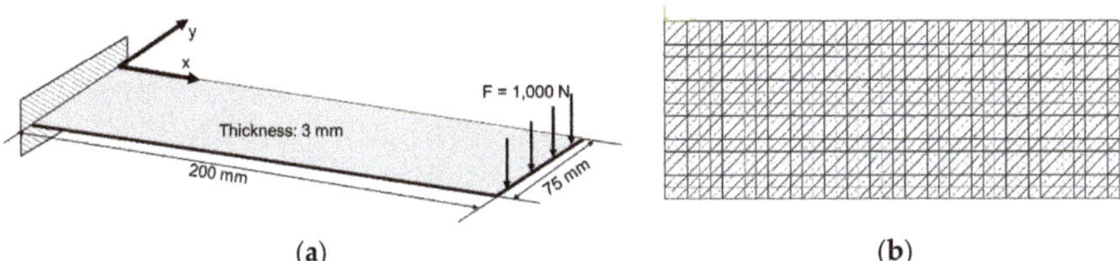

(a) (b)

Figure 1. Minimisation of the plate's maximum absolute deflection: Plate geometry, boundary conditions, and applied load (**a**). Discretisation of plate geometry with 1200 S3 elements (**b**).

Discretisation of the geometry into a structured shell mesh was conducted in Abaqus 2020, Dassault Systèmes, Vélizy-Villacoublay, France, using 1200 S3 elements (Figure 1b). This shell element type was chosen based on compatibility with the process simulation procedure, and subsequently, it was also used in structural simulation and metamodelling.

2.1. Explicit Modelling

Calculation of the fibre orientation probability distribution function (FOD) $\psi(p)$ resulting from SMC flow during compression moulding was conducted using Express 6.0, M-Base Engineering + Software GmbH, Aachen, Germany. This software was developed in close collaboration with the Institute for Plastics Processing (IKV) (Aachen, Germany) and it has been used for a 2.5D process simulation of thermoplastic and thermoset compression moulding alike [36–38]. It is based on the control volume approach as described by Osswald, which has been shown to accurately predict the filling pattern in compression moulding of thin geometries under the assumption of planar flow [12]. The governing equations implemented in the software and used in this paper are described briefly. The material data used are shown in Section 2.2.

To predict the static pressure distribution p within the mould cavity during the compression moulding of SMC, Folgar and Tuckers' application of the generalised Hele–Shaw model is used [12,15,17]:

$$\frac{\partial}{\partial x}\left(S\frac{\partial p}{\partial x}\right) + \frac{\partial}{\partial y}\left(S\frac{\partial p}{\partial y}\right) - \dot{h} = 0 \tag{1}$$

The flow conductivity S is derived from the flow gap height h and the shear and temperature-dependent viscosity η:

$$S = \frac{h^3}{12\eta} \tag{2}$$

From the pressure distribution, gap-wise average velocities $\overline{U}$ and $\overline{V}$ are derived, from which the fibre orientation probability distribution function (FOD) is subsequently calculated [12,15]:

$$\overline{U} = -\frac{S}{h}\frac{\partial p}{\partial x} \tag{3}$$

$$\overline{V} = -\frac{S}{h}\frac{\partial p}{\partial y} \tag{4}$$

Anisotropy of temperature (and therefore viscosity) is taken into account by simultaneously simulating the filling of the geometry with five shell geometries (assumed as each having one-fifth of the thickness of the plate geometry and being stacked on top of each other), for which the momentary temperature change due to conductive heat transfer is considered individually.

From the layer-wise average velocities, fibre orientations are determined for each layer. Jeffery developed a procedure that enables the prediction of change in orientation of a single ellipsoidal particle due to the flow of a surrounding fluid [39]. For the calculation of the FOD of fibre-reinforced materials, Folgar and Tucker supplemented this model with a phenomenological diffusion term, with which fibre–fibre interactions are taken into account by the fibre interaction coefficient C_I [13]:

$$\frac{\partial \psi}{\partial t} = C_I \frac{\partial^2 \psi}{\partial \Phi^2} - \frac{\partial}{\partial \Phi}\left(\psi\left(-\sin\Phi\cos\Phi\frac{\partial v_x}{\partial x} - \sin^2\Phi\frac{\partial v_x}{\partial y} + \cos^2\Phi\frac{\partial v_x}{\partial x} + \sin\Phi\cos\Phi\frac{\partial v_y}{\partial y}\right)\right) \tag{5}$$

In the past, extensive work has been conducted in enhancing this orientation model (e.g., by including fibre–matrix interaction), which has resulted in the development of new models such as Anisotropic Rotary Diffusion (ARD), Reduced Strain Closure (RSC), ARD-RSC, and an improved ARD model combined with the Retarding Principal Rate model (iARD-RPR) [40–45].

In this work, the Folgar–Tucker model is used as recent investigations conducted by Li, Chen et al. continue to show good agreement with the experimentally determined material properties of SMC and the model having been used in prior work on SMC part thickness optimisation by Kim et al. [22,46].

FOD were transferred to MATLAB for calculation of the mechanical properties of each element in each layer. The mechanical properties are calculated by methods described by Advani and Tucker, which are suitable for use in thin-walled compression moulding (thus assuming planar fibre orientation) and subsequently implemented and validated by Oldembo et al. for SMC materials [5,10]. The calculation of mechanical properties is divided into three successive steps. Initially, the fourth-order stiffness tensor C_{ii} (Voigt notation) is calculated under an assumption of unidirectional fibre orientation in the SMC using the governing equations of Halpin and Tsai [47]. Fibre orientation tensors of second order a_{ij} and fourth order a_{ijkl} are subsequently derived from the FODs [5,48]:

$$a_{ij} = \oint p_i p_j \psi(p) dp \tag{6}$$

$$a_{ijkl} = \oint p_i p_j p_k p_l \psi(p) dp. \tag{7}$$

Then, the FOD-dependent stiffness tensors T of each element are calculated by orientation tensor averaging [5]. Scalars B_i are derived from the unidirectional stiffness tensor (governing equations in Appendix A), where δ is the Kronecker-delta [5]:

$$T_{ijkl} = B_1\left(a_{ijkl}\right) + B_2\left(a_{ij}\delta_{kl} + a_{kl}\delta_{ij}\right) + B_3\left(a_{ik}\delta_{jl} + a_{il}\delta_{jk} + a_{jl}\delta_{ik} + a_{jk}\delta_{il}\right) + B_4\left(\delta_{ij}\delta_{kl}\right) + B_5\left(\delta_{ik}\delta_{jl} + \delta_{il}\delta_{jk}\right) \tag{8}$$

For final simulation of part deformation behaviour, the resulting stiffness tensor components were exported to an Abaqus 2020 input-file (.inp) as individually defined materials for each element (thus creating 6000 individual materials) using an automated script. The 5 layers were treated as plies of a composite shell section, which was also created using the script. As fibre orientations are provided in local coordinate systems in the control volume approach, these were also supplied to the .inp file for each element [12].

2.2. Metamodelling

As has been shown in the introduction, a range of different metamodelling procedures such as kriging and NN have been implemented in the field of SMC processing and SMC material description, each exhibiting individual problem-dependent benefits and drawbacks [8,34]. Jin et al. and Simpson et al. evaluated a range of metamodelling procedures, with Simpson et al. recommending the use of NN when dealing with highly nonlinear or large problems containing many parameters, which (as will be shown) is the case in the proposed procedure [30,32,49]. Furthermore, the use of FF-NN has been shown by Twu and Lee et al. to be beneficial in comparison to alternative approaches for metamodelling in the field of SMC optimisation; thus, this approach is used [30]. The cited work is greatly recommended for a more in-depth description of general procedures in setting up and training FF-NN.

Contrary to prior papers presented in the introduction, which use a small number of non-binary input variables, the initial filling state of each element (which may be completely filled or unfilled, thus, binary) is proposed for defining the geometry and position of the preform and for use as input variables (thus, 1200 binary input variables are used in total). However, this results in a high number of network connections, even without considering the further setup of the FF-NN. Training an FF-NN with a high number of connections may lead to a loss in accuracy for out-of-sample data commonly known as overfitting [50]. This is especially prone when using a limited sample size as is the case in simulation-based optimisation of the SMC process [51].

Using a large number of binary input variables is a known procedure in the field of Optical Character Recognition (OCR), in which neural networks are used to detect printed or handwritten letters in black and white images [49,52,53]. Cybenko and Hornik et al. have shown that an FF-NN with a single hidden layer can, when using sigmoid transfer functions, describe a continuous function to an arbitrary degree of accuracy [30,54,55]. However, the necessary number of nodes in the hidden layer and sample sizes have been part of an intensive debate. While general procedures for finding the most optimal parameters (commonly known as hyperparameters) for an FF-NN, such as the robust design methodology proposed by Taguchi, have been used, these parameters are usually set based on experience and trial and error procedures [56,57].

Furthermore, larger sample sizes are preferred, and in the field of OCR of handwriting, extensive databases have been formed [53]. However, sample size is limited in the discussed application on compression moulding due to the calculation effort and general practicability. Therefore, overfitting is assumed as given and may occur regardless of the chosen number of neurons in the hidden layer. However, there are metamodelling approaches that mitigate the effects of overfitting, which will be shown to be successful. Hastie et al. propose the use of ensemble modelling approaches, in which the outputs of multiple FF-NN with identical architecture (but which may each exhibit a different form of overfitting e.g., due to differences in training procedures) are combined to increase the accuracy of the model as a whole [51,58]. In this paper, bootstrap aggregation ("bagging") is implemented, and the mean of the outputs of 100 FF-NN trained with random starting weights is used to predict the maximum absolute deflection of the plate. Thus, only a limited number of neurons are used, and the focus is set on showing the general applicability of this procedure and its accuracy in prediction.

In total, 36 samples consisting of a unique rectangular preform geometry and position and affiliated maximum absolute deflection were created by simulation procedures shown in Section 2.1 for training of the FF-NN (see Figure 2 for a representation of the preform samples). Definition of the samples was based on the following principles:

- Each element should be included in a similar number of samples
- Preform geometries and positions typically used in processing are included
- Preforms have to cover at least 5% of the mould surface
- Preforms maximising flow lengths in the x and y-direction are included

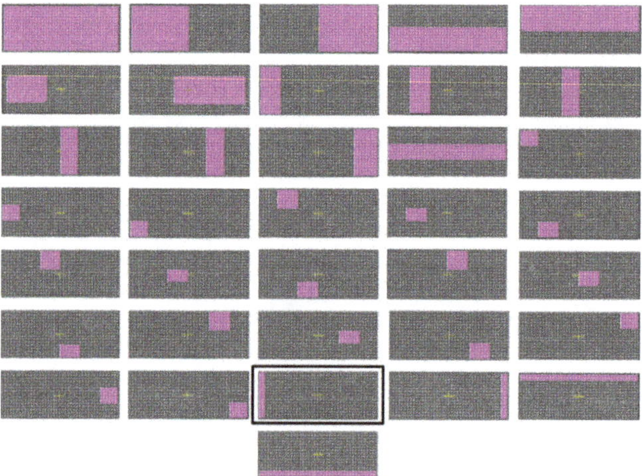

Figure 2. Geometries of preform samples used in feed forward NN (FF-NN) training. Preform resulting in the lowest maximum absolute deflection of the sample set is outlined.

Thus, limiting the samples to only a single geometry was chosen for reasons of consistency. Including alternative geometries based on these principles may be beneficial for increasing the accuracy of the trained FF-NN. However, this would necessarily increase the number of samples, and including all elements in the sample set may not be always possible (e.g., inclusion of corner elements when using round preform geometries).

The material and processing parameters applied are shown in Tables 1 and 2 and Figure 3. Material data correspond with SMC0400 of Menzolit Srl., Turate, Italy [59].

Table 1. Material parameters of SMC0400 [59].

Parameter	Values
Thermal conductivity	0.555 W/(mK)
Heat transfer coefficient (Tool/SMC)	2000 W/(m^{2}K)
Fibre weight fraction	30%
Fibre interaction coefficient C_I	0.070
Elastic modulus fibre	73,000 N/mm^2
Elastic modulus matrix	6250 N/mm^2
Poisson ratio fibre	0.220
Poisson ratio matrix	0.250
Fibre aspect ratio (length/diameter)	3000
Initial fibre orientation	isotropic

Table 2. Process settings [59]. SMC: Sheet Moulding Compounds.

Parameter	Values
Initial SMC temperature	30 °C
Temperature upper mould cavity	150 °C
Temperature lower mould cavity	145 °C
Delay time	45 s
Initial compression speed	10 mm/s
Max. compression force	1000 kN

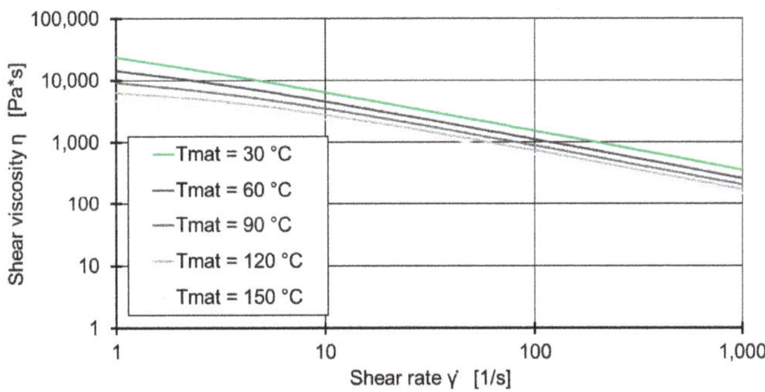

Figure 3. Shear rate and temperature-dependent viscosity [59].

Training of the FF-NN is conducted using the "fitnet" function implemented in the MATLAB Deep Learning Toolbox by Levenberg–Marquardt backpropagation, as this has been shown to be the fastest method in training the FF-NN [60–62]. FF-NN architecture and training parameters are summarised in Table 3. For a complete description of Levenberg–Marquardt backpropagation, refer to the cited paper by Hagan and Menhaj [60].

For initial validation of the metamodel, a comparison of the maximum deflection predicted by the metamodel and calculation by FEM for three preforms not used in FF-NN training (Figure 4) are compared. Further validation is conducted on preform geometries obtained using the optimisation procedure.

Table 3. Architecture of the neural networks (NN) and training parameters.

Variable	Value
Inputs	1200 (binary initial filling states of each element. 1: completely filled and 0: empty)
Outputs	1 (maximum absolute deflection under static load in mm)
Hidden layers	1
Neurons in hidden layer	6
Connection type	Fully connected
Training type	Levenberg–Marquardt
Transfer function input layer to hidden layer	Hyperbolic tangent sigmoid
Transfer function hidden layer to output	Linear
Loss function	Mean squared error
Training epochs (maximum)	1000
Drop-out	none

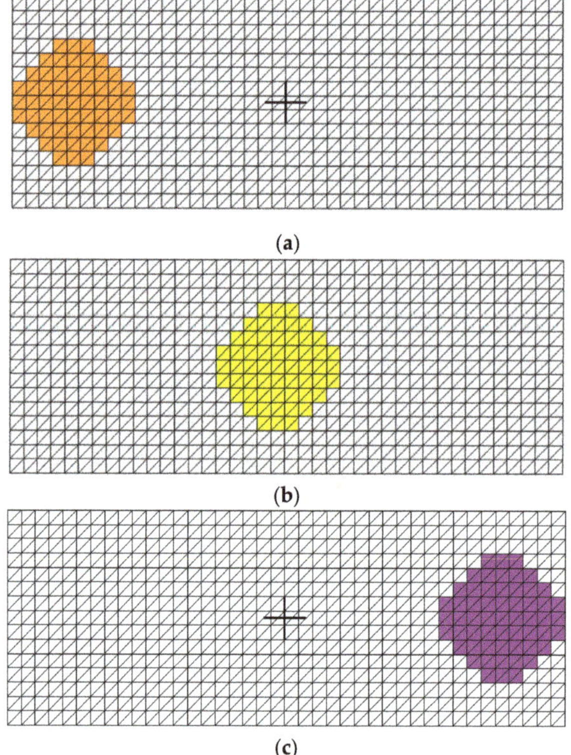

Figure 4. Preform geometries used in initial metamodel validation: validation geometry 1 (**a**), validation geometry 2 (**b**), validation geometry 3 (**c**).

2.3. Optimisation

The optimisation of a performance metric of an SMC-based component can be defined as a multivariable optimisation problem (MVO). The mathematical description of an MVO is [63–65]:

$$min(f(x)), x \in S \tag{9}$$

where $f : R^n \to R$ is the objective function and $x = (x_1, x_2, \ldots, x_n)^T$ is the decision vector belonging to the nonempty feasible region $S \subset R^n$ [63]. In this case, maximum deflection of the plate in z-direction

D_z is treated as the objective function, which is defined by the maximum absolute deflection of all nodes N observed in this direction (compare Figure 5; notation derived from Islam et al.) [66]:

$$f(x) = \max\left|(D_z)_i\right|, \ i = 1, 2, 3, \ldots N. \tag{10}$$

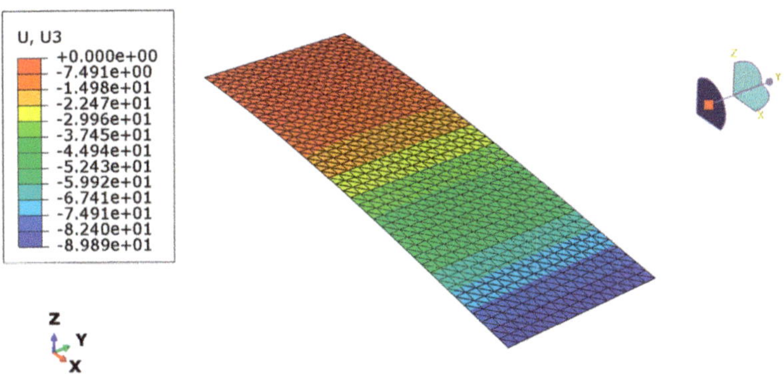

Figure 5. Exemplary deflection result of structural simulation. In this case, the maximum absolute deflection D_z is 89.89 mm.

The components of the decision vector shown in Table 4 correspond with the input variables of the metamodel described previously (initial filling state of each element, which can only be completely filled or empty).

Table 4. Description of the design variables.

Design Variable	Definition	Unit	Lower Bound	Upper Bound
x_1	Initial filling state of element 1	-	0	1
x_2	Initial filling state of element 2	-	0	1
$x_{1,200}$	Initial filling state of element 1200	-	0	1

A challenge in solving MVO is detecting the global minimum of f, for which evolutionary algorithms (EA) such as genetic algorithms (GA) have been used successfully [67–69]. These algorithms may include multiple different operators such as crossover and mutation. Here, a non-standard mutation procedure is implemented, which is based on evaluation of the metamodel (Section 2.2).

The setup of the developed optimisation routine is shown in Figure 6. Using an iterative method presented in the following, a preform geometry and position that minimises the objective function is sought.

(1) Mutation of preform and evaluation of objective function:

Starting elements (which can be a single element or a group of adjacent elements) are mutated iteratively by applying the eight subsequent procedures summarised in Table 5. During these procedures, the preform geometry is increased (or decreased) by the R adjacent, randomly chosen elements (R initially being one) in the specified direction. Decrease procedures are initialised after the minimum preform size (5% of the part surface area coverage) is reached.

After each procedure, the resulting plate deformation of the new preform geometry is evaluated using the metamodel. Mutation is retained if a decrease of the objective function is predicted. To decrease the likelihood in reaching a local minimum (thus not being the most optimal, global solution for the optimisation), R is increased to three if no decrease in deflection is reached during 10 iterations. The mutation is terminated after a total of 125 iterations.

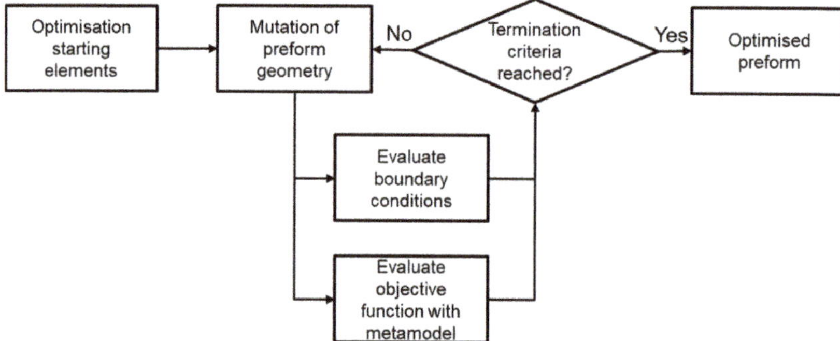

Figure 6. Flow chart of the optimisation procedure.

Table 5. Mutation procedures conducted in each optimisation iteration.

Mutation Procedure	Relative Area Change	Change Direction
1	Increase by R Elements	Positive x-direction
2	Increase by R Elements	Negative x-direction
3	Increase by R Elements	Positive y-direction
4	Increase by R Elements	Negative y-direction
5	Decrease by R Elements	Positive x-direction
6	Decrease by R Elements	Negative x-direction
7	Decrease by R Elements	Positive y-direction
8	Decrease by R Elements	Negative y-direction

(2)   Evaluation of boundary conditions:

In prior conducted studies, problem-dependent constraints and constraint handling techniques had to be implemented [25,31]. As the description of the preform in the presented approach is based on the discretisation of the geometry also used in process and structural simulation, typical constraints such as limiting the preform to the inside of the mould are not necessary.

Two problem-independent constraints (e.g., independent of the part geometry) are implemented, with which the typical processing defects and limitations of the compression moulding process are addressed:

(a)   After each iteration, the mutated preform is automatically checked for the absence of enclosed, empty elements, which can lead to part defects such as air pockets [70]. If this was detected during preform optimisation, mutation was limited to the first four procedures until the enclosed elements were eliminated.

(b)   Preform size needs to exceed at least 5% of the part surface area, thus limiting the height of the preform.

As starting elements, five evenly spaced elements contained inside the FF-NN training sample preform resulting in the smallest maximum deflection are used, as an optimal solution is presumed in this area (see Figure 2 and Figure 9). One of the following two results are expected after successfully running the optimisation procedure:

- Training sample preform is global optima: Training sample is reached regardless of starting element
- Training sample is local optima: Alternative preform is determined, which results in lower maximum plate deflection. This may vary depending on the starting element.

3. Results and Discussion

3.1. Metamodel Validation

In Figure 7, maximum absolute deflections attained by the metamodel and FEM for the validation geometries are compared. Standard deviations and outliers of the 100 individual FF-NN outputs of which the metamodel is composed are also shown. Maximum plate deflections predicted by the metamodel differ by 2.67% (validation geometry 1), 0.26% (validation geometry 2), and 0.82% (validation geometry 3) from values obtained by FEM, respectively. Therefore, plate deflections are predicted accurately by the metamodel.

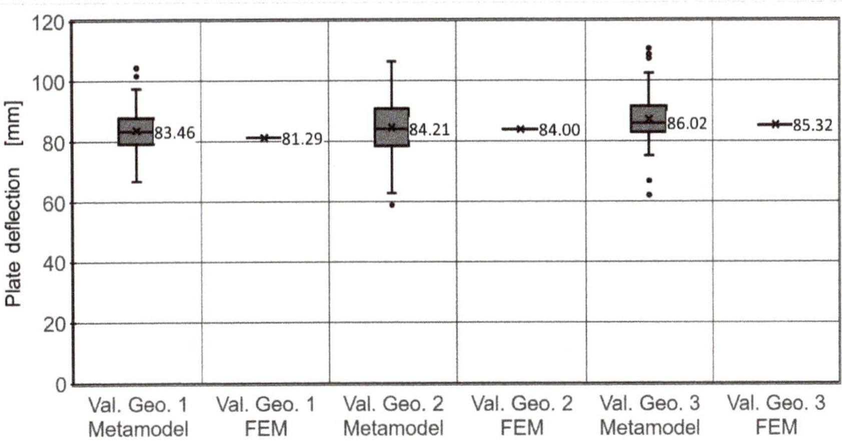

Figure 7. Comparison of plate deflections for the three preform validation geometries obtained by the metamodel and FEM.

As expected, the individual FF-NN included in the metamodel exhibit a high spread in outputs, exceeding 50% of the metamodel output value (e.g., total spread in predicted deflections for validation geometry 3: 48.47 mm), which is attributed to overfitting during the training process (see Section 2.2). No significant influence of the preform position on the spread of the individual FF-NN outputs can be detected. Potentials for decrease in spread include increasing the sample set size and implementing NN validation procedures; however, these would significantly increase the computational effort.

To further evaluate the decrease in plate deflection from 85.32 mm to 81.29 mm which results from decreasing the charge distance from the clamping location, fibre orientations are compared. Fibre orientation tensor component a_{xx}, which is visualised in Figure 8, describes the probability of fibre orientation in the x-axis direction [5]. For a decrease in the charge distance from clamping location results, a decrease of this tensor component in the left half of the plate is observed, which reduces the local flexural modulus and therefore the overall bending stiffness of the plate [71]. Although validation geometries used are symmetric relative to the principal axis of the plate, fibre orientations calculated by FEM are not symmetric relative to this axis, which may result from the non-symmetry and coarseness of the mesh used [72].

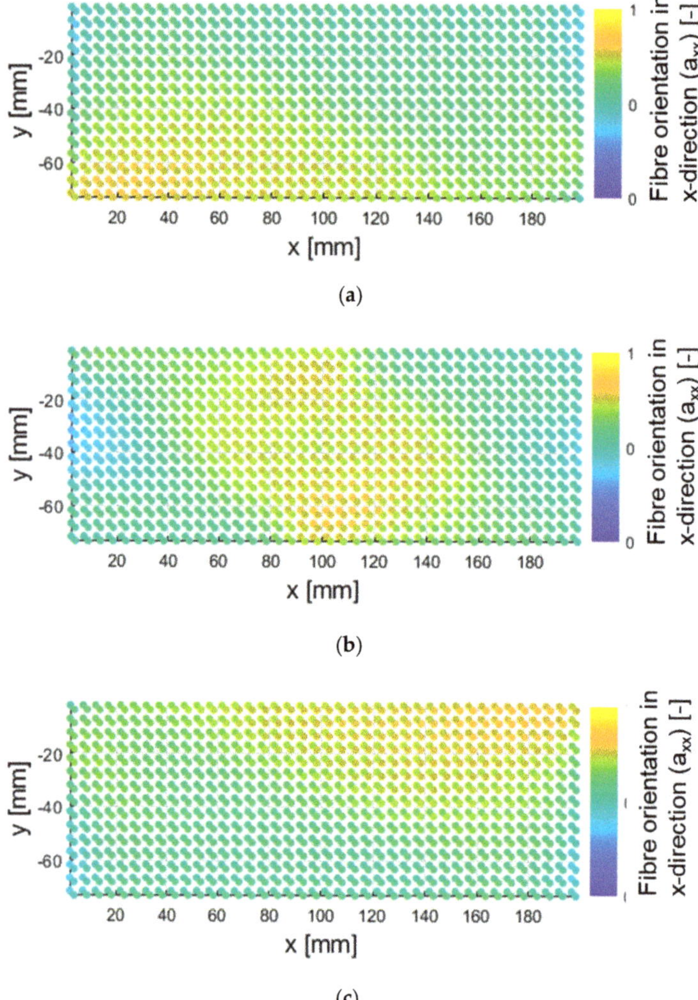

Figure 8. Comparison of fibre orientation in the x-direction for the three preform validation geometries obtained by process simulation. (**a**) = Validation geometry 1; (**b**) = Validation geometry 2; (**c**) = Validation geometry 3.

3.2. Preform Optimisation

The starting elements of the performed preform optimisations and resulting preform geometries are shown in Figure 9. Preform geometries resulting from adjacent starting points show a similarity in size and geometry; however, these are not identical in any case. Convergence of the objective function is presented in Figure 10. However, it has to be clear that the early generations do not represent valid solutions, as the minimum preform size (Section 2.3, constraint 2b) is only reached during the final generations.

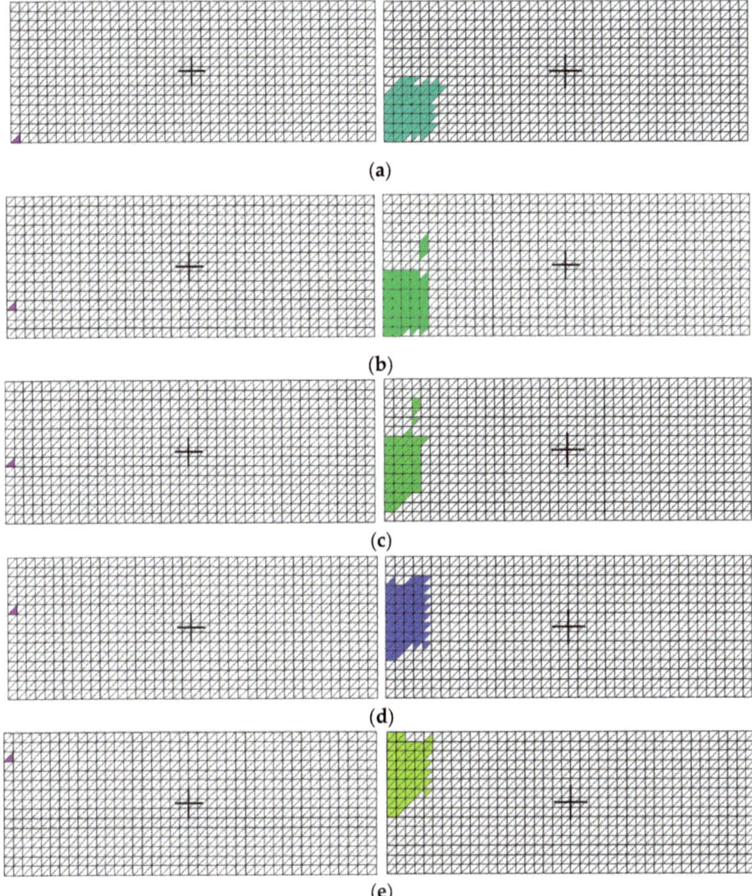

Figure 9. Initial starting elements (left) and resulting optimised preform geometries (right) of the performed optimisations. (**a**) Starting element 1, (**b**) Starting element 2, (**c**) Starting element 3, (**d**) Starting element 4, and (**e**) Starting element 5.

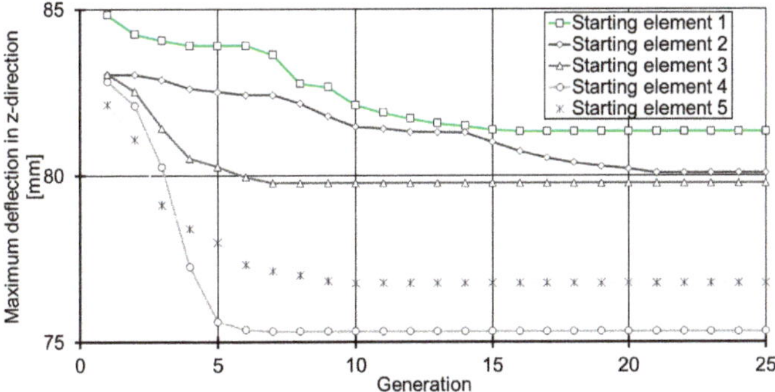

Figure 10. Convergence of objective function (maximum plate deflection in z-direction) during preform optimisation.

Although 125 iterations were conducted in each case, no decrease in the objective function value or further change in preform geometry is detected for any starting element from 25 generations onwards. The run time was under five minutes respectively. One can see a high decrease in maximum deflection during the initial iterations, with the tapering off of the attained decrease going further. The minima of the objective function range from 75.3 (Starting element 4) to 81.3 mm (Starting element 1) (Figure 9). Similar to the validation geometries, comparison of the metamodel output with FEM results again confirm accurate prediction by the metamodel, with deviations ranging from 0.46% (Starting element 4) to 2.14% (Starting element 1) (Figure 11).

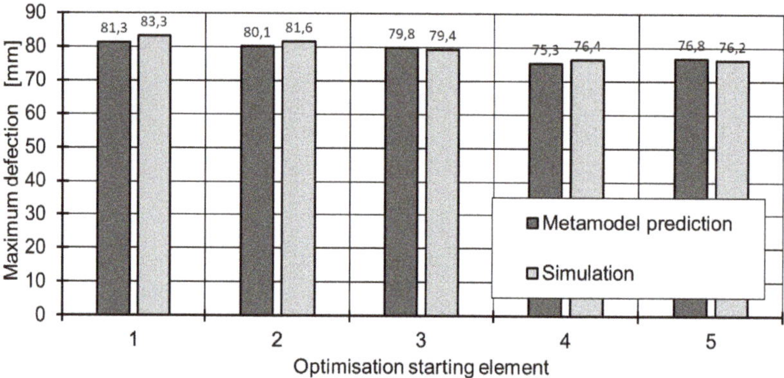

Figure 11. Comparison of deflections obtained by metamodel and FEM for optimised preforms.

As different preforms are achieved depending on the starting element and only one global minima is presumed to exist, optimised preforms represent the local minima of the MVO. Further comparison of the obtained values for the objective function (Starting element 5) with maximum deflection of the sample from which the starting elements were initially taken (Figure 2) show that the algorithm was not capable in reaching this more optimal solution (in comparison, the highest deflection of all the samples was 114.9 mm). This sample is in contact with the full length of the left plate edge, representing the highest achievable flow length while fulfilling the minimum mould coverage defined in the optimisation. While reaching the geometry of the mentioned sample may be possible when strongly increasing the number of conducted iterations, the function of the optimisation algorithm is restricted while approaching it due to it having the minimum mould coverage (5%) for conducting a valid mutation step. The sharp edges of the optimised geometries are a result of the use of S3 elements, and these could be combatted by increasing the element count, adding additional constraints to the optimisation procedure, or using alternative process simulation approaches. Additionally, an additional constraint for avoiding two preforms from forming should be implemented (as is the case in preform (b)), as these may lead to weld lines and should be avoided.

4. Conclusions

The development and refinement of metamodelling (or surrogate modelling) approaches is a beneficial step in expanding the capabilities in simulation-based SMC compression moulding process optimisation and reducing computational demand.

In this paper, an ensemble metamodelling approach is proposed, in which the spatial discretisation necessary for process and structural simulation is exploited. Hereby, the initial filling states of each element are used as input variables for the metamodel. To mitigate the effects of overfitting, an ensemble modelling approach is used in which the mean outputs of 100 FF-NN is used as output of the metamodel. Training of the FF-NN is conducted on datasets obtained by process and structural simulation with random starting weights. Contrary to metamodelling approaches successfully implemented in the past, this approach enables defining the preform without inherent geometry restrictions, as viable geometries are only dependent on the discretisation itself.

The approach is used to predict the preform geometry and position-dependent maximum deflection of a plate geometry under cantilever bending load. Maximum absolute deflection can be accurately predicted by the metamodel, with deviations between metamodel prediction and FEM validation ranging from 0.26% to 2.67%.

The usability of the metamodel in a subsequently conducted preform optimisation routine can be shown, but it is believed to be limited by the closeness of local optima to each other and chosen boundary conditions for mould coverage. For purpose of procedure development, a fairly simple plate geometry was chosen, limiting the potential in finding non-obvious solutions.

Further work will focus on the evaluation of alternative optimisation routines, which make use of the prediction accuracy of this metamodelling approach more efficiently. These could be derived from methodologies found in topology optimisation. One method that could be applicable to the metamodel is Solid Isotropic Material with Penalisation (SIMP), which was initially proposed by Bendsoe and Kikuchi [73]. An additional focus will be the application to parts with higher geometric complexity and alternative load cases, for which the potential of deriving non-apparent preform geometries and positions and thus potential for industrial applicability is higher. Subsequently, comparison with experimentally obtained part behaviour will be conducted.

Author Contributions: J.N. and R.L.P.G. conceived and implemented the methodologies and procedures. J.N. prepared the original draft and visualisations. C.H., K.F. and D.S. reviewed and edited the original draft and provided project supervision. All authors have read and agreed to the published version of the manuscript.

Funding: The investigations set out in this report received financial support from the European Regional Development Fund (No.: EFRE-0801121), to whom we extend our thanks.

Conflicts of Interest: The authors declare no conflict of interest. The funders had no role in the design of the study; in the collection, analyses, or interpretation of data; in the writing of the manuscript, or in the decision to publish the results.

Appendix A

Calculation of scalars B_i for calculation of orientation tensor averaged stiffness tensor from unidirectional stiffness tensor C_{ii} (written in compacted notion) [11]:

$$B_1 = C_{11} + C_{22} - 2C_{12} - 4C_{66} \tag{A1}$$

$$B_2 = C_{12} - C_{23} \tag{A2}$$

$$B_3 = C_{66} + \frac{1}{2}(C_{23} - C_{22}) \tag{A3}$$

$$B_4 = C_{23} \tag{A4}$$

$$B_5 = \frac{1}{2}(C_{22} - C_{23}) \tag{A5}$$

References

1. Witten, E.; Mathes, V.; Sauer, M. *Composites—Marktbericht 2018 Marktentwicklungen, Trends, Ausblicke Und Herausforderungen*; Industrievereinigung Verstärkte Kunststoffe (AVK): Frankfurt, Germany, 2018.
2. Wang, C.W.; Tsai, W.C.; Sun, S.B.; Hsu, C.H.; Chang, R. Optimizing process condition of compression molding: From material properties characterization to numerical simulation. In Proceedings of the Annual Technical Conference (ANTEC), Indianapolis, IN, USA, 23–25 May 2016; Society of Plastics Engineers: Brookfield, CT, USA; pp. 463–468.
3. Advani, S.G.; Sozer, E.M. *Process Modeling in Composites Manufacturing*; Informa UK Limited: London, UK, 2010.
4. Marjavaara, B.D.; Ebermark, S.; Lundström, T.S. Compresson moulding simulations of SMC using a multiobjective surrogate-based inverse modeling approach. *Mech. Compos. Mater.* **2009**, *45*, 503–514. [CrossRef]

5.   Advani, S.G. The Use of Tensors to Describe and Predict Fiber Orientation in Short Fiber Composites. *J. Rheol.* **1987**, *31*, 751–784. [CrossRef]

6.   Sommer, D.E.; Favaloro, A.J.; Pipes, R.B. Coupling anisotropic viscosity and fiber orientation in applications to squeeze flow. *J. Rheol.* **2018**, *62*, 669–679. [CrossRef]

7.   Sommer, D.E.; Kravchenko, O.G.; Denos, B.; Favaloro, A.J.; Pipes, R.B. Integrative analysis for prediction of process-induced, orientation-dependent tensile properties in a stochastic prepreg platelet molded composite. *Compos. Part A Appl. Sci. Manuf.* **2020**, *130*, 105759. [CrossRef]

8.   Sabiston, T.; Inal, K.; Lee-Sullivan, P. Application of Artificial Neural Networks to predict fibre orientation in long fibre compression moulded composite materials. *Compos. Sci. Technol.* **2020**, *190*, 108034. [CrossRef]

9.   Kluge, N.J.; Lundström, T.S.; Westerberg, L.G.; Olofsson, K. Compression moulding of sheet moulding compound: Modelling with computational fluid dynamics and validation. *J. Reinf. Plast. Compos.* **2015**, *34*, 479–492. [CrossRef]

10.   Oldenbo, M.; Mattsson, D.; Varna, J.; Berglund, L.A. Global Stiffness of a SMC Panel Considering Process Induced Fiber Orientation. *J. Reinf. Plast. Compos.* **2004**, *23*, 37–49. [CrossRef]

11.   Martulli, L.M.; Muyshondt, L.; Kerschbaum, M.; Pimenta, S.; Lomov, S.V.; Swolfs, Y. Carbon fibre sheet moulding compounds with high in-mould flow: Linking morphology to tensile and compressive properties. *Compos. Part A Appl. Sci. Manuf.* **2019**, *126*. [CrossRef]

12.   Osswald, T.A. Numerical Methods for Compression Mold Filling Simulation. Ph.D. Thesis, University of Illinois, Champaign, IL, USA, 1987.

13.   Folgar, F.; Tucker, C.L. Orientation Behavior of Fibers in Concentrated Suspensions. *J. Reinf. Plast. Compos.* **1984**, *3*, 98–119. [CrossRef]

14.   Osswald, T.A.; Tucker, C.L. Compression Mold Filling Simulation for Non-Planar Parts. *Int. Polym. Process.* **1990**, *5*, 79–87. [CrossRef]

15.   Tucker, C.L.; Folgar, F. A model of compression mold filling. *Polym. Eng. Sci.* **1983**, *23*, 69–73. [CrossRef]

16.   Osswald, T.A.; Tucker, C.L. A boundary element simulation of compression mold filling. *Polym. Eng. Sci.* **1988**, *28*, 413–420. [CrossRef]

17.   Hele-Shaw, H.S. The Motion of a Perfect Fluid. In *The Weekly Evening Meeting*; Wentworth Press: Sydney, Australia, 1899.

18.   Lee, C.-C.; Folgar, F.; Tucker, C.L. Simulation of Compression Molding for Fiber-Reinforced Thermosetting Polymers. *J. Eng. Ind.* **1984**, *106*, 114–125. [CrossRef]

19.   Barone, M.R.; Caulk, D.A. A Model for the Flow of a Chopped Fiber Reinforced Polymer Compound in Compression Molding. *J. Appl. Mech.* **1986**, *53*, 361–371. [CrossRef]

20.   Barone, M.R.; Caulk, D.A. Kinematics of flow in sheet molding compounds. *Polym. Compos.* **1985**, *6*, 105–109. [CrossRef]

21.   Kühn, C.; Walter, I.; Täger, O.; Osswald, T. Simulative Prediction of Fiber-Matrix Separation in Rib Filling During Compression Molding Using a Direct Fiber Simulation. *J. Compos. Sci.* **2017**, *2*, 2. [CrossRef]

22.   Li, Y.; Chen, Z.; Xu, H.; Dahl, J.; Zeng, D.; Mirdamadi, M.; Su, X. Modeling and Simulation of Compression Molding Process for Sheet Molding Compound (SMC) of Chopped Carbon Fiber Composites. *SAE Int. J. Mater. Manuf.* **2017**, *10*, 130–137. [CrossRef]

23.   Hohberg, M. *Experimental Investigation and Process Simulation of the Compression Molding Process of Sheet Molding Compound (SMC) with Local Reinforcements*; Elsevier: Amsterdam, The Netherlands, 2018.

24.   Meyer, N.; Schöttl, L.; Bretz, L.; Hrymak, A.; Kärger, L. Direct Bundle Simulation approach for the compression molding process of Sheet Molding Compound. *Compos. Part A Appl. Sci. Manuf.* **2020**, *132*, 105809. [CrossRef]

25.   Ankenman, B.; Bisgaard, S.; Osswald, T.A. Experimental Optimization of Computer Models of Manufacturing Processes. *Manuf. Rev.* **1994**, *7*, 332–344.

26.   Guo, X.; Li, W.; Iorio, F. Convolutional Neural Networks for Steady Flow Approximation. In Proceedings of the 22nd ACM SIGKDD International Conference on Knowledge Discovery and Data Mining, San Francisco, CA, USA, 13–17 August 2016; Association for Computing Machinery: New York, NY, USA, 2016; pp. 481–490.

27.   Michaeli, W.; Johannaber, F. *Handbuch Spritzgießen*, 2nd ed.; Carl Hanser Verlag: Munich, Germany, 2004; Available online: https://www.hanser-fachbuch.de/buch/Handbuch+Spritzgiessen/9783446229662 (accessed on 29 June 2020).

28. Tompson, J.; Schlachter, K.; Sprechmann, P.; Perlin, K. Accelerating Eulerian Fluid Simulation with Convolutional Networks. In Proceedings of the 34th International Conference on Machine Learning, Sunday, Australia, 10–11 August 2017; International Convention Centre: Sydney, Australia; PMLR: Sydney, Australia, 2017; Volume 70, pp. 3424–3433.

29. Ladicky, L.; Jeong, S.; Solenthaler, B.; Pollefeys, M.; Gross, M. Data-driven fluid simulations using regression forests. *ACM Trans. Graph.* **2015**, *34*, 1–9. [CrossRef]

30. Twu, J.-T.; Lee, L.J. Application of artificial neural networks for the optimal design of sheet molding compound (SMC) compression molding. *Polym. Compos.* **1995**, *16*, 400–408. [CrossRef]

31. Kim, M.-S.; Lee, W.I.; Han, W.-S.; Vautrin, A. Optimisation of location and dimension of SMC precharge in compression moulding process. *Comput. Struct.* **2011**, *89*, 1523–1534. [CrossRef]

32. Jin, R.; Chen, W.; Simpson, T. Comparative studies of metamodelling techniques under multiple modelling criteria. *Struct. Multidiscip. Optim.* **2001**, *23*, 1–13. [CrossRef]

33. Tizzard, K.; Kleijnen, J.P.C. Statistical Tools for Simulation Practitioners. *J. R. Stat. Soc. Ser. A (Stat. Soc.)* **1988**, *151*, 374. [CrossRef]

34. Chen, Z.; Huang, T.; Shao, Y.; Li, Y.; Xu, H.; Avery, K.; Zeng, D.; Chen, W.; Su, X. Multiscale finite element modeling of sheet molding compound (SMC) composite structure based on stochastic mesostructure reconstruction. *Compos. Struct.* **2018**, *188*, 25–38. [CrossRef]

35. Twu, J.-T.; Lee, L.J.; Chen, S.-C. Simulation-based design of sheet molding compound (SMC) compression molding. *Polym. Compos.* **1994**, *15*, 313–326. [CrossRef]

36. Michaeli, W.; Kremer, C. Predicting the surface waviness of sheet moulding compound parts by simulating process-induced thermal stresses. *J. Polym. Eng.* **2011**, *31*, 561–566. [CrossRef]

37. Semmler, E. Simulation of the Mechanical and Thermomechanical Behavior of Thermoplastic Fiber-reinforced Compression Moulded Parts. Ph.D. Thesis, RWTH Aachen University, Aachen, Germany, 1998.

38. Specker, O. Compression Moulding of SMC: Computer Simulations for Computer-Aided Lay-Out of the Process and for Determination of the Component Properties. Ph.D. Thesis, RWTH Aachen University, Aachen, Germany, 1991.

39. Jeffery, G.B. The motion of ellipsoidal particles immersed in a viscous fluid. *Proc. R. Soc. Lond. Ser. A Math. Phys. Sci.* **1922**, *102*, 161–179. [CrossRef]

40. Phelps, J.H.; Tucker, C.L. An anisotropic rotary diffusion model for fiber orientation in short- and long-fiber thermoplastics. *J. Non Newton. Fluid Mech.* **2009**, *156*, 165–176. [CrossRef]

41. Wang, M.-L.; Chang, R.-Y.; Hsu, C.-H.D. *Molding Simulation: Theory and Practice*; Carl Hanser Verlag: Munich, Germany, 2018.

42. Wang, J.; O'Gara, J.F.; Tucker, C.L.; Iii, C.L.T. An objective model for slow orientation kinetics in concentrated fiber suspensions: Theory and rheological evidence. *J. Rheol.* **2008**, *52*, 1179–1200. [CrossRef]

43. Gandhi, U.N.; Goris, S.; Osswald, T.A.; Song, Y.Y. *Discontinuous Fiber—Reinforced Composites—Fundamentals and Applications*; Carl Hanser Verlag: Munich, Germany, 2020.

44. Tseng, H.-C.; Chang, R.-Y.; Hsu, C.-H. Numerical prediction of fiber orientation and mechanical performance for short/long glass and carbon fiber-reinforced composites. *Compos. Sci. Technol.* **2017**, *144*, 51–56. [CrossRef]

45. Tseng, H.-C.; Chang, R.-Y.; Hsu, C.-H. An objective tensor to predict anisotropic fiber orientation in concentrated suspensions. *J. Rheol.* **2016**, *60*, 215–224. [CrossRef]

46. Kim, M.S.; Lee, W.I.; Han, W.-S.; Vautrin, A.; Park, C.H. Thickness optimization of composite plates by Box's complex method considering the process and material parameters in compression molding of SMC. *Compos. Part A Appl. Sci. Manuf.* **2009**, *40*, 1192–1198. [CrossRef]

47. Affdl, J.C.H.; Kardos, J.L. The Halpin-Tsai equations: A review. *Polym. Eng. Sci.* **1976**, *16*, 344–352. [CrossRef]

48. Chen, Z.; Tang, H.; Shao, Y.; Sun, Q.; Zhou, G.; Li, Y.; Xu, H.; Zeng, D.; Su, X. Failure of chopped carbon fiber Sheet Molding Compound (SMC) composites under uniaxial tensile loading: Computational prediction and experimental analysis. *Compos. Part A Appl. Sci. Manuf.* **2019**, *118*, 117–130. [CrossRef]

49. Simpson, T.; Poplinski, J.; Koch, P.N.; Allen, J. Metamodels for Computer-based Engineering Design: Survey and recommendations. *Eng. Comput.* **2001**, *17*, 129–150. [CrossRef]

50. Webb, G.I. Overfitting. In *Encyclopedia of Machine Learning*; Sammut, C., Webb, G.I., Eds.; Springer: Boston, MA, USA, 2010; p. 744.

51. Grant, P. *Assessment and Selection*; Springer Science and Business Media LLC: Berlin/Heidelberg, Germany, 2014.

52. Mori, S.; Nishida, H.; Yamada, H. *Optical Character Recognition*, 1st ed.; John Wiley & Sons, Inc.: Hoboken, NJ, USA, 1999.
53. Starzyk, J.A.; Ansari, N. Feedforward neural network for handwritten character recognition. In Proceedings of the 1992 IEEE International Symposium on Circuits and Systems, San Diego, CA, USA, 10–13 May 1992; Institute of Electrical and Electronics Engineers (IEEE): Miami, FL, USA, 2003; Volume 6, pp. 2884–2887.
54. Hornik, K.; Stinchcombe, M.; White, H. Multilayer feedforward networks are universal approximators. *Neural Netw.* **1989**, *2*, 359–366. [CrossRef]
55. Cybenko, G. Approximation by superpositions of a sigmoidal function. *Math. Control. Signals Syst.* **1989**, *2*, 303–314. [CrossRef]
56. Peterson, G.E.; Clair, D.S.; Aylward, S.R.; Bond, W.E. Using Taguchi's method of experimental design to control errors in layered perceptrons. *IEEE Trans. Neural Netw.* **1995**, *6*, 949–961. [CrossRef]
57. Suzuki, K. *Artificial Neural Networks—Architectures an Applications*; Suzuki, K., Ed.; InTech: London, UK, 2017.
58. Kotu, V.; Deshpande, B. *Predictive Analytics and Data Mining*; Morgan Kaufmann: Burlington, MA, USA, 2015.
59. Kremer, C. Vorhersage der Oberflächenwelligkeit von Bauteilen aus Sheet Moulding Compound durch die Simulation der Prozessinduzierten Eigenspannung. Ph.D. Thesis, RWTH Aachen University, Aachen, Germany, 2011.
60. Hagan, M.; Menhaj, M. Training feedforward networks with the Marquardt algorithm. *IEEE Trans. Neural Netw.* **1994**, *5*, 989–993. [CrossRef]
61. Kipli, K.; Muhammad, M.S.; Masra, S.M.W.; Zamhari, N.; Lias, K.; Mat, D.A.A. Performance of Levenberg-Marquardt Backpropagation for Full Reference Hybrid Image Quality Metrics. In Proceedings of the International Multiconference of Engineers and Computer Scientists, Hong Kong, China, 14–16 March 2012; pp. 704–707.
62. Kermani, B.G.; Schiffman, S.S.; Nagle, H.T. Performance of the Levenberg—Marquardt neural network training method in electronic nose applications. *Sens. Actuators B Chem.* **2005**, *110*, 13–22. [CrossRef]
63. Branke, J.; Deb, K.; Miettinen, K.; Slowiński, R. *Multiobjective Optimization—Interactive and Evolutionary Approaches*, 1st ed.; Springer: Berlin/Heidelberg, Germany, 2008.
64. Andradóttir, S. A review of simulation optimization techniques. In Proceedings of the 1998 Winter Simulation Conference (Cat. No. 98CH36274), Washington, DC, USA, 13–16 December 1998; Institute of Electrical and Electronics Engineers (IEEE): Piscataway, NJ, USA, 2002; Volume 1, pp. 151–158.
65. Luenberger, D.G.; Ye, Y. *Linear and Nonlinear Programming*; Springer Science and Business Media LLC: Berlin/Heidelberg, Germany, 2008.
66. Islam, M.; Buijk, A.; Rais-Rohani, M.; Motoyama, K. Simulation-based numerical optimization of arc welding process for reduced distortion in welded structures. *Finite Elem. Anal. Des.* **2014**, *84*, 54–64. [CrossRef]
67. Zitzler, E.; Deb, K.; Thiele, L. Comparison of Multiobjective Evolutionary Algorithms: Empirical Results. *Evol. Comput.* **2000**, *8*, 173–195. [CrossRef] [PubMed]
68. Fazilati, J.; Alisadeghi, M. Multiobjective crashworthiness optimization of multi-layer honeycomb energy absorber panels under axial impact. *Thin Walled Struct.* **2016**, *107*, 197–206. [CrossRef]
69. Sakawa, M. *Genetic Algorithms and Fuzzy Multiobjective Optimization*; Springer Science and Business Media LLC: Berlin/Heidelberg, Germany, 2002.
70. Rudd, C.D. *Composites for Automotive Applications*, 11th ed.; Rapra Technology Ltd.: Shawbury, UK, 2000.
71. Trauth, A.; Pinter, P.; Weidenmann, K.A. Investigation of Quasi-Static and Dynamic Material Properties of a Structural Sheet Molding Compound Combined with Acoustic Emission Damage Analysis. *J. Compos. Sci.* **2017**, *1*, 18. [CrossRef]
72. Katz, A.; Sankaran, V. Mesh quality effects on the accuracy of CFD solutions on unstructured meshes. *J. Comput. Phys.* **2011**, *230*, 7670–7686. [CrossRef]
73. Bendsoe, M.P.; Kikuchi, N. Generating optimal topologies in structural design using a homogenization method. *Comput. Methods Appl. Mech. Eng.* **1988**, *71*, 197–224. [CrossRef]

Journal of
Composites Science

Article

A Force-Balanced Fiber Retardation Model to Predict Fiber-Matrix-Separation during Polymer Processing

Christoph Kuhn * and Simon Wehler

Volkswagen Group, 38442 Wolfsburg, Germany; simon.wehler@volkswagen.de
* Correspondence: christoph.kuhn@volkswagen.de

Received: 22 September 2020; Accepted: 27 October 2020; Published: 1 November 2020

Abstract: The use of discontinuous fiber reinforced composites in injection and compression molding faces a number of challenges regarding process-induced changes in microstructure, which have a significant influence on the mechanical properties of the final component. The changes in final microstructure are caused by complex fiber movements, such as fiber orientation, attrition and accumulation during flow. While there are existing phenomenological prediction models for both fiber orientation and attrition, the prediction of fiber accumulation due to fiber-matrix separation is currently only possible with a complex mechanistic particle simulation, which is not applicable in industrial simulations. A simplified phenomenological model, the fiber retardation model (FRM), for the prediction of fiber-matrix separation in commercially available software tools is presented in this paper. The model applies a force balance onto an interacting two phase flow of polymer melt and fiber phase and applies a retardation factor K to calculate the slowing and accumulation of the fiber phase. The general model is successfully applied to a simple compression molding simulation.

Keywords: plastics processing; fiber reinforced plastics; composites; glass fiber; sheet molding compound; long fiber; fiber orientation; fiber content

---

1. Introduction

Discontinuous fiber reinforced polymer composites in compression molding have been applied successful in a variety of industries due to their cost-efficient processing of complex parts and their great mechanical properties to weight ratio. Their availability in a variety of polymer resins (thermoplastics and thermosets) as well as their reinforcement fibers (glass, carbon, natural, etc.) satisfy a wide range of requirements for industrial application, e.g., within the automotive and aviation industry. The design of components with fiber reinforced composites faces the challenges of anisotropic and fluctuating properties inside a component due to changes in the local microstructure during processing [1]. These changes are caused by complex fiber behavior such as fiber orientation, attrition and accumulation during the polymer flow. Simulative models are generally applied to predict these changes and improve component design, while the prediction of fiber orientation and attrition has been the main focus in the past [1–3].

With longer reinforcement fibers as traditionally found in compression molding of sheet molding compound (SMC) or long fiber thermoplastics (LFT), increasing fiber interactions lead to fiber matrix phase separation during polymer flow and hence to fluctuations in fiber content inside complex components [1–4]. Early experiments by Schmachtenberg et al. [5] showed increasing fiber-matrix separation during the compression molding with thermoset sheet molding compounds in simple plate geometries, as shown in Figure 1. Schmachtenberg's experiments display a relative fiber-free flow front followed by a section of increased fiber content.

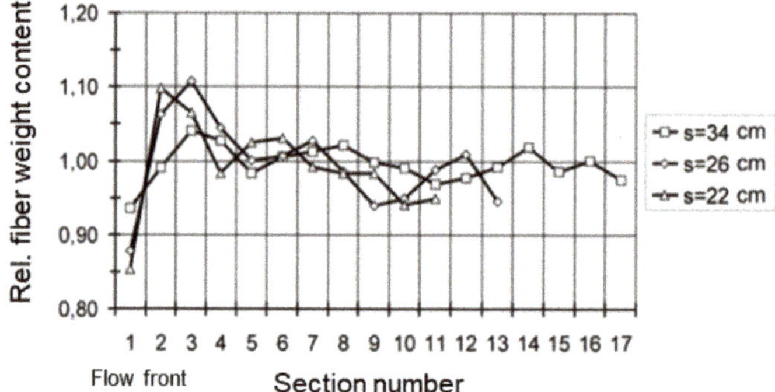

Figure 1. Relative change of fiber content distribution inside a compression molded sheet molding compound (SMC) plate from the melt front (1) to the initial charge position (17) with different flow length 's' [5].

Recent experiments by Kuhn et al. [6] show a similar behavior with glass mat reinforced thermoplastics (GMT), as shown in Figure 2. Kuhn et al. further showed the increasing fiber-matrix separation in complex rib regions.

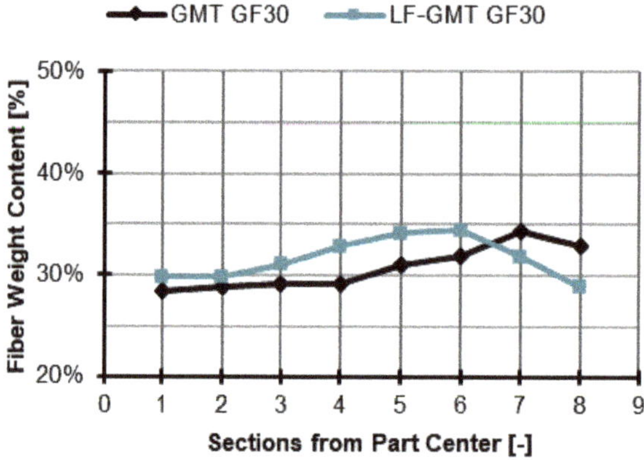

Figure 2. Change of fiber content inside a plate region with glass mat reinforced thermoplastics (GMT) GF30 and long fiber GMT GF30.

The effect of fiber-matrix separation leads to significant decreases in mechanical properties and eventually to unexpected component failure. There are simulative models available by Morris and Boulay [7] to predict local changes of fiber content through shear migration in the normal direction of a polymer flow, which are unable to predict global changes in different component regions. Currently, these global effects can only be predicted with complex particle-level simulations, such as the Mechanistic Model, as displayed in Figure 3, where the fibers are simulated as beam elements inside the polymer flow, capable of displaying complex mechanical interactions with the polymer melt, other fibers and the mold walls [8–14].

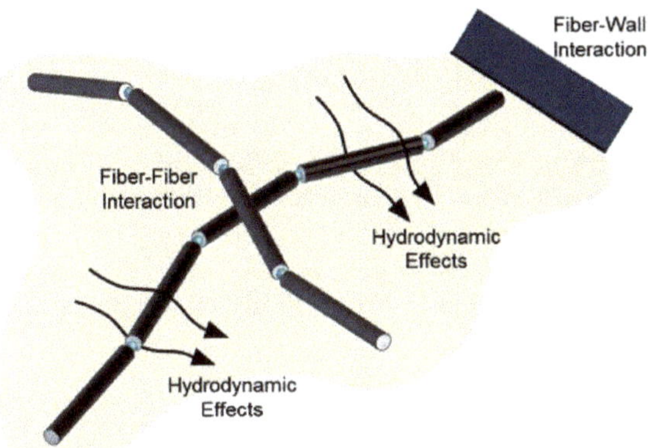

Figure 3. Mechanistic model approach with fibers displayed as beam elements inside the polymer flow [6].

Peréz et al. applied the Mechanistic Model to compression molding with long fiber reinforced composites with fiber distribution results generally complying with experiments, as shown in Figure 4 [9].

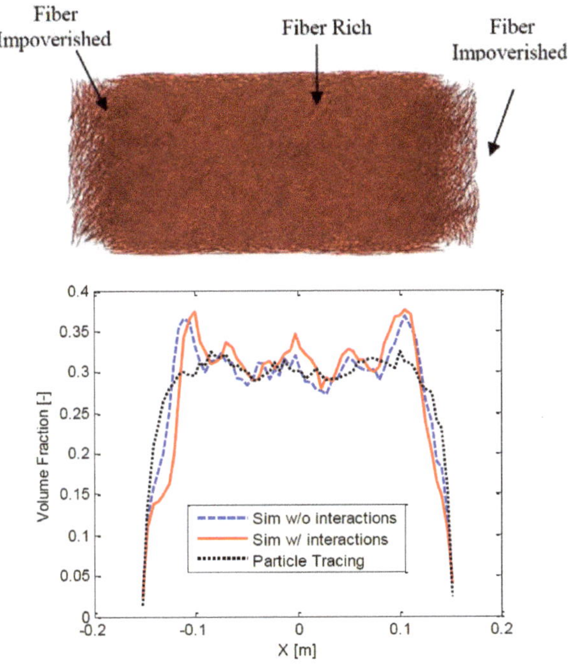

Figure 4. Mechanistic model predictions of fiber volume fraction over distance from the part center 'X' during biaxial flow in compression molding with long fiber reinforced plastics [9].

While the Mechanistic Model is capable to predict the effect of fiber matrix-separation very accurately, it is currently inapplicable to most industrial applications due to the simulative complexity, with extensive pre-and post-processing and long calculation time even with high computational capacity. Therefore, a simplified model for fiber-matrix separation is necessary to account for these effects during manufacturing, comparable to existing fiber orientation and fiber attrition models.

2. General Fiber Retardation Model

The general interactions implemented in the complex Mechanistic Model are applied to a two-phase flow model to generate a simplified fiber retardation model (FRM). Initially, a two-phase flow consisting of the polymer phase and the fiber phase is generated. In order to allow relative motion between the fiber and polymer melt phase, each phase respectively travels at its own velocity, as shown in Figure 5. For the application of the model for commercial software, it is assumed that the velocity of the melt phase u is provided.

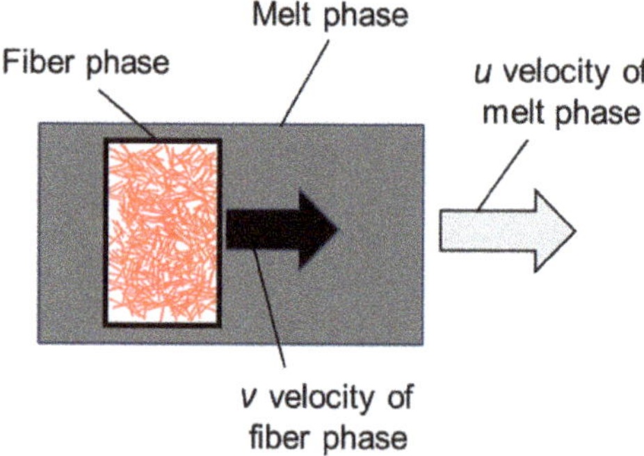

Figure 5. Two-phase model for relative motion between fiber and polymer melt phase.

Regarding the fiber phase in Figure 6, it is obvious that any movements is governed by the surrounding polymer melt. The movement of the fiber phase is based on the forces onto the fiber phase, the hydrodynamic melt forces during polymer flow F_{hydro}, and the counter-acting forces of the fiber-network (or inter-fiber) forces F_{FN} and the fiber interactions with the mold wall F_{int}. According to experimental results and Mechanistic Model simulations, the hydrodynamic forces lead to the fiber phase following the melt flow, while increasing fiber interactions lead to the slowing, or retardation, of the fiber phase.

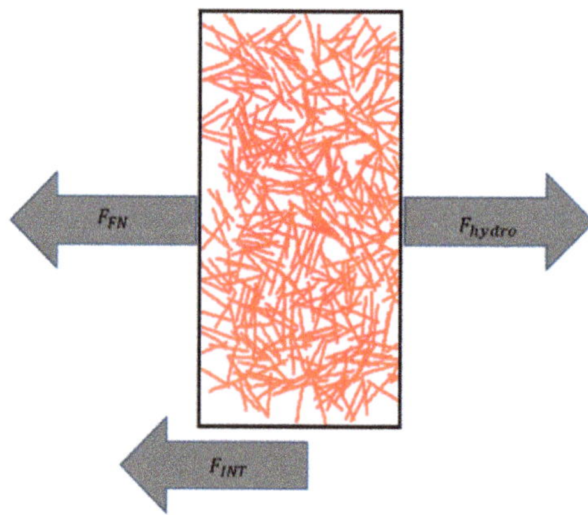

Figure 6. Fiber phase force balance.

The ratio of the interacting forces to the hydrodynamic forces can be described with the retardation factor K. For minimal fiber interaction forces and high hydrodynamic forces $K \rightarrow 0$ and no fiber matrix separation occurs and the fiber phase follows the hydrodynamic forces. With increasing fiber interaction or low hydrodynamic forces, K increases, leading to maximum fiber-matrix separation for values of K ≥ 1, which would lead to a full halt of the fiber phase movement while the polymer melt phase advances.

$$K = \frac{F_{FN} + F_{INT}}{F_{hydro}}$$

This implies that with negligible inertia of the fiber bed inside a stationary flow, the retardation factor K (for 0 < K < 1) can be used to calculate the relative velocity of the fiber bed v with a given velocity of the melt phase u.

$$v = (1 - K) \times u$$

The detailed information on the force balance calculation is described in detail in the following section.

3. Force Balance Calculation

The three main forces during processing as explained earlier are the hydrodynamic forces, leading to the initial fiber phase movement and the counteracting fiber forces–fiber interaction and fiber network forces.

The hydrodynamic forces are calculated using Stokes Equation [15] regarding the pressure gradient dp/dx through a medium with porosity P based on the fluid viscosity η and the fluid velocity u. This further complies with Darcys law.

$$\frac{dp}{dx} = -\frac{\eta}{P}u$$

Bhakarev and Tucker [16] describe the porosity of a fiber bed P with the fiber volume content ϕ and the fiber diameter d.

$$P = 0.00025\phi^{2.4}d^2$$

The fiber network forces have to be applied to the respective type of fiber network inside the composite. One solution for nonwoven felts, as generally used in glass mat reinforced thermoplastics (GMT), is the Cox model [17]. Cox describes the Youngs modulus of the fiber network with the single fiber modulus E_f, the total fiber length per unit area ρ_{geom} and the cross section area Ω.

$$E(\varepsilon = \frac{1}{3}E_f\Omega\rho_{geom})$$

The fiber network forces are important during the initial stages of compression molding, where the fiber network experiences the first deformation and fibers are pulled apart. If the hydrodynamic forces are significantly lower than the network forces, e.g., through low viscosity levels, the polymer seeps through the fiber bed without creating any fiber movement. This sponge-like phenomena is called "bleeding out". Traditionally, the general type of fiber network is matched to the viscosity of the polymer material to avoid this effect, hence fiber networks with sheet molding compounds are generally weaker than the needled felts applied in GMTs.

The fiber interaction forces with the mold walls are the main drivers for fiber-matrix separation as observed in experiments and Mechanistic Model simulations. Londono et al. [18] applied the fiber interaction forces inside a gap with the general bending forces of a fiber. The force required for fiber bending is calculated with the fiber bending constant C, deflection of the fiber δ, the Young's modulus E and momentum of inertia I and the free bending distance L.

$$F_{fiber} = C\frac{\delta EI}{L^3}$$

During molding, the fiber bending only occurs when the individual fiber is in interaction with the mold walls. With a given fiber distribution function at a specific gap height h and a fiber length l, the amount of fibers in interaction can be calculated with a critical angle of interaction ψ_{crit}.

$$\psi_{crit}(h) = \sin^{-1}\frac{h(t,\,x)}{l}$$

The geometrical bending of the fiber and the resulting fiber force is calculated on the angle and length of the fiber in Figure 7. With given fiber length l at an orientation angle ψ, a straight fiber would technically extend through the mold wall. The bending deflection of the fiber δ is then necessary so the fiber geometrically fits inside the gap.

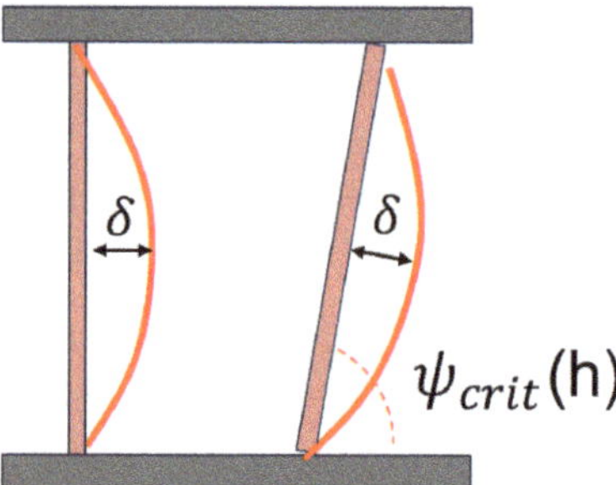

Figure 7. Fiber bending inside a gap during compression molding.

The overall fiber interaction force is then calculated by the summation of fiber interaction forces of all individual fibers.

4. Compression Molding Example

The fiber retardation model is applied in a simple compression molding simulation with a unidirectional flow front and the use of glass fibers, where a compression charge with initial dimensions h_0, x_0, constant width B and a constant glass fiber volume fraction ϕ is compressed with the constant closing speed a, as shown in Figure 8.

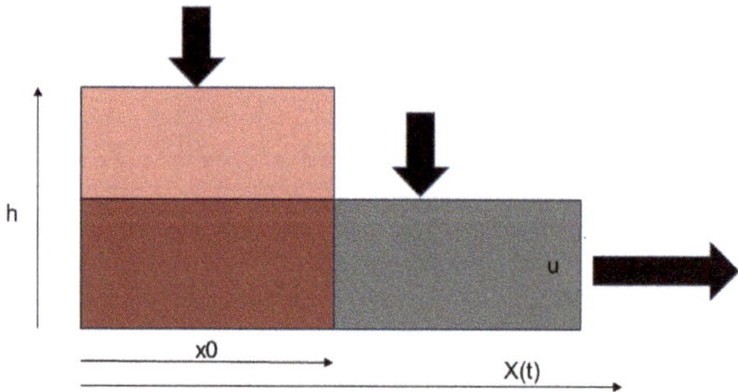

Figure 8. Compression molding setup.

During constant compression, the velocity of the melt front advances over time.

$$x'(t) = u(t) = \frac{a \times h0 \times x0}{(h0 - at)^2}$$

The resulting gap height and flow velocity development over time during compression molding are presented in Figure 9.

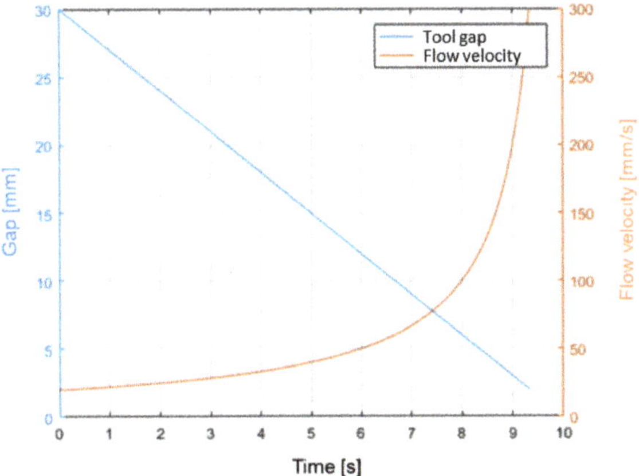

Figure 9. Gap height and flow velocity during compression molding.

The fiber orientation in commercially available molding software is calculated at every time step and therefore considered provided. For our simple demonstration, three fiber orientation distribution profiles are chosen from literature, as displayed in Figure 10, which were taken from final compression molding experiments with different initial mold coverage of 33%, 50% and 67% [19,20]. In addition, a fully random orientation distribution was added. The three orientation distribution functions are considered constant during our simulation, which of course does not comply with the realistic fiber orientation development during molding. In this first trial, the simplification is applicable.

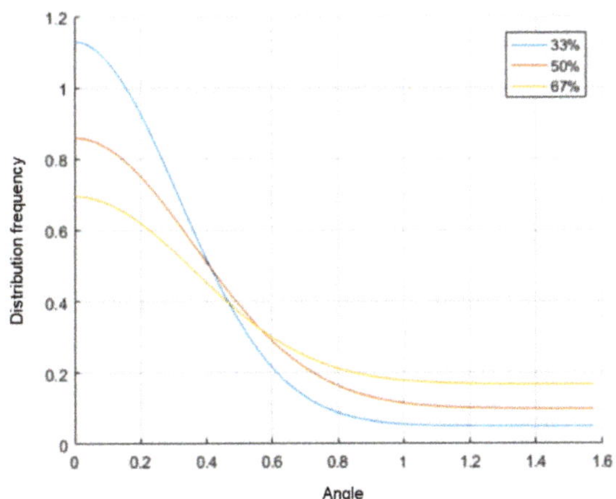

Figure 10. Fiber orientation distribution frequency profiles over fiber orientation angles in compression molded plates with 33%, 50% and 67% initial mold coverage.

The different fiber distribution functions are then applied to calculate the fiber interaction forces at every gap height h for all fibers at the specific orientation angle ψ. Fiber network forces are not considered during this simulation and will be focused on in a later publication. The hydrodynamic forces are calculated accordingly. All resulting forces are displayed in Figure 11.

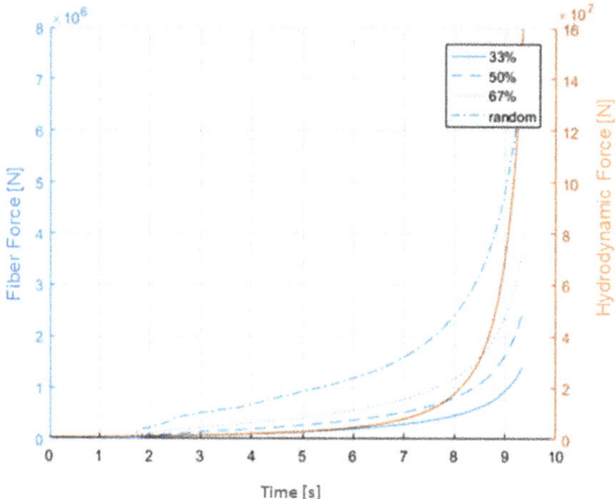

Figure 11. Development of fiber forces with different orientation distributions (blue) and hydrodynamic forces (red) during compression molding.

The development of the retardation factor K, the ratio of fiber interaction forces to hydrodynamic forces during compression molding is shown in Figure 12. It is observed that K initially remains at zero until the first fiber interactions with the mold wall occur. K increases significantly with time until the increasing molding speed leads to significantly higher hydrodynamic forces. This implies that a fiber phase of constant porosity P, would first follow the polymer melt velocity, then follow the melt phase up to 16% relatively slower than the governing polymer melt and then catch up with the melt velocity again towards the end of molding.

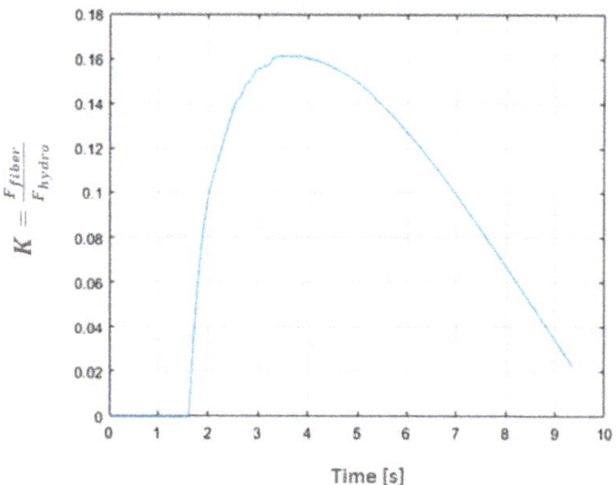

Figure 12. Fiber retardation factor K during the compression molding of a random fiber network $\phi = 0.4$.

This generally proves that the fiber retardation model is able to predict a slowing of the fiber phase caused by increasing fiber interactions during compression molding.

Additional evaluations regarding the influence of different properties have been conducted. Values which are only applied in either force calculation, hydrodynamic and fiber interaction, have a clear influence of either increasing or decreasing the forces accordingly. Other factors, which are implemented in both calculations are further evaluated in details. The influence of different fiber contents which have an influence on fiber interactions as well as hydrodynamic forces, are displayed in Figure 12. It is shown that in Figure 13, the fiber retardation factor K increases significantly with lower fiber contents and decreases with higher fiber contents. Simulations with a fiber content of 10 Vol.−% even display a K ≥ 1, which implies a complete halt of the fiber phase inside the melt flow.

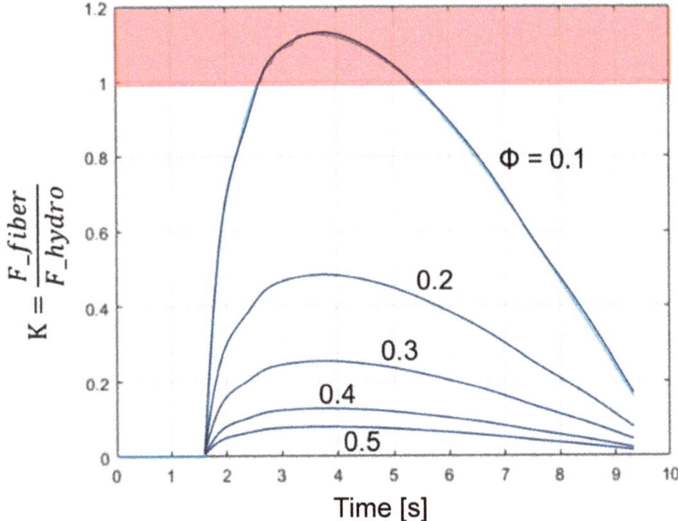

Figure 13. Development of retardation factor K in compression molding simulations with different fiber contents ϕ at random orientation.

The results on the influence of fiber content shows that the hydrodynamic forces are the leading factor regarding fiber-matrix-separation during compression molding. Furthermore, it shows that higher fiber contents display less fiber-matrix-separation, which generally agrees to earlier results from compression molding experiments with GF30 and GF40 by Kuhn.

5. Discussion and Outlook

This paper proposes a novel force-balanced approach to predict fiber-matrix-separation behavior during the processing of long fiber reinforced composites. Earlier experimental findings show increasing interaction forces of the fiber phase during flow, counteracting the hydrodynamic forces of the melt phase and eventually leading to a slowing of the fiber phase inside the melt flow. The proposed model takes these fundamental forces acting on the fiber bed into relation, enabling a slowing of the fiber phase if interaction forces increase compared to the hydrodynamic forces of the melt flow.

While the current stage of the model employs a number of generalized model assumptions and simplified mechanics, it demonstrates a general proof of concept for the realistic prediction of the "fiber-phase retardation". The model results generally comply with experimental findings and literature, whereas increasing fiber interaction forces during processing lead to a force counteracting the hydrodynamic forces of the melt. This increase in counteracting force reduces the fiber bed velocity relative to the melt velocity, eventually leading to fiber accumulation inside complex component regions and, in drastic cases, to a complete stop of the fiber bed and a bleeding out of the resin. As mentioned above, the governing factors are the fiber and melt properties, which also complies with literature findings.

While the current state of the fiber retardation model has shown its general suitability to predict fiber-matrix-separation effects, it has not been fully evaluated with experimental studies regarding the choice of simulative boundary conditions and assumed parameters, which is planned for future publications. In addition, the first calculations were applied to a single fiber region with constant fiber content in order to show the slowing of the fiber phase relative to the melt flow. The next steps regarding the fiber retardation model is to apply it to a model of multiple fiber regions, in which fiber transport and accumulation due to retardation is possible. This will make the application of the fiber retardation model more suitable for industrial application in phenomenological models in compression molding.

The authors are aware of the current simplicity of the model, with specifically chosen simple model assumptions and simplified boundary conditions. It is clear that extensive work both on the simulative and experimental side is required to ultimately create an accurate fiber content distribution model for commercial implementation. In this context, the model's current simplicity of employing a simple force-balance between fiber and melt phase is a great starting point for future improvements. Within the force balance equation, it is effortlessly possible to include more complex models, boundary conditions and formulations to utilize existing process information of for example, the surrounding flow field, local fiber orientation, fiber content and fiber length settings. Hence, it is possible to increase the predictability and efficiency of the force balance model significantly. Comparable to the evolution of fiber orientation models in the last decades, new improvements can be added to the force-balanced fiber retardation model step by step.

Author Contributions: Data curation, S.W.; Investigation, C.K.; Writing–original draft, C.K. and S.W.; Writing–review & editing, C.K. All authors have read and agreed to the published version of the manuscript.

Funding: This research received no external funding.

Conflicts of Interest: The authors declare no conflict of interest.

References

1. Thomason, J. The influence of fibre length and concentration on the properties of glass fibre reinforced polypropylene: 5. Injection moulded long and short fibre PP. *Compos. Part A Appl. Sci. Manuf.* **2002**, *33*, 1641–1652. [CrossRef]

2. Phelps, J.H.; Tucker, C.L. An anisotropic rotary diffusion model for fiber orientation in short- and long-fiber thermoplastics. *J. Non-Newton. Fluid Mech.* **2009**, *156*, 165–176. [CrossRef]

3. Folgar, F.; Tucker, C.L. Orientation behavior of fibers in concentrated suspensions. *J. Reinf. Plast. Compos.* **1984**, *3*, 98–119. [CrossRef]

4. Fan, X.-J.; Phan-Thien, N.; Zheng, R. Simulation of fibre suspension flow with shear-induced migration. *J. Non-Newton. Fluid Mech.* **2000**, *90*, 47–63. [CrossRef]

5. Goris, S.; Osswald, T. Process-induced fiber matrix separation in long fiber-reinforced thermoplastics. *Compos. Part A Appl. Sci. Manuf.* **2018**, *105*, 321–333. [CrossRef]

6. Kuhn, C.; Ton, Y.; Taeger, O.; Osswald, T. *Experimental Study on Fiber Matrix Separation during Compression Molding of Fiber Reinforced Rib Structures*; ANTEC: Orlando, FL, USA, 2018.

7. Kuhn, C.; Walter, I.; Täger, O.; Osswald, T. Experimental and numerical analysis of fiber matrix separation during compression molding of long fiber reinforced thermoplastics. *J. Compos. Sci.* **2017**, *1*, 2. [CrossRef]

8. Kuhn, C.; Walter, I.; Täger, O.; Osswald, T. Simulative prediction of fiber-matrix separation in rib filling during compression molding using a direct fiber simulation. *J. Compos. Sci.* **2017**, *2*, 2. [CrossRef]

9. Schmachtenberg, E.; Lippe, D.; Skrodolies, K. Faser-Matrix-Entmischung Während des Fliesspressens von SMC. Available online: https://www.kunststoffe.de/fachinformationen/zeitschrift-kunststofftechnik/artikel/fasermatrix-entmischung-waehrend-des-fliesspressens-von-smc-548082.html?error_description=Invalid+request+URI&state=cc-session-cookie-redirect&error=invalid_request_uri (accessed on 28 October 2020).

10. Kuhn, C. *Analysis and Prediction of Fiber Matrix Separation during Compression Molding of Fiber Reinforced Plastics*; Friedrich-Alexander-Universität Erlangen-Nürnberg (FAU): Erlangen, Germany, 2018.

11. Morris, J.F.; Boulay, F. Curvilinear flows on Noncolloidal suspensions: The role of normal stresses. *J. Rheol.* **2006**, *43*, 1213–1237. [CrossRef]

12. Londoño-Hurtado. *Mechanistic Model for Fiber Flow*; University of Wisconsin-Madison: Madison, WI, USA, 2009.

13. Peréz, C.; Ramírez, D.; Osswald, T.A. *Mechanistic Model Simulation of a Compression Molding Process: Fiber Orientation and Fiber-Matrix-Separation*; ANTEC: Orlando, FL, USA, 2015.

14. Ramírez, D. *Study of Fiber Motion in Molding Processes by Means of a Mechanistic Model*; University of Wisconsin-Madison: Madison, WI, USA, 2014.
15. Davis, B.; Gramann, P.; Osswald, T.; Rios, A. *Compression Molding*; Carl Hanser Verlag: München, Germany, 2003.
16. Bakharev, A.S.; Tucker, C.L. *Technical Papers XLII*; Society of Plastic Engineers: Bethel, CT, USA, 1996.
17. Cox, H.L. The elasticity and strength of paper and other fibrous materials. *Br. J. Appl. Phys.* **1952**, *3*, 72–79. [CrossRef]
18. Londoño-Hurtado, A.; Osswald, T.A. Fiber jamming and fiber matrix separation during compression molding. *J. Plast. Technol.* **2006**, *15*, 109.
19. Advani, S.G.; Tucker, C.L. The use of tensors to describe and predict fiber orientation in short fiber composites. *J. Rheol.* **1987**, *31*, 751–784. [CrossRef]
20. Jeffery, G.B. The motion of ellipsoidal particles immersed in a viscous flow. *R. Soc.* **1923**, *102*. [CrossRef]

Publisher's Note: MDPI stays neutral with regard to jurisdictional claims in published maps and institutional affiliations.

Journal of
Composites Science

Article

Significance of Model Parameter Variations in the pARD-RSC Model

Armin Kech [1],*, Susanne Kugler [1] and Tim Osswald [2]

[1] Corporate Sector Research and Advance Engineering, Robert Bosch GmbH, Robert-Bosch-Campus 1, 71272 Renningen, Germany; susanne.kugler2@de.bosch.com

[2] Polymer Engineering Center, University of Wisconsin-Madison, Madison, WI 53706, USA; tosswald@wisc.edu

* Correspondence: armin.kech@de.bosch.com

Received: 10 July 2020; Accepted: 29 July 2020; Published: 7 August 2020

Abstract: This study aims to evaluate how fiber orientation results are dependent on fluctuations in input parameters, such as the average fiber length, fiber volume content, and initial alignment of fibers. The range of parameters is restricted to deviations within one specific short fiber reinforced thermoplastic and is not set up to investigate the differences between materials. The evaluation was conducted by a virtual shear cell based on a mechanistic modeling approach. The fiber orientation prediction model discussed is the pARD-RSC (principal anisotropic rotary diffusion-reduced strain closure) model implemented as a user routine in AUTODESK MOLDFLOW INSIGHT® (AMI®). The material investigated was discontinuous short glass fiber reinforced PBT (polybutylene-terephthalate), which is often used for housings in various industries. It is shown that variation in the input parameters, although having an influence on the fiber orientation model parameters, only affects the final orientation moderately.

Keywords: pARD-RSC; fiber orientation; short fiber reinforced; mechanistic modelling

---

1. Introduction

The demand for lightweight structures to reduce the amount of energy required to transport a vehicle from A to B is increasing. This leads to the need for advanced models and methods to predict the performance of such components. Discontinuous short glass fiber reinforced thermoplastic components are, thus, very interesting, as their material properties combine lightweight performance with thermo–mechanical strength, such as temperature dependent ultimate tensile or flexural strength. For example, 90% of all technical thermoplastic materials purchased each year at Robert Bosch GmbH are of a short glass fiber reinforced grade.

A simulative prediction of failure and the lifetime of the glass fiber reinforced product can decrease the development time and, therefore, reduce costs significantly. To do so, an integrative simulation chain is necessary. The integrative simulation chain combines the results from process simulation, e.g., the fiber orientation distribution, with structural mechanics models. It is important in the integrative simulation chain that the fiber orientation is accurately predicted. The stiffness and strength of polymers depend on the fiber orientation significantly. Fiber orientation differs in its position based on thickness, with fibers aligned in the flow direction in the shell region and perpendicular to the flow in the core region. Models are required that are capable of predicting this orientation correctly, especially for mechanical strength and thermo–mechanical deformation. The models presently used to predict fiber orientation are derived from Jeffery [1], and the parameters introduced are purely empirical. A review of current fiber orientation models can be found in [2]. The company ALTAIR [3] recently proposed a different approach. ALTAIR is using a discrete element method approach by placing some discrete fibers in the flow field, which will be used for the orientation evolution of discontinuous short fiber reinforced thermoplastics.

This work focuses on the simulative prediction of empirical fiber orientation parameters in a shear flow. With the help of a mechanistic model, the influence of various input parameters is studied. First, we study the statistical variance caused by different cells with identical inputs. Since these cells are determined based on the second order fiber orientation, they are under constraints. Consequently, statistical variance is expected. Second, we investigate the influence of various material parameters within one short fiber reinforced material. We vary the fiber length, fiber volume content, and initial orientation. The gained parameters are then used for an injection molding simulation in Autodesk Moldflow®, and the influence on the final fiber orientation in the part is assessed.

2. Theory

The orientation of a single fiber can be characterized by a unit vector p (see Figure 1 [4]).

$$p_1 = \sin\Theta \, \cos\phi \tag{1}$$

$$p_2 = \sin\Theta \, \sin\phi \tag{2}$$

$$p_3 = \cos\Theta. \tag{3}$$

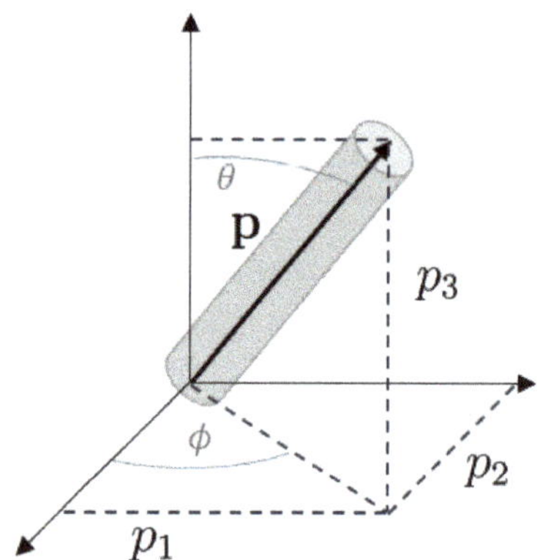

Figure 1. Definition of a single unit fiber in space.

Various models have been developed to model fiber orientation. Work was performed at Virginia Tech [5] using a sliding plate rheometer to choose a fiber orientation model—the pARD-RSC (principal anisotropic rotary diffusion-reduced strain closure) [6] model. The chosen model was implemented in AMI® using the Solver API solution within AMI® [5]. The model is given by

$$\frac{DA}{Dt} = \dot{A} = W{\cdot}A \quad -A{\cdot}W + \zeta[D{\cdot}A + A{\cdot}D - 2(A + (1-\kappa)(\mathbb{L} - \mathbb{M}:A)):D]$$
$$+\dot{\gamma}[2[C - (1-\kappa)\mathbb{M}:C] - 2\kappa(trC)A - 5(C{\cdot}A + A{\cdot}C) \tag{4}$$
$$+10[A - (1-\kappa)(\mathbb{L} - \mathbb{M}:A):C]$$

$$\mathbb{L} = \sum_{i=1}^{3} \lambda_i e_i e_i e_i e_i$$
$$\mathbb{M} = \sum_{i=1}^{3} e_i e_i e_i e_i$$

$$C = C_I R_A \begin{pmatrix} D_1 & 0 & 0 \\ 0 & D_2 & 0 \\ 0 & 0 & D_3 \end{pmatrix}; \text{ with } A = R_A \begin{pmatrix} \lambda_1 & 0 & 0 \\ 0 & \lambda_2 & 0 \\ 0 & 0 & \lambda_3 \end{pmatrix} R_A^T \tag{5}$$

$$D_1 = 1, \, D_2 = c, \, D_3 = 1 - c$$

where A is the second order orientation tensor, and $\mathbb{A}$ is the fourth order orientation tensor; R_A is the Eigen matrix, D is the strain rate, and W is the vorticity tensor. C_I is the interaction coefficient introduced in [7].

The parameters C_I, D_2, and κ were determined from a numerical shear cell experiment developed at the Polymer Engineering Center (PEC) of the University of Wisconsin—Madison [8]. For the generation of unit cells with discrete fibers, sequential addition and the migration method were used [9]. A statistical approach to determine the uncertainties of this procedure was established by varying the initial fiber cell parameters.

3. Generation of the Unit Cells

Two different setups of cells were generated. First, the influence on cell generation was investigated. The generation of the cells containing 30 wt%. in a 512 µm × 512 µm × 512 µm cubic cell resulted in 1118 discrete fibers with a diameter of 10 µm and a fiber length of 275 µm. The algorithm used assures that the fibers do not bend or overlap with each other [9]. The initial orientations, given by the fiber orientation tensor elements A_{11}, A_{22}, and A_{33}, were 0.15, 0.05, and 0.8 and correspond to the initial alignment of pellets in the sliding plate experiments. All the diagonal elements, such as A_{12}, were set to 0. Under these settings, 10 initial cells were created by the fiber generator. Notably, the second order orientation tensor does not impose a unique orientation state. Consequently, the orientation between the 10 cells was not identical.

Second, the influence of fluctuations in the initial settings for orientation, fiber length, and fiber content were examined. A Design of Experiment (DoE) scheme was used to investigate three parameters (A_{33}, fiber length, and fiber concentration) in three levels. To minimize the calculation times, a reduced DoE scheme (D-optimal using the software Cornerstone®) was applied, resulting in 15 settings to consider linear and quadratic combinations and interactions. The upper and lower bounds were chosen as the upper and lower bounds of the material properties of PBT GF 30 during injection molding, since this study aims to determine the influence of varying the parameters during the final fiber orientation of the part.

The settings applied for the D-optimal scheme with three parameters and three steps are given in Table 1.

Table 1. Settings for DoE.

Setting	Length [µm]	Volume Content [%]	A_{11}	A_{22}	A_{33}
V1	200	16.62	0.05	0.05	0.9
V2	300	19.43	0.05	0.05	0.9
V3	300	16.62	0.15	0.05	0.8
V4	300	18.02	0.05	0.05	0.9
V5	200	19.43	0.25	0.05	0.7
V6	200	18.02	0.15	0.05	0.8
V7	250	16.62	0.05	0.05	0.9
V8	300	18.02	0.25	0.05	0.7
V9	300	16.62	0.05	0.05	0.9
V10	200	19.43	0.05	0.05	0.9
V11	250	18.02	0.25	0.05	0.7
V12	300	19.43	0.25	0.05	0.7
V13	200	16.62	0.25	0.05	0.7
V14	250	19.43	0.15	0.05	0.8
V15	300	16.62	0.25	0.05	0.7

Figure 2 shows the generated fibers for V15 from Table 1 prior to the applied shear.

Figure 2. Initial fiber alignment for V15.

The mechanistic modelling approach of PEC was applied to study the orientation of the fibers in a shear flow field. The numerical shear cell applies shear in the 1, 2 plane. On the top and bottom of the shear cell, walls that prevent fibers from penetrating in two directions are modelled. For the 1 and 3-directions, periodic boundary conditions were applied to the cell. A shear rate of 1 s^{-1} was applied for 100 s and 200 s. The input parameters for the calculations include a constant viscosity (taken for a fixed temperature with the shear rate from the data sheets for an unreinforced PBT), the fiber geometry, the fiber volume content, and the cell dimensions, as well as the number of elements. The fibers are modelled as stiff rigid bodies with a single element.

In each time step, the velocity, angular velocity, and position of each fiber was calculated with the force and momentum balance. The forces/moments considered in the model are hydrodynamic forces exerted from the polymer on the fibers and interaction forces between fibers. Back coupling between fiber motion and fluid motion was not considered.

Between two approaching fibers, it is assumed that the interaction is divided into three regimes: lubrication forces at large distances, transition forces at the distance of the surface roughness of the fibers, and mechanical forces at contact [10]. The lubrication forces are dependent on the distance and approaching angle. The lubrication force model is based on the work in [10]. Since analytical solutions exist only in the parallel case [11] and in the case of infinitely long fibers [12], optimization of the lubrication force was conducted. As an experimental approach was not available, a numerical one was applied. A fully coupled simulation in COMSOL® was used to calculate the lubrication force between two fibers at varying angles, relative velocities, fiber lengths, and matrix viscosities. Based on these data, an analytic optimization of the lubrication force was performed and implemented in the mechanistic model.

4. Results and Discussion

After applying a constant shear rate of 1 1/s for times of 100 s or 200 s to the initially aligned shear cell, the evolution parameters according to Equation (4) were determined based on a generic fitting algorithm in MATLAB®. Figure 3 shows a graph of the fiber orientation evolution over time for V15. This graph shows the evolution of fiber orientations for A_{11}, A_{22}, and A_{33} over strain under a constant shear rate. The parameters for C_I, D_2, and κ are derived from the fitted lines according to Equation (4).

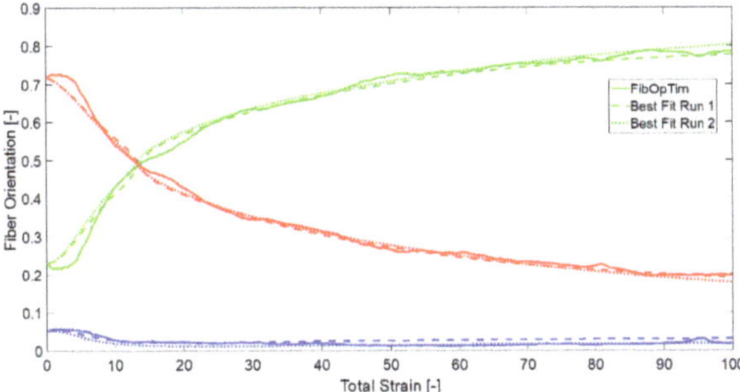

Figure 3. Fiber orientation evolution for a shear rate of 1 s^{-1} and parameters set as V15.

From the variation of cell generation with identical input parameters, it can be found that C_I varies significantly with the generation of cells and model parameter fitting. The mean values determined over all ten generations can be found together with the standard deviation in Figure 4.

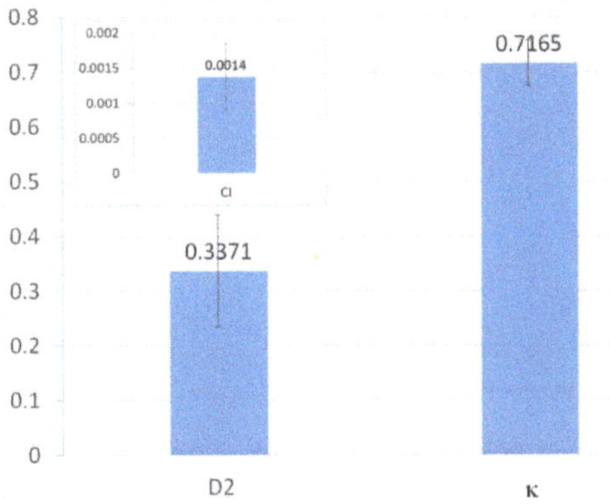

Figure 4. Average value for ten cells with identical inputs.

To determine whether such deviations play an important role, DoE calculations were performed. Table 2 shows the resulting parameters for each run.

Table 2. Results for the parameter fitting to the model (4 for DoE settings).

Setting	C_I	D_2 (=c)	κ
V1	0.00470	0.469	0.513
V2	0.00314	0.980	0.381
V3	0.00200	0.740	0.430
V4	0.00040	0.926	0.536
V5	0.09922	0.871	0.068
V6	0.05782	0.512	0.126
V7	0.00470	0.949	0.459
V8	0.00940	0.880	0.339
V9	0.00196	0.875	0.575
V10	0.00196	0.984	0.447
V11	0.01017	0.918	0.238
V12	0.01915	0.875	0.234
V13	0.01056	0.438	0.397
V14	0.03000	0.920	0.150
V15	0.00235	0.676	0.501

A statistical evaluation of the three factors (F1 = length, F2 = content, F3 = orientation) and the three response data (R1 = C_I, R2 = D_2, R3 = κ) was performed and is plotted in Figure 5a,b. For the chosen parameter range, no clear correlation could be found. The response (R1 to R3) was independent from the input parameters (factors F1 to F3).

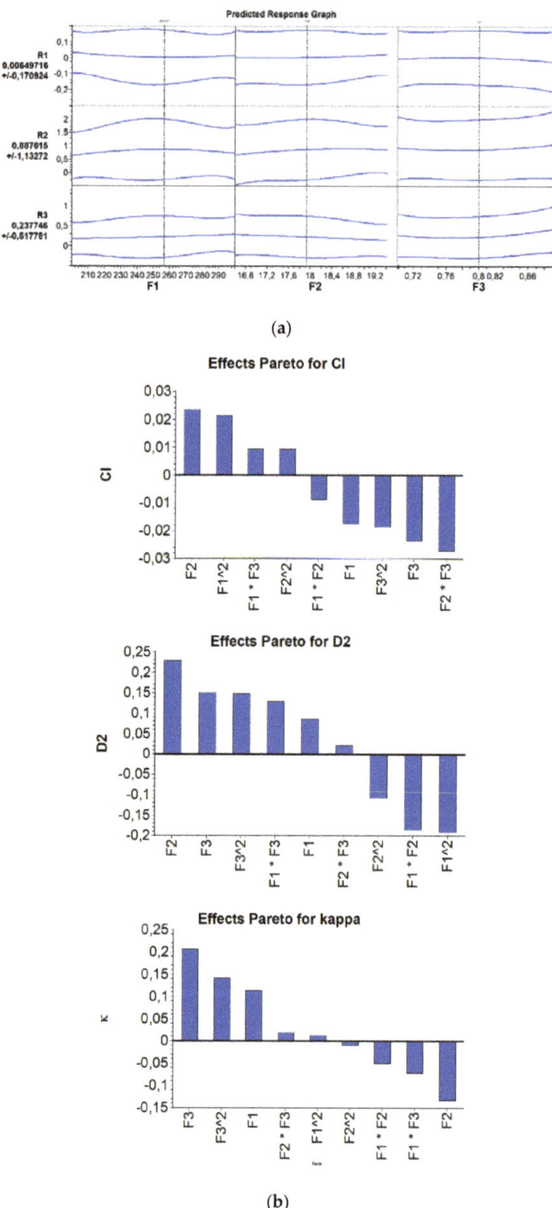

(a)

(b)

Figure 5. (**a**) Predicted Response Graph for R1 to R3 vs. F1 to F3. (**b**) Influence of factors on the response C_I (top), D_2 (middle), and κ (bottom).

In the Pareto graphs, for C_I and D_2, the fiber content has the largest positive effect, and κ is the initial orientation. Nevertheless, the effect, in general, is rather small.

5. Numerical Simulation of the Fiber Orientation in Moldflow

The influence of these varying parameters on the calculated fiber orientation in AMI® 2019 was investigated with seven parameter sets from V3, V6, V8, V9, V10, V13, and V14. These seven settings were chosen because they represent the extreme values from all valid parameter sets (min/max).

Simulations were done using the Solver API functionality in AMI® 2019 in a pre-release version. The geometry used for these simulations was a center-gated box with a non-homogeneous wall thickness distribution (see Figure 6).

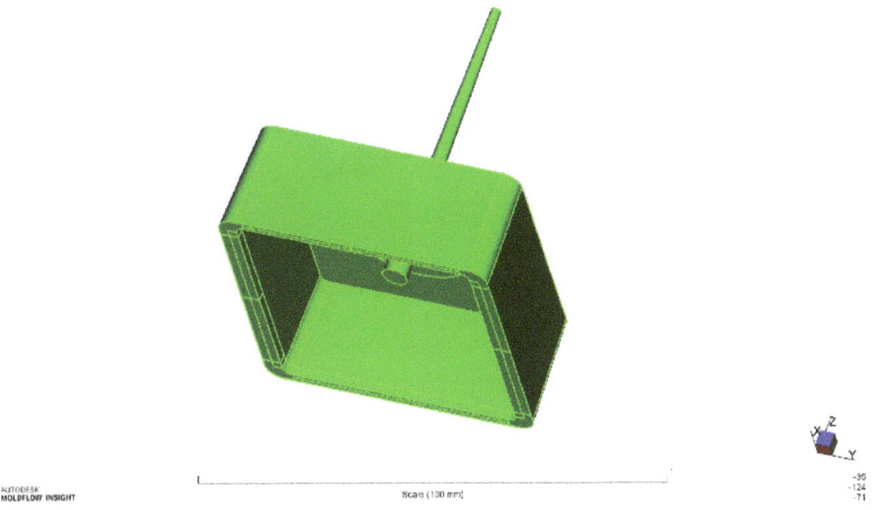

Figure 6. Geometry of the simulated center gated box.

For the simulation, material data from the AMI® 2019 data base for Ultradur B4300 G6 were chosen. The injection parameters used are given in Table 3.

Table 3. Settings in AMI® 2019.

Parameter	Value	Unit
Mold temperature	40	°C
Mass temperature	260	°C
Flow rate	8	ccm/s
Switch over	99	%
Packing pressure profile	77.8 (step 1) 8 (step 2)	MPa
Packing time profile	8 (step 1) 7 (step 2)	s
Cycle time	47	s

The cooling system of the mold was also taken into account (see the blue cylinders in Figure 7).

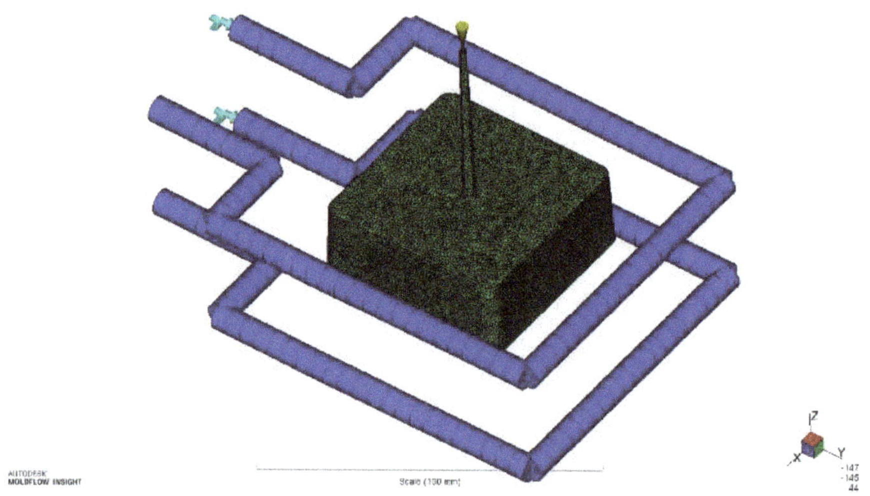

Figure 7. Model in AMI® 2019 with the cooling system (blue) and melt entrance (yellow cone).

To make a quantitative comparison between the seven chosen model parameter sets, each was simulated with identical processing parameters given in Table 3. Two element stacks of hexaeder elements over thickness were generated. The calculated fiber orientations were then mapped on these stacks. The positions of each stack are given in Figure 8 (identified by red circles).

Figure 8. Positions of the element stacks to evaluate fiber orientation.

For each calculation, the deviation to the first (V3) model parameter set was determined according to the root mean square deviation (RMSD) method.

$$\text{RMSD} = \sqrt{\frac{\sum_{n=1}^{N}\left(x_{1,n} - x_{2,n}\right)^2}{N}} \tag{6}$$

Table 4 gives a summary of the RMSD determined for each position.

Table 4. RMSD (root mean square deviation) values with reference to V3 for A_{11}, A_{22}, and A_{33}.

	Position 1				Position 2			
	A_{11}	A_{22}	A_{33}	Overall	A_{11}	A_{22}	A_{33}	Overall
V3-V6	0.0513	0.1327	0.1443	0.117	0.1456	0.0375	0.157	0.1255
V3-V8	0.0278	0.0281	0.0293	0.0284	0.0349	0.0333	0.0165	0.0294
V3-V9	0.0245	0.0139	0.0128	0.0179	0.0045	0.0179	0.0141	0.0134
V3-V10	0.024	0.0164	0.0155	0.019	0.0022	0.0151	0.0142	0.012
V3-V13	0.0313	0.0434	0.0469	0.0411	0.0348	0.022	0.0412	0.0336
V3-V14	0.0384	0.0801	0.0808	0.0693	0.0942	0.0381	0.0771	0.0737

With a maximum orientation of 0.8 in the flow direction, for the chosen input parameters and the determined model parameters of the DoE, the change in fiber orientation is 10–20%.

6. Conclusions

This study showed that varying the input parameters (average fiber length, fiber volume content, and initial alignment) within one material range for fiber orientation modeling does not have a large influence on the final orientation. Although the fiber volume content and the average fiber length varied over the flow, these changes did not affect the fiber orientation model parameters. Thus, it can be concluded that one constant set of parameters is sufficient for one material. This reduces the needed time for parameter identification significantly.

For materials with different viscosities, the average fiber length and fiber content parameters need to be determined again. To evaluate the dependence of the input on the model parameters and, thus, the fiber orientation results, the variations in the parameters must be wider.

Author Contributions: Conceptualization, S.K., A.K., and T.O.; methodology, A.K. and S.K.; software: S.K.; validation, A.K.; writing—original draft preparation, A.K.; writing—review, S.K.; visualization, A.K. All authors have read and agreed to the published version of the manuscript.

Funding: This research received no external funding.

Acknowledgments: The authors would like to express their thanks to Mohamed Besher Baradi (Robert Bosch GmbH) for providing the code for the RMSD evaluation.

Conflicts of Interest: The authors declare no conflict of interest.

Abbreviations

The following abbreviations are used in this manuscript:

AMI®	AUTODESK MOLDFLOW INSIGHT®
API	Application Programming Interface
PEC	Polymer Engineering Center
DoE	Design of Experiments
PBT	Polybutylene terephthalate
μCT	micro Computer Tomography
RSC	Reduced strain closure
ARD	Anisotropic rotary diffusion
pARD	Principal anisotropic rotary diffusion
RMSD	Root mean square deviation

References

1. Jeffery, G.B. The motion of ellipsoidal particles immersed in a viscous fluid. *Proc. R. Soc. Lond. Ser. A Math. Phys. Eng. Sci.* **1922**, *102*, 161–179.

2. Kugler, S.K.; Kech, A.; Cruz, C.; Osswald, T.A. Fiber Orientation Predictions—"A Review of Existing Models". *J. Compos. Sci.* **2020**, *4*, 69. [CrossRef]

3. Mayavaram, R.; Celigueta, M.A.; Becker, P.A.; Venkatesan, G.; Singh, N. *Predicting Fiber Orientation in Short Fiber Reinforced Injection Molding Process Using Discrete Element Method. ANTEC®2020 Papers: The Virtual Edition*; ANTEC®: Danbury, CT, USA, 2020.

4. Advani, S.G.; Tucker, C.L. The use of tensors to describe and predict fiber orientation in short fiber composites. *J. Rheol.* **1987**, *31*, 751–784. [CrossRef]

5. Kugler, S.K.; Lambert, G.M.; Cruz, C.; Kech, A.; Osswald, T.A.; Baird, D.G. Macroscopic fiber orientation model evaluation for concentrated short fiber reinforced polymers in comparison to experimental data. *Polym. Compos.* **2020**, *41*, 2542–2556. [CrossRef]

6. Tseng, H.C.; Chang, R.Y.; Hsu, C.H. The use of principal spatial tensor to predict anisotropic fiber orientation in concentrated fiber suspensions. *J. Rheol.* **2017**, *62*, 313–320. [CrossRef]

7. Folgar, F.; Tucker, C.L. Orientation behavior of dibers in concentrated suspensions. *J. Reinf. Plast. Compos.* **1984**, *3*, 98–119. [CrossRef]

8. Pérez, C. The Use of a Direct Particle Simulation to Predict Fiber Motion in Polymer Processing. Ph.D. Thesis, University of Wisconsin-Madison, Madison, WI, USA, 2017.

9. Schneider, M. The sequential addition and migration method to generate representative volume elements for the homogenization of short fiber reinforced plastics. *Comput. Mech.* **2017**, *59*, 247–263. [CrossRef]

10. Lindström, S.; Uesaka, T. Simulation of the motion of flexible fibers in viscous fluid flow. *Phys. Fluids* **2007**, *19*, 113–307. [CrossRef]

11. Yamane, Y.; Kaneda, Y.; Dio, M. Numerical simulation of semi-dilute suspensions of rodlike particles in shear flow. *J. Non-Newton. Fluid Mech.* **1994**, *54*, 405–421. [CrossRef]

12. Kromkamp, J.; Van der Ende, D.T.; Kandhai, D.; Van der Sman, R.G.; Boom, R.M. Shear-induced self-diffusion and microstructure in non-Brownian suspensions at non-zero Reynolds numbers. *J. Fluid Mech.* **2005**, *529*, 253–278. [CrossRef]

Journal of
Composites Science

Article

Direct Fiber Simulation of a Compression Molded Ribbed Structure Made of a Sheet Molding Compound with Randomly Oriented Carbon/Epoxy Prepreg Strands—A Comparison of Predicted Fiber Orientations with Computed Tomography Analyses

Jan Teuwsen [1,*], Stephan K. Hohn [1] and Tim A. Osswald [2]

[1] Volkswagen AG, Group Innovation, 38436 Wolfsburg, Germany; stephan.hohn@volkswagen.de
[2] Polymer Engineering Center (PEC), University of Wisconsin-Madison, Madison, WI 53706, USA; tosswald@wisc.edu
* Correspondence: jan.teuwsen@volkswagen.de

Received: 21 October 2020; Accepted: 22 October 2020; Published: 31 October 2020

Abstract: Discontinuous fiber composites (DFC) such as carbon fiber sheet molding compounds (CF-SMC) are increasingly used in the automotive industry for manufacturing lightweight parts. Due to the flow conditions during compression molding of complex geometries, a locally varying fiber orientation evolves. Knowing these process-induced fiber orientations is key to a proper part design since the mechanical properties of the final part highly depend on its local microstructure. Local fiber orientations can be measured and analyzed by means of micro-computed tomography (μCT) and digital image processing, or predicted by process simulation. This paper presents a detailed comparison of numerical and experimental analyses of compression molded ribbed hat profile parts made of CF-SMC with 50 mm long randomly oriented strands (ROS) of chopped unidirectional (UD) carbon/epoxy prepreg tape. X-ray μCT scans of three entire CF-SMC parts are analyzed to compare determined orientation tensors with those coming from a direct fiber simulation (DFS) tool featuring a novel strand generation approach, realistically mimicking the initial ROS charge mesostructure. The DFS results show an overall good agreement of predicted local fiber orientations with μCT measurements, and are therefore precious information that can be used in subsequent integrative simulations to determine the part's mesostructure-related anisotropic behavior under mechanical loads.

Keywords: discontinuous fiber composites (DFC); compression molding; sheet molding compound (SMC); carbon fiber sheet molding compound (CF-SMC); randomly oriented strands (ROS); fiber orientation; computed tomography (CT); process simulation; direct fiber simulation (DFS); prepreg platelet molding compound (PPMC); tow-based discontinuous composite (TBDC)

---

1. Introduction

1.1. Motivation and Materials

Fiber reinforced composites have gained importance in the automotive industry due to their potential for lightweight applications. During compression molding of discontinuous fiber composites (DFC) the long fibers undergo low shear stresses and can flow in the melt without significant fiber breakage that is common during plastification and mold filling in the injection molding process [1,2]. The shorter the required flow length to fill the mold, the lower is the possibility of fiber attrition [3]. Since typical sheet molding compounds (SMC) charge coverages lead to comparably short flow paths, the reinforcing fibers maintain their length [3,4] and compression molded DFCs show more advanced mechanical properties with high mass-specific stiffness and strength than compared to injection molded long fiber reinforced polymers [3,5].

The low tooling costs (compared to steel processing) and fast cycle times make compression molding a cost-efficient method to manufacture large DFC parts in a one-shot high volume production process, enabling DFC parts to replace automotive metal components for mass-reduction purposes [6,7]. As this material class and manufacturing method allow a high freedom of design [5], part integration (e.g., fasteners or inserts) [7], and can be used to mold complex three-dimensional (3D) shaped structural and non-structural components with corrugations, ribs and domes, it is widely used among the automotive industry and a key aspect for the endeavor to reduce the vehicle weight as much as possible [5,8]. Among this type of DFC, carbon fiber sheet molding compounds (CF-SMC) have been extensively used for interior and exterior, structural and non-structural composite applications in the automotive and aerospace industry [7,9–12].

The focus of this work is on a DFC consisting of transversely chopped unidirectional (UD) carbon fiber prepreg tows, so called 'strands', 'chips' or 'platelets', which are randomly distributed into a mat. SMCs with those randomly oriented strands (ROS) show a high degree of heterogeneity (variability in intra- and inter-part structure on the meso- and macro-scale) yet seek to reach quasi-isotropic mechanical properties [13–15]. This material class is also called prepreg platelet molding compound (PPMC) [4,16–27] or tow-based discontinuous composite (TBDC) [28–33]. The material system is characterized by a high fiber volume fraction with good impregnation, comparable to continuous fiber layups, and therefore higher mechanical properties are reachable than with traditional SMCs [18]. High performance CF-SMCs, such as the epoxy-based material used in this study (see Section 2.1), are further characterized by a high delamination resistance, near quasi-isotropic in-plane stiffness, high out-of-plane strength and stiffness, and low notch sensitivity [13]. Moreover, state of the art resin systems enable very short curing times, leading to an 84% shorter molding time and an overall process time reduction of 44% for a one-piece inner monocoque compared to the same part produced in a resin transfer molding (RTM) process [34].

However, due to the part design together with the high fiber volume fraction and fiber length in the CF-SMC material, complex anisotropic material flow conditions occur [21], which induce a characteristic microstructure in compression molded components [5,31]. This process-induced microstructure is mainly characterized by locally varying fiber orientations and fiber concentrations. Besides defects such as voids (air entrapments), swirls or resin rich pockets in between strands [35–40], the flow-induced strand alignments have the biggest impact on the mechanical performance of DFC parts [21,31,41]. Therefore, obtaining realistic 3D information of the local strand orientations in a CF-SMC part is key for a better understanding of its related mechanical behavior and a sophisticated part design. 3D representations of the morphology of entire CF-SMC parts help engineers to grasp the compression molding process related fiber alignment and the gained fiber orientation information can be mapped to a structural simulation mesh and used as a digital twin in an integrative (coupled) Finite Element Analysis (FEA) [42,43]. This microstructure information can be acquired by precise and reliable process simulations or laborious micro-computed tomography (µCT) measurements. CT scans that allow fiber orientation measurements of complete parts may be used to compare 'real' with numerically predicted fiber orientations coming from molding simulations in order to evaluate the quality of the simulation results. However, especially for parts made of carbon fiber reinforced plastics, this is so far a costly challenge due to the low contrast between the carbon fibers and the polymer matrix system in CT scans.

This paper shows the application of a direct fiber simulation (DFS) tool featuring a novel strand generation approach to simulate the compression molding of ROS-based CF-SMC parts. The strand generation feature is a pre-processing step to initialize multi-bundle UD strands in the initial numerical material charge. In order to validate the filling and fiber orientation prediction accuracy on real parts, automotive-related demonstrator parts with a complex ribbed structure are molded with a high-performance ROS-based CF-SMC material. As the accuracy of the numerical flow prediction is crucial for fiber orientation predictions, it is compared with the real part filling behavior observed in short shot experiments. Subsequently, simulated fiber orientations are compared with true process-induced strand orientations inside the molded parts, which are determined by CT scan analyses. The micro-, or more appropriately, mesostructures of three entire parts are non-destructively analyzed using low-resolution CT scans. The determined fiber orientations of all three CT scanned parts are averaged to eliminate local mesostructure differences between the parts and to gain a representative image of the general fiber orientations in the ribbed structure. In order to eliminate further influencing factors when comparing measured and predicted fiber orientations, both the process simulation and the CT scan analyses are conducted with exactly the same tetra mesh, enabling a one-to-one comparison. The size of the molded, simulated and CT scanned CF-SMC parts and the conducted detailed analyses surpass previously published studies dealing with this material class. The objective of this paper is to provide a good understanding of the process-induced mesostructure in complex ribbed DFC parts and to show the application of a commercially available DFS tool in a DFC part development process to avoid costly trial and error molding experiments and part design loops.

1.2. Compression Molding Simulations

Compression molded DFCs show distinct long fiber effects such as complex fiber orientations and fiber matrix separation (FMS) due to fiber entanglements and accumulations leading to fiber content irregularities and consequently to significant changes in mechanical properties [44,45]. As this process-induced microstructure has a strong effect on the part performance, molding simulation results such as fiber orientations and fiber content distributions are nowadays mapped to FEA meshes and used in integrative (coupled) simulations to account for the created anisotropy and to obtain realistic mechanical part behavior predictions. Therefore, accurate fiber configuration predictions are of major importance for the quality of subsequent structural simulations. Furthermore, high-quality virtual process chains help to avoid cumbersome trial-and-error molding experiments and design loops, which decreases the associated costs for the trials and mold changes and also accelerates the part development process. In conclusion, accurate numerical predictions are crucial to keep costs as low as possible and enable efficient material use and safe lightweight part design.

Process simulation tools are widely used in the automotive industry to predict the flow front advancement during molding and to indicate adverse effects like short shots or knit lines [6]. These tools are further used for the calculation of changing fiber configurations during molding of fiber reinforced plastics. Process simulations of CF-SMC parts can be done with traditional numerical tools that use phenomenological models or with more elaborate methods like smoothed particle hydrodynamics (SPH) or direct fiber simulations (DFS).

1.2.1. Statistical Fiber Orientation Models

Common commercially available molding simulation tools show good results predicting fiber orientation and fiber length changes during processing of short fiber reinforced plastics using standard empirical calculation models designed for short fibers [6,46–48]. Based on Jeffery's hydrodynamic model, the fibers are modeled as ellipsoidal shaped rigid bodies rotating in a viscous flow [47]. Leveraging the work on short fiber orientation behavior in concentrated suspensions by Folgar and Tucker [48], the change of fiber orientations during processing can be described by an evolution equation introduced by Advani and Tucker [49] stated in Equation (1):

$$\frac{\mathrm{D}a_{ij}}{\mathrm{D}t} = -\frac{1}{2}\left(\omega_{ik}a_{kj} - a_{ik}\omega_{kj}\right) + \frac{1}{2}\lambda\left(\dot{\gamma}_{ik}a_{kj} + a_{ik}\dot{\gamma}_{kj} - 2\dot{\gamma}_{kl}a_{ijkl}\right) + 2C_{\mathrm{I}}\dot{\gamma}\left(\delta_{ij} - 3a_{ij}\right). \tag{1}$$

This evolution equation for the second order orientation tensor a_{ij} is often used in contemporary molding software providing a compact expression for the fiber orientation state at reasonable calculation speed [2]. It includes the vorticity tensor ω_{ij}, the rate of deformation $\dot{\gamma}_{ij}$, the scalar magnitude of the rate of deformation tensor $\dot{\gamma}$, and the identity tensor δ_{ij} ($\delta_{ij} = \mathrm{I}$). λ denotes the particle shape parameter (for long fibers $\lambda \to 1$), which is related to the fiber aspect ratio. The fiber interaction term $2C_I\dot{\gamma}\left(\delta_{ij} - 3a_{ij}\right)$ results from the isotropic rotary diffusion (IRD) model by Folgar and Tucker and includes the empirical fiber interaction coefficient C_I [48,50]. In order to determine the fourth order tensor a_{ijkl} in the evolution equation, a closure approximation is needed and various are proposed in the literature to achieve good simulation accuracy at reasonable calculation times [51–57].

In addition to this fundamental equation for isotropic rotary diffusion and fiber dynamics [48,49], some enhancements by successor models that are implemented in conventional software improved the ability to take fiber–fiber interactions into account by implementing various anisotropic rotary diffusion (ARD) models [50,58–61]. The observed orientation delay between the measured fiber orientation and the theoretical orientation evolution is considered through strain reduction models [62–64]. Excluded volume effects can be considered to describe the dependency of the fiber interaction and the volume fraction of long fiber filled materials [65]. Recently the flow and fiber orientation prediction was coupled using anisotropic viscosity models to analyze the behavior of concentrated fiber suspensions where the fiber orientation state affects the viscosity [66–71].

These statistical fiber orientation models strongly depend on empirical coefficients, which are cumbersome to determine experimentally [72,73]. Furthermore, the model simplifications and boundary condition assumptions made in these models are not valid for flexible fibers significantly longer than structural features of the part (scale separation) [21,45,73]. In addition, phenomenological models do not consider segments of one fiber being in different flow conditions (at different positions inside the material flow) at the same time, which is highly probable for long fibers [45]. Consequently, velocity and strain rate distributions along the fiber axes cannot be taken into account and the prediction of fiber bending is not feasible [2,74]. Since interactions of long bendable fibers lead to fiber entanglements and accumulations influencing the fiber movement, the capability to model fiber bending is a crucial element [15].

In conclusion, the used empirical models implemented in conventional molding simulation tools, such as Moldflow (Autodesk, Inc., San Rafael, CA, USA) or Moldex3D (CoreTech System Co., Ltd., Zhubei City, Taiwan), are designed for rigid particles with low aspect ratios and are therefore not suited for highly filled long fiber materials. As expected and seen in the literature, even with modified evolution equations, flow-induced long fiber specific effects like fiber bending, fiber–fiber interactions, and changes in fiber content distribution (especially FMS) cannot be accurately predicted by traditional short fiber models [2,6,45,46].

1.2.2. Stochastic Particle-Based Simulation Method (Purdue University)

A particle-based flow simulation method with a stochastic approach was recently developed at the Purdue University (West Lafayette, IN, USA) to investigate ROS materials with flow-induced fiber alignment [4,17,21,25,75,76]. In contrast to conventional flow simulations typically executed in Eulerian frameworks [25], a smoothed particle hydrodynamics (SPH) method with a variable user-defined material subroutine (VUMAT) is applied in Abaqus/Explicit achieving a Lagrangian solution that allows large deformations required by mold filling [17,21,22,25,76]. The 3D flow molding simulation method is based on a stochastically generated random planar initial strand orientation distribution [17]. To predict the complex 3D strand deformations and the resulting orientations in the final part, it calculates the strand orientation evolution based on the assumption of affine motion (equivalent to Jeffery's equation) coupled with an anisotropic viscosity model [17,21,22,25,76].

The initially introduced variability of the charge orientation state is generated with a pseudo-random number generator (strand generation scheme) [25]. Coded in Python, the strand generator randomly defines in-plane strand centroid locations with random strand orientation angles as long as all SPH particles in the layer are assigned to one strand set [22,25]. Accordingly, the strands are modeled as a set of interconnected SPH particles ('groupings' [4,17]) forming a rectangular strand shape with one particle through the thickness [25]. This leads to a low resolution of the orientation distribution through the thickness of the flow simulation [17]. However, the particle size cannot be set to be equal to the strand thickness, since the computation time would be unreasonably high [17]. This is due to the fact that the particle size equally reduces in all three spatial directions so that a huge amount of particles would be needed to describe one strand.

During the flow simulation the initial charge is extended and the strand sets deform and progressively align along their flow direction tending to disaggregate [17]. Since an anisotropic (transversely isotropic) viscosity model is used, strands that are initially oriented in flow direction tend to translate, while those oriented in cross-flow direction widen [17]. Both effects are qualitatively consistent with physical observations [17,31]. Moreover, strand–strand interfaces [17] and interactions are only implicitly taken into account through the anisotropic viscosity model [21].

1.2.3. Direct Fiber Simulations (DFS)

In order to overcome the deficiencies of the described phenomenological models, many single fiber based models were devolved. Single fiber simulations, also called direct fiber simulations (DFS), connect a number of particles or rigid rods (beams) to model long flexible fibers inside the polymer flow [2,45,74,77–84]. Because of the fiber segmentation DFSs can consider the strain rate distribution along the fiber axis and are therefore suited to predict long fiber behavior like bending [74]. These models use the lubrication theory combined with small flexible inextensible threads [85], stretchable, bendable, and twistable chains of bonded spheres [78] or rigid spheres connected by ball and socket joints [86–89]. Contact formulations describe the fiber–fiber interactions [90,91] and fiber–fluid interactions are modeled by hydrodynamic drag forces [92–94].

Two types of DFS techniques can be distinguished—the velocity-based method and the mechanistic model [45]. Kuhn et al. show notable amelioration in prediction precision when using DFS tools compared to commercial phenomenological simulation tools predicting long fiber behavior in a small rib structure [45]. However, modeling a large amount of connected particles leads to long computation times if applied to practical/industrial parts, whereas the duration can be decreased by using connected rigid rods to model long fibers with a velocity-based method [74]. The same observations are made by Kuhn et al. who describe decreased preparation efforts and calculation durations when using a commercial velocity-based non-interaction DFS instead of a non-commercial, research-focused mechanistic DFS [45]. However, due to the enormous amount of calculated fibers necessary, and the therefore very numerically expensive task of single fiber simulations, this method is often restricted to predict the resulting fiber configuration in small volumes [45].

Mechanistic Model (University of Wisconsin-Madison)

Kuhn et al. use a bead chain model for a DFS of a single rib of a molded component using a reduced amount of fibers [6,45,46]. The presented study uses a particle level simulation approach based on a mechanistic model to simulate fiber bending, fiber interactions, and especially FMS during compression molding of long fiber reinforced plastics [6,45,46]. The mechanistic model simulation approach begins with a traditional mold filling calculation with a standard tool like Moldex3D [46]. The calculated flow field is then extracted and used as basis for the mechanistic model simulation. After a stack of fibers is randomly inserted into the cavity volume, the movement of all single fibers is determined by interaction with the flow field. In the mechanistic model each fiber is individually modeled as a chain of rigid rods connected by ball-and-socket joints [6,46,95,96]. The fiber discretization numerically enables fiber bending at the joints (nodes) between the rods, which is important for the prediction of long fiber behavior in component areas that are smaller than the fiber length [46,97]. Furthermore, the inside of the rods is modeled as a chain of balls to include fiber interactions and hydrodynamic effects [6]. Accordingly, the fibers are simulated as a collection of equidistant nodes where the connecting rods experience elastic deformation, hydrodynamic drag forces and excluded volume effects in the polymer flow field [46,97]. The contact and hydrodynamic forces are used in force and momentum balance equations to calculate the rotations and movements of each modeled fiber. Additionally, complex interactions with the enclosing melt flow, other fibers and mold walls are considered [45,46,96,98–100]. The excluded volume forces act as repulsive forces to inhibit fibers to interpenetrate or overlap one another or with the mold wall and are used to model fiber–fiber as well as fiber–mold interactions [46,97,98].

Due to the high fiber stiffness and the more significant impact of fiber bending, extensional fiber deformation is neglected [46]. In order to decrease the tremendous calculation times required, fiber–fiber and fiber–mold interactions are only computed once every 50 integrations [46]. However, due to the still limited amount of calculable fibers or rather the overall low simulation speed, the mechanistic model is currently limited to small part volumes [45]. In sum, the mechanistic model calculates the motion, bending, and rotation of single fibers considering all their complex interactions with the fluid, the mold walls, and other fibers. Although the mechanistic model describes only a one-way coupling from the fluid flow to the fibers, the direct fiber simulation approach results in more accurate fiber orientation and fiber content distribution predictions when compared to common process simulation software [6,45].

Direct Bundle Simulation (DBS)

For a DFS of a small compression molded cross-rib-shaped SMC part in LS-DYNA (Livermore Software Technology Corporation (LSTC), Livermore, CA, USA) Hayashi et al. use beam elements constrained in highly deformable solid elements representing the matrix [101]. In order to enable large deformations of the matrix elements, the 3D adaptive Element-Free-Galerkin (EFG) method is applied. Recently, this tool was enhanced to handle ROS composites. The fibers within one strand are modeled by multiple connected elastic beam elements in a row [102]. Since the real amount of fibers in ROS composites and even within a strand is very large, the calculated number of fibers per strand was drastically reduced, while the beam thickness was increased to ensure maintaining the nominal fiber volume fraction of the material [102]. This simplification method reduces the numerical calculation time by representing thousands of fibers as fiber bundles. Considering microstructure studies reported in the literature [4,73,103,104], it can be seen that most of the fiber bundles or strands widen and flatten, but do not disperse. Only the highly sheared strands are more likely to fan out. This observation justifies the simplifying assumption to represent thousands of carbon fibers within a strand as one bundle in order to shorten the calculation time and to enable full component scale simulations [73].

Recently, Meyer et al. [73] presented a DBS method that reproduces Jeffery's equation for single fiber bundles in shear flow, which are described as a chain of truss elements. Since the bundles are represented as one-dimensional instances, they can move independently from the matrix material, flow and interact through contact forces and with hydrodynamic drag forces of the surrounding flow. Since bundle–wall and bundle–bundle interactions are considered, there is no need for empirical interaction parameters, unlike in the commonly used statistical descriptors of fiber orientation changes (Folgar–Tucker-based models). Since this DBS approach additionally considers two-way coupling it allows for a more accurate calculation of fiber volume fraction distributions, knit lines and FMS at a component level with reasonable computation times [73].

1.2.3.3. 3D TIMON CompositePRESS

Fiber Orientation Simulation in 3D TIMON CompositePRESS

A commercially available compression molding simulation tool with a similar DFS approach is 3D TIMON CompositePRESS (Toray Engineering D Solutions Co., Ltd., Ōtsu, Shiga, Japan) [105]. Here, each fiber within the initial fiber cluster is modeled using a flexible, non-elastic chain of rods (rigid bodies, constant rod length) connected by hinge nodes [2], as shown in Figure 1. The velocity-based DFS tool begins with the filling simulation determining the flow field, assuming homogeneous isotropic resin properties (isotropic viscosity) and using a one-way coupling from the fluid flow to the fibers. Subsequently, it calculates the motion of each single fiber based on the flow velocities [2,45]. The position and curvature of each single fiber are changed according to the strain rate distribution along the fiber length [74]. Thereby, the nodes function as universal joints allowing the rods to rotate in three independent axes and thus enabling the calculation of movement, bending, and rotation of each single long fiber [2,106]. In this velocity-based direct fiber simulation approach the fiber node position u_i is a function of the enclosing fluid velocity v_{fluid} at every time step n [2]. The melt velocity v_{fluid} is interpolated from the node positions in the velocity field [2]. Since the motion u_i of node i directly follows the surrounding melt flow velocity, the temporary new node position for the next time step $n+1$ can be calculated by Equation (2) [45,107]:

$$u_i^{n+1} = u_i^n + v_{\text{fluid},i}^{\,n} \mathrm{d}t. \tag{2}$$

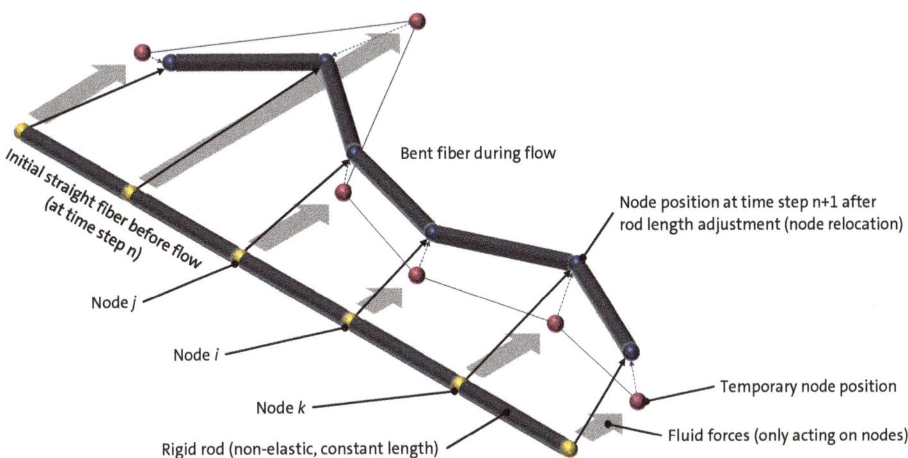

Figure 1. Schematic illustration of a flexible long fiber modeled with 6 hinge nodes and 5 rods in 3D TIMON shown in its initial straight form and bent after moved by fluid forces and applied rod length adjustments (node relocations).

This leads to an unrealistic and impermissible stretching of the fibers, which is numerically corrected by rod length adjustments (cf. Figure 1) or node relocations, respectively. At each time step the change of the rod length L is reviewed to a maximum elongation threshold ε (Equation (3)):

$$\max\left|1 - \frac{L_{ij}^{n+1}}{L_{ij}^n}\right| < \varepsilon. \tag{3}$$

If the rod length change exceeds this threshold, node i is relocated considering the neighboring nodes and rod lengths by Equation (4):

$$u_i^{n+1} = u_i^n + \mathrm{d}u_i, \tag{4}$$

with

$$\mathrm{d}u_i = \left(u_i^{n+1} - u_i^n\right) + \frac{1}{2}\left(1 - \frac{L_{ij}^{n+1}}{L_{ij}^n}\right)\left(u_j^{n+1} - u_i^{n+1}\right) + \frac{1}{2}\left(1 - \frac{L_{ik}^{n+1}}{L_{ik}^n}\right)\left(u_k^{n+1} - u_i^{n+1}\right), \tag{5}$$

where L_{ij} is the rod length between node i and node j and L_{ik} is the rod length between node i and node k. Equation (5) is iteratively solved until the rod length changes are smaller than ε and the node positions are updated by time integration based on an explicit Euler scheme [2].

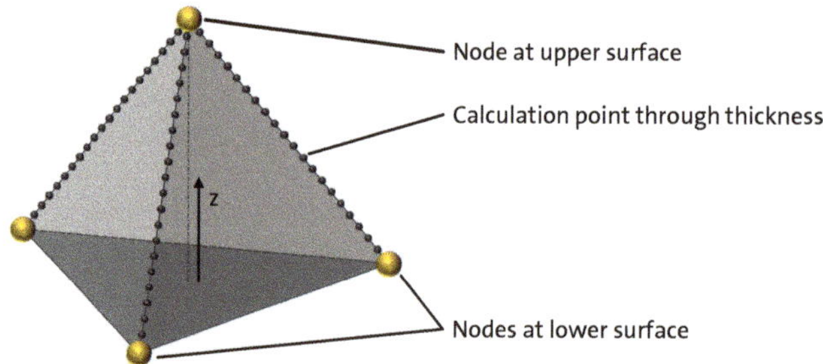

Figure 2. Schematic illustration of a tetra element in the morphing method with one node at the top, 3 nodes at the bottom, and 20 calculation points through the thickness for Light 3D analyses in 3D TIMON CompositePRESS.

Outperforming conventional fiber orientation models, the velocity-based DFS method applied in 3D TIMON CompositePRESS enables the prediction of local fiber orientations and fiber content distributions in compression molded SMC parts where long fibers (typically 25–50 mm) are used [2]. Moreover, fiber breakage can be simulated through detachment of rod–rod connections [2]. However, due to the fact that fiber attrition is of minor importance during SMC compressing molding, this effect is typically neglected. Fiber–fiber interactions are not considered in the current R6.0 version of 3D TIMON 10 to keep the calculation time reasonable [2,73,105]. Furthermore, the software does not account for anisotropic viscosity and two-way coupling [73]. Compared to phenomenological models though, the long fiber behavior is superiorly predicted, despite being based on non-colliding, velocity-following nodes [45]. However, it is reported that fiber content distribution simulation results have some insufficiencies when compared with experiments. Kuhn et al. conclude that these discrepancies are induced by extensive fiber interactions causing FMS in reality, which cannot be accurately predicted by either phenomenological (statistical) models based on Morris and Boulay or the velocity-based DFS by 3D TIMON [45].

Due to the current status of the presented long fiber orientation models, Kuhn et al. recommend to use velocity-based DFSs like 3D TIMON for the fiber orientation simulation of large SMC components [45]. However, for smaller volumes, where FMS is pronounced and of interest, they suggest to use the mechanistic model to be able to predict fiber agglomeration more accurately [45]. Kuhn et al. further propose to develop and implement a phenomenological or simplified model for FMS in common computational tools like Moldflow, Moldex3D or 3D TIMON [45].

Flow Simulation Simplifications in 3D TIMON CompositePRESS

In order to simulate the compression molding process, the flow analysis in 3D TIMON is based on the standard equations of traditional fluid dynamics: the continuity equation, the momentum equation, and the energy equation. These governing equations for the filling phase are given in the following. Theoretical background and a more detailed deduction of the general and special forms of the equations are given in Appendix A.

Continuity equation:

$$\frac{\partial \rho}{\partial t} + \frac{\partial}{\partial x}(\rho u_x) + \frac{\partial}{\partial y}\left(\rho u_y\right) + \frac{\partial}{\partial z}(\rho u_z) = 0, \tag{6}$$

where ρ is the density, t is time and u_i stands for the fluid velocity vector in direction x, y, and z, respectively. Assuming incompressibility and a steady state, neglecting inertia and gravity, and considering the fluid motion in z-direction, the governing equations for non-Newtonian polymer melt flow can be simplified to a volume continuity equation:

$$\frac{\partial u_x}{\partial x} + \frac{\partial u_y}{\partial y} + \frac{\partial u_z}{\partial z} = 0. \tag{7}$$

Momentum equation in x-direction:

$$\rho\left(\frac{\partial u_x}{\partial t} + u_x\frac{\partial u_x}{\partial x} + u_y\frac{\partial u_x}{\partial y} + u_z\frac{\partial u_x}{\partial z}\right) = -\frac{\partial p}{\partial x} + \mu\left(\frac{\partial^2 u_x}{\partial x^2} + \frac{\partial^2 u_x}{\partial y^2} + \frac{\partial^2 u_x}{\partial z^2}\right) + \rho g_x, \tag{8}$$

where p: pressure, μ: viscosity of Newtonian fluids, g_i: body force acting on the continuum, for example, gravity. This equation can also be expressed in terms of deviatoric stress τ and is then commonly called the Cauchy momentum equation:

$$\rho\left(\frac{\partial u_x}{\partial t} + u_x\frac{\partial u_x}{\partial x} + u_y\frac{\partial u_x}{\partial y} + u_z\frac{\partial u_x}{\partial z}\right) = -\frac{\partial p}{\partial x} + \left(\frac{\partial \tau_{xx}}{\partial x} + \frac{\partial \tau_{yx}}{\partial y} + \frac{\partial \tau_{zx}}{\partial z}\right) + \rho g_x. \tag{9}$$

Energy equation for a fluid with constant properties is given by:

$$\rho c_p \frac{DT}{Dt} = k\left(\frac{\partial^2 T}{\partial x^2} + \frac{\partial^2 T}{\partial y^2} + \frac{\partial^2 T}{\partial z^2}\right) + \dot{Q}_{viscous\ heating} + \dot{Q}, \tag{10}$$

where ρ: density, c_p: specific heat, T: temperature, t: time, k: thermal conductivity, $\dot{Q}_{viscous\ heating}$: viscous dissipation and $\dot{Q}$: arbitrary heat source (i.e., exothermic reaction).

For polymer flows, the fundamental governing equations given above are non-linear, non-unique, complex, do not have a general solution, and are thus difficult to solve [108]. To gain analytical solutions and to reduce the required CPU time the balance equations must be simplified. 3D TIMON CompositePRESS avoids solving the Navier–Stokes equations to realize a 3D flow analysis of non-Newtonian polymer melts, and instead uses the traditional Hele-Shaw approximation for a 2.5D flow simulation (called Light 3D) conducted on a one layer tetra mesh when using the morphing method [109]. The Light 3D solver virtually divides each mesh element (tetra or hexa) into 20 calculation points across the thickness (between the nodes on the surfaces) in order to predict the flow front advancement realistically [109] (see Figure 2). This procedure allows the calculation of temperatures, velocities, shear rates, and viscosity changes through the thickness by using a single layer mesh [109]. The Light 3D method reduces the number of necessary elements for accurate flow predictions for large and thin parts drastically [109].

The Hele-Shaw simplification, which is commonly used in injection molding simulations, was also applied by Folgar and Tucker to solve the compression molding process [110,111]. By assuming a flow through a narrow cavity gap h between lower and upper platen an order-of-magnitude analysis shows that the flow rate in thickness direction is small compared to the x- and y-directions. Furthermore, it shows that the in-plane velocity gradient can be neglected [109]. Therefore, in a thin layer flow field the shearing stresses $\frac{\partial \tau_{zx}}{\partial z}$ across the narrow cavity gap h are dominant [112]. These assumptions allow simplifying the volume continuity and the momentum equation to:

$$\frac{\partial u_x}{\partial x} + \frac{\partial u_y}{\partial y} = 0, \tag{11}$$

$$\frac{\partial p}{\partial x} = \frac{\partial \tau_{zx}}{\partial z}, \tag{12}$$

$$\frac{\partial p}{\partial y} = \frac{\partial \tau_{zy}}{\partial z}. \tag{13}$$

The 2.5D Hele-Shaw simplification method applied in 3D TIMON CompositePRESS aims at solving the momentum equation with the continuity equation. Therefore, the shear stresses must be derived from the three unknown variables in a 2D flow: pressure p and the flow velocities u_x, and u_y. As the flow field through the part thickness is neglected in the Hele-Shaw model, the velocity across the thickness h is integrated from the lower platen to the upper platen to compute the gap-wise average velocities $\bar{u}_x$ and $\bar{u}_y$. When the continuity equation is integrated over the thickness it takes the form:

$$\frac{\partial}{\partial x} \int_0^h u_x \mathrm{d}z + \frac{\partial}{\partial y} \int_0^h u_y \mathrm{d}z = 0. \tag{14}$$

This integrated continuity equation further reduces to:

$$\frac{\partial}{\partial x}(h\bar{u}_x) + \frac{\partial}{\partial y}\left(h\bar{u}_y\right) = 0. \tag{15}$$

In order to solve the continuity equation (Equation (11)) 3D TIMON CompositePRESS uses an assumption of potential viscous flow to express the flow velocity components $\bar{u}_x$ and $\bar{u}_y$, which can be written as:

$$\bar{u}_x = -\frac{S}{h}\left(\frac{\partial p}{\partial x}\right), \tag{16}$$

$$\bar{u}_y = -\frac{S}{h}\left(\frac{\partial p}{\partial y}\right), \tag{17}$$

where h is the gap height between the mold halves and S is the viscosity-dependent flow conductance for 2.5D flow analysis of thin parts, defined by:

$$S = \int_0^h \frac{(z - \lambda)^2}{\eta} \mathrm{d}z, \tag{18}$$

where λ is the local value of z at which the shear stresses are zero. Since most problems are symmetric, $\lambda = h/2$.

Equations (16) and (17) show that the flow rate in each direction is proportional to its pressure gradient. Substituting the unknown flow velocities u_x and u_y in Equation (11) with the gap-wise average velocities from Equations (16) and (17) reduces the number of unknowns at each node in the calculation from three to one (pressure only) in Equation (19), which is the classic Hele-Shaw model:

$$\frac{\partial}{\partial x}\left(S\frac{\partial p}{\partial x}\right) + \frac{\partial}{\partial y}\left(S\frac{\partial p}{\partial y}\right) = 0. \tag{19}$$

However, for the compression molding process the z-velocity component cannot be fully neglected and therefore the mold closing speed $\dot{h}$ is included as an extra term in the continuity equation:

$$\frac{\partial}{\partial x}(h\bar{u}_x) + \frac{\partial}{\partial y}\left(h\bar{u}_y\right) + \dot{h} = 0, \tag{20}$$

which, when implementing Equations (16) and (17), leads to the Hele-Shaw model for compression molding:

$$\frac{\partial}{\partial x}\left(S\frac{\partial p}{\partial x}\right) + \frac{\partial}{\partial y}\left(S\frac{\partial p}{\partial y}\right) - \dot{h} = 0. \tag{21}$$

Since the flow conductance S depends on the temperature-dependent viscosity η, the temperature field must be calculated. In order to calculate the temperature distribution in the thermoset molding compound, 3D TIMON CompositePRESS assumes the heat conduction in thickness direction is dominant [105] and therefore simplifies the energy equation (Equation (10) and Equation (A10) in Appendix A) to the energy equation for thermoset materials in 2.5D flow, written as

$$\rho c_p\left(\frac{\partial T}{\partial t} + u_x\frac{\partial T}{\partial x} + u_y\frac{\partial T}{\partial y} + u_z\frac{\partial T}{\partial z}\right) = k\frac{\partial^2 T}{\partial z^2} + \tau_{yz}\left(\frac{\partial u_y}{\partial z}\right) + \tau_{zx}\left(\frac{\partial u_x}{\partial z}\right) + \dot{Q}, \tag{22}$$

where ρ: density, c_p: specific heat, T: temperature, u_i: fluid velocity vector in direction x, y, and z, respectively, k: thermal conductivity, τ_{ij}: deviatoric stress, and $\dot{Q}$: heat generation due to the exothermic reaction in the thermoset curing process.

Using the simplification $\tau_{yz} = \eta\left(\frac{\partial u_y}{\partial z}\right)$ and $\tau_{zx} = \eta\left(\frac{\partial u_x}{\partial z}\right)$ (assumption of simple shear flow) yields to the final energy equation used in Light 3D heat transfer analyses in 3D TIMON CompositePRESS:

$$\rho c_p\left(\frac{\partial T}{\partial t} + u_x\frac{\partial T}{\partial x} + u_y\frac{\partial T}{\partial y} + u_z\frac{\partial T}{\partial z}\right) = k\frac{\partial^2 T}{\partial z^2} + \eta\left(\frac{\partial u_y}{\partial z}\right)^2 + \eta\left(\frac{\partial u_x}{\partial z}\right)^2 + Q_0\frac{d\alpha}{dt}, \tag{23}$$

where Q_0 denotes the total amount of generated heat in the exothermic curing reaction and $\frac{d\alpha}{dt}$ is the curing reaction rate.

In conclusion, the presented simplification method of determining the flow conductance facilitates the pressure field calculation. Subsequently, the temperature field is calculated and the flow front advancement during the 2.5D filling simulation can be predicted. However, in more complex parts with thicker ribs the narrow gap assumption of this simplified approach has its limits. Furthermore, the model does not account for the plug flow with a slip boundary condition at the mold surface, so that a proper choice of the flow conductance is important. For a more detailed explanation of the balance equations and their transformations, the simplifications made, and general background knowledge, the reader is referred to Appendix A and [1,108,110,111,113–116]. For more information about the flow analysis in 3D TIMON CompositePRESS the reader is referred to [105,109].

Viscosity Models in 3D TIMON CompositePRESS

During the compression molding process the thermoset resin is continuously heated up by the hot mold walls and the self-generating curing reaction heat leading to a viscosity drop. Simultaneously, due to the temperature increase, the curing reaction accelerates causing a conversion rate and viscosity increase. In order to describe these dependencies between viscosity, shear rate, temperature, and the curing reaction rate of the polymer melt, 3D TIMON combines three models: the Arrhenius-like temperature-dependent Andrade model (Equation (24)) [117], the curing reaction rate-dependent Castro–Macosko model (Equation (25)) [118,119], and the shear rate-dependent Cross model (Equation (26)) [120].

Andrade model:

$$\eta_0(T) = A\,\exp\!\left(\frac{B}{T}\right), \tag{24}$$

with η_0: zero shear viscosity (initial resin viscosity before curing), A: empirical model constant (fitted), B: material specific constant in Kelvin, T: melt temperature in Kelvin.

The **Castro–Macosko model** in Equation (25) is a widely used model in molding simulation tools, which describes the viscosity of thermoset materials as a function of temperature T and degree of cure α that can be expressed as follows:

$$\eta_1(T,\alpha) = \eta_0(T)\left(\frac{\alpha_{gel}}{\alpha_{gel} - \alpha}\right)^{(D+E\alpha)}. \tag{25}$$

Here, η_1 is the viscosity of the resin at a given degree of cure (conversion) α, η_0 is the initial resin viscosity before curing, α_{gel} is the degree of cure at the gel point, and D and E are empirical model constants that fit the experimental data.

Modified Cross model:

$$\eta\left(T,\alpha,\dot{\gamma}\right) = \frac{\eta_1\left(T,\alpha\right)}{1 + \left(\frac{\eta_1\left(T,\alpha\right)\dot{\gamma}}{\tau^*}\right)^{\left(1-n\right)}},$$ (26)

with $\dot{\gamma}$: shear rate, τ^*: critical shear stress (stress level at which the viscosity is during the transition between zero shear region (Newtonian plateau) and the shear thinning (power law) region of the viscosity curve), n: power law index. This model considers the effect of shear rate and temperature on the viscosity and describes the shear thinning behavior by the power law index n.

The combination of all three viscosity models extends the Castro–Macosko model with a power-law type shear rate dependence and leads to a modified **Cross–Castro–Macosko model:**

$$\eta = \frac{A\,\exp\left(\frac{B}{T}\right)\left(\frac{\alpha_{gel}}{\alpha_{gel}-\alpha}\right)^{\left(D+E\alpha\right)}}{1 + \left(\frac{\left[A\,\exp\left(\frac{B}{T}\right)\left(\frac{\alpha_{gel}}{\alpha_{gel}-\alpha}\right)^{\left(D+E\alpha\right)}\right]\dot{\gamma}}{\tau^*}\right)^{\left(1-n\right)}},$$ (27)

which is the finally used viscosity model in 3D TIMON CompositePRESS [105]. This model considers the dependence of the thermoset resin viscosity on temperature, degree of cure (conversion), and shear rate.

Curing Model in 3D TIMON CompositePRESS

The widely used semi-empirical Kamal model [121], which is a further development of a model proposed earlier by Kamal and Sourour [122], is applied in 3D TIMON CompositePRESS in order to describe the temperature-driven, n^{th}-order autocatalytic reaction of thermoset resins. Equation (28) expresses the curing reaction velocity as a product of a temperature term following the Arrhenius temperature dependence and a function of the degree of cure α. Typically, the resin curing behavior is measured by differential scanning calorimetry (DSC) analyses and the six unknown constants in Equation (28) can be fitted to the DSC data.

Kamal model:

$$\frac{d\alpha}{dt} = \left[A_1\exp\left(-\frac{E_1}{RT}\right) + A_2\exp\left(-\frac{E_2}{RT}\right)\alpha^m\right]\left(1-\alpha\right)^n,$$ (28)

where $\frac{d\alpha}{dt}$ stands for the curing reaction rate, A_1, and A_2 for pre-exponential factors, E_1 and E_2 for activation energies, R for the universal gas constant and m and n are reaction orders.

During the flow simulation this curing reaction model calculates the reaction rate, which is then used in the Cross–Castro–Macosko model to obtain the current viscosity of the flowing and simultaneously curing material. This coupling allows for a viscosity model that considers the effect of the curing kinetics on the conversion rate and the corresponding viscosity.

1.2.4. Computed Tomography of ROS Materials

In order to collect experimental 3D information about fiber orientation distributions of heterogeneous fiber reinforced materials, X-ray computed tomography (CT) is widely used in industry as a non-destructive measuring method due to the fact that it is easy to prepare samples and only requires a difference in the density-dependent linear X-ray attenuation coefficients of the matrix and the reinforcement [123]. The morphology of inhomogeneous materials like CF-SMCs can be investigated three-dimensionally by micro-CT (μ-CT), which is a high resolution X-ray CT method, allowing an in-depth material characterization [123].

To gain microstructural information with this experimental technique the sample is exposed to X-rays while incrementally rotating on a platform between the X-ray source and the detector (cf. Figure 11a). For each angle increment the sample is penetrated with radiation for a specific exposure time. Denser parts of the sample (fibers) absorb more radiation and therefore appear brighter in CT images than less dense material (resin). Since the X-ray radiation is proportionally attenuated as a function of the material's density, the detector can record a shadow projection of the sample revealing its inner structure [124]. The sample is fractionally rotated and irradiated until radiographic projections of a full 180° sample rotation are recorded. This acquired set of angular projections is sufficient to be reconstructed into a 3D model consisting of a large number of parallel micro-slice images by applying specific mathematical algorithms [124]. In order to improve the scan quality, a 360° rotation is usually carried out and several recordings per increment can be taken and averaged. The achievable scan quality (resolution and image sharpness) further depends on the spot/focal size of the X-ray tube, geometrical magnification, detector quality, vibrations during the image recording and the chosen combination of scan parameters [125].

In the reconstructed volumetric representation of the internal sample structure the scanned material is defined and visualized by its grayscale values (high density = high grayscale value). Therefore, composites show two maxima in the grayscale value histogram—one for the fibers and one for the matrix material. In order to analyze the fiber constitution, especially the fiber orientation, a threshold between both maxima is set by the user to define the surface of the fibers or strands, respectively. However, as this is a user-dependent manual act, analyzed fiber volume content distributions should only be used normalized.

In order to ensure a proper component design regarding mechanical requirements and for quality assurance of manufactured composite parts, the industry's endeavor is to determine the material's microstructure for large areas or ideally for entire parts [43]. However, due to the lack of contrast between polymer resins and carbon fibers (similar linear X-ray attenuation coefficients) in X-rayed composites, μ-CT scans of entire CF-SMC components are so far limited in size when receiving useful data for fiber orientation analyses is required. Normally, attaining fiber orientations by CT scan data analysis of fiber reinforced polymer parts requires a finer scan resolution than the fiber's diameter to distinguish between individual fibers, which by implication limits the scan volume size [126]. Obtaining CT data for a larger 3D part is therefore always a trade-off between scan volume size, or part size, respectively, achievable resolution (voxel size) and the required scanning time. In conclusion, with increasing sample size, the achievable scan resolution decreases, which makes it very time-consuming and costly or even impossible to resolve fiber-scale details in CT scans of entire composite parts [126].

Useful μ-CT scans for CF-SMC parts, however, just need a sufficient resolution so that it is clearly distinguishable between strands and resin by grayscale value differences related to local relative fiber volume fraction variations enabling a fiber orientation analysis with a common CT scan analysis software (e.g., the commercially available software package VGSTUDIO MAX 3.3., Volume Graphics GmbH, Heidelberg, Germany). The analysis algorithm within VGSTUDIO MAX 3.3 (VG hereafter) is indeed intended to be used for the orientation analysis of discretely visible fibers [42,127–130]. Yet with correctly set parameters and with local relative density gradients between resin and fibers the image analysis principles are suited to be used for scans with mesoscale resolutions, where no single fibers, only coarser structures like strands and fiber bundles, respectively, are visible [43,123]. Since microscopy shows that during compression molding the fibers within one strand mostly flow and orient together, deforming yet remaining as an intact strand with locally highly aligned fibers, the density gradients of CT scanned ROS-based materials are sufficient for a determination of local average strand orientations even at a coarser scan resolution [4,42,43,123]. In a CT scan of a ROS-based material the smallest density gradient is present in fiber direction, intermediate density differences are visible transverse to the strand orientation, and the highest density gradient occurs normal to the strand plane and at strand–strand boundaries (matrix rich strand interfaces), respectively [4,20,21,42,43,123].

Due to these intra- and inter-strand density changes the determination of strand orientations from mesoscale resolution CT scan data is feasible [4,20,21,43].

Denos et al. use a CT scan with a mesoscale resolution of 53 µm (voxel edge length) to determine the heterogeneous internal microstructure of a prepreg carbon/polyetherketoneketone (PEKK) UD strand-based T-bracket part with maximum dimensions of 65 × 65 × 45 mm [20,42,43]. However, at that resolution and with the applied CT scan settings it is not possible to distinguish between single carbon fibers ($\varnothing \approx$ 5–7 µm) and the matrix or to discern strand boundaries (~100 µm thick) in thickness direction [42]. Although Denos' CT scan configuration is not able to represent distinct strand boundaries, it is still possible to receive a mean local fiber orientation due to sufficiently pronounced relative density gradients [20,42].

Another common method to achieve bigger CT scan volumes at a reasonable resolution is to merge several scan volumes generating a digital twin of the scanned part and its microstructure. Sommer et al., Kravchenko, and Denos et al. merged 8 partial scans of a prepreg carbon/PEKK UD strand-based tensile test specimen with a size of 30 × 30 × 5 mm each, using a scan resolution of 15 µm [17,123,131]. At that resolution the CT scan quality is high enough to discern between strands and suitable for precise fiber orientation analyses. An analysis mesh size of 0.7 × 0.7 × 0.1 mm is used to determine a single orientation tensor from each of the measured orientation vectors by a grayscale analysis [123,131]. The finer analysis resolution better resolves the thin strands in thickness direction and enables gathering more detailed information about the local strand orientation changes.

The spectrum generated by the X-ray source significantly depends on the elemental composition of the used target material. Tungsten (W) is widely used as target material for microfocus X-ray sources. However, depending on the applied X-ray voltage and the absorption behavior of the sample material, alternative target materials might deliver beneficial spectrum characteristics that can improve CT measurement quality concerning the separation capability of fiber and matrix for fiber orientation analysis [132]. A higher µ-CT scan quality, by means of a higher contrast between fiber and matrix, also enables to scan bigger composite part volumes, fulfilling industry demands, where the fiber orientation within a whole component is of interest.

2. Material and Methods

2.1. Compression Molding

The high-performance CF-SMC used in this study is Hexcel's HexMC® (Hexcel Corporation, Stamford, CT, USA). This DFC material is designed for compression molding of complex 3D shaped parts in a heated metal tool. HexMC consists of prepreg carbon/epoxy UD tapes that are longitudinally slit and transversely chopped into strands and then randomly distributed into a mat [14]. Those ROS have nominal in-plane dimensions of 50 × 8 mm and a thickness of approximately 0.15 mm (Figure 3). The strands contain high strength carbon fibers impregnated by fast-curing Hexcel HexPly® M77 epoxy resin [14,133]. The carbon fiber content amounts to 62% by weight in the raw material, corresponding to 57% fiber volume content in a molded part and giving a material density of 1.55 g/cm^3 (Table 1) [14,15].

For the compression molding trial a research tool with a ribbed hat profile tool insert designed for compression molding of CF-SMCs is used (Figure 4a). The tool is made to mold a complex ribbed structure with ribs of varying heights and non-symmetrically alternating wall thicknesses with an attached plate (Figure 4b). The molded part has outer dimensions of 450 × 450 mm and a nominal wall thickness of 2 mm. The ribbed hat profile spans an area of 150 × 450 mm and has a hat height of 52 mm.

The ribbed hat profile parts analyzed in this study were manufactured with a mold coverage of approximately 80% using a 1000 ton compression molding machine (Dieffenbacher DCL-S 1000, Dieffenbacher GmbH, Eppingen, Germany) at a temperature of 140 °C, a pressure of 200 bar, a closing speed of 5 mm/s, and a closing time of 480 s (Table 2). The mold coverage would be high for standard SMCs, yet for ROS-based high-performance CF-SMCs it is rather low giving the material a sufficiently long flow path to create a flow-induced fiber mesostructure especially in the hat end brim section, which is wanted in this study.

Material flow simulations in 3D TIMON CompositePRESS were used to develop an SMC charge pattern (Figure 5), which allows for a 100% filling of the complex ribbed structure without defects, such as FMS. Predicted pressure distributions, shear rates, flow velocities and flow front advancements are analyzed for various charge patterns with the goal to reach uniform filling of the cavity. Due to the slightly asymmetrical design of the ribbed structure with very thin (0.8 mm) and very thick ribs (8.2 mm) of different rib heights, the cavity volume in the hat profile area changes locally. Therefore, the finally identified charge pattern used in this study consists of several smaller single charge packages optimized in size and position in order to achieve a balanced raw material distribution in the cavity and to facilitate optimal rib filling. Furthermore, the edges of some charge packages are placed underneath some rib entries to ease the flow of the UD strands into the ribs. All smaller charge packages are stacked on one 440 × 440 mm base layer to ease the charge placement in the tool.

In order to reach a lower viscosity for optimal flow behavior, the material charge was pre-heated for 17 s. This pre-heating time was found in a previous flowability study for HexMC plates. For the pre-heating procedure the charge is placed in the bottom mold half and the upper mold half is fast lowered to the pre-heating position leaving a small gap between the cavity wall and the material surface. Heat conduction from the lower mold half and heat radiation from the upper mold half decrease the material viscosity enabling better flowability when the mold is closed after the pre-heating phase.

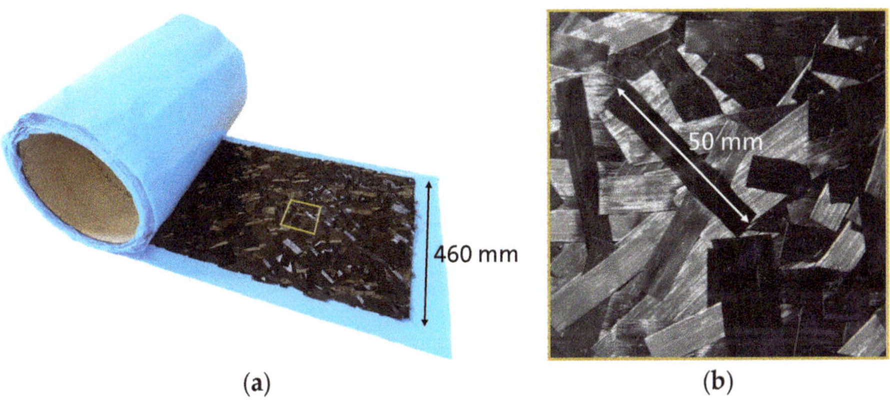

(a)   (b)

Figure 3. (**a**) HexMC raw material roll; (**b**) HexMC mesostructure with randomly in-plane oriented prepreg carbon/epoxy unidirectional (UD) strands.

Table 1. HexMC material properties.

Material Property	Value/Type	Unit
Fiber	High strength carbon	-
Fiber length	50	mm
Fiber density	1.80	g/cm^3
Resin	M77 epoxy	-
Resin density	1.22	g/cm^3
UD strand dimensions	$50 \times 8 \times 0.15$	mm
Material density	1.55	g/cm^3
Nominal fiber weight content	62	%
Nominal fiber volume content	57	%
Areal weight	2000	g/m^2

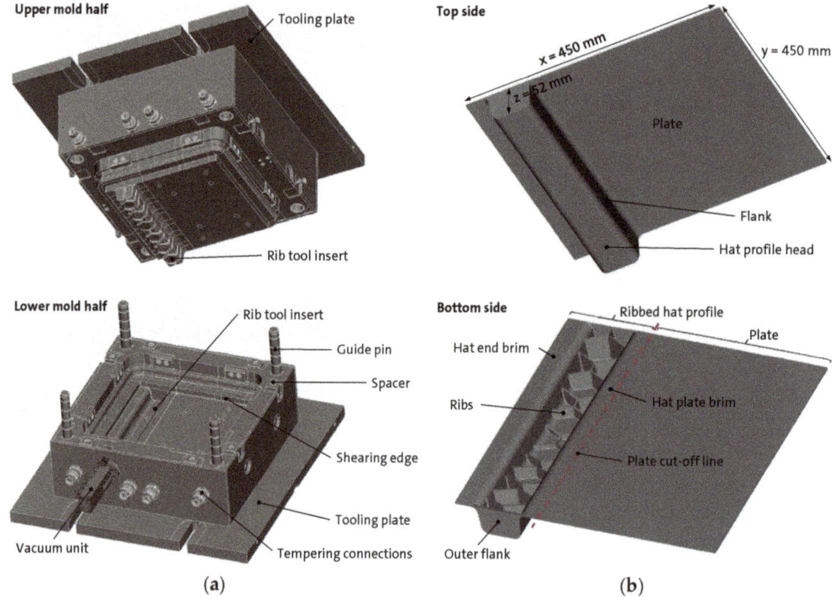

Figure 4. (**a**) CAD images of the upper and lower mold half of a research tool with a ribbed hat profile tool insert designed for compression molding of carbon fiber sheet molding compounds (CF-SMCs); (**b**) CAD geometry of the ribbed hat profile part with varying rib heights and thicknesses.

Table 2. Compression molding processing conditions.

Molding Parameter	Value	Unit
Mold temperature	140	°C
Preheating time	17	s
Pressure	200	bar
Closing speed	5	mm/s
Curing time	480	s
Charge weight	~1000	g
Mold coverage	~80	%

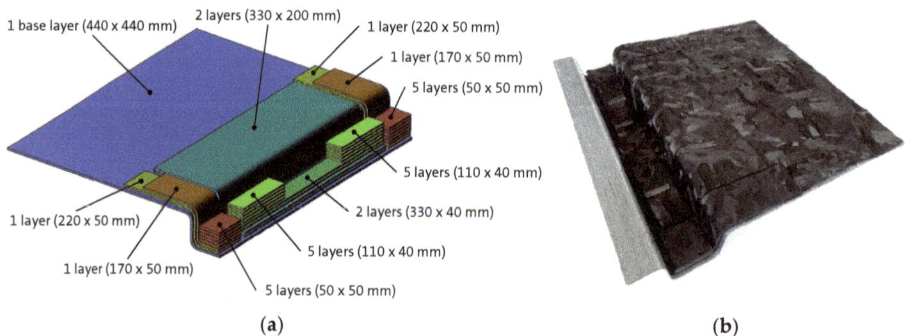

Figure 5. HexMC charge pattern for ribbed structure part; (**a**) schematic illustration with dimensions and number of layers per charge package (for visualization purposes the charge package are slightly separated in thickness direction); (**b**) prepared HexMC charge on a metal preform.

2.2. Short Shots

For accurate fiber orientation predictions with DFS tools it is crucial to obtain as precise filling predictions as possible. In order to receive a first understanding about the simulation quality, short-shot experiments were conducted with a shimmed mold for a comparison of the true material flow behavior and the numerical flow front advancement inside the mold cavity. For the shimming technique, two different shims are used as spacers between both mold halves preventing a complete mold closure but allowing the material to undergo the real pressure applied in the unshimmed mold. The used spacers have dimensions so that the press contacts the shims at intermediate positions leaving a gap between the mold halves of 8.65 and 4.00 mm, respectively. The schemes in Figure 6 illustrate the shimming technique. For the short shot trials the material charge pattern and the molding parameters remain the same as for the normal press experiments (cf. Figure 5 and Table 2).

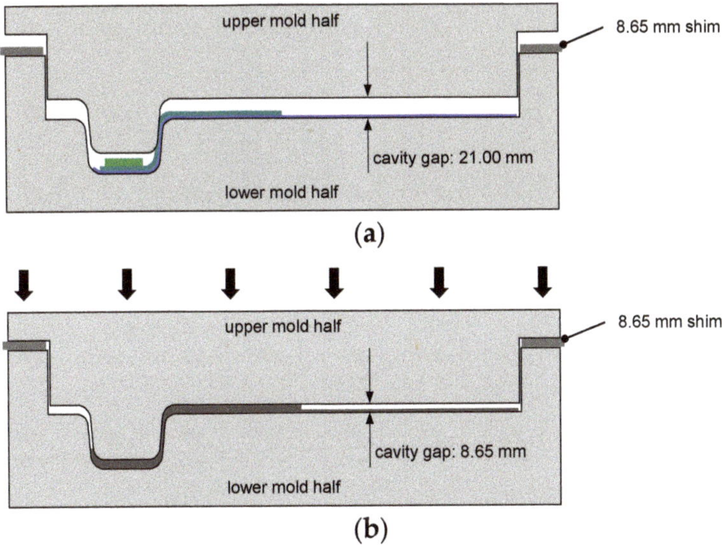

Figure 6. Schematic illustration of the shimming technique for short shot experiments in a cross-sectional view; (**a**) initial pre-heating position of the upper mold half; (**b**) final position of the upper mold half touching the 8.65 mm shims and showing the short shot of the pressed HexMC charge.

2.3. Compression Molding Simulation with 3D TIMON CompositePRESS

2.3.1. Filling Simulation

The 3D TIMON software module CompositePRESS is used to simulate the press process of the ribbed structure. It calculates the cavity filling and the resulting fiber orientations for the thermoset sheet-shaped HexMC charge. In 3D TIMON CompositePRESS there are two analysis methods available to simulate the compression molding process—the 'Euler method' and the 'morphing method' [105]. The Euler method uses a multi-layer voxel mesh to discretize the part by several elements over the thickness, whereas the morphing method utilizes a one-layer tetra mesh to represent the final part geometry (mold-closed shape). Advantages of the Euler method are the ease of 3D expressions for the charge definition and its flow as well as the possibility to discretize the part by several elements over the thickness. The accuracy of the movement, rotation and bending prediction of each single fiber improves when the number of elements across the thickness increases. However, this ultimately leads to a huge number of elements for more complex and/or larger components. Furthermore, due to the cubic-shaped voxel elements, a 3D-shaped component cannot be meshed without terraced steps in the contour.

In order to calculate the material flow during the compression molding process with the morphing method, the mesh is initially morphed in negative pressing direction, which means that the tetra elements are artificially stretched in z-direction to represent the open mold condition (cf. Figure 7). During the mold closure the morphed elements are then pressed back into their initial form (mold-closed shape) enabling the material flow calculation. The biggest advantage of the morphing method is that a tetra mesh can be used. Tetra elements allow easy meshing of complex geometries without terraced steps. This enables a high quality mesh consisting of a much smaller number of elements than a comparable voxel mesh. Therefore, the morphing method allows for faster analyses than the Euler method.

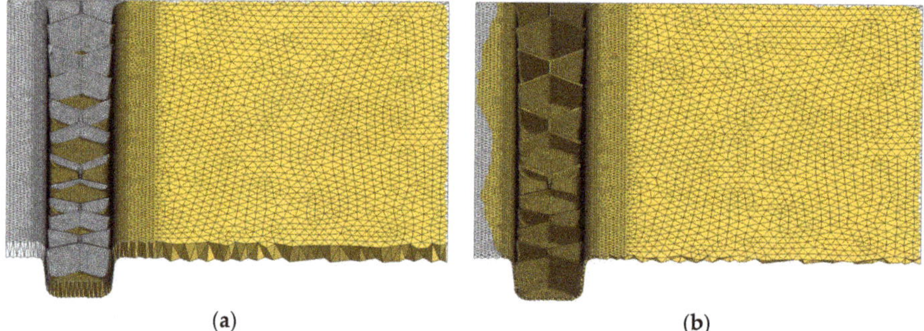

(a)(b)

Figure 7. Mold filling simulation in 3D TIMON CompositePRESS using the morphing method; (**a**) cross sectional view of the morphed tetra mesh in the initial open-mold position; (**b**) pressed back tetra mesh close to end of compression showing the proceeding flow front in the hat brim.

Considering the size and complexity of the ribbed part, the morphing method is the method of choice in this study since a high-quality tetra mesh without terraced steps can be used with a reasonable calculation time. As the applied Light 3D solver virtually divides each tetra element into 20 calculation points across the thickness, realistic flow predictions are possible even with a one-layer mesh. For the discretization of the entire part 141,385 tetra elements are used. Element sizes are varied to reduce the total amount of elements. Simple geometry areas of less interest, like the plate area, are meshed coarser with a maximal element edge length of 14 mm, whereas the hat area is discretized finer with a minimal element edge length of 0.4 mm in some rib fillets (cf. Figure 7) giving precise prediction results at a still reasonable calculation time. In total, 203 output steps are defined in order to gain a finely resolved filling analysis. Each charge package used in the physical moldings trials is also defined in 3D TIMON CompositePRESS by selecting all elements in the respective charge area and assigning the real charge thickness to those elements. Stacked charge packages are considered as one combined volume so that the actual thickness of each charge stack can be taken into account by the 20 virtual calculation points across the thickness of the stretched mesh (open-mold position). This enables a realistic modeling of the total charge volume.

Besides the chosen curing and viscosity models, good filling simulation quality depends on the material properties deposited in the material card. The interaction between the fibers and the fluid flow (two-way coupling) and the resulting anisotropic viscosity is not considered. Typical polymer characterization techniques, such as viscosity measurements by means of rotational viscometers, differential scanning calorimetry (DSC), and dielectric analysis (DEA) measurements are used in order to calibrate the HexMC material parameters for the applied Cross–Castro–Macosko viscosity model (Table 3) and to find fitting Kamal parameters (Table 4) to model the curing behavior appropriately.

Figure 8 shows three DSC curing curves at relevant molding temperatures of 150 °C, 140 °C, and 130 °C, one exemplary DEA measurement at 145 °C, and the corresponding fitted Kamal model curves. Additionally, three Kamal model curves at 155 °C, 135 °C, and 125 °C are plotted in gray to show the modeled curing behavior at slightly varied temperatures. The generated Kamal model curves fit well to the experimentally observed curing behavior of HexMC.

Table 3. Cross–Castro–Macosko viscosity model constants.

Constant	Value	Unit
a	1.000×10^{-11}	Pa s
b	1.100×10^{4}	K
D	1.500×10^{1}	-
E	-4.000×10^{0}	-
n	6.000×10^{-1}	-
τ^*	2.000×10^{2}	Pa
α_{gel}	8.500×10^{-1}	-

Table 4. Kamal curing model constants.

Constant	Value	Unit
m	6.100×10^{-1}	-
n	1.580×10^{0}	-
A_1	1.280×10^{7}	1/s
A_2	6.750×10^{10}	1/s
E_1	1.187×10^{4}	K
E_2	1.159×10^{4}	K

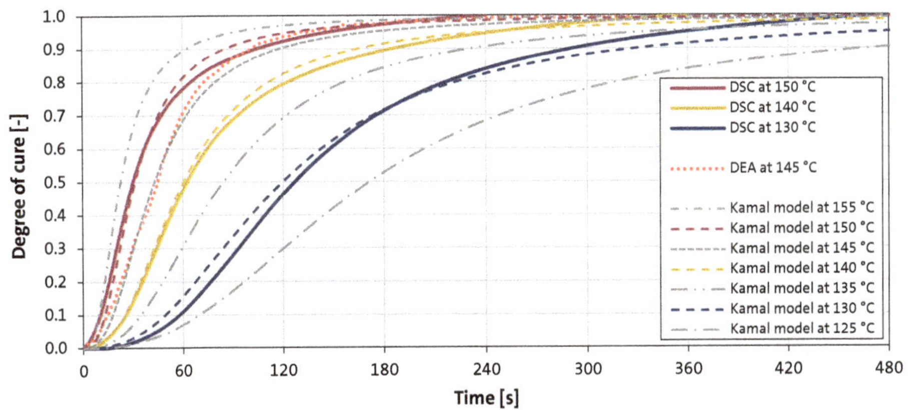

Figure 8. Isothermal differential scanning calorimetry (DSC) measurements of curing HexMC samples at different temperatures and fitted Kamal model curing curves using the identified parameters given in Table 4.

2.3.2. Direct Fiber Simulation (DFS) with Strand Generation Feature

A novel feature recently implemented in 3D TIMON CompositePRESS enables the simulation of ROS materials. In the pre-processing the new strand generation feature generates randomly oriented multi-bundle UD strands mimicking the real initial material charge. In contrast to the DBS of Meyer et al. [73], which uses one chain of truss elements per fiber bundle, 3D TIMON CompositePRESS represents strands by a grouping of several aligned numerical fibers (cf. Figure 9). Four fibers frame the outer strand dimensions and a user-defined number of additional inner fibers can be added inside the strand volume to represent a realistic amount of fiber bundles within a strand. Each fiber is still divided into a user-defined number of divisions (rods) giving the fibers the possibility to bend naturally in the flow. If it is assumed that a single fiber represents one fiber bundle within a strand, the simulation approach can be considered to be representative and gives the opportunity to mimic the strand geometry and character properly. Nonetheless, so far all fiber bundles defining one strand are numerically handled separately without a cohesive force between them.

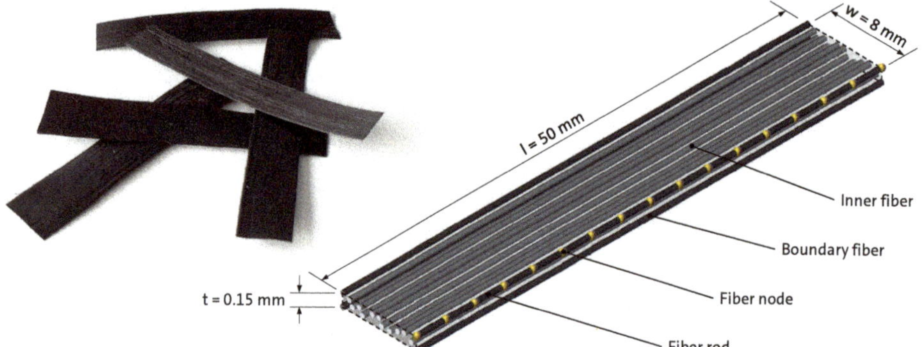

Figure 9. HexMC carbon fiber UD strands and a schematic illustration of their numerical counterpart modeled in 3D TIMON CompositePRESS with 4 boundary fibers defining the outer strand dimensions and 11 added inner fibers (here each fiber consists of 20 nodes and 19 rods, indicated on just one boundary fiber) (note: for visualization purposes length specifications are not true to scale).

More fiber divisions allow a more realistic bending behavior, especially in situations where fibers flow into small features such as ribs. Regarding the calculation effort, the number of divisions as well as the number of additional fibers should be carefully chosen. However, the more fibers are calculated, the better the simulation accuracy and the better the fiber volume fraction representation. The numerical strands have dimensions of $50 \times 8 \times 0.15$ mm, equal to the real HexMC UD strands. For this study each strand is represented by 4 boundary fibers and 11 inner fibers, all divided into 19 rods connected by 20 nodes (cf. Figure 9). Using 15 fibers to represent the designated strand volume is chosen as a compromise between the ability to visually analyze the strands' movements and deformations during the DFS and the needed calculation time.

For the DFS of the ribbed structure a random in-plane strand orientation distribution in the 3D-shaped initial charge is generated (Figure 10a). The parameters for the strand generation algorithm are set so that one strand is generated per mesh element and that the minimum distance between initial fibers is zero, which means that fibers can be generated close to each other. The strand generation approach does not model the strands as enclosed volumes, so that strands intersect each other and compaction is not considered. In Figure 10b the numerically defined individual charge packages are displayed in different colors. Both the macroscopic shape of the generated charge and the distribution of strands are comparable to the physical HexMC. Since the strands are cut at each edge of a charge package, it is possible that incomplete and/or shorter virtual strands are generated—just like in reality. Even in the almost vertically oriented cavity zones the numerical representation of the initial charge pattern is reasonably realistic when compared to the true charge configuration (cf. Figure 5b). As only straight fibers are generated, a negligible amount of fibers is cut in order to fit into the mesh in tight radii. Initially, the generated strands are not bent in order to follow the curvature of the charge in those regions.

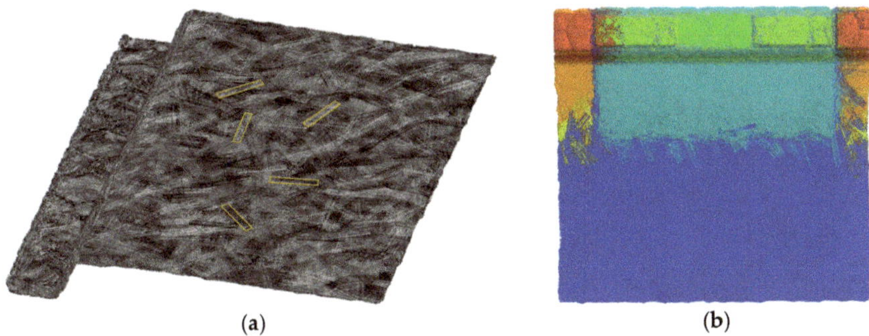

Figure 10. (**a**) Numerically modeled initial charge pattern for the ribbed hat profile part consisting of in-plane randomly oriented UD strands (5 UD strands are highlighted); (**b**) bottom view of the numerical charge definition where each charge package is colored differently for better discernibility.

2.4. Micro-CT Measurements

Three entire ribbed hat profile parts were μCT scanned with a CT-AlphaDuo device (Procon X-Ray GmbH, Sarstedt, Germany) operated by the Fraunhofer WKI, Hannover, Germany. The system is equipped with a 240 kV microfocus X-ray tube XWT-240-TCHE Plus (X-RAY WorX GmbH, Garbsen, Germany) with a high energy diamond/tungsten transmission target offering a JIMA (Japan Inspection Instruments Manufacturers' Association) test resolution of 0.9 μm. The X-rays are recorded by a 249 × 302 mm PaxScan® 2530DX detector (Varian Medical Systems, Inc., Salt Lake City, UT, USA) with a resolution of 1792 × 2176 pixels. In order to maximize the voxel resolution at the given sample diameter, the detector panel width was virtually extended by measuring field extension (MFE) (in situ horizontal movement of the detector panel). The focus object distance (FOD) and the focus detector distance (FDD) were 642 mm and 1500 mm, respectively. For best possible scan results, the parts' plates were removed from the hat section with a water jet cutter. This improves the CT scan quality since the penetration length is shorter and thereby induced scan defects (e.g., the resulting image noise) are kept low. Each ribbed part was mounted on a rotating table between X-ray source and detector (see Figure 11a). The whole scan setup is built on an air-damped granite base in order to minimize vibration-induced scan artifacts. For the measurements the X-ray tube was operated at a voltage of 160 kV and a current of 250 μA. The sample was rotated in 2400 rotation steps (0.15° angular increments) for a full 360° rotation with an integration time of 6 × 200 ms per angular increment. This was repeated for both detector positions. The measuring time for each partial scan accumulated to approximately 100 min.

All samples were scanned at a mesoscale resolution of 59.6 μm (voxel edge length) corresponding to approximately 16.8 pixels/mm. Each voxel contains a single 16-bit relative density-defined grayscale value between 0 and 65,535. With the applied scan resolution, which is approximately 10× the carbon fiber diameter, it is possible to clearly see in-plane strand boundaries. However, the strands are only resolved with ~2.5 voxels in thickness direction (minimum fiber bundle dimension), which is not enough for clear boundary detections in that strand dimension. Further, a 1 mm thick aluminum filter is used in front of the X-ray target in order to reduce the effect of beam hardening. As a result of different penetration lengths of low and high energy X-ray photons beam hardening causes grayscale value gradients inside the sample, which can especially affect grayscale value sensitive analyses such as the fiber orientation analysis. By using a filter, low-energy X-ray photons are suppressed and thereby the X-ray spectrum is shifted to higher energies. In previous experiments this specific CT scan configuration has proven to be conducive for reasonable fiber orientation analyses of larger CF-SMC components. Reconstruction of the 3D volumes was performed with the software 'Offline CT' (Fraunhofer Institute for Integrated Circuits IIS, Erlangen, Germany). Table 5 summarizes the applied μCT scan parameters.

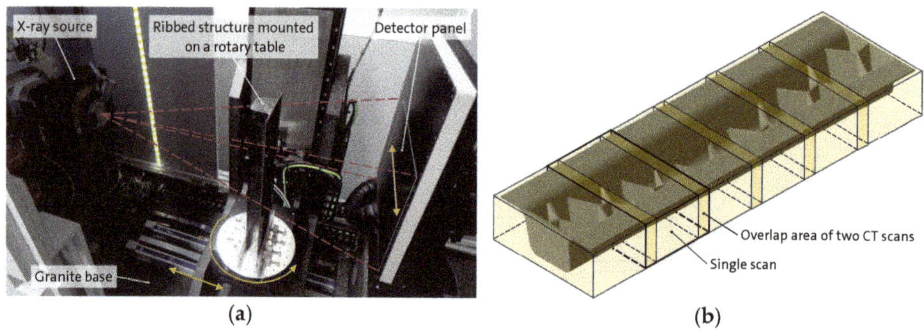

Figure 11. (**a**) Exemplary CT measurement setup (note: here, focus object distance (FOD) and the focus detector distance (FDD) are not the ones used for the scans); (**b**) schematic illustration of merged CT scans with overlap areas.

Table 5. Micro-CT measurement parameters.

Scan Parameter	Value	Unit
Voxel resolution	59.6	μm
X-ray voltage	160	kV
X-ray current	250	μA
Focus object distance (FOD)	642	mm
Focus detector distance (FDD)	1500	mm
Integration time	6×200	ms
Projections	2×2400	-
Measuring time per partial scan	100	min

To capture the entire ribbed structure at the necessary resolution, seven partial scans were performed and afterward virtually stitched together using the CT scan visualization and processing software VGSTUDIO MAX 3.3. As the individual CT scans slightly vary in their grayscale value distribution, for each 3D data set the average grayscale values of background and material were determined from the grayscale value histograms. Based on these values the volumes were imported into VG applying the 'histogram calibration' function to create a homogeneous grayscale value profile over all scans. Each individual scan captures a sample section of about 150 mm in width, 52 mm in height, and 90 mm in axial direction (Figure 11b) leading to an overlap of at least 15 mm between adjacent scans that eases the manual alignment and merging of the imported partial volumes. The alignment and merging procedure was done with utmost care so that any negative effect on the final 3D data set and the subsequent CT scan analysis is ruled out. Regions of interest (ROIs) were created defining the area of two adjacent volumes to be merged. The filter 'merge volumes' in mode 'mean' was applied to combine two adjacent volumes based on these ROIs. By repeating this procedure, one continuous 3D volume of the entire sample was composed of the partial scans.

The surface determination was conducted on the merged volume so that the scanned parts' thicknesses match with the average true thicknesses of the molded parts measured with an electronic external measuring gauge. The final object volume is then used as ROI where the background (surrounding air) is removed. This reduces the 3D volume data to be analyzed leading to faster fiber orientation calculations. The final volume of each stitched ribbed part has outer dimensions of $150 \times 450 \times 52$ mm and an analyzed volume of approximately 350,000 mm^3 captured by almost 1.78 billion voxels.

All three scanned ribbed structures are analyzed regarding their second order fiber orientation tensors (hereafter FOTs) in several sample areas using the fiber composite material analysis (FCMA) tool implemented in VG. The same tetra mesh that is used for the 3D TIMON CompositePRESS process simulation is imported to VG and is utilized as an integration mesh for the CT scan analysis.

Therefore, each 3D volume is registered to the tetra mesh's coordinate system, so that the scanned ribbed structures are perfectly superimposed with the tetra mesh. With this procedure a one-to-one comparison with the process simulation results is possible since the same tetra elements used in the process simulation can be utilized for the fiber orientation analyses in VG. For each element of this analysis mesh the FOTs, eigenvectors, and eigenvalues are determined and evaluated.

3. Results

3.1. Compression Molded Ribbed Structure

Typically, long fiber reinforced plastics (LFRP) are predestined to interlock and accumulate for instance at the entrance of ribs causing FMS. This effect is even more pronounced for SMC that consist of ROS since the UD strands are stiffer than single long fibers, which impedes the flow into narrow structures such as ribs. Filling the complex ribbed part with 50 mm long UD strands is therefore challenging as FMS is highly likely. Simple charge patterns and non-optimal compression molding parameters led to incomplete part filling, as depicted on the left hand side in Figure 12. Simulating several different charge configurations in 3D TIMON CompositePRESS showed uneven pressure distributions and finally helped to develop a complex charge pattern with a more even pressure distribution. The virtually developed final charge pattern (see Figure 5) allows a balanced flow to all cavity areas for an even part filling without visible FMS. The cut strands at the edges and the size and position of each charge package help to fully fill the part. Lowering the material viscosity by the previously described pre-heating procedure increases the flowability when the material is compression molded, which further prevents FMS. Moreover, the charge pattern ensures a good venting, so that no air traps occur. The incompletely filled part in Figure 12 is juxtaposed with a fully filled ribbed part after applying the pre-heating technique and using the final process simulation optimized charge configuration.

Figure 12. Ribbed structure compression molded with a simple quadratic HexMC charge pattern and unsuited molding parameters showing extensive fiber matrix separation (FMS) (left) compared to a part molded with a pre-heated and optimized charge pattern developed by means of filling simulation studies using 3D TIMON CompositePRESS (right) (plate areas are removed from the ribbed hat profiles).

Figure 13 shows a picture of three randomly picked ribbed hat profiles that were pre-heated and then compression molded with the optimized charge pattern. These three parts are CT scanned for a comparison with the DFS results of 3D TIMON CompositePRESS. The detailed view on the right hand side shows the middle section of sample #19 in bottom view. In the plate brim of the hat profile just slightly deformed strands can be seen. In contrast, the strands are exposed to complex flow conditions and undergo high shear forces when they flow from the initial charge edge in the hat to the end of the cavity in the end brim. The flow path length is approximately 100 mm. Therefore, the end brim is characterized by highly deformed strands split into fiber bundles. At the flow path end these bundles align along the cavity wall.

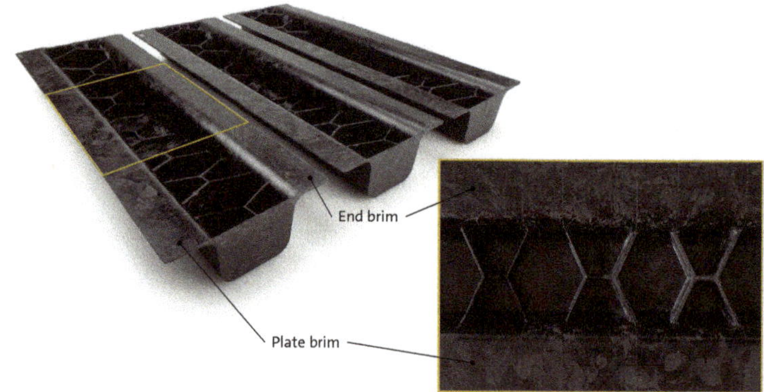

Figure 13. Three compression molded ribbed structures made of HexMC (plate areas are removed from the ribbed hat profiles) picked for CT scanning and a detailed view on the right hand side showing the middle section of sample #19 in bottom view.

3.2. Filling Simulation Results Compared to Short Shot Experiments

In velocity-based DFSs good fiber orientation predictions depend on the accuracy of the filling simulation. In order to check and evaluate the flow prediction accuracy in 3D TIMON CompositePRESS short shots were conducted and are used as first quality indicator in direct comparison with the simulation results. Figure 14 shows the predicted mold filling in 3D TIMON CompositePRESS and qualitatively compares it with the real flow front advancement in the corresponding short shot. When the mold closing is interrupted at a cavity gap of 8.65 mm the ribs have begun to fill (dark grey colored) and the numerical filling status is in very good agreement with the experiment. Due to its higher thickness the charge material in the ribbed area is pressed first, whereas the base layer material in the plate area has no contact to the upper mold half yet (light grey colored). Moreover, at a remaining cavity gap of 4.00 mm the proceeding flow front in the end brim of the hat profile is almost identical in experiment and simulation (see Figure 15). In the plate area the numerical flow front is slightly faster than in reality. In both, in reality and in the simulation the last unfilled rib is the left outer flank, whereas the right outer flank is already filled. Since these numerical filling results are in such a good agreement with the experimentally observed flow front advancement in the short shots, especially in the complex geometry of the hat area, it is assumed to have an accurate flow field prediction as basis for the subsequent DFS. In total, the flow simulation takes only 21 min (wall clock time) using four cores of a standard tower PC (Intel® Xeon® E-2246G CPU @ 3.60 GHz, 32 GB RAM).

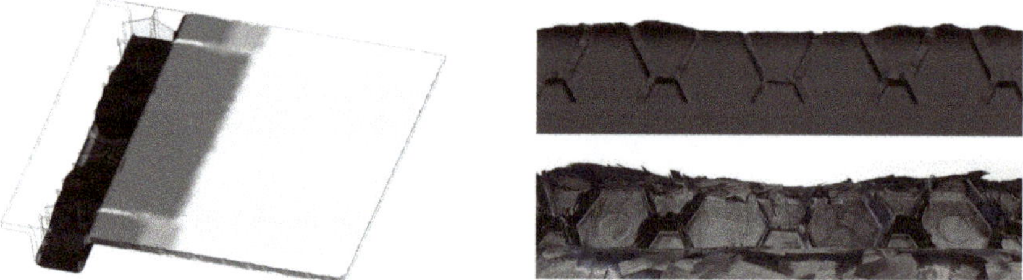

Figure 14. 3D TIMON CompositePRESS fill simulation status compared to a real HexMC short shot using 8.65 mm shims.

Figure 15. 3D TIMON CompositePRESS fill simulation status compared to a real HexMC short shot using 4.00 mm shims (numerical and real short shot are shown from different viewing angles).

3.3. Direct Fiber Simulation (DFS) Results

For the DFS of the ribbed part 1,193,306 fibers are generated in the initial charge corresponding to approximately 62,800 strands. The DFS simulation for this number of fibers, the amount of tetra elements and the number of output steps takes 112.41 h or 4.68 days (wall clock time, corresponding to 202.96 h or 8.46 days in CPU time), respectively, using four cores of a standard tower PC (Intel® Xeon® E-2246G CPU @ 3.60 GHz, 32 GB RAM). Since fiber attrition is insignificant during compression molding of CF-SMC materials, fiber breakage is not simulated in this work.

Figure 16 shows the initial charge with the generated UD strands at the start of the DFS and the flowing strands close to the end of the press process. At the end of the end brim the fiber bundles orient parallel to the cavity wall and the material flow stops. The proceeding flow front in the plate area is characterized by blurriness due to the deformed fibers. This effect can also be observed in Figure 17, where the movement and deformation of one highlighted UD strand in the lower right corner of the plate is depicted. The images clearly show how the UD strand flows and slightly rotates. From the images it can be seen that the fibers building one strand (cf. the purple colored fibers of one individual strand) are following the flow field as a grouping, although there is no cohesive force between the fibers of one strand. The initially straight fibers deform into a zigzag shape causing the blurriness. This behavior corresponds to the observations made in reality, where the UD strands stay intact and flow together in the plate area. For a one-to-one comparison, images of the initial strand configuration in the real HexMC charge and of flown strands in a molded part are given in Figure 18. Under more complex flow conditions, like in the ribbed structure, a strand splitting can be seen (cf. Figure 19).

The DFS shows how the fiber bundles of one blue highlighted strand flow into and through a rib. After exiting the rib the fibers separate. This effect is also visually notable on the surface of molded parts (for example in the end brim area in Figure 13) and in CT scans of the same area (see Figure 21).

Figure 20 shows the predicted fiber orientations in the entire ribbed hat profile displayed as vectors. After flowing through the ribs the fibers align in flow direction. Between the rib exits a more random orientation is visible. Highly oriented and random orientation area alternate. At the end brim it comes to a distinct alignment with the cavity wall. Following the melt flow direction, the fibers point into the left and right end brim corners, which are the last filled areas of the ribbed structure. Due to the more even flow conditions in the plate brim a uniform fiber orientation distribution can be discerned there. The two connecting rib bridges at the left hand side and the right hand side of the ribbed structure exhibit a complex fiber orientation, whereas the three smaller connecting rib bridges, shown in the cross section in Figure 20b, are characterized by almost homogeneous horizontal fiber alignment.

(a)   (b)

Figure 16. (**a**) Progressing flow front and deformed UD strands during the direct fiber simulation (DFS) with 3D TIMON CompositePRESS at an intermediate time step; (**b**) DFS result close to the end of compression.

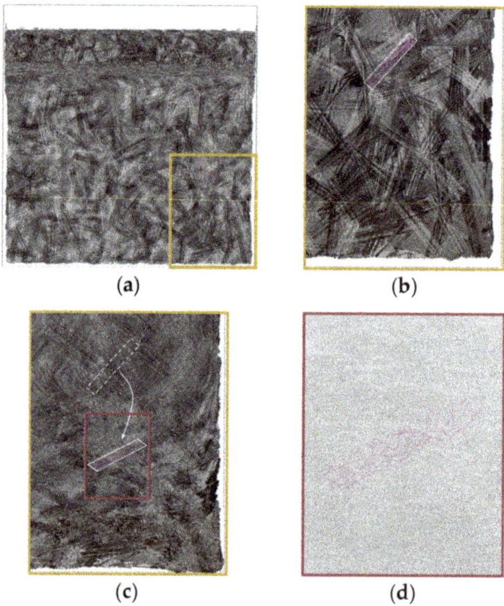

(a)   (b)

(c)   (d)

Figure 17. (**a**) Initial virtual charge configuration with randomly oriented UD strands consisting of 15 fibers each; (**b**) highlighted UD strand (purple colored) at the plate surface before molding; (**c**) same UD strand near to the end of its flow path; (**d**) detailed view showing the 15 slightly deformed single fibers of the flown UD strand (for visualization purposes all other strands are colored in light gray).

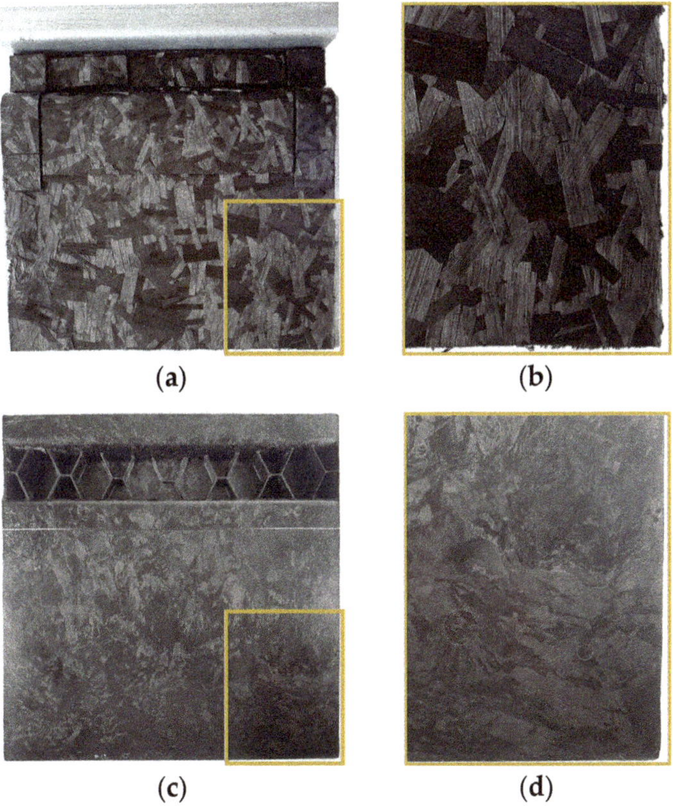

Figure 18. (**a**) Photo of the initial HexMC charge configuration with randomly oriented UD strands in several charge packages on a metal preform; (**b**) detailed view of the UD strands in the lower right corner of the base charge; (**c**) compression molded part; (**d**) detailed view of the lower right corner of the plate showing the flown and slightly deformed UD strands.

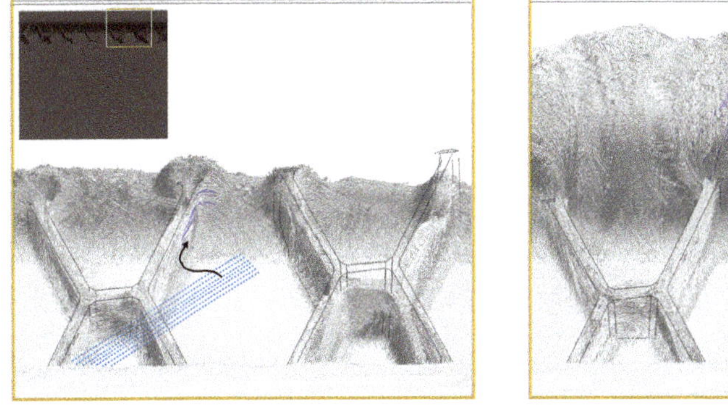

Figure 19. Virtual UD strand (colored in blue; initial position blue dotted) in the hat profile area deforming and splitting after flowing into and through a rib shown at two different time steps (all other fibers are colored in light gray).

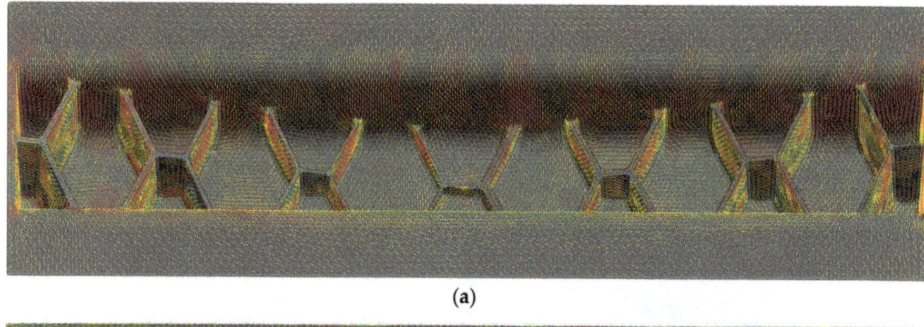

(a)

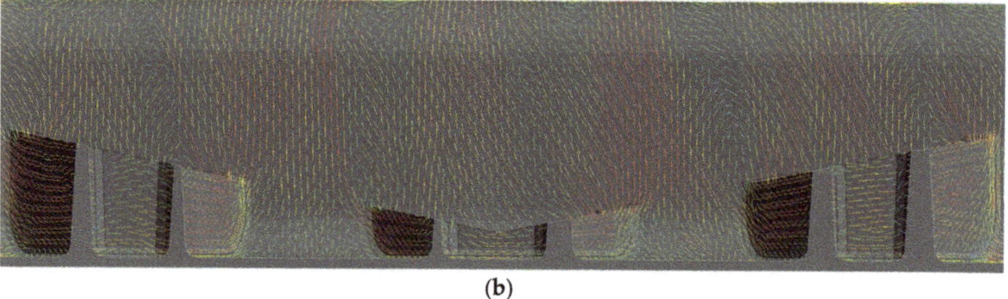

(b)

Figure 20. (a) Predicted fiber orientation vectors displayed at the part surface of the hat profile; (b) detailed view of a cross section (front view) showing the fiber orientation vectors in the three middle ribs.

3.4. Fiber Orientation Measurements (VGSTUDIO MAX 3.3)

For this study a working combination of CT scan hardware and scanning parameters is found that allows for full-sized analyses of fiber orientations in carbon fiber composite parts. Three randomly picked samples of a series of ribbed HexMC parts are CT scanned. A first example of the achieved scan quality is given in Figure 21. The detailed view of the right end brim section of sample #20 shows a flow-induced mesostructure, where individual fiber bundles can be clearly discerned. This determined mesostructure also resembles the observed strand splitting in the end brim predicted in 3D TIMON CompositePRESS. In the grayscale image the denser carbon fiber bundles are defined by white voxels, whereas the pure low-density epoxy is represented by black voxels. Consequently, gray pixels indicate the homogenized mesoscale density variations. These variations span a range from the density at maximal fiber volume fraction (nominal 57 vol.%) in tightly compacted strands (bright voxels) to the epoxy density occurring at strand boundaries, strand intersections, and in pure resin areas (dark voxels) due to FMS.

The CT images of all three scanned samples in Figure 22a allow a visual analysis of the fiber bundle orientations in the middle section of the hat profile and also a comparison among the scans. The plate brims show less deformed and more randomly oriented UD strands, whereas in the end brims clear alignments of split fiber bundles can be seen. In some areas slight indications of FMS are visible. In the head area of the hat profiles the ribs stand out. This is due to the material flow into the ribs leading to highly oriented areas underneath and within each rib.

For the fiber orientation analysis with the FCMA tool in VG, fiber bundles and matrix material are distinguished by appropriate thresholding in each of the three merged scan volumes. Although the applied scan resolution of 59.6 µm is not fine enough to see discrete carbon fibers or to exactly differentiate between individual strands in thickness direction (~150 µm strand thickness), it is sufficient to determine strand orientations, proven in Figure 22b. This is possible due to the large strand scale and since the direction of least density change within a strand stack is aligned with the strand's longitudinal axis. Instead of identifying single fibers for the orientation analysis, the FCMA algorithm detects inter- and intra-strand density gradients and uses them as indicators for the local fiber bundle orientation. The local relative density gradients in the recorded voxel data can thus be used to determine the mean orientations for each element volume of the integration mesh without even capturing all strand boundaries. In Figure 22b the determined fiber orientations in the brim areas are displayed as ellipsoids in the end brim area and as compass needles representing the 1st eigenvectors in the plate brim area, which is a typical second method to visualize fiber orientations in composite parts. The rounder and flatter the green ellipsoids get, the more random in-plane are the determined bundle orientations. This is especially visible in the middle section of the end brim, whereas the outer sections of the end brim are characterized by more distinct bundle alignments, shown as red elongated ellipsoids. The measured orientations indicate reasonable bundle orientations proving that the local mesoscale density variations in the scanned parts can be used to analyze the fiber orientations by VG.

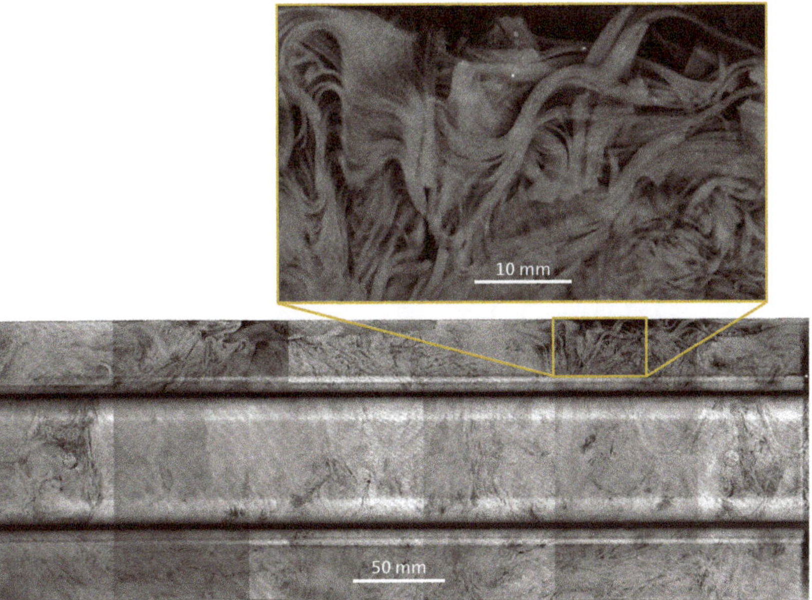

Figure 21. CT scan of sample #20 showing fiber bundles of split UD strands on the surface of the hat profile's end brim after a flow length of approximately 100 mm.

In Figure 23a a detailed 3D view of the hat profile middle section of sample #20 shows fiber bundle orientations at the part surface and inside the slightly removed brim areas. The FOTs visible as ellipsoids in Figure 23b are determined using the process simulation tetra mesh also used for the fill simulation and DFS in 3D TIMON CompositePRESS. For better visibility the CT scan data are set to 100% transparency so that only the ellipsoids are visible. Highly oriented fiber bundles, displayed as red elongated ellipsoids, characterize the ribs and also some areas in the head and in the end brims. Especially the rib exits shows high bundle alignments in flow direction. At the flow path end at the outer edge of the end brim it comes to a clearly visible alignment with the cavity wall. The plate brim shows a more random orientation status. Overall, the image color is more green than red, which indicates that most of the hat profile has a random in-plane fiber bundle orientation, displayed as green flattened ellipsoids. Only in areas with strong material flow the initially random strand orientation of the raw material changes into a process-induced mesostructure with areas of distinct bundle orientations (red elongated ellipsoids). Since the process simulation mesh has only one element across the part thickness, it averages the bundle orientations over the thickness for each element volume. However, it still delivers reasonable FOTs when visually compared with the CT scan grayscale image. Therefore, an easy one-to-one comparison with the DFS results on the same mesh in order to evaluate the simulation accuracy is possible.

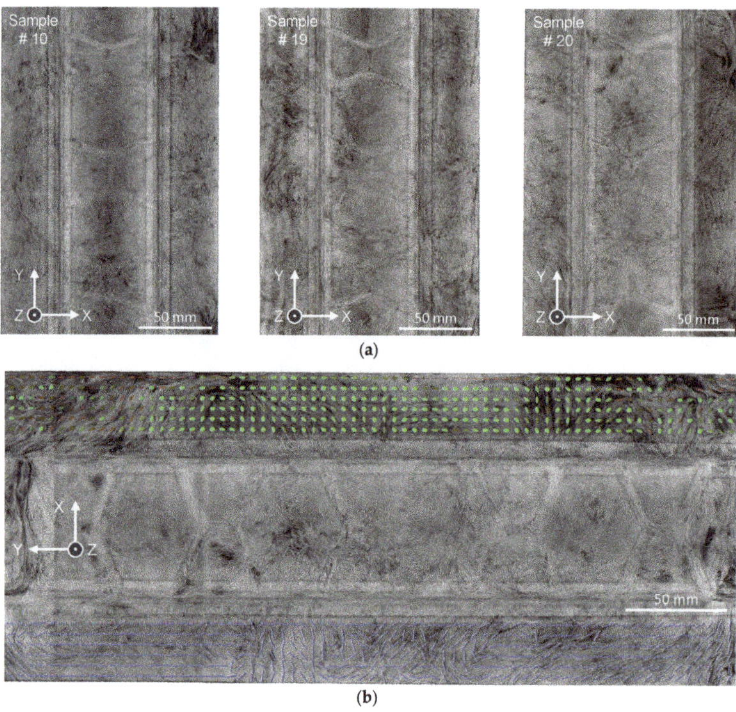

Figure 22. (**a**) CT images of the hat profile middle sections of all three scanned parts (top view; 1 mm of the each part surface is removed to see the inner strand bundle orientations); (**b**) fiber bundle orientations determined by VGSTUDIO MAX 3.3 on a 5 × 5 mm integration mesh in a wider section of sample #20 displayed as ellipsoids in the end brim and as vectors (1st eigenvector) in the plate brim.

Figure 23. (**a**) Detailed 3D view of the CT scanned ribbed structure #20 (displayed are density gradients at the part surface; the surface of the hat brims is slightly removed to show the carbon fiber bundle orientations inside the brims); (**b**) determined FOTs of the same part displayed as ellipsoids in VGSTUDIO MAX 3.3 (analysis based on the process simulation tetra mesh; CT scan data are set to 100% transparency) (red elongated ellipsoids: fibers highly oriented in one direction, green flattened ellipsoids: planar fiber orientation, blue spherical ellipsoids: 3D random fiber orientation).

3.5. Comparison of Predicted and Measured Fiber Orientations

All three CF-SMC hat profiles show visually similar fiber bundle orientations. This observation can be proved by comparing determined FOTs, eigenvectors, and eigenvalues of all three completely scanned parts. In order to eliminate the expectable local fiber orientation differences between the scanned parts, certain areas in the hat profile are used for the analysis. The determined orientation values for all tetra elements within these analysis zones are averaged to give a representative orientation status in that area. The analysis areas with their designations are given in Figure 24. The analysis boxes are superimposed with the tetra mesh and the box transparency makes it possible to discern which elements are used for the analysis.

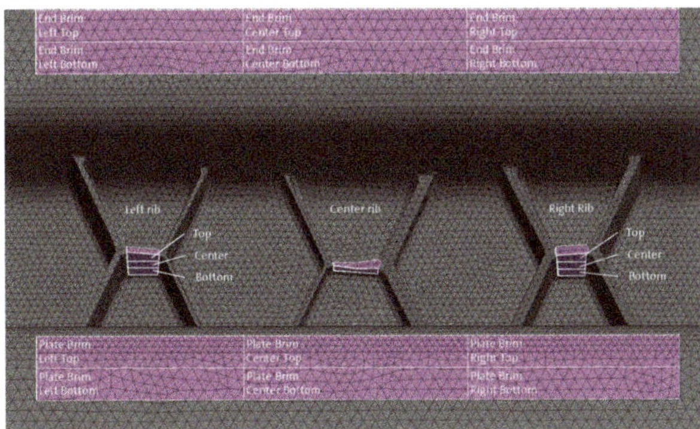

Figure 24. CAD image of the ribbed structure's middle section (3D view, superimposed tetra mesh) with all analysis areas used for the comparison of the averaged fiber orientations of all three CT scans with the fiber orientations predicted with 3D TIMON CompositePRESS.

The orientation angles for the rib analysis areas are determined by calculating angle θ (see Figure 25). The conversion equation for spherical coordinates given in Equation (29) is applied on the 1st eigenvectors in x-, y-, and z-direction and subsequently the calculated angle is projected onto the yz-plane. For the brims and head analysis areas in the xy-plane Equation (30) is used to calculate angle φ (see Figure 25) based on the 1st eigenvectors in x and y-direction.

$$\theta = \arccos\frac{z}{\sqrt{x^2 + y^2 + z^2}}. \tag{29}$$

$$\varphi = \arctan2(x, y) = \begin{cases} \arctan\left(\frac{y}{x}\right), & \text{if } x > 0, \\ \arctan\left(\frac{y}{x}\right) + \pi, & \text{if } x < 0 \wedge y \geq 0, \\ \arctan\left(\frac{y}{x}\right) - \pi, & \text{if } x < 0 \wedge y < 0. \end{cases} \tag{30}$$

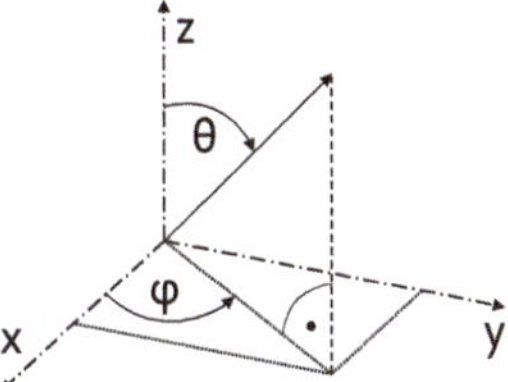

Figure 25. 1st eigenvector (displayed as arrow) of a fiber orientation in 3D space described by the polar angle θ and the azimuthal angle φ (Eulerian angles) in a Cartesian coordinate system.

The CT scan results with the determined average values and standard deviations (SD) for the seven rib analysis areas are given in Table 6. The average fiber orientation components A_{xx}, A_{yy}, and A_{zz} have mean standard deviation between all three scans of just 0.03, 0.04, and 0.03. Furthermore, the measured average orientation angles show a very low mean standard deviation of 4.1°. All measured angle SDs are below 4°, the only higher SD has the left rib center analysis area. Since there is a distinct uniform material flow in the rib areas, these low differences between the three CT scans are expectable.

Table 6. Averaged fiber orientation measurement results for the rib analysis areas of three CT scanned ribbed hat profiles.

Analysis Area		A_{xx}	A_{yy}	A_{zz}	1st Eigenvalue	2nd Eigenvalue	Measured Angle θ
Left Rib	Top	0.05	0.70	0.26	0.73	0.23	79.4°
	SD	0.02	0.04	0.02	0.03	0.01	2.6°
	Center	0.06	0.60	0.34	0.65	0.30	87.0°
	SD	0.03	0.02	0.00	0.03	0.01	15.0°
	Bottom	0.07	0.61	0.32	0.65	0.28	74.8°
	SD	0.02	0.02	0.01	0.02	0.00	1.4°
Center Rib		0.10	0.71	0.19	0.73	0.21	96.5°
SD		0.03	0.06	0.04	0.07	0.04	0.6°
Right Rib	Top	0.15	0.61	0.24	0.66	0.25	101.2°
	SD	0.06	0.09	0.04	0.09	0.05	2.5°
	Center	0.12	0.54	0.34	0.61	0.30	107.2°
	SD	0.03	0.03	0.05	0.05	0.04	3.6°
	Bottom	0.09	0.49	0.42	0.60	0.32	115.5°
	SD	0.03	0.03	0.05	0.05	0.03	3.1°
Mean SD		**0.03**	**0.04**	**0.03**	**0.05**	**0.03**	**4.1°**

* normalized to a positive angle between 0° and 180° related to the z-axis in the yz-plane.

In the brim analysis areas the mean standard deviation for the main FOT components A_{xx}, A_{yy}, and A_{zz} are comparable low (Table 7). The mean SD for the tensor component A_{zz} is even lower than in the rib areas as in the brim and head areas a low fiber orientation component in z-direction is apparent. The mean SD for the measured orientation angles is at 18.6°. This deviation is reasonably small considering the initially random in-plane orientation of the UD strands and shows that there is a measurable flow-induced mesostructure.

The consistency between the three scans, ascertainable by the low standard deviations, is justifying to average the fiber orientation results of all three samples in the hat profile section in order to get a representative depiction of the average mesostructure. Furthermore, the averaged FOTs can subsequently be used to compare the CT scan measurements with the process simulation results. This comparison is a direct method to validate the predicted fiber orientations.

In Table 8 the predicted fiber orientation results coming from the DFS in 3D TIMON CompositePRESS are given for the rib analysis areas. Here the absolute errors with the averaged CT scans in the respective analysis areas are given in order to quantify deviations. To indicate the overall degree of error for all regions of interest, the mean absolute errors (MAEs) are given. The MAE for the FOT components is 0.06 for A_{xx}, 0.15 for A_{yy}, and 0.16 for A_{zz}. The highest FOT and orientation angle deviation between CT measurement and prediction are observable in the right rib's bottom analysis area. This could be linked to the rib's base wall thickness, which is thicker than at the other two ribs. Here, the narrow gap assumption of the Hele-Shaw simplification method applied in 3D TIMON CompositePRESS might have reached its limits. The 1st and 2nd eigenvalues have a MAE of 0.12 and 0.10, respectively. These MAE values show that the DFS results are not perfect but certainly realistic. The predicted orientation angles have a comparably low MAE versus the measured bundle orientation angles in the CT scans of only 11.6°, which explicitly shows that the orientation behavior is correctly captured by the DFS.

In order to make the comparison between the measured and the predicted fiber orientations in the respective analysis areas easier, the orientations are visually superimposed as 2D ellipses in Figure 26. Fiber orientations in space can be easily visually represented by ellipsoids. However, in stationary 2D images it is easier to demonstrate FOTs as ellipses, neglecting the extension in the third room direction. For the visualization of the mean bundle orientations in those analysis areas the determined average values for the 1st and 2nd eigenvalues are used to define the length of the ellipses principal axes (outer dimensions) and the averaged orientation angle is used for the rotation of the ellipses (cf. [134]). For the rib analysis areas the angle θ is related to the z-axis and rotates the ellipses in the yz-plane. Since the 1st and 2nd eigenvalues already consider the ignored 3rd eigenvalue (representing the ellipsoid's spatial extent in the third room direction), there is no error between the 3D ellipsoid and the 2D ellipse presentation of the orientation when using a 2D front view without any viewing angle. This method enables to easily visually compare the measured and the predicted orientation values in one figure independently from the software they are coming from.

In Figure 26 the 2D ellipses calculated from the CT measurements in green are compared with the 3D TIMON CompositePRESS results in pink ellipses. There are analysis boxes with pretty good and some with less good but overall reasonable agreements between orientation measurement and prediction. The eigenvalues do not agree perfectly in all cases, yet the orientation angles capture the right orientation trends. According to Table 8 the highest absolute errors exit in the left rib top and the right rib bottom analysis areas, which can be quickly visually checked in Figure 26.

Table 7. Averaged fiber orientation measurement results for the brim and head analysis areas of three CT scanned ribbed hat profiles.

Analysis Area			A_{xx}	A_{yy}	A_{zz}	1st Eigenvalue	2nd Eigenvalue	Measured Angle φ
End Brim	**Left**	Top	0.27	0.70	0.03	0.73	0.25	103.5°
		SD	0.01	0.01	0.02	0.02	0.01	19.9°
		Bottom	0.42	0.55	0.03	0.63	0.34	113.5°
		SD	0.05	0.04	0.02	0.03	0.01	13.3°
	Center	Top	0.32	0.65	0.03	0.68	0.29	102.0°
		SD	0.03	0.01	0.02	0.01	0.01	10.6°
		Bottom	0.46	0.51	0.03	0.61	0.36	129.5°
		SD	0.01	0.01	0.02	0.03	0.01	13.0°
	Right	Top	0.28	0.70	0.02	0.73	0.26	97.8°
		SD	0.01	0.01	0.01	0.01	0.01	3.9°
		Bottom	0.40	0.57	0.02	0.65	0.33	115.9°
		SD	0.04	0.04	0.01	0.04	0.03	11.3°
Plate Brim	**Left**	Top	0.36	0.61	0.03	0.65	0.33	96.2°
		SD	0.04	0.05	0.01	0.04	0.04	23.2°
		Bottom	0.39	0.60	0.01	0.64	0.35	97.4°
		SD	0.05	0.05	0.00	0.04	0.04	4.5°
	Center	Top	0.38	0.58	0.04	0.63	0.35	85.1°
		SD	0.05	0.06	0.01	0.05	0.04	18.6°
		Bottom	0.45	0.54	0.01	0.61	0.37	92.4°
		SD	0.06	0.06	0.01	0.03	0.02	17.6°
	Right	Top	0.37	0.59	0.03	0.64	0.34	91.2°
		SD	0.04	0.05	0.01	0.05	0.04	23.1°
		Bottom	0.41	0.57	0.01	0.62	0.36	99.7°
		SD	0.04	0.04	0.01	0.05	0.04	67.6°
Head	**Left**		0.56	0.40	0.04	0.61	0.36	156.5°
	SD		0.06	0.06	0.02	0.06	0.04	14.6°
	Right		0.51	0.44	0.05	0.58	0.37	141.3°
	SD		0.07	0.05	0.03	0.06	0.04	19.3°
Mean SD			**0.04**	**0.04**	**0.01**	**0.04**	**0.03**	**18.6°**

* normalized to a positive angle between 0° and 180° related to the x-axis in the xy-plane.

Table 9 gives the predicted fiber orientation results for the brim and head analysis areas. The results from 3D TIMON CompositePRESS exhibit lower MAEs for the FOTs and the eigenvalues than compared to the rib analysis areas. The maximal FOT error is 0.09 for A_{xx} and 0.07 for the eigenvalues, respectively. These lower MAEs could arise out of the fact that here, in all analysis areas, the maximal part thickness is just 2 mm and the orientation component in the thickness direction (A_{zz}) is very low, so that it has a low significance. The mean absolute error (MAE) is in a still reasonable area around the true orientations measured in the CT scans. The tensor components A_{xx} and A_{yy} are always bigger than A_{zz} similar to the CT scan measurements. So the overall orientation tendencies are correctly captured. The predicted orientation angles' MAE is only 2.9° higher than the SD of the measured bundle orientation angles in the CT scans. This is a good indication that on average the orientation trends can be reasonably predicted. This deduction can also be visually proven in Figure 27. Here, the angle φ is related to the x-axis and rotates the ellipses in the xy-plane.

In comparison to the ribs, the brims and head analysis areas have lower MAEs for the FOT components and the eigenvalues. This manifests in better agreement of the ellipses outer dimensions, which in most areas fit well. Additionally, in some areas the deviation between the measured and the predicted orientation angles is very small, so that visually a very good agreement can be quickly found, for example, in the right end and plate brim section (cf. Figure 27).

Table 8. Predicted fiber orientation results for the rib analysis areas of the ribbed hat profile using 3D TIMON CompositePRESS.

	Analysis Area	A_{xx}	A_{yy}	A_{zz}	1st Eigenvalue	2nd Eigenvalue	Predicted Angle θ
Left Rib	Top	0.12	0.53	0.34	0.54	0.34	95.8°
	Absolute error vs. CT scans	0.08	0.16	0.09	0.19	0.11	16.4°
	Center	0.09	0.67	0.23	0.67	0.24	88.4°
	Absolute error vs. CT scans	0.03	0.07	0.11	0.03	0.06	1.4°
	Bottom	0.10	0.78	0.13	0.79	0.12	82.6°
	Absolute error vs. CT scans	0.03	0.16	0.19	0.14	0.16	7.8°
Center Rib		0.21	0.60	0.18	0.61	0.22	84.2°
	Absolute error vs. CT scans	0.12	0.11	0.01	0.12	0.01	12.3°
Right Rib	Top	0.15	0.71	0.15	0.71	0.15	94.1°
	Absolute error vs. CT scans	0.00	0.10	0.10	0.05	0.10	7.1°
	Center	0.15	0.76	0.09	0.77	0.15	94.5°
	Absolute error vs. CT scans	0.03	0.22	0.25	0.15	0.15	12.7°
	Bottom	0.19	0.74	0.07	0.74	0.19	92.4°
	Absolute error vs. CT scans	0.09	0.25	0.35	0.14	0.13	23.2°
MAE vs. averaged CT scans		**0.06**	**0.15**	**0.16**	**0.12**	**0.10**	**11.6°**

* normalized to a positive angle between 0° and 180° related to the z-axis in the yz-plane.

Figure 26. CAD image of the ribbed structure's middle section (front view) with the rib analysis areas (white boxes) showing the determined averaged fiber orientations of all three CT scans in green superimposed with the fiber orientations predicted with 3D TIMON CompositePRESS in pink (both displayed as 2D ellipses on the used tetra mesh).

Table 9. Predicted fiber orientation results for the brim and head analysis areas of the ribbed hat profile using 3D TIMON CompositePRESS.

	Analysis area		A_{xx}	A_{yy}	A_{zz}	1st Eigenvalue	2nd Eigenvalue	Predicted Angle φ
End Brim	Left	Top	0.18	0.81	0.01	0.81	0.19	91.2°
		Absolute error vs. CT	0.08	0.10	0.02	0.08	0.06	12.3°
		Bottom	0.29	0.70	0.01	0.70	0.29	85.7°
		Absolute error vs. CT	0.13	0.15	0.02	0.07	0.05	27.8°
	Center	Top	0.45	0.54	0.02	0.55	0.43	112.1°
		Absolute error vs. CT	0.13	0.12	0.01	0.13	0.14	10.1°
		Bottom	0.70	0.29	0.01	0.70	0.29	1.9°
		Absolute error vs. CT	0.24	0.22	0.02	0.10	0.07	52.5°
	Right	Top	0.24	0.75	0.01	0.76	0.23	97.4°
		Absolute error vs. CT	0.05	0.05	0.01	0.03	0.02	0.4°
		Bottom	0.43	0.56	0.01	0.59	0.40	111.7°
		Absolute error vs. CT	0.02	0.01	0.01	0.05	0.07	4.3°
Plate Brim	Left	Top	0.21	0.75	0.05	0.83	0.14	109.3°
		Absolute error vs. CT	0.15	0.14	0.02	0.17	0.19	13.1°
		Bottom	0.36	0.63	0.00	0.74	0.26	117.5°
		Absolute error vs. CT	0.03	0.04	0.01	0.10	0.09	20.1°
	Center	Top	0.26	0.68	0.07	0.69	0.27	101.7°
		Absolute error vs. CT	0.13	0.09	0.03	0.06	0.08	16.6°
		Bottom	0.32	0.66	0.01	0.68	0.30	102.5°
		Absolute error vs. CT	0.12	0.12	0.00	0.07	0.07	10.1°
	Right	Top	0.29	0.64	0.07	0.67	0.29	76.1°
		Absolute error vs. CT	0.08	0.05	0.03	0.03	0.05	15.0°
		Bottom	0.38	0.61	0.01	0.61	0.38	87.1°
		Absolute error vs. CT	0.03	0.04	0.01	0.01	0.02	12.5°
Head		**Left**	0.57	0.42	0.00	0.60	0.40	20.3°
		Absolute error vs. CT	0.01	0.02	0.04	0.01	0.04	43.8° **
		Right	0.59	0.40	0.00	0.64	0.36	23.5°
		Absolute error vs. CT	0.08	0.03	0.05	0.06	0.01	62.2° **
	MAE vs. averaged CT scans		**0.09**	**0.08**	**0.02**	**0.07**	**0.07**	**21.5°**

* normalized to a positive angle between 0° and 180° related to the x-axis in the xy-plane. ** since the calculated normalized angle is over 90°, the smaller supplementary angle (adjacent angle) is taken for the deviation calculation.

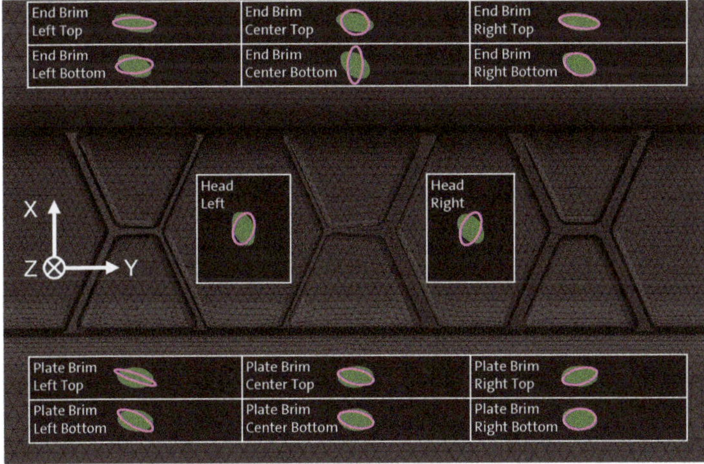

Figure 27. CAD image of the ribbed structure's middle section (bottom view) with the brims and head analysis areas (white boxes) showing the determined averaged fiber orientations of all three CT scans in green superimposed with the fiber orientations predicted with 3D TIMON CompositePRESS in pink (both displayed as 2D ellipses on the tetra mesh).

4. Discussion

For this study the compression molding of ROS-based CF-SMC into complex ribbed structures without defects, such as FMS, is accomplished although the chosen coarsely structured material with the scale and stiffness of its 50 mm long and 8 mm wide UD strands is typically not suited to fill thin and high ribs like those existent in the part. This is achieved by increasing the material's flowability with a pre-heating step and an optimal charge pattern allowing an even part filling without untimely curing through long flow paths or inhomogeneous shear loads. The sizes and positions of each of the single smaller charge packages is devised by means of fill simulation studies using 3D TIMON CompositePRESS. Such filling simulations take less than half an hour on a standard tower PC so that several studies can be conducted in reasonable time frames. This simulation-aided charge pattern development process prevented long trial and error molding sessions to find a working charge pattern and thus shortened the experimental time. This molding success shows the possible capabilities of a good charge preparation technique and a proper process simulation tool.

The accuracy of the cavity filling prediction is qualitatively evaluated by a comparison with short shots. The consistency between the experimentally obtained and the predicted mold filling is conclusive so that the calculated flow field can be assumed to be a sound basis for the consecutive DFS of the part. The slightly faster numerical filling of the plate area could be traced back to the fact that only a one-way coupling is used in the flow simulation. The missing impact of the fibers on the fluid flow behavior may lead to an over prediction of the material flow velocity at longer flow path lengths. A promising approach to overcome these shortcomings could be the incorporation of a two-way coupling between fibers and fluid to achieve an anisotropic material flow. From the convincing fill simulation results in the hat profile section and the marginal over prediction of the plate area filling it can be further concluded that the developed HexMC material card with its determined Kamal parameters and isotropic viscosity properties is suited to be used. The applied simplifications for the flow prediction, such as the Hele-Shaw model for the compression molding process and the only one-layered tetra mesh, lead to accurate results and are thus valid. Apparently, the 20 calculation points per element over the part thickness are a reliable approach to use simple tetra meshes and yet deliver convincing results in comparable short calculation times.

In order to be able to validate the 3D TIMON CompositePRESS process simulation results, CT scans of three entire ribbed structures were recorded. The quality of the CT scans is remarkable for the part size and the carbon fiber reinforced material. Due to the low contrast between carbon fibers and polymer matrix systems, there is always a trade-off between the needed CT scan resolution and the maximal scannable size of carbon fiber composite samples. However, in this study an optimum interaction of the CT scan hardware and scan parameter settings enabled CT scans with sufficient scan resolution to determine local orientations in a complex ribbed CF-SMC part. Single scans were successfully merged to full-part 3D CT scan volumes. As far as is known, such high-quality CT scans of a full CF-SMC part of this size are not published elsewhere. The scans can be easily analyzed by using the local density gradients instead of discrete carbon fibers for the fiber orientation measurement with a commercial CT scan analysis software—in this study VGSTUDIO MAX 3.3. Here, the analysis benefits from scale of the strands, which is used to correlate inter- and intra-strand density gradients to local orientation states similar to findings by Denos et al. [20] and Favaloro et al. [21]. Although the scan resolution is not fine enough to discern individual fibers from each other or to differentiate between strands in thickness direction, the local orientation states can be accurately measured for the whole part volume. Due to the mesoscale character of ROS materials, the scan quality and the used analysis mesh is suited for mesostructure analyses and comparisons with process simulation results or as input data for subsequent integrative structural simulations. If more detailed information is really needed, a finer scan resolution must be applied and either more and smaller CT scan volumes have to be merged (higher expenses), or just local spots of interest can be scanned without gaining whole-part 3D fiber orientation information. Finer analysis meshes with several elements over the thickness might also be beneficial for coupled structural simulations. On the other hand, when 3D orientation information is needed for an even bigger part, the CT scan method comes to its limits if the costs may not be too extensive.

Subsequently, the measured orientations of the three scanned parts were averaged to receive a characteristic representation of the part's mesostructures in 21 analysis areas. The averaging reduces local differences between the samples and reveals low standard deviations between the measured orientations of the three CT scanned mesostructures. This consistency justifies using the averaged 3D fiber orientation information as basis for the DFS software validation. Although additional CT scans of further parts would obviously enhance the reliability of the CT scan measurement results, the detected low standard deviations are a good sign that the small number of just three scanned parts is sufficient to represent the general orientation state occurring in the ribbed structures. As the gained CT orientation information for the CF-SMC parts are the only expedient data basis to validate predicted orientations in 3D, the apparent full-part CT scan results are valuable data, without precedent.

As far as is known, there is no other commercially available software tool that enables one to simulate the compression molding of ROS-based SMCs by applying a DFS method other than 3D TIMON CompositePRESS. Therefore, 3D TIMON CompositePRESS and its novel strand setting feature is applied and evaluated in this study. In comparison to the DBS of Meyer et al., which uses one chain of truss elements per fiber bundle, 3D TIMON models each strand by several fibers representing the fiber bundles within one strand. This gives the opportunity to mimic the strand geometry and behavior, such as strand splitting, properly. Since there is so far no cohesive force between the fibers of one strand or any interaction between fibers or strands modeled, all fiber bundles are numerically handled separately and fiber spreading and strand deformation will be inevitably overestimated. Furthermore, the assumed higher stiffness of strands compared with single fibers and especially its effect on the rib filling behavior is not taken into account. This could be addressed and studied in prospective simulations by numerically increasing the fibers' stiffness in the material card. However, the numerical strands show realistic flow behavior when compared to the experiments and therefore the software's UD strand mimicking is considered as a valid and working modeling strategy. From these results it is concluded that cohesive forces and interactions may not be absolutely necessary to represent a realistic strand movement and that the additional CPU time, when implementing such feature in the software, can be saved without losing too much accuracy. As the DFS simulation takes more than four days, the number of mesh elements, fibers, and output steps should be adapted to the situation and the needed degree of detail. For very large structural CF-SMC parts, which are currently targeted by automotive engineers, the needed simulation power for DFSs is a limiting factor for the presented DFS method. Even with high performance clusters the calculation of billions of fibers will not be effective anymore at a certain part size. Possible improvements could be made by modeling the UD strands by deformable cuboids instead of a group of single fibers, ignoring the strands' splitting behavior. As commonly very high charge coverages are used for such big parts, split strands play a minor role and this modeling idea could decrease the calculation times effectively. Although it has a minor impact on the DFS results, the charge generation in tighter radii, where the initial strands are so far unrealistically cut, could be improved, so that the strands follow the charge curvature instead of being cut.

The used high-performance CF-SMC HexMC is a typical member of the ROS-SMC material class and is therefore utmost suited for the software validation. For the evaluation the averaged CT measurements of the three ribbed hat profile parts are used. With the process simulation tetra mesh imported into the CT analysis software, a one-to-one comparison with the DFS results in the defined 21 analysis areas using exactly the same elements was conducted. The average CT values for the FOTs, eigenvectors and eigenvalues in the regions of interest were juxtaposed with the predicted values. Besides the small CT database with just three scanned parts, the detected deviations between CT measurements and DFS results might also originate from the chosen analysis areas and their sizes. Appropriate sizes for the analysis areas are crucial in order to ensure that local differences between the three CT scans and also in the process simulation are not overrated by using too small analysis areas. On the other hand, the analysis areas could also be too large so that clear orientation effects, for example, at cavity walls, are averaged out. Moreover, the underlying assumptions and boundary conditions in the DFS are a very simplified approach to simulate ROS materials and deviations from the real mesostructures in the ribbed structures are expectable. However, the calculated comparable low MAEs indicate overall reasonable agreement between measurement and simulation, especially when the initially random in-plane strand orientation and the comparably short flow paths (mold coverage of ~80%) are considered, which does not inherently lead to very pronounced flow-induced orientations. For this reason, it is concluded that 3D TIMON CompositePRESS delivers sufficiently accurate fill and orientation predictions that are valuable for ROS-based SMC part design processes and to avoid trial and error molding trials to find suitable processing conditions and a working charge pattern. The similarity between CT and DFS results was also visually shown in an orientation depiction method with 2D ellipses, allowing for an easy visual comparison. This method would be also suited to quickly compare simulation results using other settings, charge patterns or coming from different software.

5. Conclusions

For this study, complex ribbed structures were compression molded with a high-performance ROS-based CF-SMC. This relatively new material class with its advanced mechanical properties is a promising candidate for large lightweight structural parts for sports cars. However, due to the characteristic material configuration with its randomly oriented strands, a flow-induced mesostructure is evolving when the compound is compression molded into complex structures. As this mesostructure can be characterized by strong fiber alignments when the material is forced to flow, it directly impacts the parts response to applied mechanical loads. Therefore, the local mesostructure with its anisotropic mechanical properties must be considered in the part development process in order to avoid structural shortcomings due to adverse fiber orientations or knit lines and also plays a paramount role in the part performance calculation. For this reason, highly accurate simulations tools that can predict these unwanted effects are needed in the automotive field and must be developed and tested.

This paper aimed to evaluate a novel DFS tool with the feature to model ROS-based materials. In order to check its simulation accuracy by comparison with fiber orientation measurements full-part CT scans of the compression molded CF-SMC parts of unprecedented size are conducted. Full-part CT scans delivered a holistic depiction of the inner mesostructure with low standard deviations between the measured orientations of the scans and enabled comparisons with simulation results at any point of the part. Scan size and quality as well as the conducted detailed analyses of the presented CT scans surpass previously published studies dealing with this material class and show the current state of the art. Besides the quality assurance aspect, these full-part CT scans are the only expedient nondestructive microstructural characterization method to evaluate the accuracy of predicted fiber orientations. Although this method is admittedly limited in size as scan time and costs would be unreasonable for noticeably bigger parts, in future work, CT scans of larger components at maintaining quality are imaginable when CT hardware further improves and with increasing knowledge about the needed resolution for the specific part of interest. Notwithstanding, this study effectively demonstrates what kind of information for entire part volumes is already available.

Trial and error molding sessions of big CF-SMC parts quickly lead to high expenses for material, molds and manpower, hence trustworthy simulation tools are crucial for the application of the new material class. The DFS tool 3D TIMON CompositePRESS is extensively tested on a complex 3D-shaped part with ribs in different thicknesses and heights. Its main limitations are the non-considered anisotropic viscosity, the simple one-way coupling, and the lacking interactions between the fibers or strands, respectively. However, the fill simulation results were remarkably precise so that the software was used to optimize the charge configuration used in this molding trial. This capability can be used in future part design processes to accelerate the development process and to prevent disadvantageous part filling leading to orientation-related weak spots.

In order to validate the accuracy of the novel process simulation approach the orientation results attained with 3D TIMON CompositePRESS were compared with the mean orientations of three averaged CT scans in certain analysis areas. For this purpose, exactly the same analysis mesh is used in the simulation and the CT scan analyses allowing a one-to-one comparison. The determined deviations (average errors) between the predicted orientations and the CT measurements were calculated for 21 analysis areas in the ribbed structure and showed a reasonable low MAE. This result indicates that the examined DFS method is capable to accurately assess orientation trends in complex parts using ROS-based materials. Furthermore, the presented DFS approach can be applied to optimize part geometries, charge patterns, processing conditions and to choose suitable material prior to the mold manufacturing or the compression molding of SMC parts avoiding several design loops and costly experiments.

The obtained high-quality microstructure information, either from CT measurements or process simulations, will be used as input data for building a digital twin incorporated into structural simulations in further studies. In such integrative simulations, where the part performance predictions are based on its process-induced microstructure, this 3D information is highly valuable. The key benefit of this integrative simulation approach is that the automotive engineer can predict the mechanical performance of CF-SMC parts even before they physically exist. Due to the fact that, for example, different variations of charge patterns can be quickly virtually analyzed, the new DFS tool can make its own contribution to the development of reliable structural parts made of high-performance ROS-based materials for automotive lightweight applications.

Author Contributions: Conceptualization, J.T.; methodology, J.T.; software, J.T., S.K.H.; validation, J.T.; formal analysis, J.T.; investigation, J.T. and S.K.H.; resources, J.T.; data curation, J.T. and S.K.H.; writing—original draft preparation, J.T.; writing—review and editing, J.T., T.A.O.; visualization, J.T.; supervision, T.A.O.; project administration, J.T. All authors have read and agreed to the published version of the manuscript.

Funding: This research received no external funding.

Acknowledgments: The authors wish to thank Katsuya Sakaba and Ryugo Tanaka (Toray Engineering D Solutions Co., Ltd., Ōtsu, Shiga, Japan) for the technical provision of 3D TIMON 6.0, their professional support with the software, and many helpful hints. A special thanks also goes to Florian Bittner (Leibniz University, Institute of Plastics and Circular Economy IKK, Hanover, Germany) and Christina Haxter (Fraunhofer Institute for Wood Research, Wilhelm-Klauditz-Institute WKI, Hanover, Germany) for the excellent CT scans. Finally, the Volkswagen AG Group Innovation would like to thank the University of Wisconsin-Madison for their ongoing support in joint research projects.

Conflicts of Interest: The authors declare no conflict of interest.

Appendix A

In order to simulate the pressing/compression molding process, the flow analysis in 3D TIMON is based on the standard equations of traditional fluid dynamics: the equation of continuity, the equation of motion, and the equation of energy. These governing equations for the filling phase are given in the general and special form of the balance equations as follows:

The mass **continuity equation** (mass balance):

$$\frac{\partial \rho}{\partial t} + \frac{\partial}{\partial x}(\rho u_x) + \frac{\partial}{\partial y}(\rho u_y) + \frac{\partial}{\partial z}(\rho u_z) = 0, \tag{A1}$$

where ρ is the density, t is time and u_i stands for the fluid velocity vector in direction x, y, and z, respectively. For incompressible flow ρ is constant and Equation (A1) can be reduced to a volume continuity equation:

$$\frac{\partial u_x}{\partial x} + \frac{\partial u_y}{\partial y} + \frac{\partial u_z}{\partial z} = 0. \tag{A2}$$

The **equation of motion** (momentum balance) in fluid flow in all three directions can be expressed in terms of deviatoric stress τ_{ij} and this form of the equation is commonly called the Cauchy momentum equation:

$$\rho \frac{Du_i}{Dt} = -\frac{\partial p}{\partial x_i} + \frac{\partial \tau_{ji}}{\partial x_j} + \rho g_i, \tag{A3}$$

which in the x-direction becomes:

$$\rho\left(\frac{\partial u_x}{\partial t} + u_x\frac{\partial u_x}{\partial x} + u_y\frac{\partial u_x}{\partial y} + u_z\frac{\partial u_x}{\partial z}\right) = -\frac{\partial p}{\partial x} + \left(\frac{\partial \tau_{xx}}{\partial x} + \frac{\partial \tau_{yx}}{\partial y} + \frac{\partial \tau_{zx}}{\partial z}\right) + \rho g_x, \tag{A4}$$

where p: pressure, g_i: body force acting on the continuum, for example, gravity. In this force balance the mass is represented by the fluid density ρ and the following bracket represents the acceleration, meaning how the velocity of a particle changes with time. Therein, $\frac{\partial u_x}{\partial t}$ stands for the change of velocity over time and the three following terms represent the speed and direction in which the fluid is moving. The right hand side of the equation shows all forces acting in the fluid, where $-\frac{\partial p}{\partial x}$ stands for the internal pressure gradient of the fluid, the second term is representing the internal stress forces acting on the fluid (viscous effects are considered), and the last term represents all external forces acting on the fluid, such as gravity.

In fluid mechanics the deviatoric stress tensor τ_{ij} is commonly defined as:

$$\tau_{ij} = \mu\dot{\gamma}_{ij}, \tag{A5}$$

with μ: viscosity and $\dot{\gamma}_{ij}$: rate of deformation tensor reducing the Cauchy momentum equation to the Navier–Stokes equation for a simple Newtonian fluid with constant density ρ and viscosity μ:

$$\rho \frac{Du_i}{Dt} = -\frac{\partial p}{\partial x_i} + \mu\frac{\partial^2 u_i}{\partial x_j \partial x_j} + \rho g_i, \tag{A6}$$

which in the x-direction looks like:

$$\rho\left(\frac{\partial u_x}{\partial t} + u_x\frac{\partial u_x}{\partial x} + u_y\frac{\partial u_x}{\partial y} + u_z\frac{\partial u_x}{\partial z}\right) = -\frac{\partial p}{\partial x} + \mu\left(\frac{\partial^2 u_x}{\partial x^2} + \frac{\partial^2 u_x}{\partial y^2} + \frac{\partial^2 u_x}{\partial z^2}\right) + \rho g_x. \tag{A7}$$

The **equation of energy** (energy balance) for a Newtonian fluid with constant properties is given by:

$$\rho c_p \frac{DT}{Dt} = k\left(\frac{\partial^2 T}{\partial x^2} + \frac{\partial^2 T}{\partial y^2} + \frac{\partial^2 T}{\partial z^2}\right) + \dot{Q}_{\text{viscous heating}} + \dot{Q}, \tag{A8}$$

where ρ: density, c_p: specific heat, T: temperature, t: time, k: thermal conductivity, $\dot{Q}_{\text{viscous heating}}$: viscous dissipation and $\dot{Q}$: arbitrary heat source (i.e., exothermic reaction).

The viscous heating for a Newtonian material is written as:

$$\dot{Q}_{\text{viscous heating}} = \mu\dot{\gamma}_{ij}^2 = \mu\left(\frac{\partial u_i}{\partial x_j} + \frac{\partial u_j}{\partial x_i}\right)^2. \tag{A9}$$

Inserting the viscous heating term into Equation (A8) gives the energy equation in Cartesian coordinates:

$$\rho c_p\left(\frac{\partial T}{\partial t} + u_x\frac{\partial T}{\partial x} + u_y\frac{\partial T}{\partial y} + u_z\frac{\partial T}{\partial z}\right)$$
$$= k\left(\frac{\partial^2 T}{\partial x^2} + \frac{\partial^2 T}{\partial y^2} + \frac{\partial^2 T}{\partial z^2}\right) + 2\mu\left[\left(\frac{\partial u_x}{\partial x}\right)^2 + \left(\frac{\partial u_y}{\partial y}\right)^2 + \left(\frac{\partial u_z}{\partial z}\right)^2\right]$$
$$+\mu\left[\left(\frac{\partial u_x}{\partial y} + \frac{\partial u_y}{\partial x}\right)^2 + \left(\frac{\partial u_x}{\partial z} + \frac{\partial u_z}{\partial x}\right)^2 + \left(\frac{\partial u_y}{\partial z} + \frac{\partial u_z}{\partial y}\right)^2\right] + \dot{Q}. \tag{A10}$$

With the convention to denominate the viscosity of Newtonian fluids with μ and the viscosity of non-Newtonian fluids as η Equation (A9) can be rewritten as:

$$\dot{Q}_{\text{viscous heating}} = \eta\dot{\gamma}^2. \tag{A11}$$

In the compression molding process the polymer melt can be considered as a viscous fluid. When the material is pressed through the cavity to fill the mold, the melt is assumed to behave as a generalized Newtonian fluid (GNF). Therefore, the non-isothermal 3D flow can be described by the following simplified variant of the energy equation:

$$\rho c_p\left(\frac{\partial T}{\partial t} + u_x\frac{\partial T}{\partial x} + u_y\frac{\partial T}{\partial y} + u_z\frac{\partial T}{\partial z}\right) = k\left(\frac{\partial^2 T}{\partial x^2} + \frac{\partial^2 T}{\partial y^2} + \frac{\partial^2 T}{\partial z^2}\right) + \eta\dot{\gamma}^2. \tag{A12}$$

During the exothermic curing reaction of thermosets, heat generates. This effect is considered with an additional energy source terms on the right hand side in Equation (A13), where the energy balance for thermosets in a mold is shown.

$$\rho c_p\left(\frac{\partial T}{\partial t} + u_x\frac{\partial T}{\partial x} + u_y\frac{\partial T}{\partial y} + u_z\frac{\partial T}{\partial z}\right) = k\left(\frac{\partial^2 T}{\partial x^2} + \frac{\partial^2 T}{\partial y^2} + \frac{\partial^2 T}{\partial z^2}\right) + \eta\dot{\gamma}^2 + \frac{dQ}{dt}. \tag{A13}$$

Therein, the left terms stands for the temperature change and the convective heat transfer (more precisely thermal advection), whereas the terms on the right hand side of the equation represent the thermal conduction, shear heat generation (viscous heating), and the heat generation due to an exothermic reaction. Equation (A14) shows the relation between the heat generation velocity $\frac{dQ}{dt}$ and the total amount of generated heat Q_0 in the exothermic curing reaction affected by the curing reaction rate $\frac{d\alpha}{dt}$:

$$\frac{dQ}{dt} = Q_0\frac{d\alpha}{dt}. \tag{A14}$$

With the viscosity standardly defined as:

$$\mu = \frac{\tau}{\dot{\gamma}} \text{ or } \eta = \frac{\tau}{\dot{\gamma}}, \tag{A15}$$

and substituting μ in Equation (A9) the viscous dissipation can also be written as:

$$\dot{Q}_{\text{viscous heating}} = \tau_{ij}\dot{\gamma}_{ij}. \tag{A16}$$

For generalized Newtonian fluids (incompressible viscous fluids), the deviatoric stress τ can be expressed by:

$$\tau_{ij} = \eta\dot{\gamma}_{ij} = \eta\left(\nabla u + (\nabla u)^{\text{T}}\right), \tag{A17}$$

where η is the non-Newtonian viscosity that depends both on temperature T and the rate of deformation tensor $\dot{\gamma}_{ij}$, ∇u is the velocity gradient tensor and $(\nabla u)^{\text{T}}$ is the transposed velocity gradient tensor.

Since η is a scalar it must depend only on scalar invariants of $\dot{\gamma}$. Therefore, another notation for the symmetric rate of deformation tensor $\dot{\gamma}_{ij}$ in the Generalized Newtonian Fluid (GNF) model in Equation (A17) is the scalar strain rate $\dot{\gamma}$:

$$\dot{\gamma} = \left|\dot{\gamma}_{ij}\right| = \sqrt{\frac{1}{2}\left(\dot{\gamma}_{ij} : \dot{\gamma}_{ji}\right)} = \sqrt{\frac{1}{2}\mathrm{II}} = \sqrt{\frac{1}{2}\sum_i \sum_j \dot{\gamma}_{ij}\dot{\gamma}_{ji}} \text{ with } i, j = x, y, z, \tag{A18}$$

where $\dot{\gamma}$ is the magnitude of the rate of deformation tensor $\dot{\gamma}_{ij}$ and II is the second invariant of the rate of deformation tensor.

The rate of deformation tensor components in Equation (A18) are defined by:

$$\dot{\gamma}_{ij} = \frac{\partial u_i}{\partial x_j} + \frac{\partial u_j}{\partial x_i}, \tag{A19}$$

which yields:

$$\dot{\gamma} = \sqrt{2\left(\frac{\partial u_x}{\partial x}\right)^2 + 2\left(\frac{\partial u_y}{\partial y}\right)^2 + 2\left(\frac{\partial u_z}{\partial z}\right)^2 + \left(\frac{\partial u_x}{\partial y} + \frac{\partial u_y}{\partial x}\right)^2 + \left(\frac{\partial u_y}{\partial z} + \frac{\partial u_z}{\partial y}\right)^2 + \left(\frac{\partial u_z}{\partial x} + \frac{\partial u_x}{\partial z}\right)^2}, \tag{A20}$$

respecting the flow field in thickness direction.

The stress tensor components for GNFs in Cartesian coordinates are consequently given by:

$$\begin{aligned}
\tau_{xx} &= 2\eta\frac{\partial u_x}{\partial x} & \tau_{xy} = \tau_{yx} &= \eta\left(\frac{\partial u_x}{\partial y} + \frac{\partial u_y}{\partial x}\right) \\
\tau_{yy} &= 2\eta\frac{\partial u_y}{\partial y} & \tau_{yz} = \tau_{zy} &= \eta\left(\frac{\partial u_y}{\partial z} + \frac{\partial u_z}{\partial y}\right) \\
\tau_{zz} &= 2\eta\frac{\partial u_z}{\partial z} & \tau_{zx} = \tau_{xz} &= \eta\left(\frac{\partial u_z}{\partial x} + \frac{\partial u_x}{\partial z}\right)
\end{aligned} \tag{A21}$$

$$\tau_{ij} = \begin{bmatrix} \tau_{xx} & \tau_{xy} & \tau_{xz} \\ \tau_{yx} & \tau_{yy} & \tau_{yz} \\ \tau_{zx} & \tau_{zy} & \tau_{zz} \end{bmatrix} = \eta \begin{bmatrix} 2\frac{\partial u_x}{\partial x} & \frac{\partial u_x}{\partial y} + \frac{\partial u_y}{\partial x} & \frac{\partial u_z}{\partial x} + \frac{\partial u_x}{\partial z} \\ \frac{\partial u_x}{\partial y} + \frac{\partial u_y}{\partial x} & 2\frac{\partial u_y}{\partial y} & \frac{\partial u_y}{\partial z} + \frac{\partial u_z}{\partial y} \\ \frac{\partial u_z}{\partial x} + \frac{\partial u_x}{\partial z} & \frac{\partial u_y}{\partial z} + \frac{\partial u_z}{\partial y} & 2\frac{\partial u_z}{\partial z} \end{bmatrix}. \tag{A22}$$

With Equation (A16) the stress tensor components are implemented into Equation (A10), which results in the energy equation for thermoset materials, written as:

$$\begin{aligned}
\rho c_p\left(\frac{\partial T}{\partial t} + u_x\frac{\partial T}{\partial x} + u_y\frac{\partial T}{\partial y} + u_z\frac{\partial T}{\partial z}\right) \\
= k\left(\frac{\partial^2 T}{\partial x^2} + \frac{\partial^2 T}{\partial y^2} + \frac{\partial^2 T}{\partial z^2}\right) + \tau_{xx}\frac{\partial u_x}{\partial x} + \tau_{yy}\frac{\partial u_y}{\partial y} + \tau_{zz}\frac{\partial u_z}{\partial z} \\
+ \tau_{xy}\left(\frac{\partial u_x}{\partial y} + \frac{\partial u_y}{\partial x}\right) + \tau_{yz}\left(\frac{\partial u_y}{\partial z} + \frac{\partial u_z}{\partial y}\right) + \tau_{zx}\left(\frac{\partial u_z}{\partial x} + \frac{\partial u_x}{\partial z}\right) + \dot{Q},
\end{aligned} \tag{A23}$$

where $\dot{Q}$ denotes the heat generation due to the exothermic reaction in the thermoset curing process. In order to calculate the temperature distribution in the thermoset molding compound, 3D TIMON, however, assumes the heat conduction in thickness direction is dominant and therefore simplifies Equation (A23) to:

$$\rho c_p\left(\frac{\partial T}{\partial t} + u_x\frac{\partial T}{\partial x} + u_y\frac{\partial T}{\partial y} + u_z\frac{\partial T}{\partial z}\right) = k\frac{\partial^2 T}{\partial z^2} + \tau_{yz}\left(\frac{\partial u_y}{\partial z}\right) + \tau_{zx}\left(\frac{\partial u_x}{\partial z}\right) + Q_0\frac{d\alpha}{dt}, \tag{A24}$$

where $\dot{Q}$ is substituted with Equation (A14).

Using the simplification $\tau_{yz} = \eta\left(\frac{\partial u_y}{\partial z}\right)$ and $\tau_{zx} = \eta\left(\frac{\partial u_x}{\partial z}\right)$ (assumption of simple shear flow) yields to the final energy equation applied in 3D TIMON's Light 3D Heat Transfer:

$$\rho c_p\left(\frac{\partial T}{\partial t} + u_x\frac{\partial T}{\partial x} + u_y\frac{\partial T}{\partial y} + u_z\frac{\partial T}{\partial z}\right) = k\frac{\partial^2 T}{\partial z^2} + \eta\left(\frac{\partial u_y}{\partial z}\right)^2 + \eta\left(\frac{\partial u_x}{\partial z}\right)^2 + Q_0\frac{d\alpha}{dt}. \tag{A25}$$

References

1. Osswald, T.A.; Hernández-Ortiz, J.P. *Polymer Processing-Modeling and Simulation*; Carl Hanser Verlag: Munich, Germany, 2006; ISBN 978-3-446-41286-6.
2. Nakano, R.; Sakaba, K. Development of CAE software for injection and BMC/SMC molding including short/long fibers reinforcement. In Proceedings of the SAMPE Technical Conference, Seattle, WA, USA, 2–5 June 2014.
3. Song, Y.; Gandhi, U.N.; Sekito, T.; Vaidya, U.K.; Hsu, J.; Yang, A.; Osswald, T.A. A novel CAE method for compression molding simulation of carbon fiber-reinforced thermoplastic composite sheet materials. *J. Compos. Sci.* **2018**, *2*, 33. [CrossRef]
4. Denos, B.R.; Kravchenko, S.G.; Sommer, D.E.; Favaloro, A.J.; Pipes, R.B.; Avery, W.B. Prepreg platelet molded composites process and performance analysis. In Proceedings of the 33rd Technical Conference of the American Society for Composites 2018, Seattle, WA, USA, 24–26 September 2018.
5. Schemmann, M.; Görthofer, J.; Seelig, T.; Hrymak, A.; Böhlke, T. Anisotropic meanfield modeling of debonding and matrix damage in SMC composites. *Compos. Sci. Technol.* **2018**, *161*, 143–158. [CrossRef]
6. Kuhn, C.; Walter, I.; Taeger, O.; Osswald, T.A. Experimental and numerical analysis of fiber matrix separation during compression molding of long fiber reinforced thermoplastics. *J. Compos. Sci.* **2017**, *1*, 2. [CrossRef]
7. Automobili Lamborghini, S.p.A. Technical Data Sheet–Forged Composites. Available online: https://www.lamborghini.com/sites/it-en/files/DAM/lamborghini/forged/Forgedpresentation_EN.pdf (accessed on 17 April 2020).
8. Görthofer, J.; Meyer, N.; Pallicity, T.D.; Schöttl, L.; Trauth, A.; Schemmann, M.; Hohberg, M.; Pinter, P.; Elsner, P.; Henning, F.; et al. Virtual process chain of sheet molding compound: Development, validation and perspectives. *Compos. Part B Eng.* **2019**, *169*, 133–147. [CrossRef]
9. NORDAM Group Inc. Boeing 787 features composite window frames. *Reinf. Plast.* **2007**, *51*, 4. [CrossRef]
10. Feraboli, P.; Gasco, F.; Wade, B.; Maier, S.; Kwan, R.; Masini, A.; De Oto, L.; Reggiani, M. Lamborgini "forged composite" technology for the suspension arms of the sesto elemento. In Proceedings of the 26th Annual Technical Conference of the American Society for Composites 2011: The 2nd Joint US-Canada Conference on Composites, 6–28 September; Curran Associates, Inc.: Red Hook, New York, USA; Montreal, QC, Canada, 2011; Volume 2, pp. 1203–1215.
11. Gardiner, G. Is the BMW 7 Series the Future of Autocomposites? Available online: https://www.compositesworld.com/articles/is-the-bmw-7-series-the-future-of-autocomposites (accessed on 17 April 2020).
12. Bruderick, M.; Denton, D.; Shinedling, M. Applications of carbon fiber SMC for the 2003 Dodge Viper. In Proceedings of the Society of Plastics Engineers (SPE) Automotive Composites Conference & Exhibition (ACCE), Troy, MI, USA, 12–13 September 2002.
13. Tuttle, M.E.; Shifman, T.J.; Boursier, B. Simplifying certification of discontinuous composite material forms for primary aircraft structures. In Proceedings of the International SAMPE Symposium and Exhibition, Seattle, WA, USA, 17–20 May 2010.
14. Hexcel Corporation. HexMC®User Guide. Available online: https://www.hexcel.com/user_area/content_media/raw/HexMC_UserGuide.pdf (accessed on 27 August 2020).
15. Hexcel Corporation. HexMC®-i Moulding Compound-Product Data Sheet. Available online: https://www.hexcel.com/user_area/content_media/raw/HexMCi_C_2000_M77_RA_DataSheet.pdf (accessed on 27 August 2020).
16. Sommer, D.E. *Anisotropic Flow and Fiber Orientation Analysis of Preimpregnated Platelet Molding Compounds*; Purdue University: West Lafayette, IN, USA, 2018.

17. Sommer, D.E.; Kravchenko, S.G.; Denos, B.R.; Favaloro, A.J.; Pipes, R.B. Integrative analysis for prediction of process-induced, orientation-dependent tensile properties in a stochastic prepreg platelet molded composite. *Compos. Part A Appl. Sci. Manuf.* **2020**, *130*, 105759. [CrossRef]

18. Cutting, R.A.; Rios-Tascon, F.; Goodsell, J.E. Experimental investigation of the crush performance of prepreg platelet molding compound tubes. *J. Compos. Mater.* **2020**. [CrossRef]

19. Favaloro, A.J.; Sommer, D.E. On the use of orientation tensors to represent prepreg platelet orientation state and variability. *J. Rheol.* **2020**, *64*, 517–527. [CrossRef]

20. Denos, B.R.; Sommer, D.E.; Favaloro, A.J.; Pipes, R.B.; Avery, W.B. Fiber orientation measurement from mesoscale CT scans of prepreg platelet molded composites. *Compos. Part A Appl. Sci. Manuf.* **2018**, *114*, 241–249. [CrossRef]

21. Favaloro, A.J.; Sommer, D.E.; Denos, B.R.; Pipes, R.B. Simulation of prepreg platelet compression molding: Method and orientation validation. *J. Rheol.* **2018**, *62*, 1443–1455. [CrossRef]

22. Favaloro, A.J.; Sommer, D.E.; Pipes, R.B. Manufacturing simulation of composites compression molding in Abaqus/Explicit. In Proceedings of the Science in the Age of Experience (SIMULIA Global User Meeting), Boston, MA, USA, 18–21 June 2018.

23. Kravchenko, S.G.; Denos, B.R.; Sommer, D.E.; Favaloro, A.J.; Avery, W.B.; Pipes, R.B. Analysis of open hole tensile strength in a prepreg platelet molded composite with stochastic morphology. In Proceedings of the 33rd Technical Conference of the American Society for Composites 2018, Seattle, WA, USA, 24–27 September 2018; Volume 3, pp. 1881–1896.

24. Kravchenko, S.G.; Pipes, R.B. Progressive failure analysis in discontinuous composite system of prepreg platelets with stochastic meso-morphology. In Proceedings of the Science in the Age of Experience (SIMULIA Global User Meeting), Boston, MA, USA, 18–21 June 2018.

25. Sommer, D.E.; Favaloro, A.J.; Kravchenko, S.G.; Denos, B.R.; Byron Pipes, R. Stochastic process modeling of a prepreg platelet molded composite bracket. *Tech. Conf. Am. Soc. Compos.* **2018**, *4*, 2159–2169. [CrossRef]

26. Kravchenko, S.G.; Sommer, D.E.; Denos, B.R.; Favaloro, A.J.; Tow, C.M.; Avery, W.B.; Pipes, R.B. Tensile properties of a stochastic prepreg platelet molded composite. *Compos. Part A Appl. Sci. Manuf.* **2019**, *124*. [CrossRef]

27. Kravchenko, S.G.; Sommer, D.E.; Denos, B.R.; Avery, W.B.; Pipes, R.B. Structure-property relationship for a prepreg platelet molded composite with engineered meso-morphology. *Compos. Struct.* **2019**, *210*, 430–445. [CrossRef]

28. Li, Y.; Pimenta, S.; Singgih, J.; Nothdurfter, S.; Schuffenhauer, K. Experimental investigation of randomly-oriented tow-based discontinuous composites and their equivalent laminates. *Compos. Part A Appl. Sci. Manuf.* **2017**, *102*, 64–75. [CrossRef]

29. Li, Y. *The Effect of Variability in the Microstructure of Tow-Based Discontinuous Composites on Their Structural Behaviour*; Imperial College London: London, UK, 2018.

30. Li, Y.; Pimenta, S. Development and assessment of modelling strategies to predict failure in tow-based discontinuous composites. *Compos. Struct.* **2019**, *209*, 1005–1021. [CrossRef]

31. Martulli, L.M.; Muyshondt, L.; Kerschbaum, M.; Pimenta, S.; Lomov, S.V.; Swolfs, Y. Carbon fibre sheet moulding compounds with high in-mould flow: Linking morphology to tensile and compressive properties. *Compos. Part A Appl. Sci. Manuf.* **2019**, *126*, 105600. [CrossRef]

32. Martulli, L.M.; Creemers, T.; Schöberl, E.; Hale, N.; Kerschbaum, M.; Lomov, S.V.; Swolfs, Y. A thick-walled sheet moulding compound automotive component: Manufacturing and performance. *Compos. Part A Appl. Sci. Manuf.* **2020**, *128*, 105688. [CrossRef]

33. Alves, M.; Carlstedt, D.; Ohlsson, F.; Asp, L.E.; Pimenta, S. Ultra-strong and stiff randomly-oriented discontinuous composites: Closing the gap to quasi-isotropic continuous-fibre laminates. *Compos. Part A Appl. Sci. Manuf.* **2020**, *132*, 105826. [CrossRef]

34. De Oto, L. Carbon fibre innovation for high volumes: The forged composite. In Proceedings of the IICS JEC, Paris, France, 29–31 March 2011.

35. Feraboli, P.; Cleveland, T.; Ciccu, M.; Stickler, P.; De Oto, L. Defect and damage analysis of advanced discontinuous carbon/epoxy composite materials. *Compos. Part A Appl. Sci. Manuf.* **2010**, *41*, 888–901. [CrossRef]

36. Landry, B.; Hubert, P. Experimental study of defect formation during processing of randomly-oriented strand carbon/PEEK composites. *Compos. Part A* **2015**, *77*, 301–309. [CrossRef]

37. Selezneva, M.; Lessard, L. Characterization of mechanical properties of randomly oriented strand thermoplastic composites. *J. Compos. Mater.* **2016**, *50*, 2833–2851. [CrossRef]

38. Stelzer, P.S.; Plank, B.; Major, Z. Mesostructural simulation of discontinuous prepreg platelet based carbon fibre sheet moulding compounds informed by X-ray computed tomography. *Nondestruct. Test. Eval.* **2020**, *35*, 342–358. [CrossRef]

39. Sommer, D.E.; Kravchenko, S.G.; Pipes, R.B. A numerical study of the meso-structure variability in the compaction process of prepreg platelet molded composites. *Compos. Part A Appl. Sci. Manuf.* **2020**, *138*, 106010. [CrossRef]

40. Martulli, L.M.; Kerschbaum, M.; Lomov, S.V.; Swolfs, Y. Weld lines in tow-based sheet moulding compounds tensile properties: Morphological detrimental factors. *Compos. Part A Appl. Sci. Manuf.* **2020**, *139*, 106109. [CrossRef]

41. Pipes, R.B.; McCullough, R.L.; Taggart, D.G. Behavior of discontinuous fiber composites: Fiber orientation. *Polym. Compos.* **1982**, *3*, 34–39. [CrossRef]

42. Denos, B.R.; Pipes, R.B. Local mean fiber orientation via computer assisted tomography analysis for long discontinuous fiber composites. In Proceedings of the American Society for Composites (ASC) 2016 – 31st Technical Conference on Composite Materials, Williamsburg, VA, USA, 19–22 September 2016.

43. Denos, B.R. *Fiber Orientation Measurement in Platelet-Based Composites via Computed Tomography Analysis*; Purdue University: West Lafayette, IN, USA, 2017.

44. Kuhn, C.; Ton, Y.; Taeger, O.; Osswald, T.A. Experimental study on fiber matrix separation during compression molding of fiber reinforced rib structures. In Proceedings of the Society of Plastics Engineers (SPE) Annual Technical Conference (ANTEC), Orlando, FL, USA, 7–10 May 2018.

45. Kuhn, C.; Körner, E.; Täger, O. A simulative overview on fiber predictions models for discontinuous long fiber composites. *Polym. Compos.* **2020**, *41*, 73–81. [CrossRef]

46. Kuhn, C.; Walter, I.; Täger, O.; Osswald, T.A. Simulative prediction of fiber-matrix separation in rib filling during compression molding using a direct fiber simulation. *J. Compos. Sci.* **2018**, *2*, 2. [CrossRef]

47. Jeffery, G.B. The motion of ellipsoidal particles immersed in a viscous fluid. *Proc. R. Soc. London A Math. Phys. Eng. Sci.* **1922**, *102*, 161–179. [CrossRef]

48. Folgar, F.; Tucker, C.L., III. Orientation behavior of fibers in concentrated suspensions. *J. Reinf. Plast. Compos.* **1984**, *3*, 98–119. [CrossRef]

49. Advani, S.G.; Tucker, C.L., III. The use of tensors to describe and predict fiber orientation in short fiber composites. *J. Rheol.* **1987**, *31*, 751–784. [CrossRef]

50. Phelps, J.H.; Tucker, C.L., III. An anisotropic rotary diffusion model for fiber orientation in short- and long-fiber thermoplastics. *J. Nonnewton. Fluid Mech.* **2009**, *156*, 165–176. [CrossRef]

51. Bay, R.S. *Fiber Orientation in Injection-Molded Composites: A Comparison of Theory and Experiment*; University of Illinois at Urbana-Champaign: Urbana, IL, USA, 1991.

52. Advani, S.G.; Tucker, C.L., III. Closure approximations for three-dimensional structure tensors. *J. Rheol.* **1990**, *34*, 367–386. [CrossRef]

53. Cintra, J.S., Jr.; Tucker, C.L., III. Orthotropic closure approximations for flow-induced fiber orientation. *J. Rheol.* **1995**, *39*, 1095–1122. [CrossRef]

54. Chung, D.H.; Kwon, T.H. Improved model of orthotropic closure approximation for flow induced fiber orientation. *Polym. Compos.* **2001**, *22*, 636–649. [CrossRef]

55. Chung, D.H.; Kwon, T.H. Invariant-based optimal fitting closure approximation for the numerical prediction of flow-induced fiber orientation. *J. Rheol.* **2002**, *46*, 169–194. [CrossRef]

56. Montgomery-Smith, S.; Jack, D.A.; Smith, D.E. The fast exact closure for Jeffery's equation with diffusion. *J. Nonnewton. Fluid Mech.* **2011**, *166*, 343–353. [CrossRef]

57. Montgomery-Smith, S.; He, W.; Jack, D.A.; Smith, D.E. Exact tensor closures for the three-dimensional Jeffery's equation. *J. Fluid Mech.* **2011**, *680*, 321–335. [CrossRef]

58. Wang, J.; Jin, X. Comparison of recent fiber orientation models in Autodesk Moldflow Insight simulations with measured fiber orientation data. In Proceedings of the Polymer Processing Society 26th Annual Meeting, Banff, AB, Canada, 4–8 July 2010.

59. Tseng, H.-C.; Chang, R.-Y.; Hsu, C.-H. An objective tensor to predict anisotropic fiber orientation in concentrated suspensions. *J. Rheol.* **2016**, *60*, 215–224. [CrossRef]

60. Tseng, H.-C.; Chang, R.-Y.; Hsu, C.-H. The use of principal spatial tensor to predict anisotropic fiber orientation in concentrated fiber suspensions. *J. Rheol.* **2018**, *62*, 313–320. [CrossRef]

61. Bakharev, A.; Yu, R.; Ray, S.; Speight, R.; Wang, J. Using new anisotropic rotational diffuion model to improve prediction of short fibers in thermoplastic injection molding. In Proceedings of the Society of Plastics Engineers (SPE) Annual Technical Conference (ANTEC), Orlando, FL, USA, 7–10 May 2018.

62. Huynh, H.M. *Improved Fiber Orientation Predictions for Injection-Molded Composites*; University of Illinois: Urbana-Champaign, IL, USA, 2001.

63. Wang, J.; O'Gara, J.F.; Tucker, C.L., III. An objective model for slow orientation kinetics in concentrated fiber suspensions: Theory and rheological evidence. *J. Rheol.* **2008**, *52*, 1179–1200. [CrossRef]

64. Tseng, H.-C.; Chang, R.-Y.; Hsu, C.-H. Phenomenological improvements to predictive models of fiber orientation in concentrated suspensions. *J. Rheol.* **2013**, *57*, 1597–1631. [CrossRef]

65. Latz, A.; Strautins, U.; Niedziela, D. Comparative numerical study of two concentrated fiber suspension models. *J. Nonnewton. Fluid Mech.* **2010**, *165*, 764–781. [CrossRef]

66. Favaloro, A.J.; Sommer, D.E.; Pipes, R.B. Anisotropic viscous flow simulation in Abaqus. In Proceedings of the Science in the Age of Experience, Chicago, IL, USA, 15–18 May 2017; pp. 115–125.

67. Sommer, D.E.; Favaloro, A.J.; Pipes, R.B. Coupling anisotropic viscosity and fiber orientation in applications to squeeze flow. *J. Rheol.* **2018**, *62*, 669–679. [CrossRef]

68. Favaloro, A.J.; Tseng, H.-C.; Pipes, R.B. A new anisotropic viscous constitutive model for composites molding simulation. *Compos. Part A Appl. Sci. Manuf.* **2018**, *115*, 112–122. [CrossRef]

69. Tseng, H.; Favaloro, A.J. The use of informed isotropic constitutive equation to simulate anisotropic rheological behaviors in fiber suspensions. *J. Rheol.* **2019**, *63*, 263–274. [CrossRef]

70. Li, T.; Luyé, J.-F. Flow-fiber coupled viscosity in injection molding simulations of short fiber reinforced thermoplastics. *Int. Polym. Process.* **2019**, *34*, 158–171. [CrossRef]

71. Wittemann, F.; Maertens, R.; Kärger, L.; Henning, F. Injection molding simulation of short fiber reinforced thermosets with anisotropic and non-Newtonian flow behavior. *Compos. Part A Appl. Sci. Manuf.* **2019**, *124*. [CrossRef]

72. Kugler, S.K.; Lambert, G.M.; Cruz, C.; Kech, A.; Osswald, T.A.; Baird, D.G. Efficient parameter identification for macroscopic fiber orientation models with experimental data and a mechanistic fiber simulation. *AIP Conf. Proc.* **2020**, *2205*, 1–6. [CrossRef]

73. Meyer, N.; Schöttl, L.; Bretz, L.; Hrymak, A.N.; Kärger, L. Direct bundle simulation approach for the compression molding process of sheet molding compound. *Compos. Part A Appl. Sci. Manuf.* **2020**, *132*, 1–12. [CrossRef]

74. Kobayashi, M.; Dan, K.; Baba, T.; Urakami, D. Compression molding 3D-CAE of discontinuous long fiber reinforced polyamide 6: Influence on cavity filling and direct fiber simulations of viscosity fitting methods. In Proceedings of the ICCM International Conferences on Composite Materials, Copenhagen, Denmark, 19–24 July 2015.

75. Favaloro, A.J. *Rheological Behavior and Manufacturing Simulation of Prepreg Platelet Molding Systems*; Purdue University: West Lafayette, IN, USA, 2017.

76. Favaloro, A.J.; Sommer, D.E.; Pipes, R.B. Process simulation of compression molding of prepreg platelet molding systems. In Proceedings of the 14th International Conference on Flow Processing in Composite Materials, Luleå, Sweden, 30 May–1 June 2018.

77. Yamamoto, S.; Matsuoka, T. Dynamic simulation of fiber suspensions in shear flow. *J. Chem. Phys.* **1995**, *102*. [CrossRef]

78. Yamamoto, S.; Matsuoka, T. A method for dynamic simulation of rigid and flexible fibers in a flow field. *J. Chem. Phys.* **1993**, *98*, 644–650. [CrossRef]

79. Switzer, L.H.; Klingenberg, D.J. Rheology of sheared flexible fiber suspensions via fiber-level simulations. *J. Rheol.* **2003**, *47*, 759–778. [CrossRef]

80. Joung, C.G.; Phan-Thien, N.; Fan, X.J. Direct simulation of flexible fibers. *J. Nonnewton. Fluid Mech.* **2001**, *99*, 1–36. [CrossRef]

81. Qi, D. Direct simulations of flexible cylindrical fiber suspensions in finite Reynolds number flows. *J. Chem. Phys.* **2006**, *125*, 1–10. [CrossRef]

82. Lindström, S.B.; Uesaka, T. Simulation of the motion of flexible fibers in viscous fluid flow. *Phys. Fluids* **2007**, *19*. [CrossRef]

83. Wu, J.; Aidun, C.K. A method for direct simulation of flexible fiber suspensions using lattice Boltzmann equation with external boundary force. *Int. J. Multiph. Flow* **2010**, *36*, 202–209. [CrossRef]

84. Yamamoto, S.; Matsuoka, T. Dynamic simulation of flow-induced fiber fracture. *Polym. Eng. Sci.* **1995**, *35*, 1022–1030. [CrossRef]

85. Hinch, E. The distortion of a flexible inextensible thread in a shearing flow. *J. Fluid Mech.* **1976**, *74*, 317–333. [CrossRef]

86. Skjetne, P.; Ross, R.F.; Klingenberg, D.J. Simulation of single fiber dynamics. *J. Chem. Phys.* **1997**, *107*, 2108–2121. [CrossRef]

87. Phan-Thien, N.; Fan, X.; Tanner, R.I.; Zheng, R. Folgar–Tucker constant for a fibre suspension in a Newtonian fluid. *J. Nonnewton. Fluid Mech.* **2002**, *103*, 251–260. [CrossRef]

88. Ross, R.F.; Klingenberg, D.J. Dynamic simulation of flexible fibers composed of linked rigid bodies. *J. Chem. Phys.* **1997**, *106*, 2949–2960. [CrossRef]

89. Meirson, G.; Hrymak, A.N. Two-dimensional long-flexible fiber simulation in simple shear flow. *Polym. Compos.* **2016**, *37*, 2425–2433. [CrossRef]

90. Yamane, Y.; Kaneda, Y.; Dio, M. Numerical simulation of semi-dilute suspensions of rodlike particles in shear flow. *J. Nonnewton. Fluid Mech.* **1994**, *54*, 405–421. [CrossRef]

91. Sundararajakumar, R.R.; Koch, D.L. Structure and properties of sheared fiber suspensions with mechanical contacts. *J. Nonnewton. Fluid Mech.* **1997**, *73*, 205–239. [CrossRef]

92. Fan, X.; Phan-Thien, N.; Zheng, R. A direct simulation of fibre suspensions. *J. Nonnewton. Fluid Mech.* **1998**, *74*, 113–135. [CrossRef]

93. Ausias, G.; Fan, X.; Tanner, R.I. Direct simulation for concentrated fibre suspensions in transient and steady state shear flows. *J. Nonnewton. Fluid Mech.* **2006**, *135*, 46–57. [CrossRef]

94. López, L.; Ramírez, D.; Osswald, T.A. Fiber attrition and orientation productions of a fiber filled polymer through a gate - A mechanistic approach. In Proceedings of the Society of Plastics Engineers (SPE) Annual Technical Conference (ANTEC), Cincinnati, OH, USA, 22–24 April 2013.

95. Pérez, C. *The Use of a Direct Particle Simulation to Predict Fiber Motion in Polymer Processing*; University of Wisconsin-Madison: Madison, WI, USA, 2016.

96. Ramírez, D. *Study of Fiber Motion in Molding Processes by Means of a Mechanistic Model*; University of Wisconsin-Madison: Madison, WI, USA, 2014.

97. Walter, I.; Goris, S.; Teuwsen, J.; Tapia, A.; Pérez, C.; Osswald, T.A. A direct particle level simulation coupled with the folgar-tucker RSC model to predict fiber orientation in injection molding of long glass fiber reinforced thermoplastics. In Proceedings of the Society of Plastics Engineers (SPE) Annual Technical Conference (ANTEC), Anaheim, CA, USA, 8–10 May 2017; pp. 573–579.

98. Pérez, C.; Ramírez, D.; Osswald, T.A. Mechanistic model simulation of a compression molding process: Fiber orientation and fiber-matrix separation. In Proceedings of the Society of Plastics Engineers (SPE) Annual Technical Conference (ANTEC), Orlando, FL, USA, 23–25 March 2015.

99. Londoño-Hurtado, A.; Hernandez-Ortiz, J.P.; Osswald, T.A. Mechanism of fiber–matrix separation in ribbed compression molded parts. *Polym. Compos.* **2007**, *28*, 451–457. [CrossRef]

100. Londoño-Hurtado, A. *Mechanistic Models for Fiber Flow*; University of Wisconsin-Madison: Madison, WI, USA, 2009.

101. Hayashi, S.; Chen, H.; Hu, W. Development of new simulation technology for compression molding of long fiber reinforced plastics. In Proceedings of the 15th International LS-DYNA® Users Conference, Detroit, MI, USA, 10–12 June 2018.

102. Hayashi, S. New simulation technology for compression molding of long fiber reinforced plastics: Application to randomly-oriented strand thermoplastic composites. In Proceedings of the ECCM 2018 – 18th European Conference on Composite Materials, Athens, Greece, 24–28 June 2018.

103. Motaghi, A. *Direct Sheet Molding Compound Process (D-SMC)*; University of Western Ontario: London, ON, Canada, 2018.

104. Le, T.-H.; Dumont, P.J.J.; Orgéas, L.; Favier, D.; Salvo, L.; Boller, E. X-ray phase contrast microtomography for the analysis of the fibrous microstructure of SMC composites. *Compos. Part A Appl. Sci. Manuf.* **2008**, *39*, 91–103. [CrossRef]

105. Toray Engineering Co., Ltd. *3D TIMON 10 — CompositePRESS*; Toray Engineering Co., Ltd.: Ōtsu, Shiga, Japan, 2019.

106. Kim, H.-S.; Chang, S.-H. Simulation of compression moulding process for long-fibre reinforced thermoset composites considering fibre bending. *Compos. Struct.* **2019**, *230*, 111514. [CrossRef]

107. Kuhn, C. *Analysis and Prediction of Fiber Matrix Separation during Compression Molding of Fiber Reinforced Plastics*; Friedrich-Alexander-Universität Erlangen-Nürnberg: Erlangen, Germany, 2018.

108. Dantzig, J.A.; Tucker, C.L., III. *Modeling in Materials Processing*; Cambridge University Press: New York, NY, USA, 2001; ISBN 9781139175272.

109. Toray Engineering Co., Ltd. *3D TIMON 10 – Reference Manual*; Toray Engineering Co., Ltd.: Ōtsu, Shiga, Japan, 2019.

110. Tucker, C.L., III; Folgar, F. A model of compression mold filling. *Polym. Eng. Sci.* **1983**, *23*, 69–73. [CrossRef]

111. Lee, C.C.; Folgar, F.; Tucker, C.L., III. Simulation of compression molding for fiber-reinforced thermosetting polymers. *J. Manuf. Sci. Eng. Trans. ASME* **1984**, *106*, 114–125. [CrossRef]

112. Osswald, T.A. *Numerical Methods for Compression Mold Filling Simulation*; University of Illinois at Urbana-Champaign: Urbana-Champaign, IL, USA, 1987.

113. Osswald, T.A.; Menges, G. *Materials Science of Polymers for Engineers*, 3rd ed.; Carl Hanser Verlag: Munich, Germany, 2012; ISBN 978-1-569-90514-2.

114. Osswald, T.A.; Rudolph, N.M. *Polymer Rheology-Fundamentals and Applications*; Carl Hanser Verlag: Munich, Germany, 2015; ISBN 9781569905173.

115. Osswald, T.A. *Understanding Polymer Processing-Processes and Governing Equations*; Carl Hanser Verlag: Munich, Germany, 2017; ISBN 9781569906477.

116. Bay, R.S.; Tucker, C.L., III. Fiber orientation in simple injection moldings. Part I: Theory and numerical methods. *Polym. Compos.* **1992**, *13*, 317–331. [CrossRef]

117. Andrade, E.N. da C. The viscosity of liquids. *Nature* **1930**, *125*, 309–310. [CrossRef]

118. Castro, J.M.; Macosko, C.W. Kinetics and rheology of typical polyurethane reaction injection molding systems. In Proceedings of the Society of Plastics Engineers (SPE) Annual Technical Conference (ANTEC), New York, NY, USA, 5–8 May 1980; pp. 434–438.

119. Castro, J.M.; Macosko, C.W. Studies of mold filling and curing in the reaction injection molding process. *AIChE J.* **1982**, *28*, 250–260. [CrossRef]

120. Cross, M.M. Rheology of non-Newtonian fluids: A new flow equation for pseudoplastic systems. *J. Colloid Sci.* **1965**, *20*, 417–437. [CrossRef]

121. Kamal, M.R. Thermoset characterization for moldability analysis. *Polym. Eng. Sci.* **1974**, *14*, 231–239. [CrossRef]

122. Kamal, M.R.; Sourour, S. Kinetics and thermal characterization of thermoset cure. *Polym. Eng. Sci.* **1973**, *13*, 59–64. [CrossRef]

123. Kravchenko, S.G. *Failure Analysis in Platelet Molded Composite Systems*; Purdue University: West Lafayette, IN, USA, 2017.

124. Centea, T.; Hubert, P. Measuring the impregnation of an out-of-autoclave prepreg by micro-CT. *Compos. Sci. Technol.* **2011**, *71*, 593–599. [CrossRef]

125. Willems, F.; Beerlink, A.; Metayer, J.; Kreutzbruck, M.; Bonten, C. Bestimmung der Faserorientierung langglasfaserverstärkter Thermoplaste mittels bildoptischer Analyse und Computertomografie. In Proceedings of the DGZfP-Jahrestagung, Leipzig, Germany, 7–9 May 2018.

126. Garcea, S.C.; Wang, Y.; Withers, P.J. X-ray computed tomography of polymer composites. *Compos. Sci. Technol.* **2018**, *156*, 305–319. [CrossRef]

127. Shen, H.; Nutt, S.; Hull, D. Direct observation and measurement of fiber architecture in short fiber-polymer composite foam through micro-CT imaging. *Compos. Sci. Technol.* **2004**, *64*, 2113–2120. [CrossRef]

128. Riedel, T. Evaluation of 3D fiber orientation analysis based on x-ray computed tomography data. In Proceedings of the 4th International Conference on Industrial Computed Tomography (iCT), Wels, Austria, 19–21 September 2012; pp. 313–320.

129. Goris, S.; Osswald, T.A. Progress on the characterization of the process-induced fiber microstructure of long glass fiber-reinforced thermoplastics. In Proceedings of the Society of Plastics Engineers (SPE) Automotive Composites Conference & Exhibition (ACCE), Novi, MI, USA, 7–9 September 2016.

130. Maier, D.; Dierig, T.; Reinhart, C.; Günther, T. Analysis of woven fabrics and fiber composite material aerospace parts using industrial CT data. In Proceedings of the 5th International Symposium on NDT in Aerospace, Singapore, 13–15 November 2013.

J. Compos. Sci. **2020**, *4*, 164

131. Denos, B.R.; Kravchenko, S.G.; Pipes, R.B. Progressive failure analysis in platelet based composites using CT-measured local microstructure. In Proceedings of the International SAMPE Technical Conference, Seattle, WA, USA, 22–25 May 2017.
132. Teuwsen, J.; Bittner, F.; Steffen, J.P. Evaluation of X-ray target materials to improve CT-based measurement of fiber orientations inside CF-SMC components. In Proceedings of the International Symposium on Digital Industrial Radiology and Computed Tomography – DIR2019, Fürth, Germany, 2–4 July 2019.
133. Hexcel Corporation. HexPly® M77-Product Data Sheet. Available online: https://www.hexcel.com/user_area/content_media/raw/HexPly_M77_EpoxyResin_DataSheet.pdf (accessed on 27 August 2020).
134. Park, C.H.; Lee, W.I.; Yoo, Y.E.; Kim, E.G. A study on fiber orientation in the compression molding of fiber reinforced polymer composite material. *J. Mater. Process. Technol.* **2001**, *111*, 233–239. [CrossRef]

Publisher's Note: MDPI stays neutral with regard to jurisdictional claims in published maps and institutional affiliations.

Journal of
Composites Science

Article

Process Chain Optimization for SWCNT/Epoxy Nanocomposite Parts with Improved Electrical Properties

Manuel V. C. Morais, Marco Marcellan, Nadine Sohn, Christof Hübner * and Frank Henning

Polymer Engineering Department, Fraunhofer Institut für Chemische Technologie (ICT),
Joseph-von-Fraunhofer-Straße 7, 76327 Pfinztal, Germany; manuel.morais@ict.fraunhofer.de (M.V.C.M.);
s182212@student.dtu.dk (M.M.); sona1014@hs-karlsruhe.de (N.S.); frank.henning@ict.fraunhofer.de (F.H.)
* Correspondence: christof.huebner@ict.fraunhofer.de

Received: 17 July 2020; Accepted: 11 August 2020; Published: 14 August 2020

Abstract: Electrically conductive nanocomposites present opportunities to replace metals in several applications. Usually, the electrical properties emerging from conductive particles and the resulting bulk values depend on the micro/nano scale morphology of the particle network formed during processing. The final electrical properties are therefore highly process dependent. In this study, the electrical resistivity of composites made from single-walled carbon nanotubes in epoxy was investigated. Three approaches along the processing chain were investigated to reduce the electrical resistivity of nanocomposites-the dispersion strategy in a three-roll mill, the curing temperature, and the application of electric fields during curing. It was found that a progressive increase in the shear forces during dispersion leads to a more than 50% reduction in the electrical resistivities. Higher curing temperatures of the nanocomposite resin also lead to a decrease of around 50% in resistivity. Furthermore, a scalable resin transfer molding set-up with gold-coated electrodes was developed and tested with different mold release agents. It has been shown that curing the material under electric fields leads to an electrical resistivity approximately an order of magnitude lower, and that the properties of the mold release agent also influence the final resistivity of different samples in the same batch.

Keywords: Carbon nanotubes; CNTs; nanocomposites; electrical resistivity; conductivity; electric fields

---

1. Introduction

With the dispersion of discontinuous micro or nanofibers, such as carbon nanotubes (CNTs), in a polymer matrix, it is possible to obtain composite materials with mechanical and electrical properties that open a vast range of applications. Depending on the composition, particle concentration and dispersion, these materials have an electrical conductivity ranging from electrical insulation (10^{-18}–10^{-7} Scm^{-1}), to semi-conductivity (10^{-7}–10^{-1} Scm^{-1}) or even metallic conductivity (above 10^{-1} Scm^{-1}), allowing for higher flexibility in applications when compared to plastics and metals that possess a unique and defined conductivity [1]. Electrically conductive polymer composites may be able to substitute metals in a variety of applications that allow for lightweight solutions and easy shaping, such as resistive or capacitive sensors, battery technology components, electromagnetic shielding or resistive heating elements [2–4].

The electrical properties in these functional composite materials arise due to a network of interconnected particles which ultimately allows for the flow of electrons. This network not only depends on the particle shape and properties, but also on all the processing steps of the material, from the compounding of matrix and particles up to the shaping of the final part. Therefore, the complete history of the material affects the final electrical properties of the composite in the final parts [5–7]. Traditionally, the best way to reduce electrical resistivity is to increase the particle concentration, which entails additional costs and affects the processability due to higher viscosity [8].

The CNTs dispersive mixing operation consists of several stages-filler incorporation and the wetting and infiltration of the epoxy matrix, followed by dispersion, distribution and flocculation [9,10]. During the stage of filler dispersion, the large initial filler agglomerates are reduced in size up to the smallest dispersible unit. This can be mainly attributed to two coexisting mechanisms: rupture (a bulk phenomenon) and erosion (a surface phenomenon). The rupture mechanism breaks down agglomerates in a short time, related to erosion that happens slower by removing single or bundles of CNTs from the surface of the agglomerate [11]. This first step is the most difficult and important one, since it determines the rate at which CNTs disperse in the polymer. The prevalence of a mechanism over the other depends on the shear stress: if it exceeds a certain threshold value (dependent on the filler) the rupture tends to be the dominant mechanism [12]. From the existing literature [13] it emerges that the ratio of applied shear stress and cohesive strength of the agglomerate defines the dominating mechanism. Dispersion by rupture is the fastest way to obtain small final agglomerates, but the sudden breaking of a whole cluster often implies a reduction in the aspect ratio of its constituent particles, resulting in a worsening of the macroscopic properties that the nanoparticles can give to the composite. To avoid the excessive breakages of single nanotubes, the mixing parameters should be ideally controlled in order to maintain the rupture mechanism at the lowest possible rate, promoting erosion that is slow but preserving the nanotube aspect ratio, and leading to a better infiltration of the epoxy matrix in the agglomerates [11].

Finally, this work presents a scalable method to investigate and produce nanocomposite samples with improved controlled electrical properties by manipulating the nanoparticles with electric fields during the shaping step [14,15]. This is possible due to the difference in dielectric and electric properties between the particles and matrix, and particularly due to the high aspect ratio of carbon nanotubes [16,17]. According to the theory of dielectrophoresis, randomly dispersed particles under an electric field become polarized and orient, interacting due to Coulomb forces and assembling in conductive chains [16–18].

The experimental set-up hereby developed is inspired in resin-transfer-molding (RTM) [19] and allows for the application of an electric field on the electrically conductive composite resin during the curing. It was designed to be easily scaled-up for different plate geometries and integrated with available composite technology, aiming at an industrial-relevant production. It is here demonstrated for a system of single-wall carbon nanotubes (SWCNTs) and epoxy. In a first step, the process of dispersion of SWCNT in epoxy is investigated in view of optimization of electrical resistivity. Then, this nanocomposite is cured under the influence of electrical fields, and the impact of different mold release agents on the final resistivity is investigated.

2. Materials and Methods

The experimental procedure for the production of bulk epoxy nanocomposites was done in the following steps:

- Dispersion of nanoparticles in the liquid resin (with a three-roll mill);
- Addition of the curing agent in the correct stoichiometric amount;
- Feeding the mold, setting the electric field and the curing temperature;
- After reaction completion, removal of the field and retrieval of the samples.

This section describes the materials and methodology used in the different stages of the sample production.

2.1. Materials

Commercial pristine SWCNTs TUBALL™ were provided in powder form by OCSiAl (Luxembourg, Luxembourg) and used as received. These were chosen due to their high aspect ratio, given their diameter of 1.6 ± 0.4 nm and length above 5 μm, according to the manufacturer.

Epikote MGS RIMR426, a low viscosity epoxy resin (500–900 mPa·s at 25 °C), usually used for fiber impregnation processes, was used together with an amine-based curing agent, Epikure RIHMH433. Both materials were provided by Lange+Ritter GmbH (Gerlingen, Germany). Epoxy resin and hardener were manually mixed in a single step for 120 seconds at room temperature (RT) and subsequently cured at the investigated curing temperature. Bar-shaped specimens were cured in a silicon mold as follows: samples cured at RT were kept in a climatized lab (23 °C and 50% humidity) for 24 h and samples cured at 80 °C were heated in an oven for 40 min. Disk-shaped samples were heated by placing the RTM set-up on top of a custom-made heating plate at 100 °C for 40 min.

2.2. Dispersion of SWNTs in Epoxy

Given the initial agglomerate state of the CNTs in powder form [5,6], the first mixing step is crucial to guarantee that the CNTs are sufficiently dispersed in the resin and ensure the reproducibility of results. This was accomplished by using a three-roll mill. Comparing to other dispersion techniques (such as ultrasonication or ball milling), it has the advantage of providing more process control in the mixing [7,20]. This is achieved by the fine tuning of the rotation speed of the rolls together with the gaps between them. Moreover, it is a scalable production method because the mixing energy does not depend on the amount of material, in contrast with other dispersion technologies such as ultrasonication, for example, and so it is more relevant for industrial applications.

The used three-roll mill model was the Exakt 80E (EXAKT Advanced Technologies GmbH, Norderstedt, Germany), whose rolls were made of chemically neutral silicon carbide and measure 80 mm of diameter and 200 mm of length. The gap between rolls can be as low as 5 μm, while the maximum throughput capacity is of 20,000 cm^3/h. As depicted in Figure 1, three rolls were coupled together to operate in alternating current at scaling velocities: the speed was set for the apron roll, while the center and feed rolls were automatically set at three- and nine-times lower speeds, respectively. The suspension was placed in the gap between the feed and center roll, and was then forced to pass through the first gap while experiencing shear forces caused by the speed difference among the two rolls. The suspension then went through the second and last gap that is usually set to be three-times smaller than the previous gap, undergoing, once again, shear forces and being collected through the use of a sharp blade in direct contact with the apron roll, concluding a full dispersion cycle that, in this work, will be referred to as "pass".

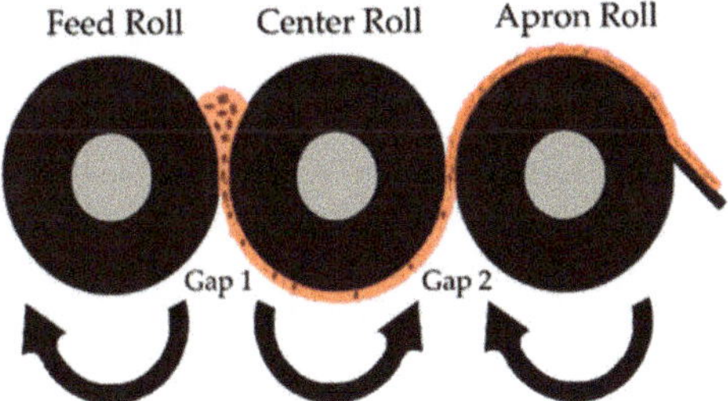

Figure 1. Three-roll mill schematic: suspension is fed between the first two rolls, passes through gap 1 and gap 2 and is finally collected completing a "pass" cycle.

Based on the literature [20,21], the chosen apron roll speed was 180 rpm, at which CNT rupture phenomena are minimized, better preserving their aspect ratio. In order to investigate the influence on the electrical resistivity of the final sample in the absence of the electric field, the gaps between rolls were varied from 60–20 μm to the minimum of 15–5 μm following two different sequences later presented. The overall number of passes was varied from 2 to 8. The aim was to obtain a stable and reproducible electrical response of the samples produced while minimizing their electrical resistivity. Suspensions with 0.001 wt.% and 0.005 wt.% CNTs were obtained by producing 0.01 wt.% masterbatches that were subsequently thinned down to the target concentration.

2.3. Set-Up for Electric Field Application

A set-up for the simultaneous production of multiple disk samples was developed, inspired by resin transfer molding, depicted in Figure 2. A silicone form for six disk samples of 20 mm diameter and 1.5 mm thickness (3 in Figure 2) was placed between two gold-coated electrodes (2 in Figure 2). Together with another layer of silicone (4 in Figure 2), which served as a sealant, the four layers were placed between two steel plates (1 in Figure 2). The outer plates were fitted with four nuts screwed together and used for clamping the structure. The inner arrangement was additionally fixed with 3 guiding pins, in order to prevent slipping or warping of the silicone form and thus avoiding the clogging of the narrow flowing channels between the disk molds.

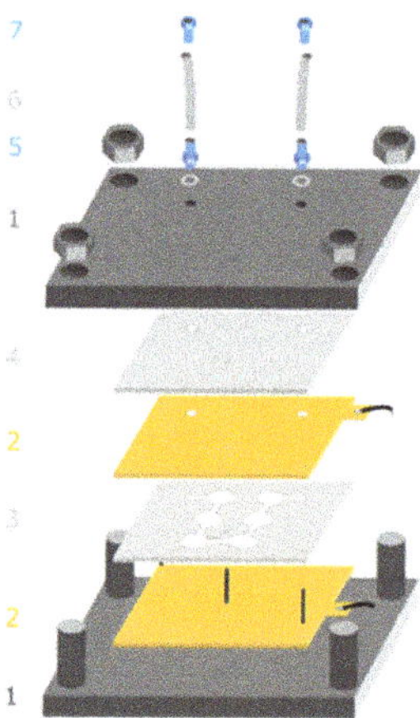

Figure 2. Exploded view of the mold used to apply the electric field during resin curing: six disc samples are produced at a time.

After mounting the system together, two hose couplings (5 in Figure 2) were screwed on the top plate, each then connecting a feeding tube (6 in Figure 2) fastened with a tightening nut (7 in Figure 2). One of the hoses was closed with a clamp, the other one connected to a syringe with double sealing ring. With the syringe, a negative pressure is generated in the mold. Subsequently, the epoxy dispersion was mixed with the curing agent and, to avoid air bubbles in the samples, degassed under vacuum in a desiccator. The composition was then filled into a second syringe, connected to the feeding tube and fed to the cavity until it emerged from the other tube. The electrodes were connected to the voltage source through a temperature-resistant silicone coated cable. Finally, the complete mold was placed on top of an aluminum heating plate set to the desired curing temperature. In order to be able to remove the samples from the electrodes, a mold release agent had to be used. The standard mold release agent is usually either silicone- or polytetrafluoroethylene-based, and so in both cases electrically nonconductive. In order to evaluate the effect of the electrical conductivity of the interface between nanocomposite and electrode, experiments were performed with three different release agents: conductive silver paint, mechanical grease and carbon black filled grease.

2.4. Electric Field

A sinusoidal voltage of 60 V_{pp} at a frequency of 10 MHz was used to generate a sinusoidal alternating electric field strength of 400 V_{pp}/cm using a function generator Agilent 33250A and a high frequency power amplifier (Tabor 9260). The voltage and the frequency were chosen based on results presented in [22], where the procedure for electric field application is further described.

2.5. Characterization of Electrical Resistivity

For the investigation of electrical properties of samples produced with the three-roll mill, SWCNT/epoxy dispersions were used to produce bar-shaped samples (5 × 2 × 40 mm) in a silicone mold. For each experimental configuration, the electrical resistivity of six samples was measured using the four-point method. In this method, four contact points are used where the two outer ones lay an electrical current and the resulting voltage is measured by the inner ones.

Disk-shaped samples produced in the mold presented in Section 2.3 were characterized with electrochemical impedance spectroscopy in an IM6 workstation from Zahner-Elektrik GmbH (Kronach, Germany). This means that a measuring setup was used where a small sinusoidal potential of 1 mV with fixed frequency is applied to the sample, the response is measured, and the impedance computed at each frequency. The starting point frequency was set at 3 MHz and then it was gradually decreased to 500 mHz where the impedance assumes a constant value corresponding with acceptable approximation to the through-plane resistance.

3. Results

3.1. Dispersion

Two mixing modes were investigated in the three-roll mill:

- Constant gap—The suspension was passed eight times with a fixed gap 1 of 15 µm (between the feed and center rolls—refer to 2 in Figure 2) and a gap 2 of 5 µm (between the center and apron rolls).
- Regressive gap—Starting with a larger gap between rolls (60–20), and then every two passes halving the distance until reaching the physical limit used on constant gap mode, 15–5 µm.

The eight total suspension pass gaps are resumed in Table 1. To avoid an excessive promotion of rupture mechanism that could damage the aspect ratio of the nanoparticles, the speed on the apron roll was always set at 180 rpm.

Table 1. Three-roll mill mixing modes investigated.

Pass	1	2	3	4	5	6	7	8
Constant gap (µm)	15–5	15–5	15–5	15–5	15–5	15–5	15–5	15–5
Regressive gap (µm)	60–20	60–20	30–10	30–10	15–5	15–5	15–5	15–5

Resistivity measurements for the two gap modes investigated are plotted in Figure 3. While the difference in first passes is still not perceivable, it reaches one order of magnitude after eight passes. Using the regressive gap mode seems to promote the lower electrical resistivity of SWCNT/epoxy composites for the same curing process parameters and without increasing the particle concentration. This is expected to result from the different agglomerate dispersion mechanisms [8], where erosion is fostered over rupture. The constant mode (gaps of 15–5 µm for each pass) imposes high shear forces to the bigger initial CNT agglomerates, promoting their rupture and, most likely, rupture of the nanotubes. On the other hand, starting with wider gaps in the regressive mode favors the erosion of the agglomerates over rupture, limiting the breakage of CNTs. A higher CNT aspect ratio for the same concentration seems to help the formation of more efficient secondary networks for electrons to flow across the bulk composite, by enhancing the chances of contact between adjacent particles.

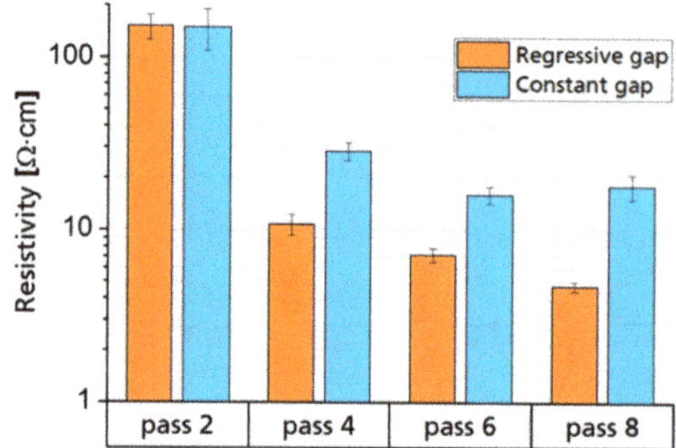

Figure 3. Electrical resistivity of SWCNT/epoxy 0.1 wt.% bar-shaped samples produced with two gap-modes in the three-roll mill and measured with the four-point method.

Additionally, Figure 4 depicts the through-plane electrical resistivity of SWCNT/epoxy nanocomposites at different concentrations. For the lowest amount of SWCNTs (0.001 wt.%) the electrical resistivity was too high to be measured by the multimeter in resistance mode (limited to 50 MΩ) and hence the value presented corresponds to the multimeter range limit. From the figure it can be concluded that the percolation concentration lies between 0.001 wt.% and 0.005 wt.%. It can be observed that the through-plane resistivity for 0.1 wt.% SWCNT/epoxy disk-shaped samples in Figure 4 is substantially higher than the value in Figure 3, which corresponds to the in-plane resistivity measured with the four-point method.

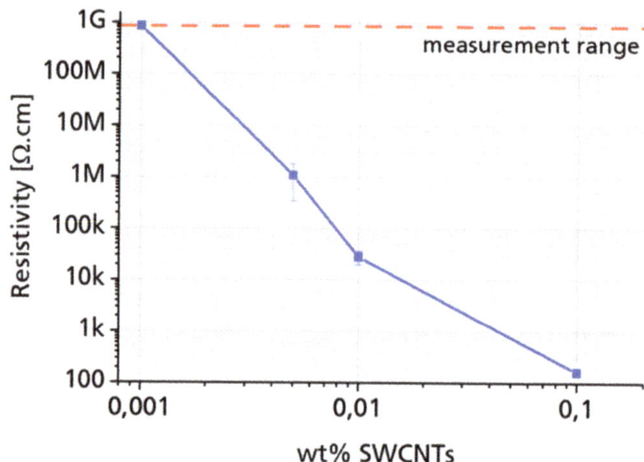

Figure 4. Through-plane electrical resistivity of SWCNT/epoxy as a function of particle concentration for eight passes.

The influence of the curing temperature of the SWCNT/epoxy composites on the final electrical resistivity was investigated for samples collected after different numbers of passes in the three-roll mill. Specimens of SWCNT/epoxy at a concentration of 0.1 wt.% (mixed in constant gap mode) and at four different passes number were cured both at room temperature and heated up at 80 °C, accordingly also varying the overall curing time. The resistivity measurements are depicted in Figure 5. The beginning of a plateau is noticeable at pass number six, after which no improvements come from additional mixing in the three-roll mill. Moreover, by increasing the curing temperature, the electrical resistivity of the SWCNT/epoxy 0.1 wt.% composites appears to decrease around 50%. This is probably due to the fact that, at 80 °C, the resin matrix is less viscous than at room temperature, as it emerges from the rheological tests performed in [22]. A more liquid suspension provides less viscous resistance for the secondary agglomerates' network formation, under Van der Waals forces or intentional applied dielectrophoresis (useful consideration for alignment experiments described further on in this paper, which in fact were performed at 80 °C).

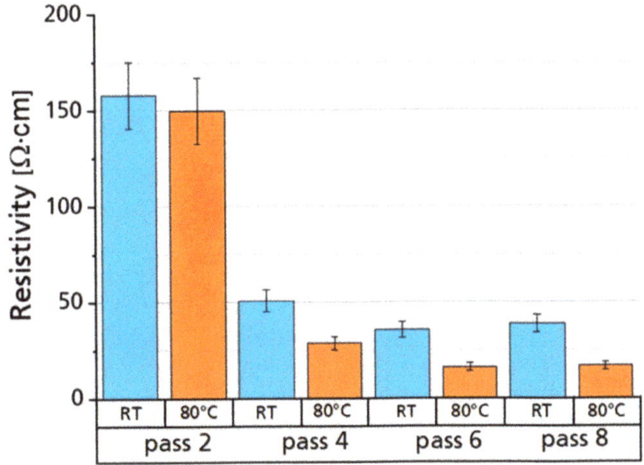

Figure 5. SWCNT/epoxy 0.1 wt.%—influence of curing temperature on electrical resistivity at different passes.

3.2. Electric Fields

Dielectrophoresis experiments were performed with SWCNT/epoxy 0.01 wt.% composites. The main goal of this set-up was to produce multiple plate specimens simultaneously and with a shape defined by the silicone mold, which is a scalable process with potential application in industry. With every experiment, six disk samples were produced (20 mm diameter and 1.5 mm thickness), as can be seen in Figure 6.

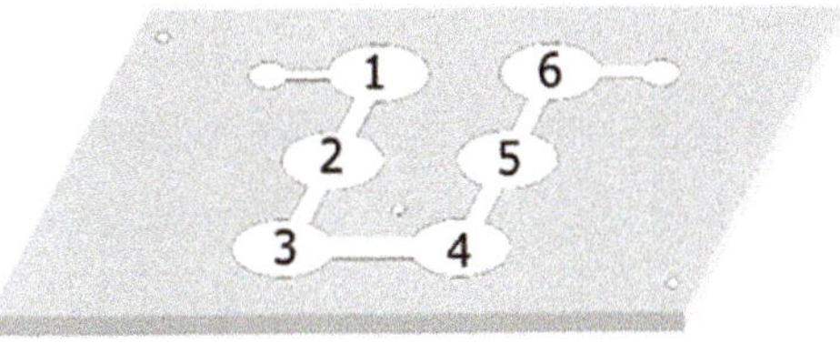

Figure 6. Silicon mold (corresponding to point 3 in Figure 2), which allows for the production of six discs of 20 mm diameter per 1.5 mm thickness.

Furthermore, the influence of the use of distinct mold release agents was investigated. Figure 7 shows the resistivity values measured between the electrodes after the curing of SWCNT/epoxy 0.01 wt.% at 100 °C, with and without electric field. The first measurements presented here were done before opening the mold by connecting the electrodes to a multimeter after sample cool down. Therefore, the values represent the total resistivity of the material connecting the electrodes (six plates and the interconnections). A one order of magnitude decrease in the resistivity was generally measured.

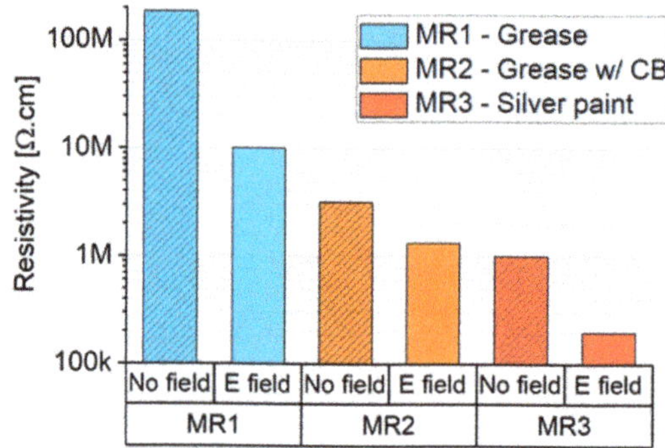

Figure 7. Impact of mold release agent (MR1,2,3) on plate resistivity measured at the electrodes (after cure) for SWCNT/epoxy 0.01 wt.% composites.

However, a clear difference was noticeable when different mold release agents were used. With the most conductive one among the three, MR3 (silver paint), both the resistivity with and without electric field were two orders of magnitude lower than that of MR1. MR2 (intermediate conductivity) appeared to be in between the two and showed a minor difference between field and no field application. This is likely related to the contact resistance of the interface between the electrodes and the SWCNT/epoxy composite, which is presumably higher for MR1 and MR2.

After specimen retrieval from the mold and thorough cleaning and gold sputtering of the individual disk samples for through-plane resistivity measurements, this property was evaluated. Figure 8 displays the average resistivity of individual samples corresponding to the results in Figure 7. This not only shows the difference between electric field and no field to be less pronounced than in Figure 7, but the resistivity values also show the opposite trend—MR3 leads to the highest average resistivity and MR1 to the lowest. This behavior can be explained by considering the properties of the individual samples after curing under the electric field, depicted in Figures 9 and 10. In these graphs, the electrical resistivity of samples is plotted according to their fixed position in the mold, numbered from 1 to 6. For samples produced without electric field, the electrical resistivity of six samples for each mold release agent does not vary considerably (Figure 10). However, when an electric field is applied, as depicted in Figure 9, it is possible to observe that MR3 and MR2 show a great disparity of results between the six samples. MR3, for example, has specimen 4 with a resistivity of only 1.7 kΩ·cm, being the most conductive sample produced. However, samples 1, 2, 5 and 6 are all above 10 kΩ·cm.

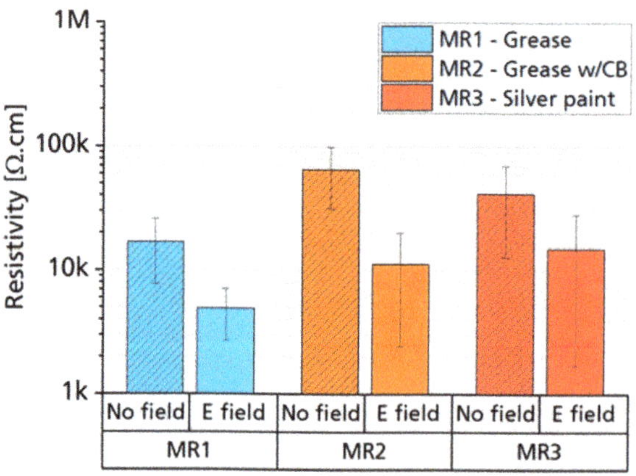

Figure 8. Impact of mold release agent on single disc average resistivity of SWCNT/epoxy 0.01 wt.%.

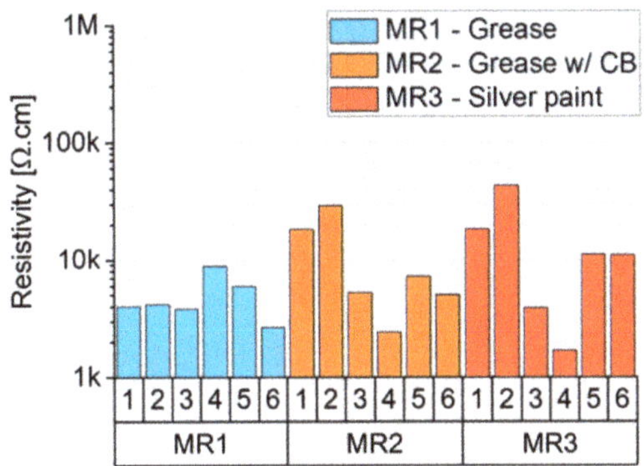

Figure 9. With electric field: Impact of mold release agent on resistivity of SWCNT/epoxy 0.01 wt.% individual samples cured at 100 °C.

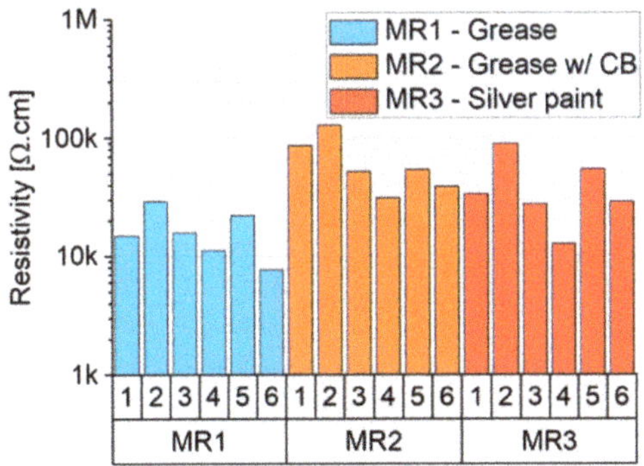

Figure 10. Without electric field: Impact of mold release agent on resistivity of SWCNT/epoxy 0.01 wt.% individual samples cured at 100 °C.

A possible explanation for the measurements of total resistivity in Figure 7 is the following phenomenon: given the very low contact resistance of silver paint (MR3), more electrical current can flow between the electrodes. Due to this micro-current, the electric field propagates in the material across the sample thickness as new conductive paths are generated in the material due to CNT dielectrophoresis. However, as soon as one of the six samples becomes more conductive (e.g., sample 4 of MR3 in Figure 9), more and more current starts flowing only through it, since it offers less resistivity, and less current through the other samples (such as 1, 2, 5 and 6 of MR3 in Figure 9). This would explain the scatter in the resistivity values of samples MR2 and MR3 (Figure 9) and the resistivity measured in the closed mold presented in Figure 7. These results highlight the importance of the contact interface (more specifically its resistivity) between the electrodes and the composite for influencing the homogeneous electrical resistivity of SWCNT nanocomposites with electric fields.

4. Discussion and Conclusions

An optimized dispersion process that limits rupture mechanisms, hence presumably ensuring a high aspect ratio and well dispersed particles, proved to be fundamental when the aim is to decrease the electrical resistivity of thermoset nanocomposites. Once a good level of matrix infiltration, dispersion and distribution in the CNT agglomerates is obtained, it was found that heat application during the curing process further decreased the electrical resistivity of the nanocomposites. This is expected to be due to a reduction in the matrix viscosity at higher temperatures, allowing more mobility of the particles and hence leading to more contacts between them, as investigated in [22]. This proved to be beneficial when combined with the application of an electric field between the electrodes, leading to a reduction in resistivity of up to one order of magnitude compared to the samples cured without the field. As explored in [22], the clear alignment of CNTs due to the electric field was not observed in scanning electron microscope (SEM) images of sample fracture surfaces. Instead, the reduction in electrical resistivity was attributed to an electric-field-induced increase in contact points between the nanoparticles. This possibility was investigated in [22] with finite element simulations of CNT networks in epoxy at different concentrations.

Furthermore, an RTM set-up for applying electric fields to nanocomposites during curing has been developed and successfully tested. When applying electric fields to the nanocomposite during curing, the conductivity of the interface between the electrodes and the nanocomposite resin was shown to play a crucial role in determining the final resistivity of the bulk materials produced. A highly conductive interface (MR3) promoted more inhomogeneous samples, leading to pronounced changes in the resistivity at a local level (significant differences between six samples produced simultaneously). On the other hand, a highly resistive interface (MR1) prevented such local differences and resulted in a smaller average resistivity in the samples produced. The use of a nonconductive mold release agent therefore seems more suitable for a scalable industrial process since it leads to composites with more homogeneous electrical properties.

Overall, the use of an electric field during processing shows promising results for enhancing the through-plane electrical conductivity of polymer composites with carbon-based nanoparticles by influencing the structure of the conductive network during the processing step. Such technology would reduce the amount of CNTs necessary to achieve similar electrical properties, hence allowing a cheaper production of composites with a broad range of industrial applications (e.g., sensors, batteries, electrical heating technologies). The aim of further research in this topic should be to understand the impact of the type of carbon nanoparticles used (or a combination of particles with different size scales), the effect of stronger electric fields with lower particle concentrations and limitations arising from matrix viscosity.

Author Contributions: Conceptualization, M.V.C.M., M.M., C.H. and F.H.; Data curation, M.V.C.M., M.M. and N.S.; Funding acquisition, C.H.; Investigation, M.V.C.M., M.M. and N.S.; Methodology, M.V.C.M. and C.H.; Supervision, C.H. and F.H.; Writing—original draft, M.V.C.M. and M.M.; Writing—review and editing, M.V.C.M., M.M., N.S., C.H. and F.H. All authors have read and agreed to the published version of the manuscript.

Funding: This research has received funding from the European Union's Horizon 2020 research and innovation program under the Marie Skłodowska-Curie Grant Agreement no. 642890 (TheLink, www.thelink-project.eu).

Conflicts of Interest: The authors declare no conflict of interest.

References

1. Kaur, G.; Adhikari, R.; Cass, P.; Bown, M.; Gunatillake, P. Electrically conductive polymers and composites for biomedical applications. *RSC Adv.* **2015**, *5*, 37553–37567. [CrossRef]
2. Yeetsorn, R.; Fowler, M.W.; Tzoganakis, C.A. Review of Thermoplastic Composites for Bipolar Plate Materials in PEM Fuel Cells. In *Nanocomposites with Unique Properties and Applications in Medicine and Industry*; InTech: London, UK, 2011.
3. Lan, Y.C.; Wang, Y.; Ren, Z.F. Physics and applications of aligned carbon nanotubes. *Adv. Phys.* **2011**, *60*, 553–678.
4. Zeng, Z.; Jin, H.; Chen, M.; Li, W.; Zhou, L.; Zhang, Z. Lightweight and anisotropic porous mwcnt/wpu composites for ultrahigh performance electromagnetic interference shielding. *Adv. Funct. Mater.* **2016**, *26*, 303–310. [CrossRef]
5. Alig, I.; Pötschke, P.; Lellinger, D.; Skipa, T.; Pegel, S.; Kasaliwal, G.R.; Villmow, T. Establishment, morphology and properties of carbon nanotube networks in polymer melts. *Polymer* **2012**, *53*, 4–28. [CrossRef]
6. Bauhofer, W.; Kovacs, J.Z. A review and analysis of electrical percolation in carbon nanotube polymer composites. *Compos. Sci. Technol.* **2009**, *69*, 1486–1498. [CrossRef]

7. Krause, B.; Potschke, P.; Haussler, L. Melt mixed SWCNT-polypropylene composites with very low electrical percolation. *Polymer* **2016**, *98*, 45–50. [CrossRef]

8. Kasaliwal, G.R.; Göldel, A.; Pötschke, P.; Heinrich, G. Influences of polymer matrix melt viscosity and molecular weight on MWCNT agglomerate dispersion. *Polymer* **2011**, *52*, 1027–1036. [CrossRef]

9. Kasaliwal, G.R.; Pegel, S.; Göldel, A.; Pötschke Petra, P.; Heinrich, G. Analysis of agglomerate dispersion mechanisms of multiwalled carbon nanotubes during melt mixing in polycarbonate. *Polymer* **2010**, *51*, 2708–2720. [CrossRef]

10. Sandler, J.K.W.; Kirk, J.E.; Kinloch, I.A.; Shaffer, M.S.P.; Windle, A.H. Ultra-low electrical percolation threshold in carbon-nanotube-epoxy composites. *Polymer* **2003**, *44*, 5893–5899. [CrossRef]

11. Peng, B.; Locascio, M.; Zapol, P.; Li, S.; Mielke, S.L.; Schatz, G.C. Measurements of near-ultimate strength for multiwalled carbon nanotubes and irradiation-induced crosslinking improvements. *Nat. Nanotechnol.* **2008**, *3*, 626–631. [CrossRef] [PubMed]

12. Iosif, D.; Rosca, S.; Hoa, V. Highly conductive multiwall carbon nanotube and epoxy composites produced by three-roll milling. *Carbon* **2009**, *47*, 1958–1968.

13. Yamamoto, K.; Akita, S.; Nakayama, Y. Orientation of carbon nanotubes using electrophoresis. *Jpn. J. Appl. Phys.* **1996**, *2*, L917. [CrossRef]

14. Yamamoto, K.; Akita, S.; Nakayama, Y. Orientation and purification of carbon nanotubes using ac electrophoresis. *J. Phys. D Appl. Phys.* **1998**, *31*, L34. [CrossRef]

15. Oliva-Avilés, A.I.; Avilés, F.; Sosa, V.; Oliva, A.I.; Gamboa, F. Dynamics of carbon nanotube alignment by electric fields. *Nanotechnology* **2012**, *23*, 465710. [CrossRef]

16. Martin, C.A.; Sandler, J.K.W.; Windle, A.H.; Schwarz, M.K.; Bauhofer, W.; Schulte, K.; Shaffer, M.S.P. Electric field-induced aligned multi-wall carbon nanotube networks in epoxy composites. *Polymer* **2005**, *46*, 877–886. [CrossRef]

17. Li, J.; Zhang, Q.; Peng, N.; Zhu, Q. Manipulation of carbon nanotubes using AC dielectrophoresis. *Appl. Phys. Lett.* **2005**, *86*, 1–3. [CrossRef]

18. Park, C.; Wilkinson, J.; Banda, S.; Ounaies, Z.; Wise, K.E.; Sauti, G.; Lillehei, P.T.; Harrison, J.S. Aligned single-wall carbon nanotube polymer composites using an electric field. *J. Polym. Sci. Polym. Phys. Ed.* **2006**, *44*, 1751–1762. [CrossRef]

19. Cheng, Q.F.; Wang, J.P.; Wen, J.J.; Liu, C.H.; Jiang, K.L.; Li, Q.Q.; Fan, S.S. Carbon nanotube/epoxy composites fabricated by resin transfer molding. *Carbon* **2010**, *48*, 260–266. [CrossRef]

20. Olowojoba, G.; Sathyanarayana, S.; Caglar, B.; Kiss-Pataki, B.; Mikonsaari, I.; Hübner, C.; Elsner, P. Influence of process parameters on the morphology, rheological and dielectric properties of threeroll-milled multiwalled carbon nanotube/epoxy suspensions. *Polymer* **2013**, *54*, 188–198. [CrossRef]

21. Olowojoba, G.; Sathyanarayana, S.; Caglar, B.; Mikonsaari, I.; Hübner, C.; Elsner, P. Influence of processing temperature, carbon nanotube agglomerate bulk density and functionalization on the dielectric and morphological properties of carbon nanotube / epoxy suspensions. In Proceedings of the Polymer Processing Society 29th Annual Meeting PPS-29, Nuremberg, Germany, 15–19 July 2013.

22. Morais, M.V.C.; Oliva-Avilés, A.I.; Matos, M.A.S.; Tagarielli, V.L.; Pinho, S.T.; Hübner, C.; Henning, F. On the effect of electric field application during the curing process on the electrical conductivity of single-walled carbon nanotubes–epoxy composites. *Carbon* **2019**, *150*, 153–167. [CrossRef]

Article

Comparative Analysis of the Impact of Additively Manufactured Polymer Tools on the Fiber Configuration of Injection Molded Long-Fiber-Reinforced Thermoplastics

Lukas Knorr [1,2,*,†], **Robert Setter** [3,†], **Dominik Rietzel** [2], **Katrin Wudy** [3] **and Tim Osswald** [4]

1   Institute of Polymer Technology, University of Erlangen-Nuremberg, 91058 Erlangen, Germany
2   Department of Additive Manufacturing, BMW Group, 85764 Oberschleissheim, Germany; Dominik.Rietzel@bmw.de
3   Department of Mechanical Engineering, Professorship of Laser-Based Additive Manufacturing, Technical University of Munich, 85748 Munich, Germany; robert.setter@tum.de (R.S.); katrin.wudy@tum.de (K.W.)
4   Mechanical Engineering Department, Polymer Engineering Center, University of Wisconsin-Madison, Madison, WI 53706, USA; tosswald@wisc.edu
*   Correspondence: knorr.lukas@gmx.de
†   Contributed equally to this work.

Received: 7 August 2020; Accepted: 11 September 2020; Published: 15 September 2020

Abstract: Additive tooling (AT) utilizes the advantages of rapid tooling development while minimizing geometrical limitations of conventional tool manufacturing such as complex design of cooling channels. This investigation presents a comparative experimental analysis of long-fiber-reinforced thermoplastic parts (LFTs), which are produced through additively manufactured injection molding polymer tools. After giving a review on the state of the art of AT and LFTs, additive manufacturing (AM) plastic tools are compared to conventionally manufactured steel and aluminum tools toward their qualification for spare part and small series production as well as functional validation. The assessment of the polymer tools focuses on three quality criteria concerning the LFT parts: geometrical accuracy, mechanical properties, and fiber configuration. The analysis of the fiber configuration includes fiber length, fiber concentration, and fiber orientation. The results show that polymer tools are fully capable of manufacturing LFTs with a cycle number within hundreds before showing critical signs of deterioration or tool failure. The produced LFTs moldings provide sufficient quality in geometrical accuracy, mechanical properties, and fiber configuration. Further, specific anomalies of the fiber configuration can be detected for all tool types, which include the occurrence of characteristic zones dependent on the nominal fiber content and melt flow distance. Conclusions toward the improvement of additively manufactured polymer tool life cycles are drawn based on the detected deteriorations and failure modes.

Keywords: additive tooling; additive manufacturing; rapid tooling; injection molding; polypropylene; long-fiber-reinforced thermoplastics; fiber length; fiber orientation; fiber concentration; stereolithography

---

1. Introduction

1.1. Motivation

In the most recent decades, the ease of processability of long-fiber-reinforced thermoplastics (LFT) has enabled their use as advanced lightweight engineering materials, particularly within the automotive sector [1,2]. As a key technology, the injection molding process is expected to take a major role in terms of value and volume [3]. With an estimated market volume percentage of 65%, polypropylene (PP) is the material with the biggest market volume of LFTs in this field [4]. Material glass is predominantly used for reinforcement due to its low cost and superior mechanical properties [5].

Additive manufacturing (AM) is becoming increasingly important in industry in general and is becoming especially important in the automotive sector [6]. Although a lot of research and development is carried out, AM is still very limited in its available range of materials, which prevents its widespread entry into large-scale industry [7]. Due to its generative layer structure, most of the materials show lower molecular cohesion in one or more load directions [8]. Consequently, anisotropic material properties that deviate from those of the injection molded (IM) pedant must be expected [9]. This counts particularly for polymers; therefore, commodity construction materials such as PP, PU, PA, PC, blends, and other compounds can only be substituted to a very limited amount [10].

Additive tooling (AT) describes a process which combines the potential of AM with the material spectrum of a traditional manufacturing process, such as IM, by replacing the molding tool unit by an additive one and keeping the existing equipment of the traditional manufacturing technology alive. In this way, it is possible to produce moldings with series properties, to reduce the cost and time-consuming tool manufacturing process, and to increase the tooling complexity and the resulting part design freedom at the same time [11,12]. For this procedure, polymer molds are currently preferred due to their superior surface quality, lower production time, and lower costs in comparison to its metal pedant [13]. For AT, the most important production method is stereolithography (SL), which provides supreme surfaces with a high geometrical accuracy and the access to cross-linked and reinforced polymers. These composites are based on a high-temperature resistant polymer matrix which is enhanced with fillers such as aluminum or ceramics particles to improve the thermal conductivity, heat deflection point, and mechanical properties [14,15]. In general, plastic molds show a significant deviation from the conventional steel tools in points of heat capacity as well as in their temperature and thermal conductivity around a factor of 10 to 100 [16]. As the properties and behavior of plastic components are mainly defined by their thermal history, the cooling process of the polymer tools is crucial for the morphology and the crystallization of the moldings.

The objective of this investigation is to qualify additively manufactured polymer tools for the fabrication of LFT parts for functional validation, spare parts, and small series production. For this purpose, the effects of three different tool materials (ceramic reinforced plastic, steel, and aluminum) on the process-induced fiber configuration are investigated and compared with each other. The iterative tool design process is performed for all materials individually and is based on Moldflow simulations and basic principles of heat transfer. Especially in terms of the cooling systems, the advantages of AT are specifically implemented. Representative for the most used LFT, the glass fiber-reinforced PP injection molding material STAMAX™ from SABIC is used in different fiber weight percentages. Tensile specimens and discs are used as sample geometries to map both a linear and a radial source flow. The process-induced fiber configuration is investigated over the flow path and over the component thickness. The central injection point at the disc mold will also allow investigation of the transition from a swirling to a pointed flow. Furthermore, the impact of the different tool materials on significant mechanical properties of the fiber-reinforced moldings are analyzed. For the test methods, CT scan, pyrolysis, tensile testing, and a fiber length measurement method modeled after that of Goris et al. [17] are carried out. To evaluate the economic efficiency, the tool life and the suitability of AT will be assessed on the degree of wear, defect patterns, and output quantity.

1.2. State of the Art

AM is ideally used to manufacture components directly without using the indirect process chain of AT. For this purpose, there are many commercially available fabrication methods (i.e., fused filament fabrication (FFF), selective laser sintering (SLS), and multi jet fusion) and materials that contain fiber reinforcements. However, their properties are not comparable with IM materials and especially not with LFTs due to their short fiber character and diverging fiber configuration. The commodity polymer PP is well known for its difficulty in processing within AM due to its shrinkage and low adhesion. Nevertheless, Sodeifian et al. [18] achieved good processing results of PP/GF with the additive POE-g-MA on a FFF platform. Still, the material is lagging behind in its mechanical properties and is deviating strongly from its compression molding pedant by 30%. To fabricate classic IM materials directly, another promising approach arose in the most recent decades. It is a modified FFF method where the filament extruder is exchanged for a granulate-fed portable extruder unit. On that base, Hertle et al. [19] achieved good results by melt extrusion of PP injection molding granulate. Though it was unlikely, the process suffers the same issue of an insufficient interlaminar bonding, which is strongly incoherent compared to the intralaminar ones and is leading into anisotropic material properties [20].

AT is the indirect process chain and a tool-bound alternative, which allows it to maintain the original manufacturing process and its material variety. Kampker et al. [21], investigated the suitability of polyjet (PJ), SL, digital light synthesis (DLS), and SLS processes for direct IM tool production. The investigations showed that the material PA 3200 GF is the most suitable for SLS, while for the SL method, composite materials such as Accura® Bluestone™ and Perform are the most promising ones. Rahmati and Dickens [22] examined the output of SL produced tools in the field of IM. They achieved an output of more than 500 parts and found that the main tool failure was due to flexural and excessive friction. Hofstätter et al. [23] achieved the same results.

For the improvement of the part properties, as well as the extension of the tool life and to decrease the cycle time, extensive research is being done on optimized cooling systems and adapted process parameters. Altaf et al. [24] demonstrate the general effectiveness of a conformal cooling system in a direct comparison with a conventional cooling design. In this example, the conformal cooling system leads to a cycle time reduction of approx. 20%. The investigations of Park et al. [25] show even better performances of up to 30%, whereby only tool areas relevant to the cycle time are replaced by SLM (Selective Laser Melting) inserts with a conformal cooling unit. However, Hopkins et al. [26] observed in a direct comparison of an aluminum and a polymer IM tool an increased cycle time and significant differences in the rheology of the injection molding material. The lower thermal conductivity of the mold materials results into longer flow paths, which in turn requires lower process pressures and melt temperatures. In earlier investigations of Martinho et al. [27], the influence on the morphology of semi-crystalline PP moldings is revealed in a direct comparison of an epoxy and steel mold material. As soon as different materials were used for core and cavity, a clear asymmetrical crystalline structure was observed. In concerns of the influence on the material properties of the moldings, the investigations of Harris et al. [28,29] should be emphasized. He states that with the aid of a modified cooling system and the adaptation of the necessary process parameters, the crystallinity can be strongly influenced, even to the extent that comparable part properties can be achieved using polymer and CNC milled steel/aluminum molds. Similar results were achieved by Fernandes et al. [30] and Volpato et al. [31]. Recent investigations by Kampkar et al. [16] proved that the materials still show a rather brittle and non-ductile fracture behavior. However, the deviations can be attributed with high probability to particle agglomeration due to the lower thermal conductivity and a greater surface roughness.

On the side of LFTs, the investigations of Kim et al. [32] show that the mechanical and the impact strength increases with the fiber length. Furthermore, they measured a reduction of the initial fiber length of 4–16 mm to a residual fiber length of about 0.5–2 mm during processing. The results are consistent with Seong et al. [33], who noted an increase in Young's modulus, melting temperature, and viscoelastic properties, and a less uniform distribution of the fiber with an increasing fiber length. Hou et al. [34] measured a reduction of the average fiber length with an increasing injection rate over the whole length of tensile specimens. Based on the work of Tadmor [35] as well as Osswald and Menges [36], there are seven characteristic regions that can be differentiated in injection molded parts, each with a specific fiber orientation. The two skin layers provide random fiber orientation, while the shell layers are aligned parallel to the melt flow direction. The fiber alignment of the core layer is perpendicular to the melt flow direction, with two randomly oriented transition layers between the core layer and the shell layers. This specific orientation pattern is well-known and caused by the fountain flow effect. However, the numerical predictions of Hou et al. [34] do not match with the empirical orientation in the skin layer. Parveen et al. [37] achieved comparable results to Hou et al. on a disc geometry (2 mm; 75 mm D) and stated that a higher fiber length leads to a wider core region with more fibers aligned transverse to the flow direction at the end of the flow path. Zhu et al. [38] compared a centered and an end-gated injecting model. It is found that the end-gated plate has a more defined shell area of 55%, where the fibers align along the main flow direction and a core region of 20%, with fibers aligned perpendicular to it. The shell area of the center-gated plates takes a 35% and the core area 35% at an equal density and fiber weight content. Goris et al. [17] detected a substantial increase in fiber length along the flow path, which is probably due to a fiber pullout effect. The same could be observed by Lafranche et al. [39], who detected a strong influence of the used gate type in addition. Furthermore, Goris [17] demonstrated in a comparative study that the so-called epoxy plug method, which is based on the investigations of Kunc et al. [40]. The new measurement method is especially suitable for LFTs with a large sample size and can be used to generate accurate results with a strong repeatability while minimizing the manual handling effort. It is shown that the taken measurements agree with the outcome of previously reported studies [12,39,41]. Aside from the epoxy plug method, the comparative study concentrated on two other fiber analysis methods: first, a full fiber analysis, which is a proprietary measurement system developed by SABIC (Geleen, Netherlands) and based on the work by Krasteva [42]; and second, the FASEP method, which is a commercially available analysis method proprietary to IDM Systems (Darmstadt, Germany).

For the analysis of the fiber orientation distribution, Sharma et al. [43] achieved sufficient results with a tensor-based marker watershed method. However, the method is relatively sensitive to improper surface preparation and image processing techniques. A better application seems to be the evaluation of industrial micro-computed tomography (μCT) images. The fiber orientation is described in accordance with the work of Adavani and Tucker [44], who assume that each fiber can be represented by a unit vector p spherical coordinates. The orientation is then described in tensor form. A schematic representation can be seen in Figure 1 (left). Mathematically, the unit vector is described with the angles Φ, θ as:

$$p = \begin{pmatrix} p_1 \\ p_2 \\ p_3 \end{pmatrix} = \begin{pmatrix} \cos(\Phi)\cdot\sin(\theta) \\ \sin(\Phi)\cdot\sin(\theta) \\ \cos(\theta) \end{pmatrix} \tag{1}$$

The orientation tensor or Advani–Tucker tensor is calculated as:

$$\alpha_{ij} = \oint p_i\cdot p_j\cdot\Psi(p)\,dp \tag{2}$$

While $p_i\cdot p_j$ is the product of the fiber orientation vector with itself, $\Psi(p)$ represents a probability function of all possible orientations. This results in the following tensor components:

$$\begin{aligned} a_{11} &= \left\langle \sin^2(\Phi)\cos^2(\theta) \right\rangle & a_{12} &= \left\langle \sin^2(\Phi)\cos(\theta)\sin(\theta) \right\rangle & a_{13} &= \left\langle \sin(\Phi)\cos(\Phi)\cos(\theta) \right\rangle \\ a_{21} &= a_{12} & a_{22} &= \left\langle \sin^2(\Phi)\sin^2(\theta) \right\rangle & a_{23} &= \left\langle \sin(\Phi)\cos(\Phi)\sin(\theta) \right\rangle \\ a_{31} &= a_{13} & a_{32} &= a_{23} & a_{33} &= \left\langle \cos^2(\Phi) \right\rangle \end{aligned} \tag{3}$$

Experimental data are conventionally illustrated using the orientation tensor form. The diagonal components of the second order orientation tensor (a_{11}, a_{22}, and a_{33}) describe the degree of orientation with respect to the defined coordinate system. Conventionally, the reference coordinates are defined so that the 1-direction represents the inflow direction, the 2-direction is the crossflow direction and the 3-direction is the thickness direction. The off-diagonal components of the orientation tensor show the tilt of the orientation tensor from the coordinate axes. Hence, they are zero only if the coordinate axes align with the principal directions of the orientation tensor [44]. Two examples for possible fiber alignment are given in Figure 1. If completely random orientation occurs, the diagonal elements are $a_{11} = a_{22} = a_{33} = 1/3$ (middle). For fiber orientation perpendicular to in-flow direction, the tensor elements are $a_{11} = 0$ $a_{22} = 1$ and $a_{33} = 0$ (right).

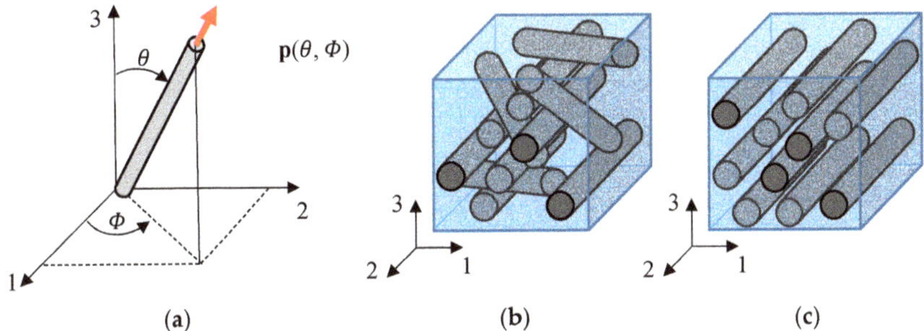

Figure 1. Representation of the orientation of a single rigid fiber by the vector $p(\Phi, \theta)$ (**a**) adapted after [44]; orientation of a fiber population within a volume: randomly oriented fibers (**b**) and aligned fibers (**c**), adapted after [36].

2. Materials and Methods

2.1. Materials and Specimens

2.1.1. Specimen Geometry and Mold Insert Design

The specimen geometries used in this work are shown in Figure 2. The test specimen geometry for tensile tests according to DIN EN ISO 527 is based on the multipurpose specimen geometry 1A according to DIN EN ISO 527-1 and DIN EN ISO 3167. For a high reproducibility of the specimen geometry, the cavity of the injection molding tool is flow-optimized in deviation to DIN ISO 294-1 to keep the wear of the AM tools as low as possible. The test specimen with its long flow path ratio is representative for a linear directed flow behavior. Due to the wall thickness of 4 mm, a good separation of a skin–shear–core–shear–skin orientation can be expected. The cavity and specimen geometry are shown in Figure 2.

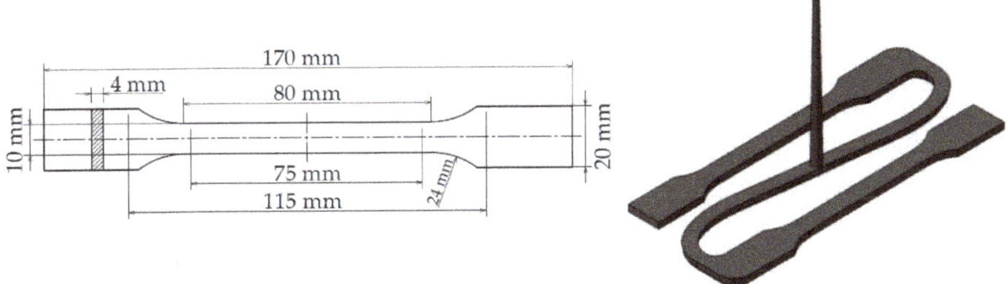

Figure 2. Melt flow path optimized tensile specimen geometry after DIN EN ISO 527 and DIN EN ISO 3167.

For the investigation of the fiber configuration, a nonstandard disc with a thickness of 3 mm and a diameter of 120 mm is used. Due to the rotation symmetry of the disc, only a partial segment of the disc needs to be examined to draw conclusions about the entire test specimen. The central injection point at the disc mold will allow investigation of the transition from a swirling to a pointed flow. Likewise, a radial geometry provides an even pressure distribution during the IM process and thus reduces the process-induced influence on the test specimens and measurement results. This plays an important role especially for AM tools, which tend to have a high deformation and mold breathing. Furthermore, the disc geometry is frequently used in literature, so that the obtained results can be compared with existing studies of Hongyu and Baird [45]. In the cases of Rohde et al. [46] and Goris et al. [17], who focused more on a plate geometry, the results can be seen as complementary and further investigation. The cavity and specimen geometry are shown in Figure 3.

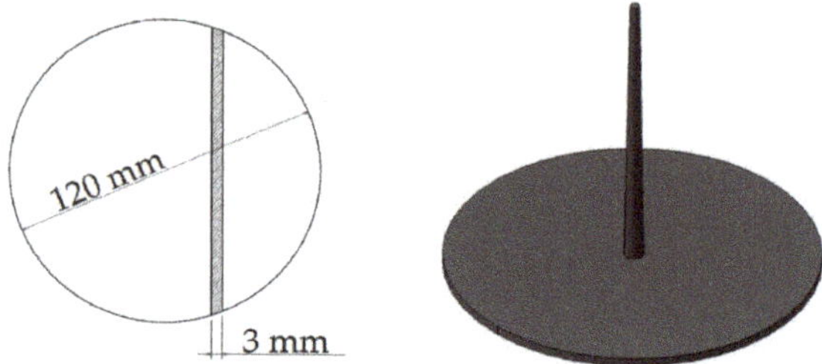

Figure 3. Disc specimen geometry with a thickness of 3 mm and a diameter of 120 mm.

2.1.2. Materials

The AM tools are made from Accura Bluestone by the company 3D Systems®, which is processed on a Viper si2 System (SLA). Accura Bluestone is a ceramic filled epoxy composite, which is characterized by relatively high heat deflection temperature of up to 284 °C and a tensile strength about 8000 MPa [47]. Due to its material characteristics, high imaging accuracy and low surface roughness, Kampkar et al. [21] recommend Accura Bluestone as an AM tool material.

The thermal conductivity coefficient of Accura Bluestone was determined for this investigation at approximately 0.781 W/mK, which is 50 times lower than the thermal conductivity of conventional steel (34.5–49.3 W/mK) [48] and 150 of aluminum (130–160 W/mK) [49]. Aside from the cooling system, all other tools are identical in their geometry and design. Table 1 shows a brief comparison of the material properties of the different tool materials. As a reference, the same tools are conventionally milled from aluminum (EN AW-7075) and steel (C45 U).

Table 1. Comparison of the material properties of the different tool materials (Bluestone, aluminum, steel).

Parameter	Unit	Bluestone	Aluminum	Steel
Thermal conductivity	W/mK	0.781	130–160	41.6–44.9
Thermal expansion coefficient	m/mK	81–98	22.5–23.4	11.1–12.1
Young's modulus	MPa	7600–11,700	71,000	210,000
Elongation at break	%	1.4–2.4	2–8	16

To increase the heat deflection temperature, the Accura Bluestone mold insert is tempered with a temperature profile as recommended by 3D Systems. This is done through thermal post curing for 2 h at 120 °C, which increases the deflection temperature from 65–66 °C to 267–284 °C [47].

Representative for the most used LFT, the glass fiber-reinforced PP injection molding material STAMAX from SABIC is used in different fiber weight percentages of 10 wt.%, 20 wt.%, 40 wt.%, and 60 wt.%. The specific fiber contents are chosen in accordance with the results of Goris [50], which showed the highest contrast and a clear differentiation for interpretation. Ratios of 30 wt.% and 50 wt.% were not considered to keep the experimental volume adequate. STAMAX is a certified product series for the automotive industry and is already available pre-mixed in many different glass fiber contents. Accordingly, the mixing ratios 20 wt.%, 40 wt.%, and 60 wt.% for the investigations can be obtained directly. The mixing ratio 10 wt.% is gravimetrically prepared from STAMAX 20 wt.% (20YM240) and pure PP, which the manufacturer SABIC uses as a base matrix material. In addition to its wide industrial usage, the material is characterized by its particularly good processability. PP has a wide processing temperature range with a low melt viscosity, melt temperature, and adhesivity. This is beneficial to reduce the thermal and mechanical stress on the polymer tools and to ensure the possibility to produce enough samples for the evaluation. For the specimen fabrication, a Boy 25 E injection molding machine from Dr. Boy GmbH & Co. KG is chosen.

2.2. Experiment Methodolgy

2.2.1. Mechanical Properties

To analyze the mechanical properties, tensile testing after DIN EN ISO 527 is performed. The results are analyzed toward tensile strength, elongation, and Young's modulus. For the calculation of the standard deviation, five samples of each specimen type are analyzed as recommended by DIN EN ISO 527. Nominal fiber contents of 10 wt.%, 20 wt.%, 40 wt.%, and 60 wt.% are analyzed. In accordance with DIN EN ISO 527, a testing speed of 1 mm/min is used for determination of the elastic properties and, more specifically, the determination of the Young's modulus. For the deformation properties, the testing speed is increased to 50 mm/min for the practical purpose of decreasing testing time. Although not in accordance with the norm, this is a common approach in the field of tensile testing. However, this effect must be considered for the discussion of the test results.

2.2.2. Fiber Length Analysis: Epoxy Plug Method

The epoxy plug method is centered around a down-sampling step after Kunc [40], which reduces the fiber count from <1,000,000 to a representative amount around 15,000–60,000 fibers. The detailed experiments steps are depicted in Figure 4.

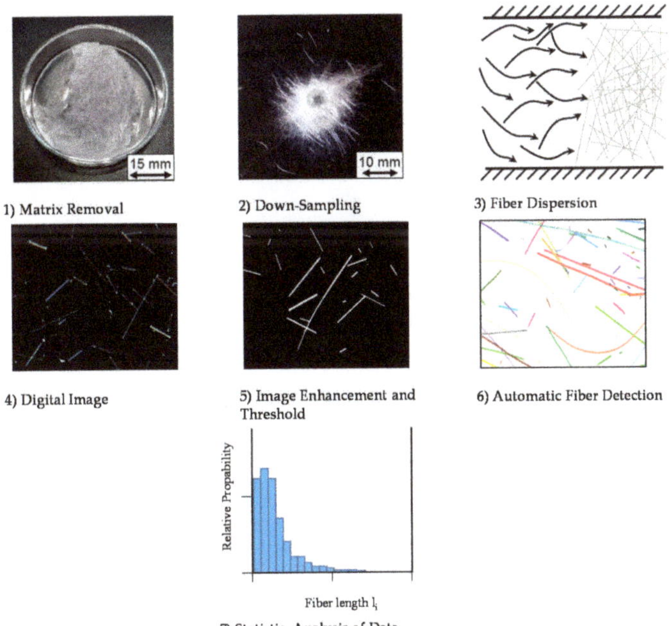

Figure 4. Overview of the developed methodology (Epoxy plug method) for fiber length analysis after Goris adapted from [50].

As can be seen in Figure 4, the matrix material of the sample is initially removed. A sample diameter of at least twice the initial fiber length must be chosen to avoid measuring fibers that crossed the cutting plane during extraction. An initial fiber length of 15 mm results in a diameter of 30 mm. The matrix removal is performed by pyrolysis at 500 °C for 8 h in an industrial oven. Rohde et al. [46] studied the impact of matrix removal by pyrolysis or chemical decomposition on the morphology of single fibers. The results show that performing pyrolysis at 500 °C for 2 h is optimal for a PP sample. However, initial test runs showed that the pyrolysis time needed to be extended to 8 h to sufficiently remove the matrix.

For this investigation, the selection of the experimental fiber analysis test method was based on two factors: to provide sufficient repeatability and comparability toward previous investigations, as well as being able to generate an adequate output with a reasonable experiment duration. Therefore, the developed method by Goris et al. [17] (compare Section 1.2) was chosen. Based on the investigations of Rohde et al. [46], a thermal matrix removal is chosen instead of a chemical removal, since the fiber length could be negatively altered. Next, the down-sampling step is performed. A defined amount of UV-activated epoxy B0027N07MM liquid-glue by the company BONDIC (Aurora, Canada) is injected in the exposed fiber bed. The diameter of the injected epoxy varies from ca. 4–7 mm. After UV-curing of the glue and careful removing of non-attached fibers, a second pyrolysis is performed at 500 °C for 8 h. The subsequent fiber dispersion step is performed within a dispersion chamber, which can be seen in Figure 5. The turbulent dispersion is performed through small amounts of pressured air, performed 3–4 times at 1 bar for around 0.5 s. The fibers are than dispersed on a 210 mm × 255 mm × 4 mm glass plate, which is positioned on an EPSON Perfection V800 Photo scanner of the company Seiko Epson (Suwa, Japan) to create a digital image at 2400 dpi. For image enhancement and threshold, Adobe® Photoshop™ is used. Threshold levels are approximately 40 at the image center and 60 at the edges. This variation is necessary due to inhomogeneous illumination. Then, a MATLAB-based algorithm is used for fiber detection, which was developed at the Polymer Engineering Center (Madison, WI, USA) [51] and is based on the work of Wang [52]. The fiber detection is automated and even detects stacked and bent fibers.

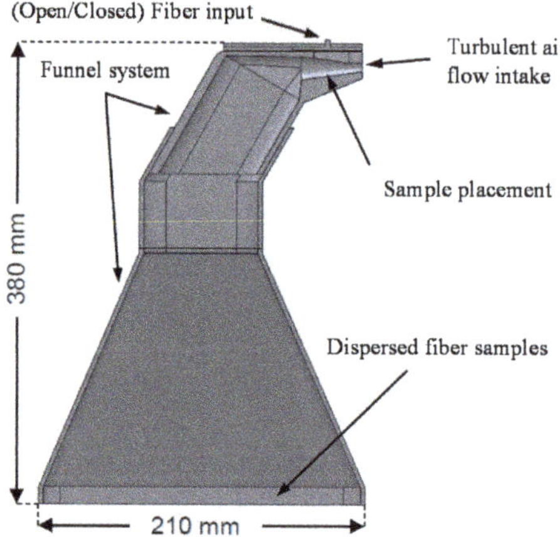

Figure 5. Dispersion chamber after Goris adapted from [50].

Two different averages are calculated for investigation of the fiber length: number average and length average. The number average is calculated using (compare nomenclature Table 2)

$$L_N = \frac{\sum_{i=1}^{n}(N_i \cdot l_i)}{\sum_{i=1}^{n}(N_i)} \tag{4}$$

and the length average (compare nomenclature Table 2)

$$L_W = \frac{\sum_{i=1}^{n}\left(N_i{\cdot}l_i^2\right)}{\sum_{i=1}^{n}\left(N_i{\cdot}l_i\right)} \tag{5}$$

Table 2. Nomenclature.

Parameter	Symbol	Unit
Fiber length	l_i	mm
Number of bins	n	-
Fiber frequency	N_i	-
Diameter of down-sampling	d	mm
Measured fiber frequency	θ	-

Long fibers have a more significant impact on the length average than short fibers. This effect is described as nonuniformity, of which high values are usually preferable in technical applications. Another effect that must be considered during the epoxy plug method is the preferred pickup of long fibers during the down-sampling step. This phenomenon can lead to a distorted average fiber length and is schematically represented in Figure 6. Five fibers are visible, from which only four are picked up by the down-sampling region. The fifth does not contribute to the experimental analysis aside from a similar alignment of the center of mass of each fiber. Therefore, Kunc et al. [40] introduced the so-called Kunc correction, which determines a corrected fiber frequency N_i in favor of shorter fibers. It is calculated as:

$$N(L_i) = \theta(L_i){\cdot}\left(1 + \frac{4{\cdot}L_i}{\pi d}\right) \tag{6}$$

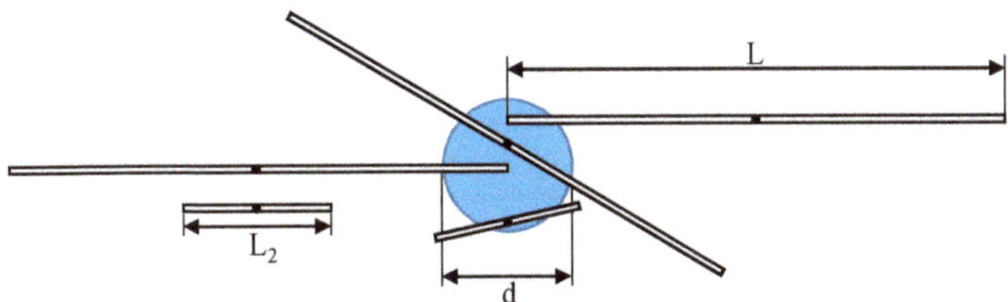

Figure 6. Illustration of the Kunc correction procedure, adapted after [40].

For this investigation, the analysis of the fiber length focuses on the analysis of the fiber-reinforced discs. Three segments—A, B, and C—with a diameter of 30 mm ± 0.5 mm are therefore removed by sawing from discs as depicted in Figure 7. The diameter is chosen based on the initial fiber length of 15 mm for STAMAX pellets. For the calculation of the standard deviation, three samples were analyzed for each location. Nominal fiber contents of 10 wt.%, 20 wt.%, 40 wt.%, and 60 wt.% are analyzed.

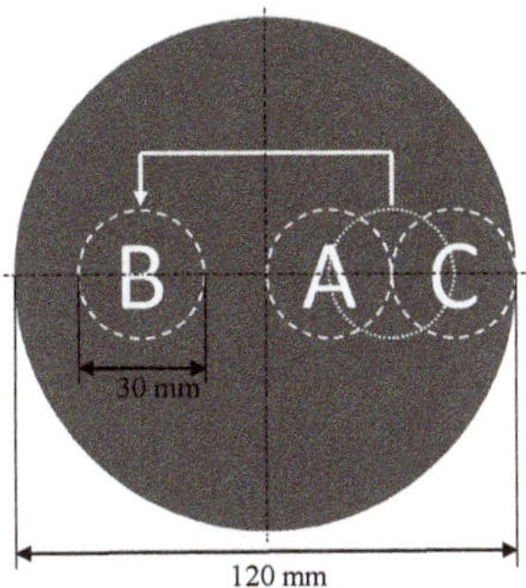

Figure 7. Disc segments for fiber length analysis.

2.2.3. Fiber Content Analysis: Pyrolysis

The analysis of the fiber content is determined gravimetrically. For this, a pyrolysis in accordance with DIN EN ISO 3451 is performed for each sample within an industrial oven. Each sample is kept at 100 °C for 30 min before the matrix is removed at 625 °C for 3 h. This represents a less gentle and therefore faster temperature program compared to the pyrolysis steps of the epoxy plug method, since the fiber quality is not important for this experiment. The fiber content is then calculated in accordance with DIN EN ISO 3451 as (compare nomenclature Table 3):

$$\alpha = \frac{m_F}{m_{Total}} \cdot 100 \ wt\% \tag{7}$$

Table 3. Nomenclature.

Parameter	Symbol	Unit
Fiber content	α	wt.%
Total sample mass	m_{Total}	mg
Fiber mass	m_F	mg

For this investigation, the analysis of the fiber content focuses on the analysis of the fiber-reinforced discs. The discs are therefore quartered and segmented in 33 squares with an edge length of 11.5 mm ± 0.5 mm as depicted in Figure 8. The samples were then cut out with scissors, resulting in 22 segments with identical in volume for each nominal fiber content, as well as 11 segments from the disc edges with varying volume and sample shape. The varying shape does not constitute a problem for later calculation of the fiber content, since the results are normalized by the individual total sample mass. For the calculation of the standard deviation, three samples were analyzed for each location. Nominal fiber contents of 10 wt.%, 20 wt.%, 40 wt.%, and 60 wt.% are analyzed.

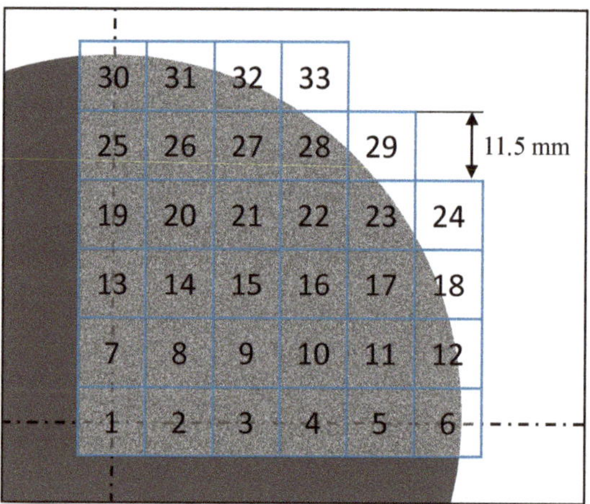

Figure 8. Segmentation of fiber-reinforced discs for fiber content analysis.

2.2.4. Fiber Orientation Analysis: Micro-Computed Tomography

For fiber orientation analysis, μCT scans are used, which are performed with a GE v|tome|x m 240/180 by the company General Electric (Boston, MA, USA). The scan parameters can be seen in Table 4.

Table 4. μCT scan parameters GE v|tome|x m 240/180.

Parameter	Unit	Value
Voltage	kV	80
Current	μA	140
Voxel size	μm	14.4
Projections	-	2000

The analysis of the μCT data is performed with VGSTUDIO MAX 3.3.0 64 Bit by the company Volume graphics GmbH (Heidelberg, Germany). The software was chosen in accordance with a comparative study by Goris [17], which compared results of the conventional method of ellipses to μCT-data analyzed with three different algorithms: VGSTUDIO MAX 3.3.0 64 Bit, slit method (SM) algorithm, and Mimics (proprietary to SABIC and Materialise MV). The results showed that a minimal resolution of 19 ± 1 μm could be used to sufficiently analyze the fiber orientation using VGSTUDIO MAX. For this investigation, the analysis of isotropic fiber orientation is performed with plane projection. The direction of the normal vector is identical to the thickness direction. Table 5 shows the detailed analysis parameters. As can be seen, the fiber material is defined by specific threshold, which is identical for all samples.

Table 5. Analysis parameters VGSTUDIO MAX 3.3.0 64 Bit.

Parameter	Unit	Value
Resolution	μm	5
Radius of integration	μm	5
Gradient threshold	-	7
Threshold for definition of the fiber material	-	116
Mode of integration for plane projection	-	isotrop

For this investigation, the analysis of the fiber orientation focuses on the analysis of the fiber-reinforced discs. As depicted in Figure 9, a rectangular sample of 18 mm × 70 mm (±0.5 mm) is removed with a saw from each disc. Three squares—A, B, and C—with an edge length of 12 mm ± 0.5 mm are then analyzed within VGSTUDIO MAX. Nominal fiber contents of 10 wt.%, 20 wt.%, 40 wt.%, and 60 wt.% are analyzed. Due to limitations for the experimental amount, only one sample of each fiber content and tool type is analyzed.

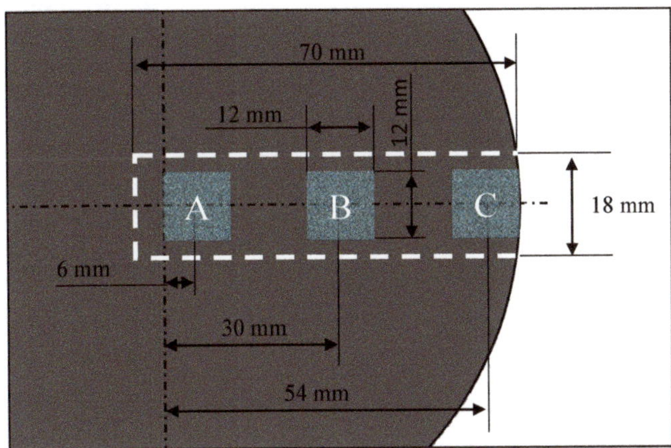

Figure 9. Segmentation of the fiber-reinforced discs for fiber orientation analysis; Segments A, B, and C (blue) are analyzed within VGSTUDIO MAX.

2.3. Processing Parameters and Tool Design

The processing parameters for STAMAX are based on the processing guidelines [53] recommended by SABIC. Within these guidelines, a distinction is made between minimum, moderate, and maximum parameters. For this investigation, moderate parameters are targeted to generate comparability to the results of Goris [50]. In accordance with the guidelines, the injection pressure is chosen at 800 bar with linear decline down to 700 bar for steel tooling, with moderate injection speeds of 70 ccm/s to 60 ccm/s. The holding pressure is recommended at 50–80% of the injection pressure. Therefore, a holding pressure of 400 bar is chosen with a linear decline to 350 bar. Since aluminum and Bluestone provide lower mechanical properties than steel, the processing parameters including the clamping force for these tools were lowered accordingly. Table 6 gives a detailed overview of the processing parameters of each tooling type.

Table 6. Processing parameters after [53].

	Steel	Aluminum	Bluestone
Melt temperature in °C	250	250	250
Mold temperature in °C	30	30	30
Injection pressure in bar	800 → 700	500 → 400	400 → 250
Injection speed in ccm/s	70 → 60	75 → 70	75 → 60
Holding pressure in s	400 → 350	300 → 240	125 → 100
Cooling time in s	65	65	85
Clamping force in kN	220	180	170

The melt temperature is chosen at 250 °C with a flat temperature profile within the screw cylinder. As can be seen, the mold temperature is set to 30 °C. The cooling process is accomplished through a water-cooling cycle of the same temperature. As soon as the implemented temperature sensors near the cavity reach the targeted temperature, a new cycle can be started.

As described before, geometrical limitations for the cooling channels are small due to additive manufacturing. This is represented in Figure 10, which shows the difference in possible cooling cycle alignment for ejector half metal disc tools (a) and additively manufactured disc tools (b).

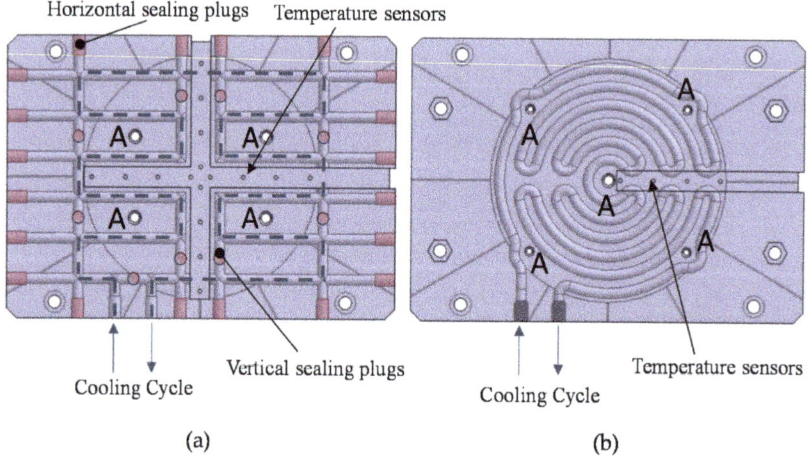

Figure 10. Ejector half disc tool designs: aluminum and steel (**a**) and Bluestone (**b**). Ejector pins are marked (A).

As can be seen, sealing plugs were used to close the boreholes of the metal tools. For the fixed mold half, the cooling path is analogue to the ejector half. The cooling channel geometry was determined through iterative calculation steps based on fundamental theorems of heat transfer and through calculation of the divergence in mold temperature. Moldflow simulations were used to guarantee a homogenic average mold temperature of the developed concepts, to minimize the effect of different cooling channel geometries on the results. As an example, the simulative results of Bluestone tensile specimen-tools are depicted in Figure 11.

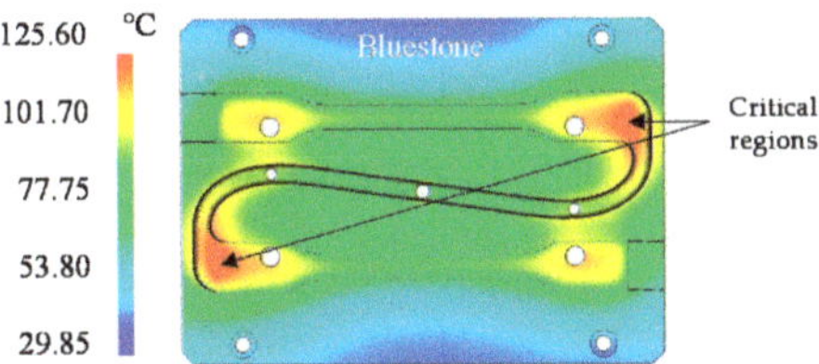

Figure 11. Moldflow simulation results of the average temperature for tensile specimen-tools out of Bluestone at 30 °C coolant temperature.

As can be seen, the average temperature ranges from approximately 30 °C up to 125 °C. The highest temperatures can be seen in the stagnation points of the melt at the tensile specimen shoulders. Since for Bluestone, a uniform temperature could not be guaranteed within the whole part, the focus lied on creating a homogenous average temperature within the functional regions, which is the gauge for tensile specimens. The stagnation points (red) were defined as critical regions, which will be discussed further in the following chapter. In case of the tensile specimens, parabolic runners with a diameter of 5 mm and a non-specified gate are used only on the ejector half. For the disc tools, the melt enters the cavity directly through the nozzle without the use of runners or a gate. The striking point of the melt also represents a critical region, especially for disc tools.

3. Results

3.1. Cycle Times Part Output and Failure-Modes

As can be seen in Figure 12, the recorded temperature near the critical region of tensile specimen Bluestone, aluminum, and steel tools are depicted. The final cycle times are determined through the critical mold temperature of 30 °C. As soon as the critical temperature is reached, a new cycle is started.

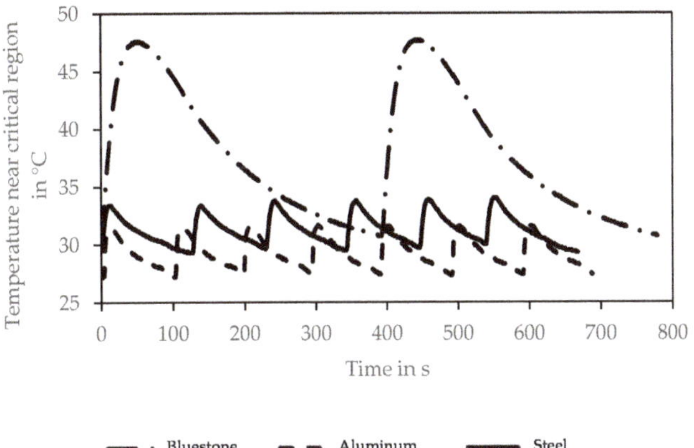

Figure 12. Recorded temperatures near critical region of cavity within Bluestone, aluminum, and steel tools for tensile specimens.

As can be seen, through the immense thermal conductivity of aluminum tools, the temperature in the depicted region fell even lower than 30 °C, close to room temperature, before a new cycle could be started. The results show that for every two Bluestone tool cycles, around seven steel cycles and eight aluminum cycles could be run. This directly translates to the increased thermal conductivity of steel and especially aluminum compared to Bluestone. In general, for every Bluestone tool type over 100 parts could be manufactured for all represented nominal fiber contents which includes tool trial runs. However, after a varying number of cycles depending on the tool type, different failure modes could be detected which were assessed as discard criteria or could be temporarily overhauled. The most prominent failure modes and deteriorations are depicted in Figure 13.

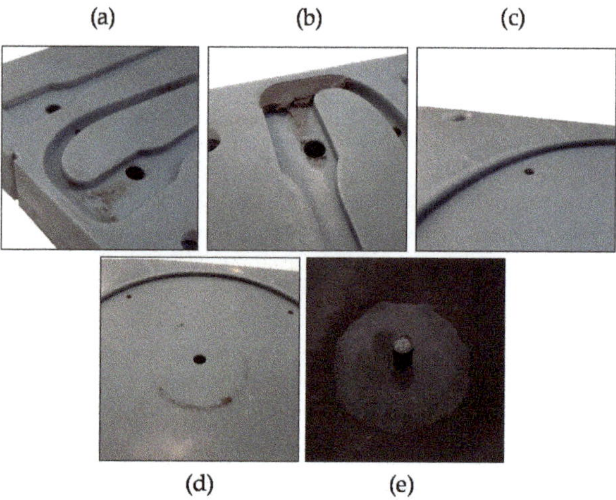

Figure 13. Failure modes and tool deterioration during specimen manufacturing: demolishing of mold at areas with strong staircase effect (**a**); clogging of channels at fiber contents of 60 wt.% (**b**); abrasion of mold material at the edges resulting in flash on parts (**c**); penetration of mold through fibers and PP at regions of high impact (**d**); contamination of parts through scraped mold material (**e**).

A typical challenge with additive manufactured components is the occurring staircase effect for steep geometries especially around angles of 45° paired with high temperatures. Figure 13 shows the critical regions with increased staircase effect (a), which caused demolishing of the mold especially at high fiber contents. For fiber contents of 60 wt.%, the demolishing eventually leads to clogging of channels at critical regions (Figure 13b). A typical deterioration for all nominal fiber contents is the abrasion of the mold material at the cavity edges (Figure 13c), which causes flashes on manufactured parts. This is either caused through underestimation of the clamping force, the flexibility of the mold or as an effect of repeated demolishing through material abrasion and heat exposure. A similar effect can be seen near the center of the mold and the melt entry point. After a non-specified number of cycles, these regions of high impact showed penetration of the mold through fibers and plain PP (Figure 13d). In extreme cases, the scraped mold material was found in the injected molded components (Figure 13e). Parts which showed signs of foreign particles and contamination were declared as rejects and the mold was eventually discarded. For aluminum and steel tools, none of the aforementioned failure modes or deteriorations could be detected.

3.2. Mechanical Properties

Generally, the results for the Young's modulus (Figure 14) and the tensile strength (Figure 15) are increasing for rising nominal fiber contents, while the elongation at tensile strength (Figure 16) is declining. Typical for brittle materials, the elongation at tensile strength and the elongation at break are basically equal. The highest variation of the results is visible for nominal fiber contents of 60 wt.%.

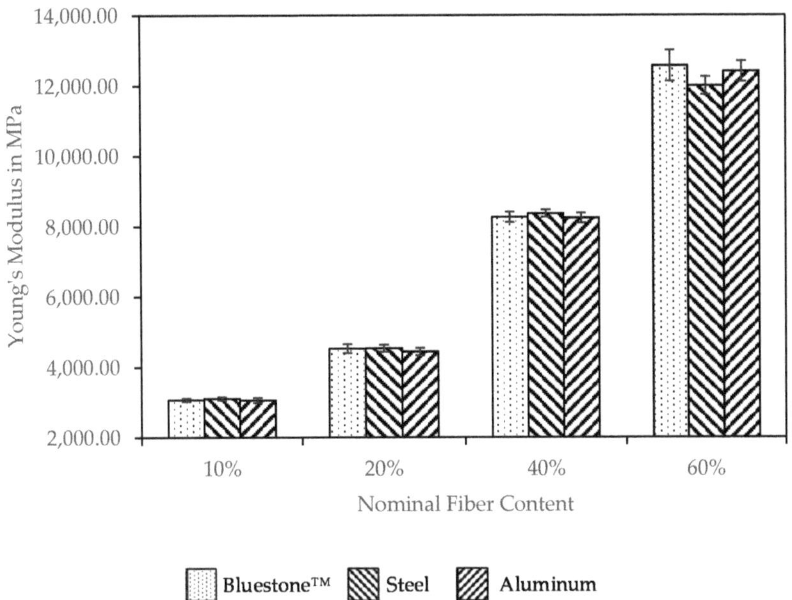

Figure 14. Young's modulus for tensile specimens from Bluestone, steel, and aluminum tools for nominal fiber contents of 10 wt.%, 20 wt.%, 40 wt.%, and 60 wt.%.

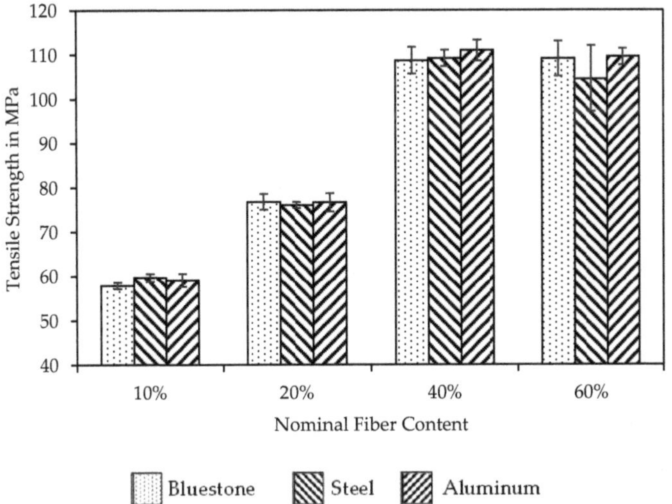

Figure 15. Tensile strength for tensile specimens from Bluestone, steel, and aluminum tools for nominal fiber contents of 10 wt.%, 20 wt.%, 40 wt.%, and 60 wt.%.

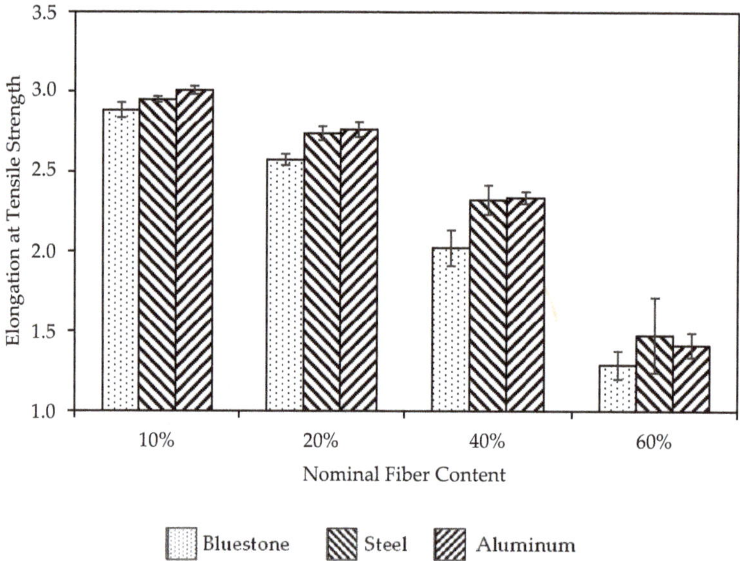

Figure 16. Elongation at tensile strength for tensile specimens from Bluestone, steel, and aluminum tools for nominal fiber contents of 10 wt.%, 20 wt.%, 40 wt.%, and 60 wt.%.

The results of the Young's modulus show that comparable values toward experimental data supported by the manufacturer are reached for every tool type and nominal fiber content. Furthermore, Bluestone tool specimens of 60 wt.% nominal fiber content are showing slightly higher values than steel and aluminum tools, although the increased variation of the results must be considered. An almost linear increase of the Young's modulus can be detected with increasing nominal fiber contents, with maximum values around 12,000 MPa.

The results of the tensile strength also meet nearly identical values for each tool type and nominal fiber content. However, there is no detectable increase in tensile strength from 40 wt.% toward 60 wt.%, with maximum values around 110 MPa at 60 wt.%. As described before, the deformation analysis is performed at higher testing speeds than the analysis of the elastic behavior. This must be considered for the comparison of the tensile strength as well as the elongation at tensile strength toward the manufacturers' data.

The elongation at tensile strength shows a constant decline toward higher nominal fiber contents since the material gets increasingly brittle. Aluminum and steel tools show almost identical results. However, test specimens of Bluestone tools show lower values than aluminum and steel tools, especially at nominal fiber contents of 40 wt.% and 60 wt.%. In further analysis, μCT data were able to show an increased number of voids for these nominal fiber contents, which are possibly caused due to limited processing conditions (Figure 17). The polymer matrix is responsible for an improved elongation behavior since stresses in lateral direction can be minimized. Through voids, the elongation and elastic behaviors are locally reduced.

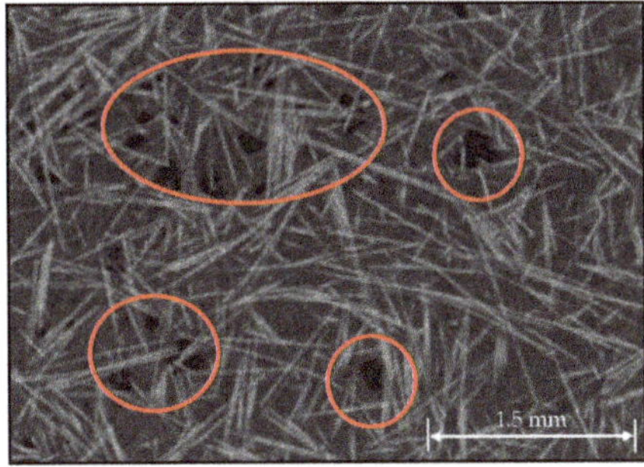

Figure 17. µCT image from spattered voids in Bluestone tool specimens for a nominal fiber content and 60 wt.%.

3.3. Fiber Length

The results of the fiber length analysis for a nominal fiber content of 40 wt.% are shown in Figure 18. All other results for 10 wt.%, 20 wt.%, and 60 wt.% are represented in Appendix A. Analogue to the results of Goris [17] for steel tools, the length average of the fiber length reduces with increasing nominal fiber contents. The same phenomenon is visible for aluminum and Bluestone tools. Processing causes severe fiber breakage, which causes the number of long fibers to reduce. With increasing fiber contents, the shear stress and fiber interlocking increases, which causes increased early state fiber breakage. Therefore, the reached average values are far below the initial fiber length of 15 mm, with higher values for the length average compared to the number average. The highest length average variations are visible close to the gate and the entry point of the melt (Location A). This is most prominently visible for steel and aluminum tools at all nominal fiber contents. A possible explanation is the highly disoriented state in which the melt enters the cavity, thus creating a broader variety in fiber length. This theory can be further supported by the length average values, which show higher varieties than number average values. As described before, the length average values tend to be more highly impacted by the existence of a small number of relatively long fibers.

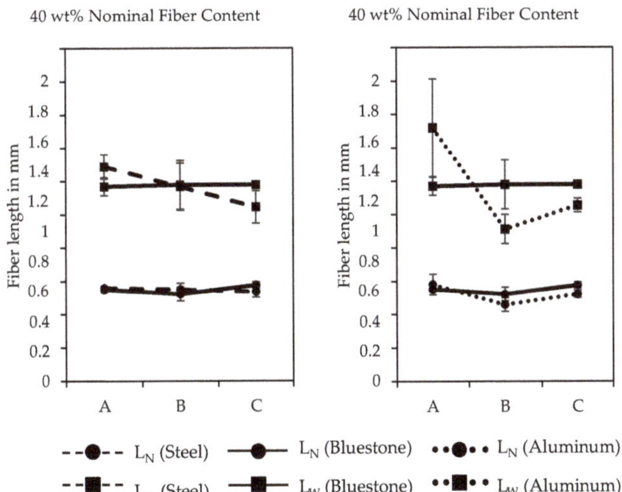

Figure 18. Fiber length (N = number average; W = weight or length average) for different tool types (for better representation in two diagrams) with 40 wt.% nominal fiber content at different specimen locations A, B, and C in accordance with Figure 7.

Depending on the locations A, B, and C, the depicted fiber contents show different behaviors with an increasing melt path. In general, the length average values of each tool type show higher fluctuations over the melt path than the number average values, which show a nearly stable length around 0.6 mm for all nominal fiber contents and tool types. This value directly translates to the results of Kim et al. [32], who declared a fiber length between 0.5–2 mm as typical values for long fibers after injection molding processing. Prominently stable results for the length average of each location are visible for Bluestone tools at nominal fiber contents of 40 wt.% and 60 wt.%, with a slight increase over the melt path (location A to C). Steel and aluminum tools show contrarian behaviors for nominal fiber contents of 40 wt.% and 60 wt.%, with maximum values at location A, followed by a decrease over location B to C. For a nominal fiber content of 10 wt.%, the samples of Bluestone tools show a decline in length average from location A to B, while increasing to maximum values of around 1.8 mm in location C. This behavior is also present for steel and aluminum tools, although to a lesser extent and with maximum values for the length average at location A. For 20 wt.% nominal fiber content, samples of Bluestone tools show contrarian behavior compared to 10 wt.%, with high length average values at locations A and B and a prominent decline in location C. A similar behavior is visible for aluminum tool samples, while samples from steel tools show almost identical behavior for 20 wt.% compared to 10 wt.%.

Despite the discussed differences in length average for different tool types, the results show that globally similar values for the average fiber length are reached depending of each tool material. Therefore, Bluestone tools can be declared viable for the use in functional validation of steel tools.

3.4. Fiber Concentration

The results of the fiber content analysis for a nominal fiber content of 40 wt.% are shown in Figure 19. All results for 10 wt.%, 20 wt.%, 40 wt.%, and 60 wt.% are represented in Figure 20. The schematic results for 10 wt.%, 20 wt.%, and 60 wt.% are shown in Appendix A. As can be seen, Figure 19 schematically depicts the results for an initial nominal fiber content of 40 wt.%, which show a characteristic increase in fiber content for all three tool types. This phenomenon can be seen for all initial nominal fiber contents. The lowest fiber concentration can be seen around the disc center with values mostly below and around the initial nominal fiber concentration. With increasing flow path, the fiber content increases until reaching above average values within the disc edges. This general increase of the fiber content with increasing melt path is visible for all nominal fiber contents, which is identical to the results of Goris [50]. A possible hypothesis for this behavior is the increase of the shear stress between liquid melt and solidified regions. Two effects occur in these regions: fiber breakage and fiber pullout. Through shear forces in the border region, the fibers are carried away by the melt flow and are shortened in this process. In retrospective to the results of the fiber length in Section 3.3, this phenomenon is especially visible on aluminum and steel tools. However, the results of Section 3.3 often show a limited or contrarian fiber shortening behavior for Bluestone tools toward metal tools. This is possibly caused by the slowed solidification and cooling process, as well as the limited processing conditions for Bluestone (injection speed, pressure, and clamping force). The fibers are more likely to be pulled out and carried off, with limited fiber breakage due to a more gradually transition between solidified material and the flowing liquid melt.

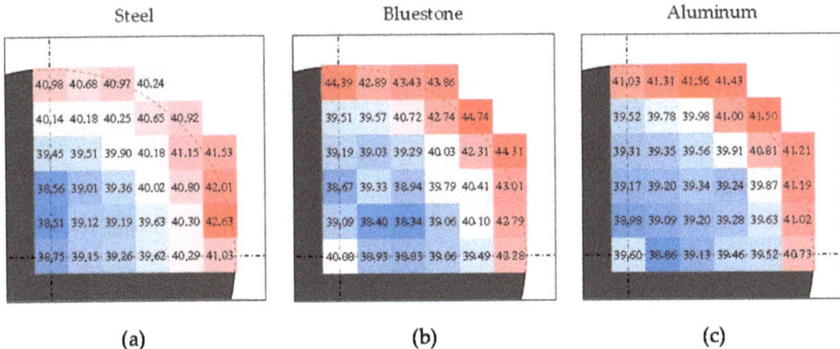

Figure 19. Characteristic fiber content distribution pattern for disc segments of 40 wt.% initial nominal fiber content: steel tool (**a**), Bluestone tool (**b**), and aluminum tool (**c**). The color scheme describes values below (blue) up to values near (white) and values above (red) the initial nominal fiber content (in accordance with Figure 8).

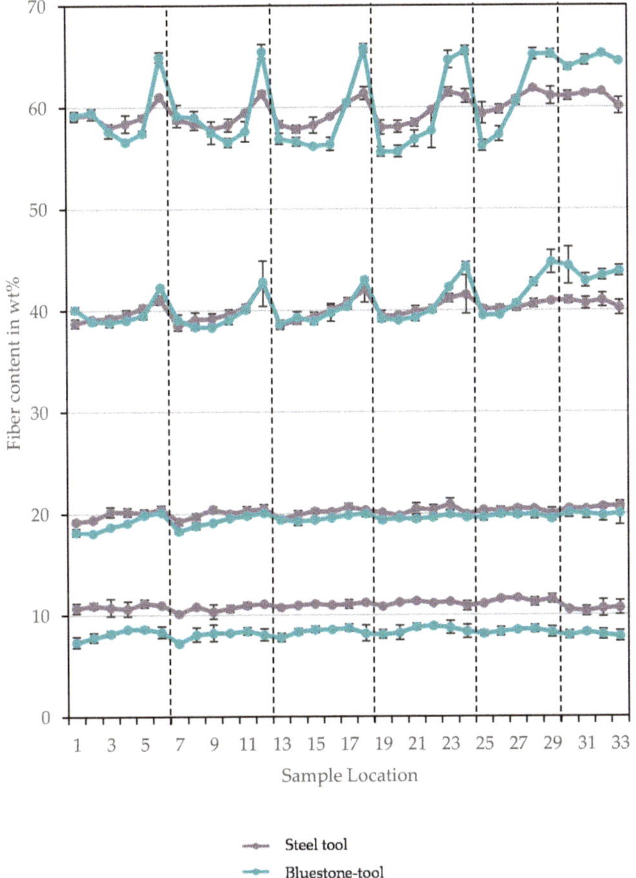

Figure 20. Fiber content distribution for disc segments of 10 wt.%, 20 wt.%, 40 wt.%, and 60 wt.% initial nominal fiber content at different locations (1–33 in accordance with Figure 7) for steel tools and Bluestone tools.

The comparison between Bluestone tools and steel tools for all initial fiber contents is represented in Figure 20. For the comparison of steel and aluminum tools, see Appendix A. As can be seen, for an initial nominal fiber content of 10 wt.%, the overall values for Bluestone tool samples are below the values of steel tools. This can most likely be explained through inaccuracy in the before discussed gravimetric mixture of equal parts 20 wt.% fiber-reinforced PP and plain PP. Regarding this, Bluestone tools and steel tools show nearly identical values and behaviors up to 40 wt.% initial nominal fiber content, with low standard variation. 60 wt.% samples from Bluestone tools show a higher spread toward minimum and maximum values compared to steel tools. This phenomenon is possibly caused by the gentler processing conditions for Bluestone tools. In comparison to the results of aluminum tools (see Appendix A), the results show that the best representation of steel tools for functional validation is through aluminum tools. However, the global fluctuation of the Bluestone tool samples is still low enough to be declared viable for functional validation of steel tools, especially for fiber contents up to 40 wt.%. The characteristic zones are identical for all tool types.

3.5. Fiber Orientation

As described in Section 1.2, seven characteristic regions can be determined in injection molded parts. Figure 21 shows the orientation tensor components over the relative sample thickness exemplified for a nominal fiber content of 60 wt.% at location B of a Bluestone tool specimen segment (in accordance with Figure 9). As can be seen, the characteristic zones can be identified (1, 2 and 3).

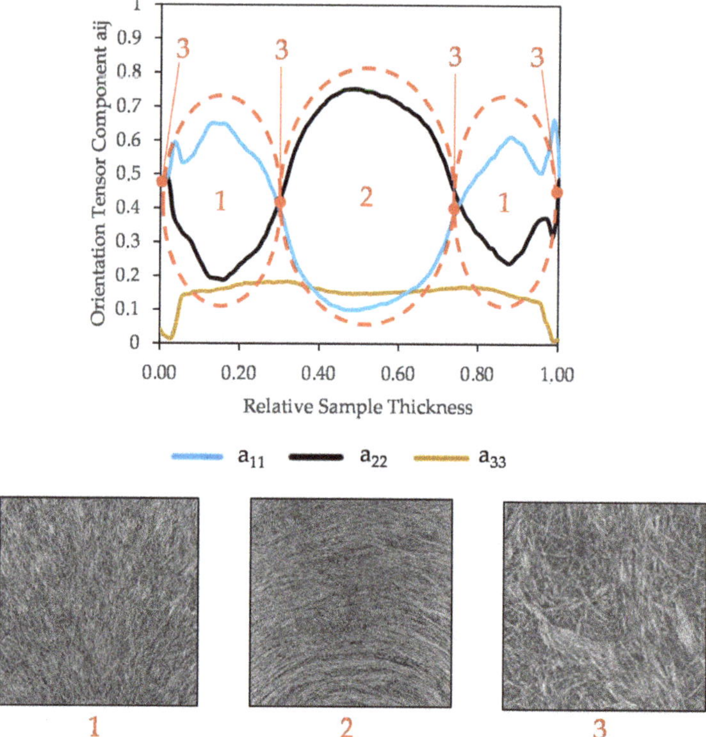

Figure 21. Orientation tensor components (a_{11}, a_{22}, a_{33}) over relative sample thickness for a nominal fiber content of 60 wt.% (Bluestone tool; segment B), with characteristic zones after Osswald and Menges [36] (shell layers = 1, core layer = 2, transition layers/skin layers = 3).

In addition to the identification of the characteristic zones, it is visible that the tensor component a_{33} is constant over the entire relative sample thickness. This is valid for all examined segments. For further analysis of the fiber orientation, only components a_{11} and a_{22} are described because of their relevance toward the identification of the characteristic zones. Figure 22 shows the results for a nominal fiber content of 60 wt.% for Bluestone tools, steel tools and aluminum tools at locations A, B, and C. The results of all other nominal fiber contents can be seen in Appendix A.

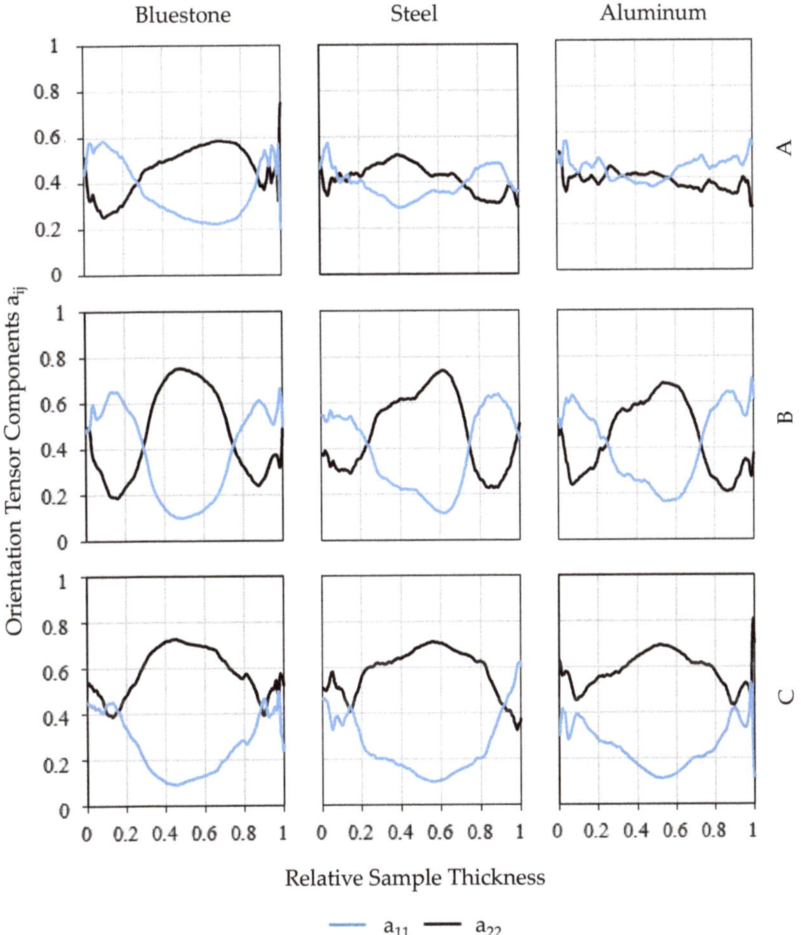

Figure 22. Orientation tensor components (a_{11}, a_{22}; without a_{33} for better overview) and relative sample thickness fora nominal fiber content of 60 wt.%. Specimens were created with Bluestone, steel, and aluminum tools, with samples from regions A, B, and C (in accordance with Figure 9).

As can be seen, the quality of the results is highly dependent on the sample location. Since B provides the most developed, unhindered, and oriented state of the melt flow, the characteristic zones are identifiable without further complication. Location A provides the most inconsistent transitions compared to locations B and C. A possible explanation for this behavior can be given based on the melt flow state. Since the melt flow is in a highly disoriented state when it enters the cavity, it is most likely that this region shows the highest inconsistencies toward fiber orientation. This behavior is present for all tool types and nominal fiber contents. Location C shows more consistent results than location A. However, due to the material stagnation at the edges of the disc, slight inconsistencies that hinder clear identification of the characteristic zones are still visible. Therefore, the decision was made to concentrate on location B entirely for the identification of specific trends dependent on increasing nominal fiber content, as well as different tool materials. The results can be seen in Figure 23.

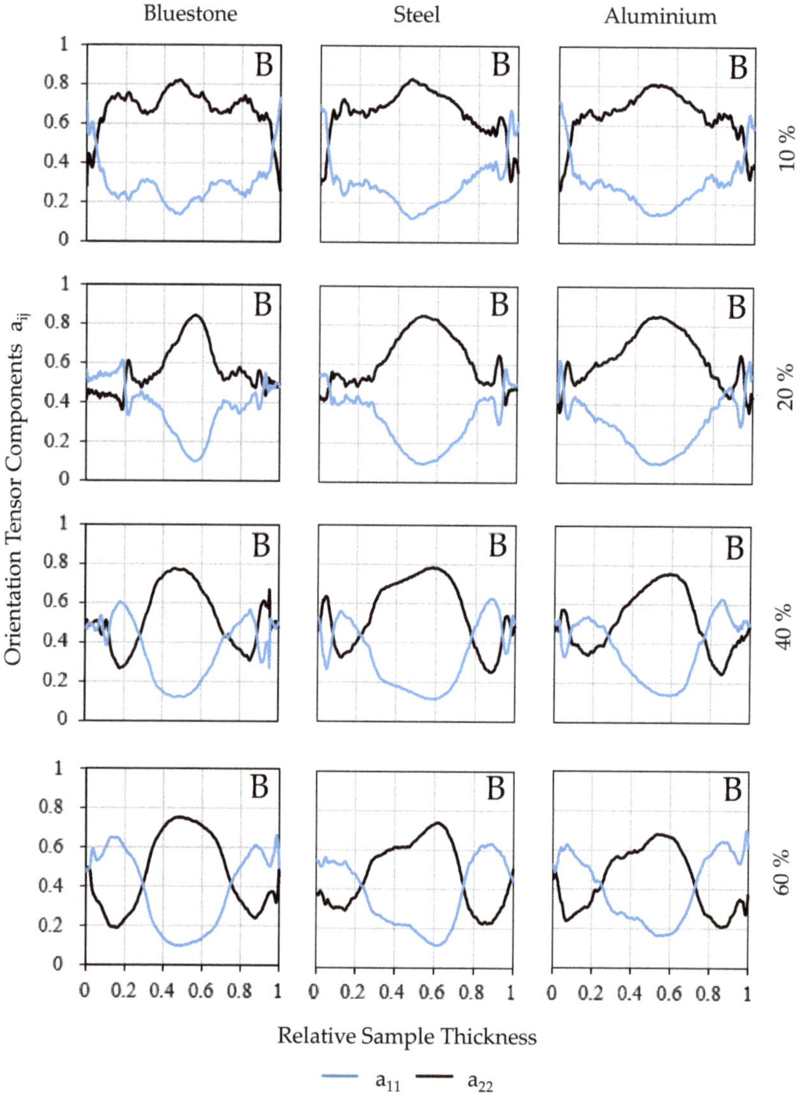

Figure 23. Orientation tensor components (a_{11}, a_{22}; without a_{33} for better overview) and relative sample thickness for nominal fiber contents of 10 wt.%, 20 wt.%, 40 wt.%, and 60 wt.%. Specimens were created with Bluestone, steel, and aluminum tools.

It can be seen that there is a global increase of the shell layer thickness with increasing nominal fiber contents. A possible hypothesis for this phenomenon is based on different crystallization behavior for increasing fiber contents. Each fiber in the melt acts as a nucleating agent during cooldown. When more fibers are included, the crystallization process and solidification take place faster, which benefits increasing shell thickness. In terms of applicability for functional validation of steel tools, the Bluestone tools show almost identical fiber orientation compared to steel and aluminum tools. The thickness of the specific zones and the range of the orientation tensors are almost identical for all used tool types, which is essential for the qualification of Bluestone tools for functional validation. However, further repeated experiments are necessary for a sufficient qualification, since only one specimen of each tool type and nominal fiber content was examined during the experiments. Furthermore, the different thermal conductivities of all tool types impact the viscosity of the melt, which must be analyzed to solidify the aforementioned hypothesis.

4. Discussion & Conclusions

The discussion of the results of this investigation concentrates on two different aspects:

(a)   The general identification of characteristic phenomena dependent on the type of tool material and the nominal fiber content.

(b)   The relevance of the results toward the overall target, to use additively manufactured Bluestone tools as an alternative for steel tools for spare part and small series productions, as well as for the functional validation of steel tools.

The results of this investigation successfully illustrate, that specimens manufactured with Bluestone tools show comparable results toward metal tools for the mechanical properties and the fiber configuration, which includes fiber length, fiber concentration, and fiber orientation. In terms of the cycle times, the part output and the tool durability, Bluestone tools are not able compete with metal tools. As expected, zones of high impact and choke points proof to be critical for the Bluestone tool durability. A possible solution would be the use of modular concepts that make use of overall usable metal inlets, which can be applied at highly stressed regions. However, for the use of Bluestone tools in spare part and small series production, the reached part output of over 100 parts proves to be sufficient. Due to the similarity in fiber configuration and the mechanical properties, the use of Bluestone tools for functional validation is nearly qualified. Further experiments toward the fiber orientation need to be conducted to solidify the existing results.

Multiple characteristic phenomena could be detected, which are applicable for polymer tools, aluminum tools, and steel tools. For all tool types, the extent of the nominal fiber content influences two aspects: the thickness of the characteristic zones defined by Osswald and Menges [36], as well as the fiber length average. The analysis of the fiber orientation shows that with increasing nominal fiber content, the thickness of the shell layers increases respectively, while the thickness of the core layer decreases. A possible hypothesis for this phenomenon is based on the altered crystallization mechanism for increasing fiber contents. When more fibers are included, the crystallization process and solidification take place faster, which benefits increasing shell thickness. The analysis of the fiber length illustrates that the length average for specimens of each tool material decreases with increasing nominal fiber contents. The number average stays nearly stable for all tool types around 0.6 mm. This value directly translates to the results of Kim et al. [32], which declared a fiber length between 0.5–2 mm as typical values for long fibers after injection molding processing. Another phenomenon is determined through the analysis of the fiber concentration. Analogous to the results for rectangular specimens created through steel tools described by Goris [50], a gradual decrease of the fiber content is visible with increasing melt flow distance. This behavior is visible for steel tool specimens and now for Bluestone tool and aluminum tool specimens as well. A characteristic maximum at the edges of the disc specimens can be declared.

All in all, additively manufactured Bluestone proofs to be a viable approach for the rapid creation of polymer tools for injection molding processing. Further investigations will be centered around the improvement of the polymer tool lifecycle, as well as improved cycle times through efficient cooling methods. The use of modular concepts that make use of overall usable metal inlets needs to be further investigated. At last, the cost efficiency of polymer tools toward metal tools needs to be analyzed for a specific example in a comparative study.

Author Contributions: This investigation is based in equal parts on the scientific work of L.K. and R.S., under the supervision of D.R., K.W. and T.O. The work was divided in the following categories: Conceptualization, L.K.; methodology, L.K. and R.S.; software, L.K. and R.S.; validation, L.K. and R.S.; formal analysis, L.K. and R.S.; investigation, L.K. and R.S.; data curation, L.K. and R.S.; writing—original draft preparation, L.K. and R.S.; writing—review and editing, L.K. and R.S.; visualization, L.K. and R.S.; supervision, D.R., K.W., T.O.; project administration, L.K.; funding acquisition, no additional/external funding. All authors have read and agreed to the published version of the manuscript.

Funding: This research received no external funding. The APC was funded by Polymer Engineering Center, University of Wisconsin-Madison, Madison, WI 53706, USA.

Conflicts of Interest: The authors declare there is no conflict of interest.

Appendix A

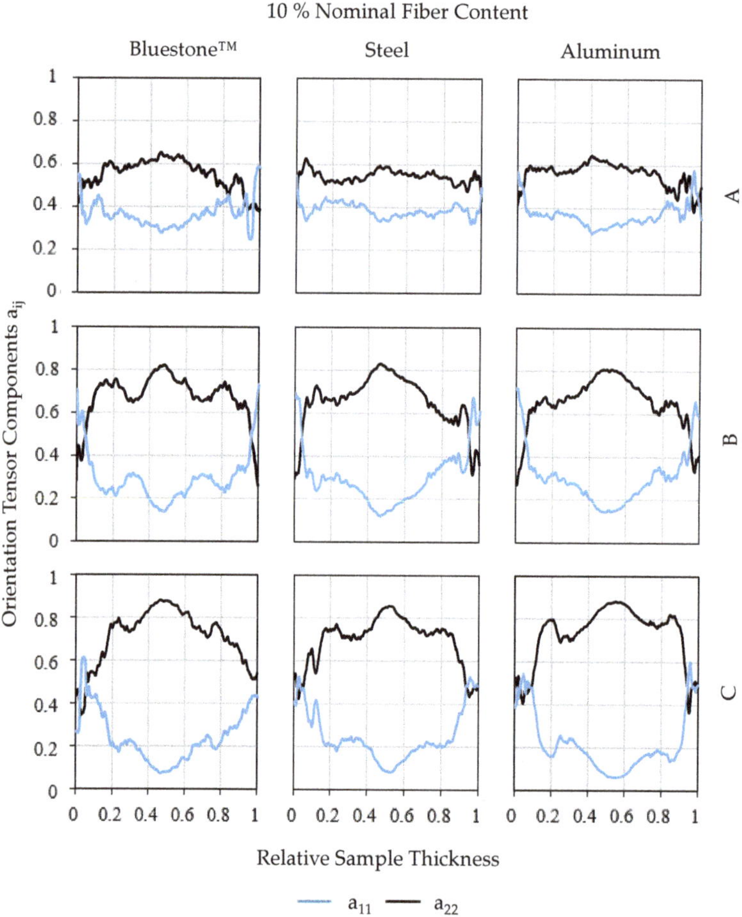

Figure A1. Orientation tensor components (a_{11}, a_{22}; without a_{33} for better overview) and relative sample thickness for a nominal fiber content of 10 wt.%. Specimens were created with Bluestone, steel, and aluminum tools, with samples from regions A, B, and C (in accordance with Figure 9).

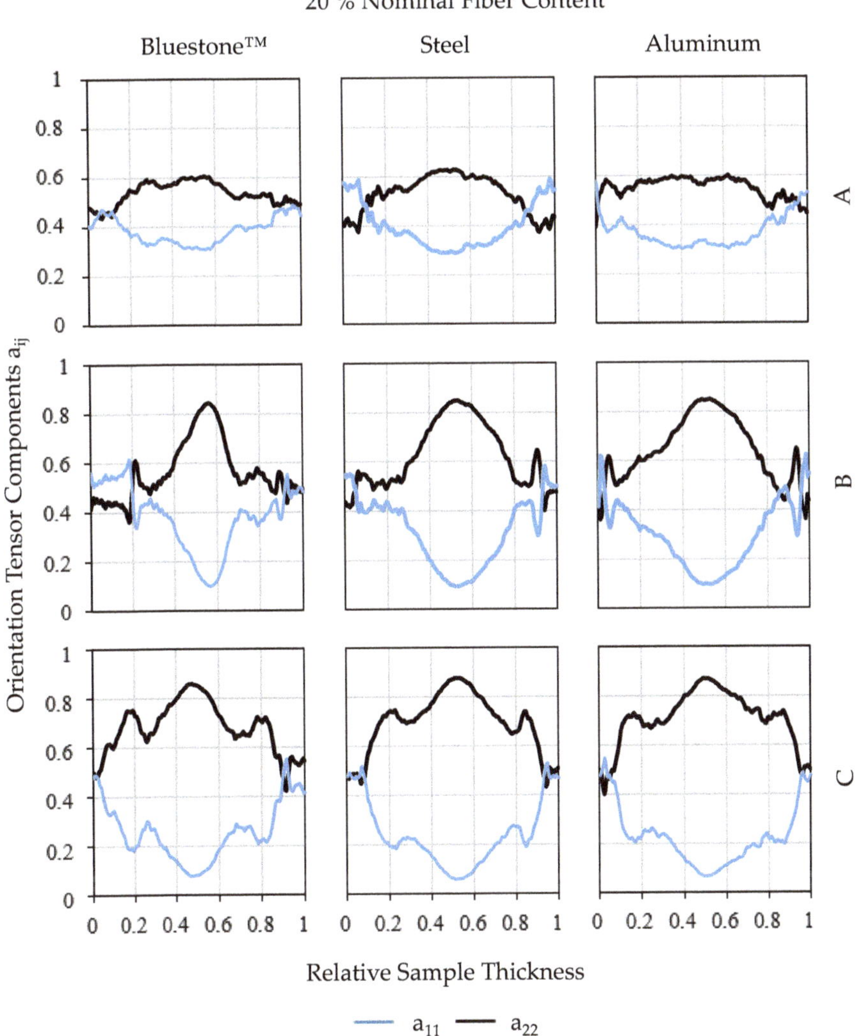

Figure A2. Orientation tensor components (a_{11}, a_{22}; without a_{33} for better overview) and relative sample thickness for a nominal fiber content of 20 wt.%. Specimens were created with Bluestone, steel, and aluminum tools, with samples from regions A, B, and C (in accordance with Figure 9).

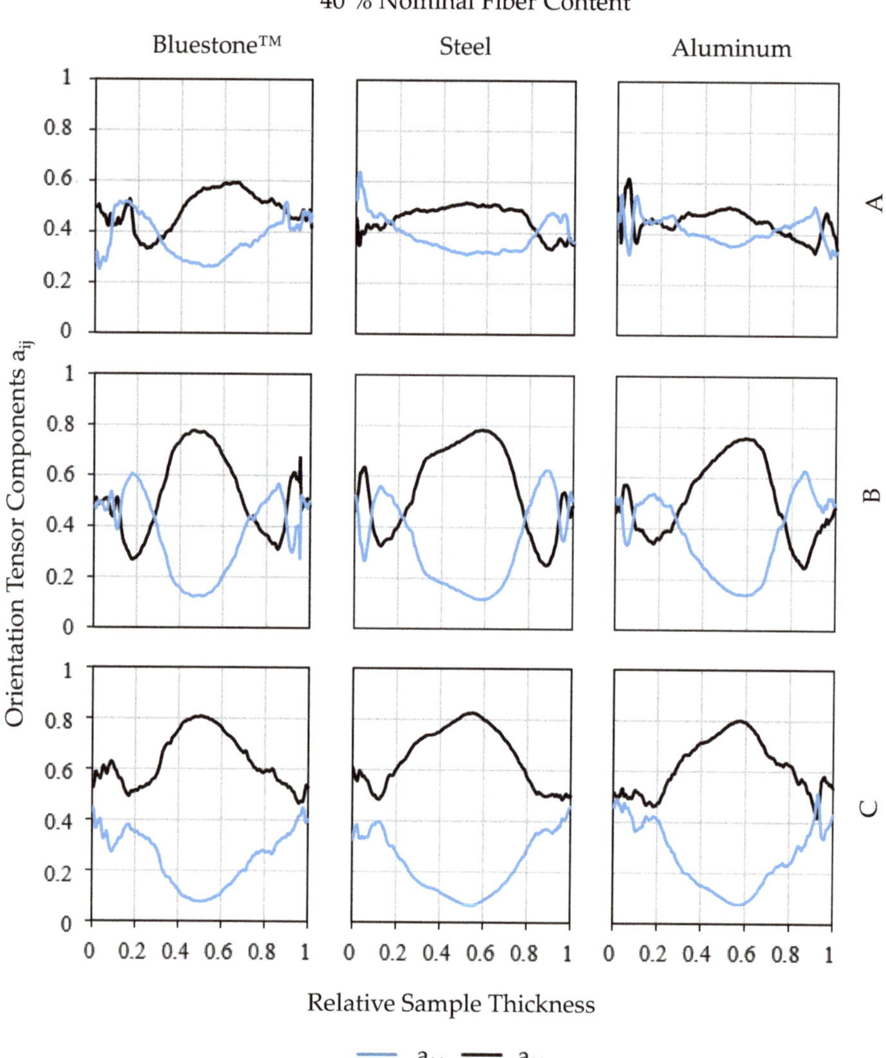

Figure A3. Orientation tensor components (a_{11}, a_{22}; without a_{33} for better overview) and relative sample thickness for a nominal fiber content of 40 wt.%. Specimens were created with Bluestone, steel, and aluminum tools, with samples from regions A, B, and C (in accordance with Figure 9).

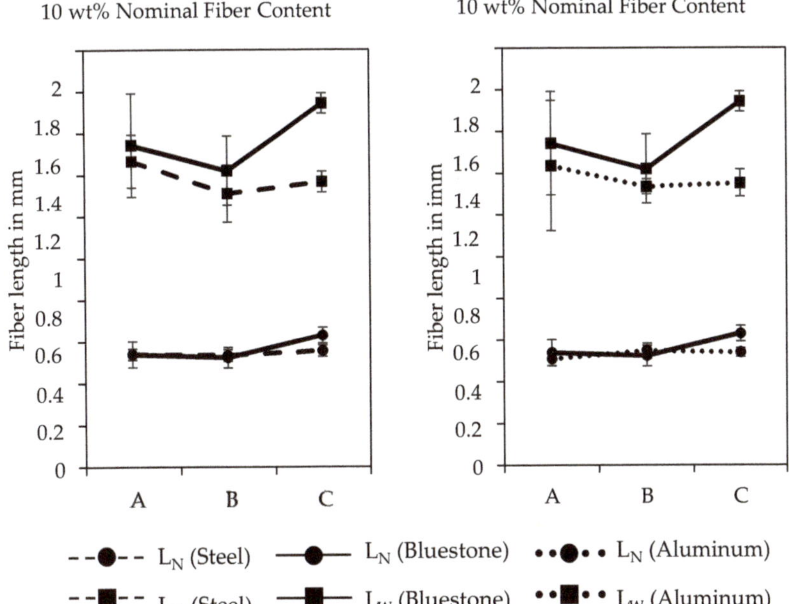

Figure A4. Fiber length (N = number average; W = weight or length average) for different tool types (for better representation in two diagrams) with 10 wt.% nominal fiber content at different specimen locations A, B, and C in accordance with Figure 7.

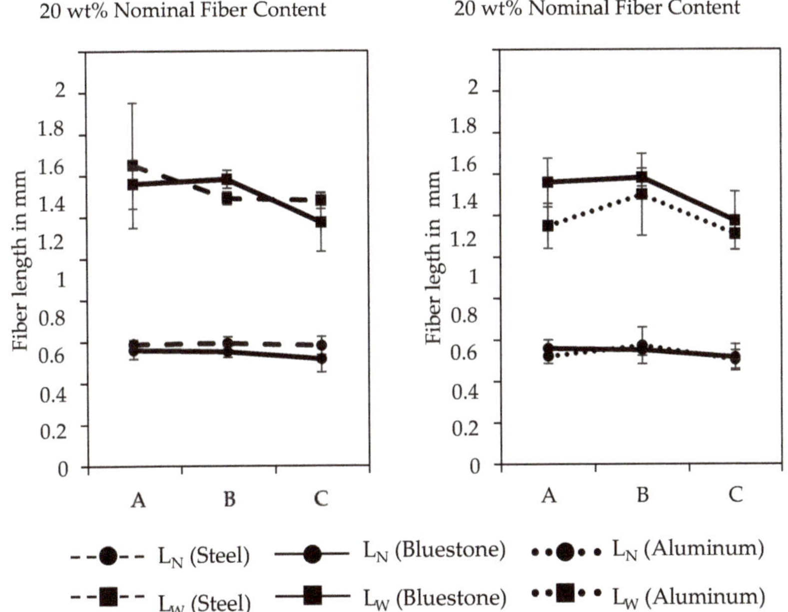

Figure A5. Fiber length (N = number average; W = weight or length average) for different tool types (for better representation in two diagrams) with 20 wt.% nominal fiber content at different specimen locations A, B, and C in accordance with Figure 7.

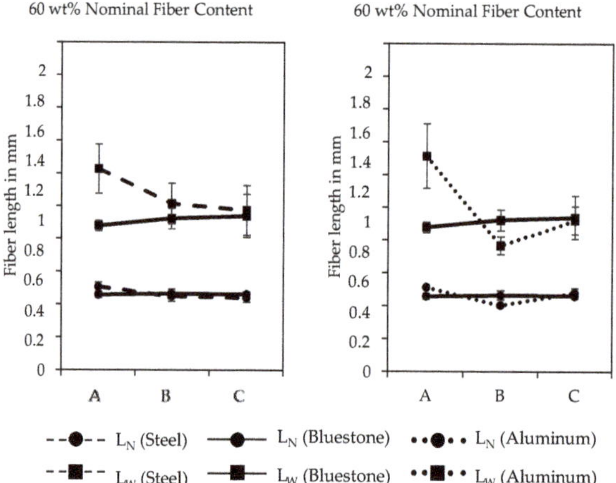

Figure A6. Fiber length (N = number average; W = weight or length average) for different tool types (for better representation in two diagrams) with 60 wt.% nominal fiber content at different specimen locations A, B, and C in accordance with Figure 7.

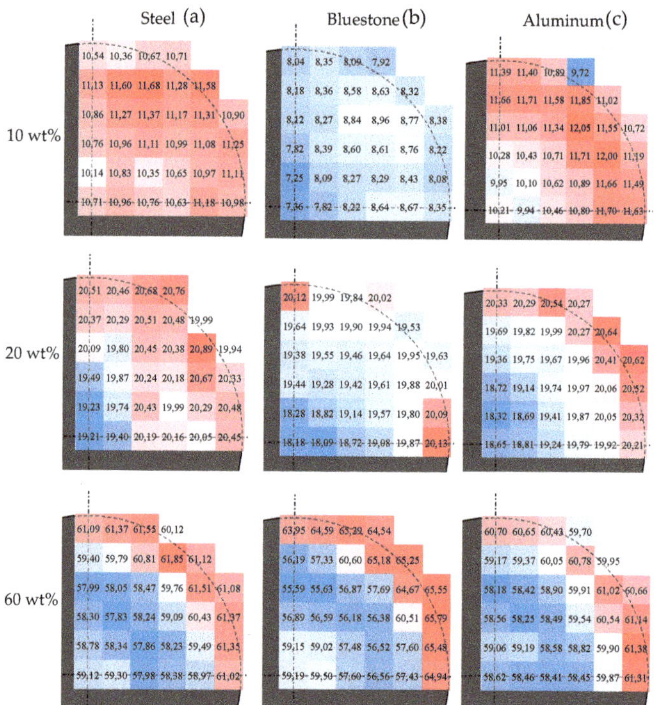

Figure A7. Characteristic fiber content distribution pattern for disc segments of 10 wt.%, 20 wt.%, and 60 wt.% initial nominal fiber content: steel tool (**a**), Bluestone tool (**b**), and aluminum tool (**c**). The color scheme describes values below (blue) up to values near (white) and values above (red) the initial nominal fiber content (in accordance with Figure 8).

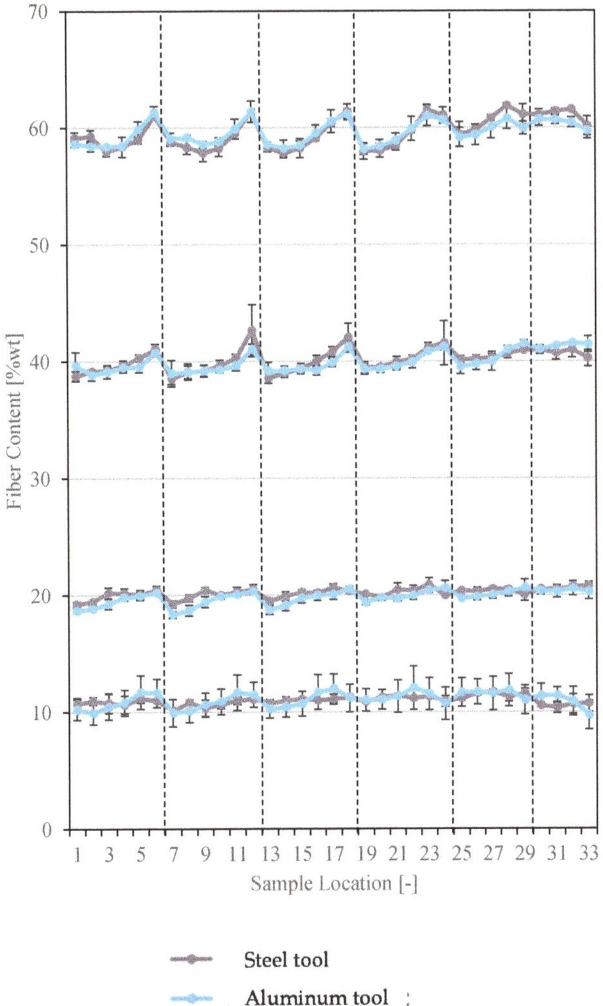

Figure A8. Fiber content distribution for disc segments of 10, 20, 40, and 60 wt.% initial nominal fiber content at different locations (1–33 in accordance with Figure 7) for steel tools and aluminum tools.

References

1. Schemme, M. LFT—Development status and perspectives. *Reinf. Plast.* **2008**, *52*, 32–39. [CrossRef]
2. Henning, F.; Ernst, H.; Brüssel, R. LFTs for automotive applications. *Reinf. Plast.* **2005**, *49*, 24–33. [CrossRef]
3. Markets and Markets Research Private Ltd. *Long Fiber Thermoplastics Market by Resin (PP, PA, PEEK, PPA), Fiber (Glass, Carbon), Manufacturing Process (Injection Molding, Pultrusion, D-LFT), End-Use Industry (Automotive, Electrical & Electronics), Region-Global Forecast to 2025;* Markets and Markets Research Private Ltd.: Magarpatta, India, 2020.
4. Markarian, J. Long fibre reinforcement drives automotive market forward. *Plast. Addit. Compd.* **2005**, *7*, 24–29. [CrossRef]
5. Gómez-Monterde, J.; Sánchez-Soto, M.; Maspoch, M.L. Microcellular PP/GF composites: Morphological, mechanical and fracture characterization. *Compos. Part A Appl. Sci. Manuf.* **2018**, *104*, 1–13. [CrossRef]

6. SmartTech Analysis. *Additive Manufacturing Market Outlook and Summary of Opportunities*; SmartTech Analysis: Charlottesville, VA, USA, 2019.
7. Ngo, T.; Kashani, A.; Imbalzano, G.; Nguyen, Q.; Hui, D. Additive manufacturing (3D printing): A review of materials, methods, applications and challenges. *Compos. Part B Eng.* **2018**, *143*, 172–196. [CrossRef]
8. Shaffer, S.; Yang, K.; Vargas, J.; Di Prima, M.; Voit, W. On reducing anisotropy in 3D printed polymers via ionizing radiation. *Polymer* **2014**, *55*, 5969–5979. [CrossRef]
9. Zou, R.; Xia, Y.; Liu, S.; Hu, P.; Hou, W.; Hu, Q.; Shan, C. Isotropic and anisotropic elasticity and yielding of 3D printed material. *Compos. Part B Eng.* **2016**, *99*, 506–513. [CrossRef]
10. Valino, A.D.; Dizon, J.R.C.; Espera, A.H., Jr.; Chen, Q.; Messman, J.; Advincula, R.C. Advances in 3D printing of thermoplastic polymer composites and nanocomposites. *Prog. Polym. Sci.* **2019**, *98*, 101162. [CrossRef]
11. Achillas, C.; Tzetzis, D.; Raimondo, M.O. Alternative production strategies based on the comparison of additive and traditional manufacturing technologies. *Int. J. Prod. Res.* **2017**, *55*, 1–13. [CrossRef]
12. Rohde, J.; Jahnke, U.; Lindemann, C.; Kruse, A.; Koch, R. Standardised product development for technology integration of additive manufacturing. *Virtual Phys. Prototyp.* **2018**, *14*, 141–147. [CrossRef]
13. Wu, T.; Jahan, S.A.; Kumaar, P.; Tovar, A.; El-Mounayri, H.; Zhang, Y.; Zhang, J.; Acheson, D.; Brand, K.; Nalim, R. A framework for optimizing the design of injection molds with conformal cooling for additive manufacturing. *Procedia Manuf.* **2015**, *1*, 404–415. [CrossRef]
14. Maravola, M.; Conner, B.; Walker, J.; Cortes, P. Epoxy infiltrated 3D printed ceramics for composite tooling applications. *Addit. Manuf.* **2019**, *25*, 59–63. [CrossRef]
15. Yuan, S.; Shen, F.; Chua, C.K.; Zhou, K. Polymeric composites for powder-based additive manufacturing: Materials and applications. *Prog. Polym. Sci.* **2019**, *91*, 141–168. [CrossRef]
16. Kampker, A.; Triebs, J.; Kawollek, S.; Ayvaz, P.; Beyer, T. Direct polymer additive tooling—Effect of additive manufactured polymer tools on part material properties for injection moulding. *Rapid Prototyp. J.* **2019**. [CrossRef]
17. Goris, S.; Back, T.; Yanev, A.; Brands, D.; Drummer, D.; Osswald, T.A. A novel fiber length measurement technique for discontinuous fiber-reinforced composites: A comparative study with existing methods. *Polym. Compos.* **2017**, *39*, 4058–4070. [CrossRef]
18. Sodeifian, G.; Ghaseminejad, S.; Yousefi, A.A. Preparation of polypropylene/short glass fiber composite as Fused Deposition Modeling (FDM) filament. *Results Phys.* **2019**, *12*, 205–222. [CrossRef]
19. Hertle, S.; Drexler, M.; Drummer, D. Additive manufacturing of poly(propylene) by means of melt extrusion. *Macromol. Mater. Eng.* **2016**, *301*, 1482–1493. [CrossRef]
20. Mulholland, T.; Goris, S.; Boxleitner, J.; Osswald, T.; Rudolph, N. Process-induced fiber orientation in fused filament fabrication. *J. Compos. Sci.* **2018**, *2*, 45. [CrossRef]
21. Kampker, A.; Triebs, J.; Ford, B.A.; Kawollek, S.; Ayvaz, P. Potential analysis of additive manufacturing technologies for fabrication of polymer tools for injection moulding—A comparative study. In Proceedings of the 2018 IEEE International Conference on Advanced Manufacturing, Yunlin, Taiwan, 16–18 November 2018.
22. Rahmati, S.; Dickens, P. Rapid tooling analysis of stereolithography injection mould tooling. *Int. J. Mach. Tools Manuf.* **2007**, *47*, 740–747. [CrossRef]
23. Summer Topical Meeting; American Society for Precision Engineering; European Society for Precision Engineering; Nanotechnology. Dimensional accuracy and surface finish in additive manufacturing. In Proceedings of the ASPE—The American Society for Precision Engineering, Raleigh, NC, USA, 27–30 June 2016.
24. Altaf, K.; Rani, A.M.A.; Raghavan, V.R. Prototype production and experimental analysis for circular and profiled conformal cooling channels in aluminium filled epoxy injection mould tools. *Rapid Prototyp. J.* **2013**, *19*, 220–229. [CrossRef]
25. Park, H.S.; Dang, X.-P. Development of a smart plastic injection mold with conformal cooling channels. *Procedia Manuf.* **2017**, *10*, 48–59. [CrossRef]
26. Hopkinson, N.; Dickens, P. A comparison between stereolithography and aluminium injection moulding tooling. *Rapid Prototyp. J.* **2000**, *6*, 253–258. [CrossRef]
27. Martinho, P.G.; Bartolo, P.J.; Pouzada, A. Hybrid moulds: Effect of the moulding blocks on the morphology and dimensional properties. *Rapid Prototyp. J.* **2009**, *15*, 71–82. [CrossRef]

28. Harris, R.; Fouchal, F.; Hague, R.; Dickens, P. Quantifying part irregularities and subsequent morphology manipulation in stereolithography plastic injection moulding. *Plast. Rubber Compos.* **2004**, *33*, 92–98. [CrossRef]

29. Harris, R.A.; Hague, R.J.; Dickens, P.M. Crystallinity control in parts produced from stereolithography injection mould tooling. *Proc. Inst. Mech. Eng. Part L J. Mater. Des. Appl.* **2003**, *217*, 269–276. [CrossRef]

30. Fernandes, A.D.C.; De Souza, A.F.; Howarth, J.L.L. Mechanical and dimensional characterisation of polypropylene injection moulded parts in epoxy resin/aluminium inserts for rapid tooling. *Int. J. Mater. Prod. Technol.* **2016**, *52*, 37. [CrossRef]

31. Volpato, N.; Solis, D.M.; Costa, C.A. An analysis of digital ABS as a rapid tooling material for polymer injection moulding. *Int. J. Mater. Prod. Technol.* **2016**, *52*. [CrossRef]

32. Kim, Y.; Park, O.O. Effect of fiber length on mechanical properties of injection molded long-fiber-reinforced thermoplastics. *Macromol. Res.* **2020**, *28*, 1–12. [CrossRef]

33. Seong, D.G.; Kang, C.; Pak, S.Y.; Kim, C.H.; Song, Y.S. Influence of fiber length and its distribution in three phase poly(propylene) composites. *Compos. Part B Eng.* **2019**, *168*, 218–225. [CrossRef]

34. Hou, X.; Chen, X.; Liu, B.; Chen, S.; Li, H.; Cao, W. Fracture and orientation of long-glass-fiber-reinforced polypropylene during injection molding. *Polym. Eng. Sci.* **2020**, *60*, 13–21. [CrossRef]

35. Tadmor, Z. Molecular orientation in injection molding. *J. Appl. Polym. Sci.* **1974**, *18*, 1753–1772. [CrossRef]

36. Osswald, T.A. *Understanding Polymer Processing: Processes and Governing Equations*; Hanser Publications: Munich, Germany, 2017.

37. Parveeen, B.; Caton-Rose, P.; Costa, F.; Jin, X.; Hine, P. Study of injection moulded long glass fibre-reinforced polypropylene and the effect on the fibre length and orientation distribution. *AIP Conf. Proc.* **2014**, *1593*, 432–435.

38. Zhu, H.; Gu, Y.; Yang, Z.; Li, Q.; Li, M.; Wang, S.; Zhang, Z. Fiber distribution of long fiber reinforced polyamide and effect of fiber orientation on mechanical behavior. *Polym. Compos.* **2020**, *41*, 1531–1550. [CrossRef]

39. Lafranche, E.; Krawczak, P. Injection moulding of long glass fibre reinforced thermoplastics (LFT): Structure/processing conditions/mechanical properties relationship. In Proceedings of the *ESAFORM* Conference on Material Forming, Glasgow, UK, 26–28 April 2006.

40. Kunc, V.; Frame, B.J.; Nguyen, B.N.; Tucker, C.L., III; Velez-Garcia, G. Fiber length distribution measurement for long glass and carbon fiber reinforced injection molded thermoplastics. *Res. Gate* **2007**, *Volume 2*, 866–876.

41. Ren, P.; Dai, G. Fiber dispersion and breakage in deep screw channel during processing of long fiber-reinforced polypropylene. *Fibers Polym.* **2014**, *15*, 1507–1516. [CrossRef]

42. Krasteva, D.L. Integrated Prediction of Processing and Thermomechanical Behavior of Long Fiber Thermoplastic Composites. Ph.D. Thesis, University of Minho, Braga, Portugal, 2009.

43. Sharma, B.N.; Naragani, D.; Nguyen, B.N.; Tucker, C.L.; Sangid, M.D. Uncertainty quantification of fiber orientation distribution measurements for long-fiber-reinforced thermoplastic composites. *J. Compos. Mater.* **2018**, *52*, 1781–1797. [CrossRef]

44. Advani, S.G. The use of tensors to describe and predict fiber orientation in short fiber composites. *J. Rheol.* **1987**, *31*, 751–784. [CrossRef]

45. Chen, H.; Baird, D.G. Prediction of young's modulus for injection molded long fiber reinforced thermoplastics. *J. Compos. Sci.* **2018**, *2*, 47. [CrossRef]

46. Rohde, M.; Ebel, A.; Wolff-Fabris, F.; Altstädt, V. Influence of processing parameters on the fiber length and impact properties of injection molded long glass fiber reinforced polypropylene. *Int. Polym. Process.* **2011**, *26*, 292–303. [CrossRef]

47. 3D Systems. Accura Bluestone Technical Data Sheet 2018. Available online: https://de.3dsystems.com/sites/default/files/2018-08/3d-systems-accura-bluestone-datasheet-uken-2018-08-21-web.pdf (accessed on 7 August 2020).

48. Wilzer, J.; Lüdtke, F.; Weber, S.; Theisen, W. The influence of heat treatment and resulting microstructures on the thermophysical properties of martensitic steels. *J. Mater. Sci.* **2013**, *48*, 8483–8492. [CrossRef]

49. Ostermann, F. *Anwendungstechnologie Aluminium*; Springer Science and Business Media LLC: Berlin, Germany, 1998.

50. Goris, S. Characterization of the Process-Induced Fiber Configuration of Long Glass Fiber-Reinforced Thermoplastics. Ph.D. Dissertation, University of Wisconsin, Madison, WI, USA, 2017.

51. Perez, C.; Osswald, T.A.; Goris, S. Study on the fiber properties of a LFT strand. *SPE ACCE* **2013**, *2*, 1115–1126.
52. Wang, H. Fiber Property Characterization by Image Processing. Master's Thesis, Texas Tech University, Lubbock, TX, USA, 2007.
53. SABIC. *Processing Guides: SABIC Stamax*; SABIC: Riyadh, Saudi Arabia, 2016.

Journal of
Composites Science

Article

Selective Laser Sintering of PA6: Effect of Powder Recoating on Fibre Orientation

Tobias Heckner [1,*], Michael Seitz [1], Sven Robert Raisch [1], Gerrit Huelder [1] and Peter Middendorf [2]

[1]   Robert Bosch GmbH, Corporate Research, 71272 Renningen, Germany;
     fixed-term.michael.seitz4@bosch.com (M.S.); svenrobert.raisch@de.bosch.com (S.R.R.);
     gerrit.huelder@de.bosch.com (G.H.)
[2]   Institute of Aircraft Design, University of Stuttgart, 70569 Stuttgart, Germany;
     middendorf@ifb.uni-stuttgart.de
*   Correspondence: tobias.heckner@de.bosch.com

Received: 30 June 2020; Accepted: 1 August 2020; Published: 6 August 2020

Abstract: In Selective Laser Sintering, fibres are strongly orientated during the powder recoating process. This effect leads to an additional increase of anisotropy in the final printed parts. This study investigates the influence of process parameter variation on the mechanical properties and the fibre orientation. A full factorial design of experiment was created to evaluate the processing parameters of recoating speed, layer thickness and laser power on the part's modulus of elasticity. Based on the mechanical testing, computed tomography was applied to selected samples to investigate the process-induced fibre microstructure, and calculate the fibre orientation tensors. The results show increasing part stiffness in the deposition direction, with decreasing layer thickness and increasing laser power, while the recoating speed only shows little effect on the mechanical performance. This complies with computed tomography imaging results, which show an increase in fibre orientation with smaller layer thickness. With thinner layers, and hence smaller shear gaps, shear stresses induced by the roller during recoating increase significantly, leading to excessive fibre reorientation and alignment. The high level of fibre alignment implies an increase of strength and stiffness in the recoating direction. In addition, thinner layer thickness under constant laser energy density results in improved melting behaviour, and thus improved fibre consolidation, consequently further increasing the mechanical properties. Meanwhile, the parameters of recoating speed and laser power do not have a significant impact on fibre orientation within their applicable process windows.

Keywords: selective laser sintering; recoating; PA6; polyamide; glass; fibres; beads; orientation; recoating speed; layer thickness; energy density

1. Introduction

In polymer and composite technologies, reinforcement fillers of various materials are commonly used to increase part strength and stiffness. In injection molding, it is well known that fibres align inside the melt flow and display characteristic fibre orientation patterns inside the center of the flow and outer shell [1]. In addition to fibre orientation, fibre content distribution and length can vary within different locations inside a part [2]. Due to the effect of fibre matrix separation, areas within a part can appear with significant deviations in filler content [3].

In Selective Laser Sintering (SLS), part strength and stiffness can also be increased with the use of fibre-reinforced powders. While the typical layer-by-layer build-up differs greatly from the injection molding process, changes in filler orientation and part anisotropy are also observed. This is mainly caused by fibre alignment during recoating.

Jansson and Pejryd [4] investigated the orientation of carbon short fibres in SLS processing. Their results show greater fibre alignment in the direction of powder recoating, while longer fibres exhibit higher levels of orientation. During recoating, the recoating blade touches longer fibres and aligns them into recoating direction, while short fibres with a length below the layer thickness align more randomly. This investigation is supported by Arai et al. [5], who determined the anisotropy of glass fibres in polybutylene terephthalate (PBT). Arai showed that the tensile strength and modulus of elasticity show their highest values in the recoating direction, and lowest mechanical properties perpendicular to it. Furthermore, Jansson and Pejryd [2] and Schmid [6] state that fibre-reinforced SLS parts can show worse mechanical properties in the perpendicular build-up direction than unreinforced parts with equal polymer materials. This is caused by reduced fibre orientation in the thickness direction, and hence lower layer-to-layer connection [4].

In addition, the influence of powder recoating and the energy density on the mechanical properties of unreinforced SLS material was investigated. Beitz [7] compared different recoating systems, such as roller and blade. He found that the recoating unit applies pressure onto the powder bed, dependent on the recoating speed and type of recoating unit. Drexler et al. [8] showed no significant effects on part density with different recoating speeds, but an impact on the mechanical properties can be observed. Tensile strength as well as elongation at break show the highest values for the lowest recoating speed. This effect is explained via the longer interaction time of the recoating unit with the powder particles. As a consequence, a more homogenous part surface results.

The influence of laser energy density was previously studied for unreinforced materials [9]. Lanzl et al. [10] conducted the first experiments to investigate the effect of the energy density on the mechanical properties of fibre-reinforced materials. In their investigations, fibre content was varied from 10 to 50 vol.%, while particularly fibre contents of up to 30 vol.% showed a high impact on the modulus of elasticity. The variation of energy density from 0.04 to 0.06 J/mm^2 had only a little influence on part performance. At very low energy density levels, the modulus of elasticity decreased significantly due to the low melting behaviour of the polymer particles. These results are supported by Arai's et al. [5] investigation. The authors investigated the orientation of glass fibres in a PBT SLS material. At a layer thickness of 100 µm, the tensor shows an orientation of 50% in the recoating direction and 35% in the cross direction, while the build-up direction Z displays only 15%. This behaviour is also observed in the mechanical properties, whereas the highest tensile strength and modulus of elasticity were found in the recoating direction, while the lowest properties were found in the build-up direction. Furthermore, it was found that the samples in the coating direction showed the highest heat deflection temperature and the lowest shrinkage effects.

Besides the process influence, the powder and fibre composition also have an influence on fibre orientation and part behaviour. Zhu et al. [11] developed a method whereby a green body manufactured with SLS is combined with epoxy resin infiltration. After the modification of carbon fibres (CF) with a Polyamide 12 (PA12) surface treatment, the PA12/CF composite powder was used for the SLS process. This method enables a better random three-dimensional distribution of the fibres.

The literature review shows that prior studies investigated the mechanical properties as a consequence of fibre orientation. The characterisation of fibre orientation is only described by different sample orientations in the build job. Therefore, this study describes the fibres' orientation by calculation of the orientation tensors in the X, Y and Z direction. In this context, the impact of the processing conditions on the fibre orientation tensor is described. The parameters of recoating speed, layer thickness and laser power are varied, and the interactions between these three parameters are identified in this study.

2. Materials and Methods

This research explores the influence of process parameters on fibre orientation. Based on a literature review, an experimental design was created to identify the influence of processing parameters on the fibre orientation. After identifying a suitable sintering window, samples were manufactured according to the experimental design and chosen parameters. The results of mechanical testing are used to identify samples with remarkable mechanical performance, to be investigated by means of X-ray computed tomography (CT).

2.1. Design of Experiment

A full factorial design of experiment (DoE) with the parameters of recoating speed (RS), layer thickness (LT) and laser power (LP) was created. The parameters were identified based on a literature review, as mentioned earlier. For each parameter, three factor levels were set: low, medium and high. The bottom and top limit values were maximised in order to observe high effects on the fibre orientation. Although going to the process limits, safe processing had to be ensured, characterised by low curling and warpage effects. The values of the DoE are summarised in Table 1. All other processing parameters were kept constant. Three parameters with three levels leads to a 3^3 DoE.

Table 1. Varied parameters for effect identification on fibre orientation.

Level	Recoating Speed in mm/s	Layer Thickness in μm	Laser Power in W
Low	168	100	16
Medium	228	120	26
High	288	140	36

2.2. Manufacturing of Samples

For the manufacturing of samples, a HT252P SLS printer made by Farsoon (Hunan, China), was applied. The printer features a maximum chamber temperature of 220 °C and is equipped with a 60 W CO_2 laser (wavelength 10.6 μm). The build chamber is heated from all side: from the top by infrared radiation (IR) heaters, and from the cylinder walls and piston plate by integrated heating bands. The surface of the powder reservoir is heated by IR heaters. In addition, the chamber is flooded with nitrogen during the complete build job, leading to a residual oxygen content below 0.3% throughout all build stages. Prior to the heat-up stage, the machine was flooded manually with nitrogen and heated up to 150 °C for 1h. After this period, the actual warm-up stage began. The warm-up stage started with a ramp from the start temperature of 150 °C up to the final build temperature. The optimum build temperature was identified prior to testing by differential scanning calorimetry (DSC) analysis, supplier recommendations and printing pre-trials. The cylinder, piston and feed temperatures had been optimised in pre-tests with the target of minimizing curling and warpage effects. After the core temperature of the part cake had cooled down to below 50 °C, samples were removed and cleaned. As the impact by blasting could lead to uneven surfaces, the removal of the remaining powder from the specimens was done manually with a brush, and additionally with a vacuum cleaner with an attached brush.

Based on the DSC analysis (Section 3.1) and knowledge from pre-trials, the parameters for the temperatures were identified as summarised in Table 2. The temperature of the cavity walls (cylinder temperature) and piston plate was set to 180 °C. The surface temperature (build temperature) of the powder after recoating was set to 210 °C. While RS, LT and LP were changed according to the DoE, the temperatures, the scanning speed and the hatch distance were kept constant.

For tensile testing, tensile bars according to DIN EN ISO 527-2 1A were printed. For CT scans, cubes with a geometry of 3 mm × 3 mm × 3 mm on a base platform were chosen. The base allowed labelling as well as easier positioning on the CT platform. Both samples are illustrated in Figure 1.

2.3. Investigated PA6 Material

A non-commercial Polyamide 6 (PA6) powder, supplied by Farsoon (FS6140GF), filled with 30 wt.% glass beads and 10 wt.% glass fibres, was used for this study. Due to the effects whereby fibres increase the anisotropy and decrease the flowability of the powder, beads facilitate the powder recoating, the isotropy and the warpage behaviour. The reinforcing fillers were added to the powder and were not integrated into the powder particles. The particle size distribution was D50 = 63 μm. Carbon black was added as colourant as well as additives for oxidation resistance. The melting point was 225 °C and the bulk density of the powder was 0.6–0.73 g/cm^3, while the density of the printed parts was 1.4 g/cm^3 at sufficient energy input.

Table 2. Processing parameters of PA6-GB-GF (Farsoon FS6140GF).

Parameter	Unit	Value		
Recoating Speed	mm/s	168	228	288
Layer Thickness	μm	100	120	140
Build Temperature	°C		210	
Cylinder Temperature	°C		180	
Piston Temperature	°C		180	
Feed Temperature	°C		165	
Scanning Speed Fill	m/s		10.16	
Hatch distance	mm		0.15	
Power Fill	W	16	26	36

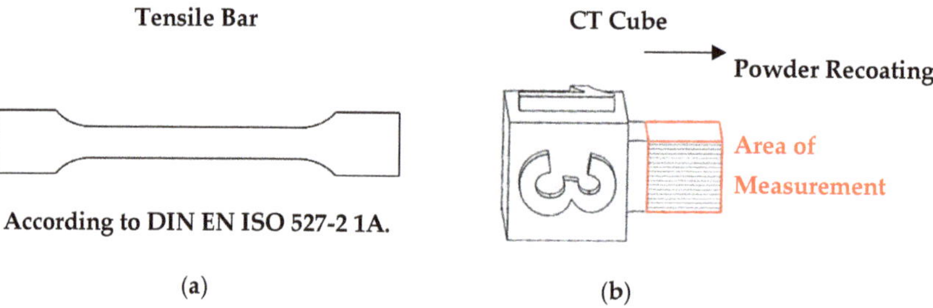

Figure 1. Overview of the manufactured samples. (**a**) Tensile bars to identify the mechanical properties and (**b**) computed tomography (CT) cubes to investigate the fibre orientation.

2.4. Mechanical Testing and Fibre Orientation Determination

Before the determination of the mechanical properties, the samples were vacuum dried at 80 °C for 168 h. The modulus of elasticity was measured with an extensometer at 1 mm/min according to DIN EN ISO 527-2/1A/5. After determining the linear-elastic strain area, the testing speed was increased to 5 mm/min. For the determination of the tensile properties, a Z050 by Zwick Roell (Ulm, Germany), with an extensometer was used. The measurements were performed under environmental conditions at room temperature of 23 °C and humidity of 50%.

X-ray computer tomography was applied for investigation of the fibre orientation and to collect images for qualitative and quantitative analysis. A v|tome|x m CT machine by GE Inspection Technologies, United States, with a resolution of 2 μm for detailed measuring range and 6 μm for overview images, was used. Figure 2 illustrates the high resolution area (green). The lower resolution is only needed for identification of the labels and positioning of the samples for calculation analysis.

The software VG Studio Max allows the detection of fibre orientation and evaluation of the differences due to different processing parameters. For the calculation of fibre orientation, a cube of 2.8 mm × 2.8 mm × 2.8 mm inside the CT cube (3 mm × 3 mm × 3 mm) was defined and analysed, as shown in Figure 3a. This way, typical SLS effects at the edge zones, such as shrink marks, are neglected, which increases the comparability of the samples. The cube (blue coloured) was further sliced into

21 layers (1 × 1 × 21, X × Y × Z mesh) as illustrated in Figure 3b. Thus, a better understanding of the fibre orientation through the single layers could be obtained.

Figure 2. Illustration of the CT resolution within the CT cube. Green represents the high detail area with a resolution of 2 μm.

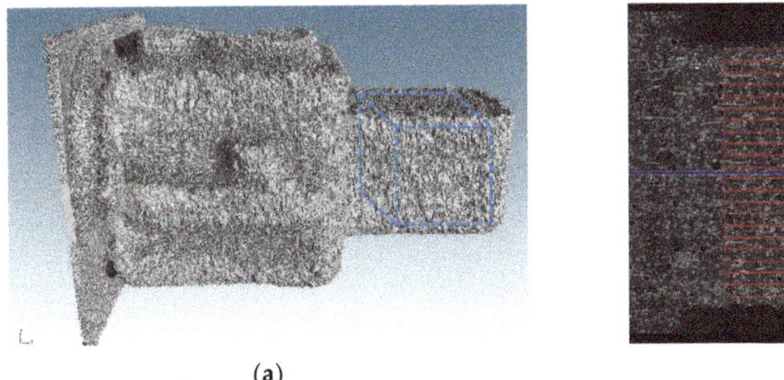

(a)   (b)

Figure 3. Illustration of the measuring cube within the sample for tensor calculation (**a**). The measuring cube is split into a 1 × 1 × 21 mesh (**b**).

3. Results

3.1. Powder Analysis

The SEM images in Figure 4 show that the fillers are added to the powder and are not integrated into the polymer particles. The length of fibres varies up to a maximum of 500 μm, while the fibre diameters have a constant value of about 15 μm. Furthermore, the diameter of the beads varies up to 80 μm. The potato-shaped geometry of the powder particles is a result of the fabrication process of the powder. Farsoon reported that the PA6 powder is fabricated by precipitation of a dissolved PA6 [12]. Additionally, small spherical additives are commonly added to the powder, which act as separating agents to decrease the particle-particle attractive forces for a better powder flowability [13]. In these SEM images, dust on the particles is identifiable, which could also support the flowability.

The dynamic DSC analysis of the powder is illustrated in Figure 5. The melting and crystallisation peaks are clearly defined, which allows one to derive the optimum processing parameters. The chosen build temperature of 210 °C is 6 °C lower than the onset melt temperature (216 °C). The fillers do not significantly influence the melting curve. The temperature of the cavity walls and piston plate was set to 180 °C to allow the melted material to cool down as slowly as possible, so as to avoid warpage effects. Furthermore, it can be observed that the crystallisation is initiated at higher temperatures (195 °C) in comparison to non-filled materials (BASF Ultrasint® PA6: 188 °C). The glass fibres, glass beads, carbon black and other additives act as crystallisation nuclei. Depending on the size and geometry of the crystallisation nucleus, the crystalline structure is influenced.

Figure 4. Powder composition of the investigated reinforced PA6 powder.

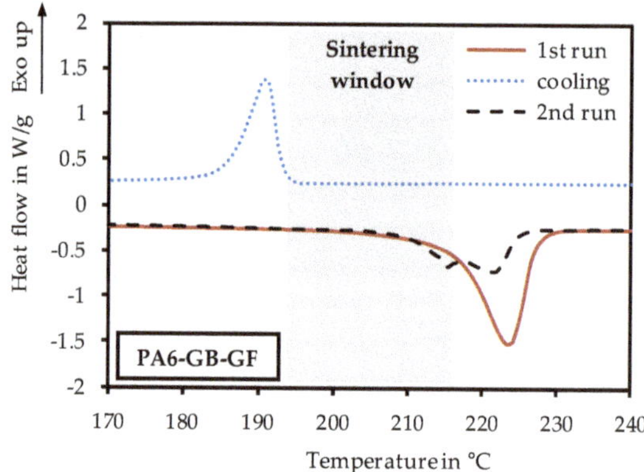

Figure 5. Characterisation of the melting and crystallisation behaviour of PA6-GB-GF powder.

3.2. Process Influence on the Mechanical Properties

For the evaluation of influences on the mechanical properties, the modulus of elasticity is chosen as a key factor. The analysis of the DoE shows that the values of the modulus of elasticity follow a normal distribution (p-value < 0.005). All factors show significant effects (significance level: 1.98). LT and LP show similar effects, with effect values above 7.2. RS shows low significance with 2.5. Furthermore, the correlation of LT and LP has an effect of 5.5. This high significance is due to the impact of both parameters on to the resulting volumetric energy density. The other correlations show no significance.

Figure 6 illustrates the results of mechanical testing. The result shows averaged values for the modulus of elasticity. As observed at effect significance, RS has only a little influence on the sample's modulus of elasticity. There is a maximum at medium setting, while the low and high setting leads to lower values. In total, the impact of RS is hardly measurable. LT and LP, on the other hand, show higher impacts. The higher LT, the lower is the modulus of elasticity. The opposite trend is observed for LP. A high increase from low to medium setting can be observed. At the medium setting, the maximum is achieved and remains on this level with a small decrease in the high setting. Both effects can be explained by the resulting volumetric energy density. The lower LT and/or the higher LP, the higher is the resulting energy. The higher modulus of elasticity can be found with the better layer-to-layer interconnection [9]. For the influence of fibre orientation on the mechanical properties, the orientation tensors in the following chapters are necessary.

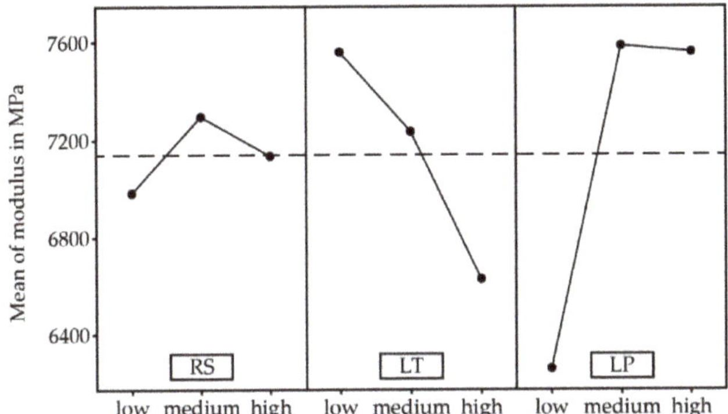

Figure 6. Effect of processing parameters on the modulus of elasticity of the produced samples.

3.3. Sample Selection for CT Investigation

The analysis of the modulus of elasticity itself does not directly allow a conclusion as regards the fibre orientation. Therefore, samples for the CT are selected, which display significant differences in their mechanical values. The selected samples are summarised in Table 3. The CT1 cube reflects the parameter combination with the highest modulus of elasticity of 8154 MPa, and is taken as the reference. The selection for the other CT cubes is reasoned to be below. In addition, the resulting volumetric energy density for each sample is calculated.

The selected samples also vary in their tensile strength and elongation at break. The values are summarised in Figure 7. The results show a correlation with the resulting laser energy density, whereas the three settings with sufficiently high laser energy density (CT1 (197 J/cm^3), CT3 (236 J/cm^3) and CT5 (197 J/cm^3)) lead to high strengths as well as elongations. Lower mechanical properties are observed for CT2 (169 J/cm^3) and CT4 (87 J/cm^3) if the laser energy density is much lower. For high layer-to-layer interconnection, the threshold value can be found between 169 J/cm^3 and 197 J/cm^3. In addition, the recoating speed seems to influence the reproducibility. The setting with the lowest recoating speed (CT5) shows the lowest standard deviation of 1.72 MPA (2.1%) for the tensile strength.

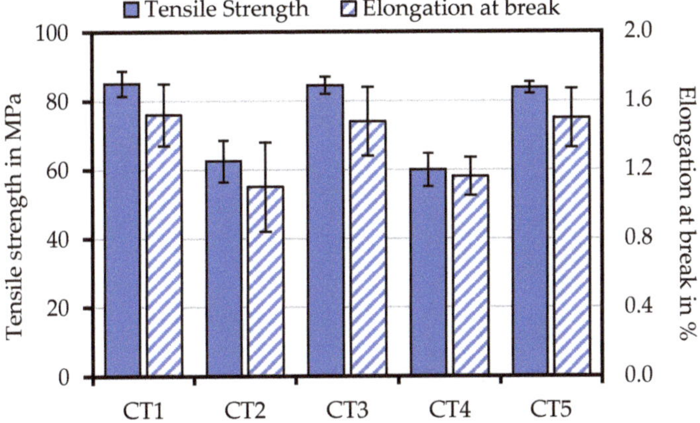

Figure 7. Effects of the varied parameters on tensile strength and elongation at break.

Table 3. Selection of samples for computed tomography scans.

Label	RS	LT	LP	Resulting Energy Density in J/cm^3	Modulus of Elasticity in MPa	Selection Criterion
CT1 (reference)	high	medium	high	197	8154	Maximum modulus of elasticity
CT2 (Appendix A: Figure A1)	high	high	high	169	6881	Impact layer thickness high
CT3 (Appendix A: Figure A2)	high	low	high	236	7695	Impact layer thickness low
CT4 (Appendix A: Figure A3)	high	medium	low	87	6417	Impact laser energy
CT5 (Appendix A: Figure A4)	low	medium	high	197	7548	Impact recoating speed

3.4. Fibre Orientation Dependent on Processing Parameters

The evaluation of CT1 shows that fibres tend to align in the powder recoating direction (Figure 8a). The SLS's typical rough surface can be detected all around the edge areas. At the right bottom edge, curling effects can be observed, which on a macroscopic view lead to a smooth transition between right side and bottom side. Next to the aligned glass fibres, glass beads of different sizes, as well as shrinkage-related voids, occur inside the part. Furthermore, the top layers show less filler particles than the bottom layers. Another difference in the comparison of the top/bottom layer is seen in the melting behaviour of the particles. The first layer of the sample displays partially melted powder particles, while particles in the top layer melted almost completely. This effect can be explained via the layer's temperature distribution before a new layer of powder is added. While in the first layer, the temperature of the applied powder corresponds to the build temperature, the existing temperature is much lower in comparison to the top layer, where a higher temperature results from the heat of previously melted layers. Beyond that, Figure 8b shows the fibre orientation in the X and Y direction. Fibres are mainly orientated in the X direction. In addition, the orientation in the Y direction is more highly detectable than orientation in the Z direction.

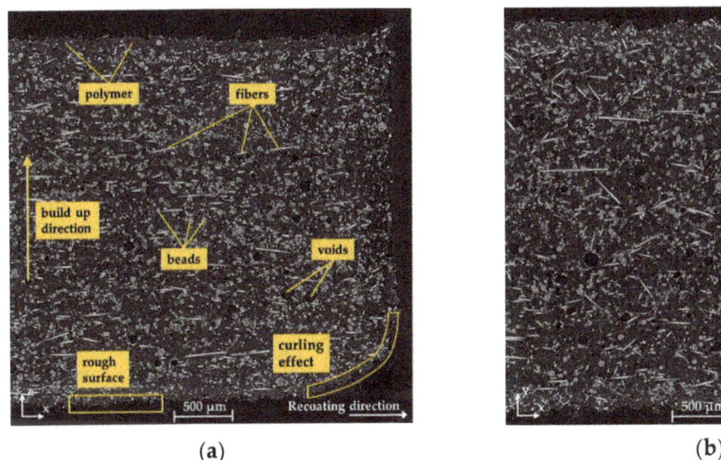

Figure 8. 2D cross-sectional slice from the CT scan of CT1. Fibre orientation in (**a**) XZ plane and (**b**) XY plane. Parameters: RS = 288 mm/s, LT = 120 μm and LP = 36 W (CT1).

The calculation of the tensors for CT1 is illustrated in Figure 9. As already seen in the qualitative results above, the X alignment is the dominating tensor, followed by Y and Z. The value of the X tensor varies from 0.49 to 0.58 along the grid elements. The Y tensors shows a value of 0.28, and Z of 0.18.

Mean values of the 21 tensor measuring ranges are calculated for each orientation tensor and CT scan sample. The mean values of all investigated tensors are summarised in Figure 10. Except CT3, the tensors show similar characteristics for the X, Y and Z directions. With a higher value of approximately 0.04, the X tensor of CT3 shows a greater orientation of fibre in the recoating direction.

The CT3 image in Figure 11 shows similar results as that of CT1. Despite the 39 J/cm^3 higher energy input, voids are still visible. The slightly increased fibre orientation is not recognisable from this single image, but could be measured by tensor evaluation across all layers. Due to the lower layer thickness, more fibres interact with the roller and an increased orientation is the result. Despite the increased laser energy and higher orientation, the modulus of elasticity decreases compared to CT1. At this point two effects overlap: the modulus of elasticity increases due to the fibre orientation in the recoating and tensile test direction, and the high energy density leads to a degradation in the polymer material.

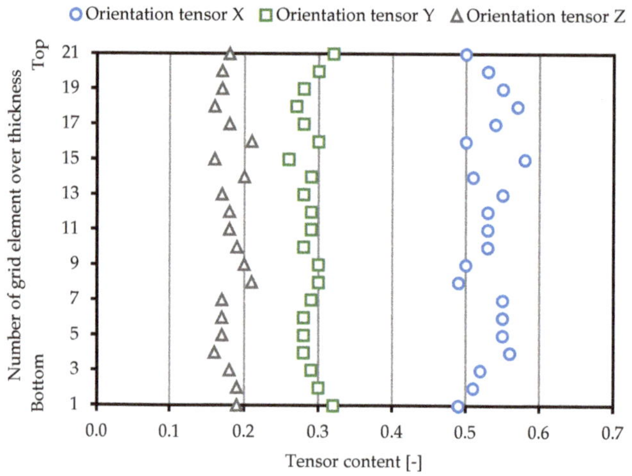

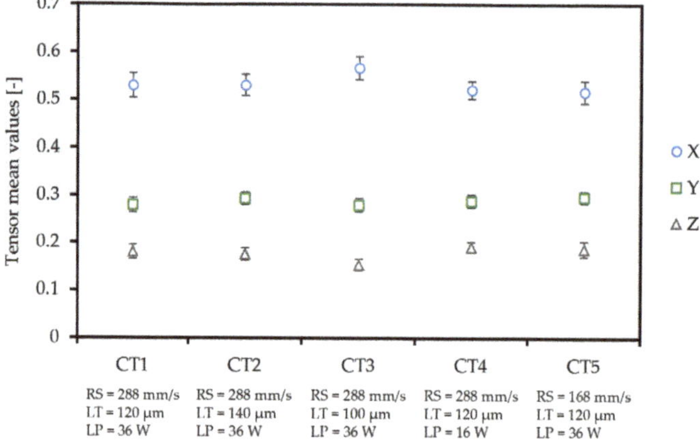

Figure 9. Fibre orientation in CT1 with the parameters RS = 288 mm/s, LT = 120 μm and LP = 36 W.

Figure 10. Comparison of all tensor values for all investigated CT scans.

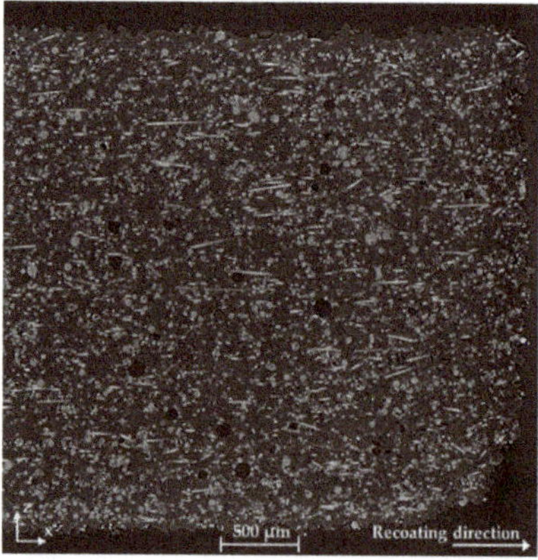

Figure 11. 2D cross-sectional slice from the CT scan of CT1. Parameters: RS = 288 mm/s, LT = 100 μm and LP = 36 W (CT3).

Figure 12 illustrates the result of insufficient laser energy in sample CT4. Many voids within the part occur, which reduces the layer-to-layer interconnection. This leads to a low tensile strength of 60 MPa and low elongation at break of 1.2%. In addition, the reduced laser energy leads to a reduction in the curling. Because of the lower melt temperature, the temperature difference between the melt and the recoated powder is comparably low. Therefore, internal stresses are reduced and less curling occurs.

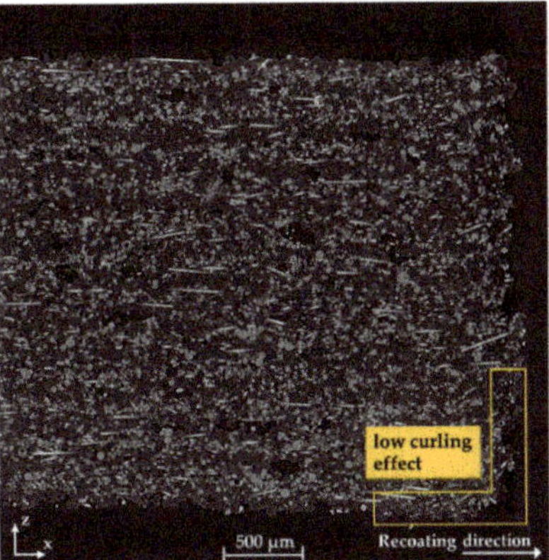

Figure 12. Insufficient laser power leads to defects in the part's material and consequently to a low layer-to-layer interconnection. Parameters: RS = 288 mm/s, LT = 120 µm and LP = 16 W (CT4).

4. Discussion

The results of the design of experiment show the high significance of the factors of layer thickness and laser power on the modulus of elasticity. The recoating speed has A low significance. Interactions between the three factors can only be identified for LT and LP. This interaction effect is the result of the dependence of these factors on the resulting energy density. The higher the energy density, the higher the mechanical properties.

The tensor analysis, for instance, shows that the varied laser power does not influence the fibre orientation, as comparing CT1 (197 J/cm^3) and CT4 (87 J/cm^3) shows similar X-tensors (0.53 and 0.52). Although the melt viscosity changes with melt temperature, which is directly related to the applied laser power, there is no external force, except gravity, acting on the fibres, which could cause changes in fibre orientation. Lower viscosity can also improve the wetting of the fibres, leading to an improved introduction of load into the fibres and higher part strength, respectively. As such, the effect of higher laser power on part strength results from an improved layer-to-layer interconnection and fibre incorporation, and not necessarily from a change of fibre orientation. This can also be seen in the CT scans, which confirm that the laser energy improves the quality of the layer-to-layer interconnection. Theoretically, perpendicular fibres are able to penetrate several layers (Z direction) more easily if the viscosity is lower. Nevertheless, as mentioned above there is no additional driving force, related to the laser power, which could cause fibre reorientation. Consequently, the CT analysis shows no influence of the energy density on fibre orientation. Thus, fibres mainly follow a 2D orientation inside a layer.

Variation of layer thickness has two effects: First, the laser energy density is proportionately dependent on the layer thickness. As a consequence, the quality of the layer-to-layer interconnection and fibre wetting is affected in the same manner as described above for the laser power. The second effect is directly related to fibre orientation, whereby lower layer thicknesses result in greater orientation in the recoating direction. By reducing the layer thickness, the recoating roller interacts with more fibres, since even smaller fibres can interact and change orientation. In addition, the shear stress between the top and bottom of each layer, which results from the roller's counter-rotation to the recoating direction, is inversely proportional to the layer thickness. Hence, the driving force causing fibre alignment in the recoating direction increases as the layer thickness decreases. For CT3 with a low layer thickness, the X tensor shows an increase of 7% in comparison to the reference with a medium setting for the layer thickness. Since spherical beads do not have an influence on the tensors, the alignment of the fibres will be higher than the calculated 7% suggests. The shown impact of the layer thickness parameter confirms Jansson's [4] findings, with the effect illustrated in Figure 13. Although the layer thickness has a significant effect on fibre orientation, the layer thickness cannot be set too low, otherwise the powder spreading process will be hindered considering the poorer flowability of the composite powders.

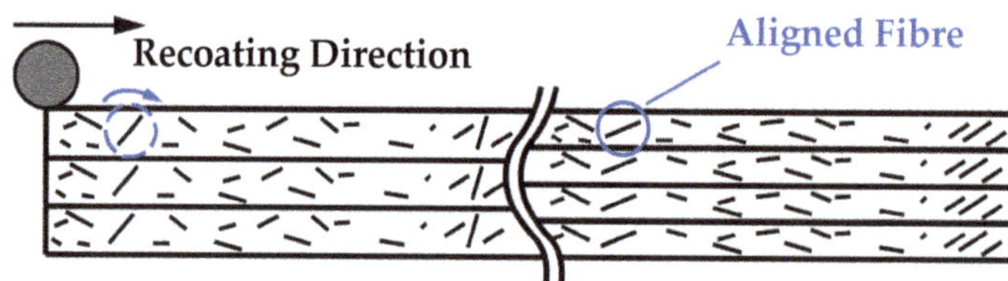

Figure 13. Orientation of fibres due to variation of the layer thickness from a high (**left**) to a low level (**right**).

As opposed to the layer thickness, the recoating speed determines the duration of the roller–fibre interaction. Results show that the factor time (interaction between roller and fibre) has little impact on the resulting modulus of elasticity within the chosen settings. The fibre orientation tensors show a slightly higher orientation for the high setting (CT1) in comparison to the low setting (CT5).

5. Conclusions and Outlook

The current work indicates that for PA6 powder, filled with 30 wt.% glass beads and 10 wt.% glass fibres, the modulus of elasticity can vary strongly (6417–8154 MPa) in dependence on the chosen process parameters. All these variations are related to the applied laser energy density, which was either varied directly by the setting of the applied laser power, or indirectly by layer thickness. The dominant effect can be found in an improved layer-to-layer interconnection and fibre wetting, which is related to the melt viscosity as a result of the varied laser energy density. In addition, reduced layer thickness leads to a slightly higher (7%) fibre alignment in the recoating direction, with smaller fibres now in interaction with the roller and thus aligning in the roller direction. In addition, a change in the melt viscosity due to variation of the energy density does not seem to affect fibre orientation.

No significant impact in terms of fibre orientation related to the parameters of recoating speed and laser power was found. With their apparently low influence on fibre orientation during the processing of fibre-filled PA6 powder, these parameters can be used to focus on process robustness. Conclusively, since the degree of anisotropy in fibre orientation can be influenced by the layer thickness, it would be possible to induce locally adjusted stiffness and anisotropic mechanical behaviour with changes in layer thickness, in order to improve part performance. This study illustrates the need to understand and optimise processing parameters so as to achieve the best mechanical properties of the laser sintered parts, especially with fibre- and bead-reinforced powders.

This study dealt with the effect of selective laser sintering on the fibre orientation in manufactured components. Based on the results shown, it can be assumed that the layer height as well as the laser power have influence on the material structure in the manufactured test specimens. Therefore, the melting process and the herein-induced properties (e.g., porosity, degree of crystallinity, temperature profile) can be investigated in detail in future work.

Author Contributions: Conceptualisation, T.H.; methodology, T.H.; validation, T.H. and M.S.; formal analysis, T.H.; investigation, T.H. and M.S.; data curation, T.H. and M.S.; writing—original draft preparation, T.H.; writing—review and editing, M.S.; S.R.R., G.H. and P.M.; visualization, T.H.; supervision, S.R.R., G.H. and P.M. All authors have read and agreed to the published version of the manuscript.

Funding: This research received no external funding.

Acknowledgments: The authors gratefully acknowledge Farsoon Europe for the technical support and provision of the investigated reinforced PA6 material (FS6140GF). Any opinions, findings, conclusions or recommendations expressed in this study are those of the authors and do not necessarily reflect the views of Farsoon Europe.

Conflicts of Interest: The authors declare no conflict of interest.

Appendix A

CT Scans and Fibre Orientations Tensors

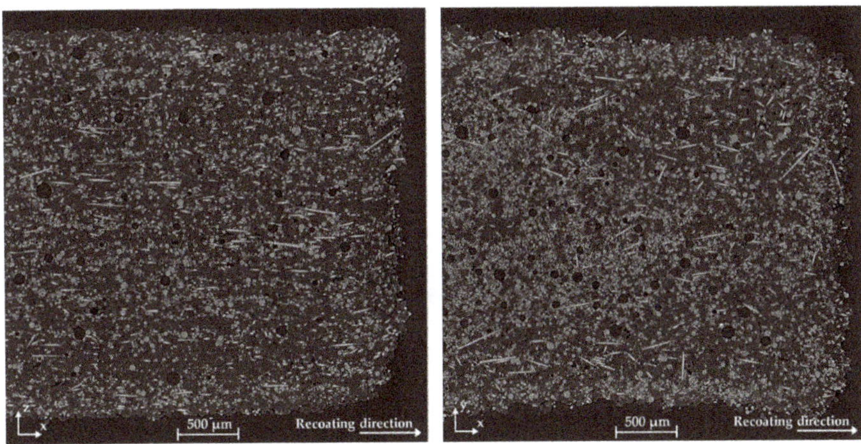

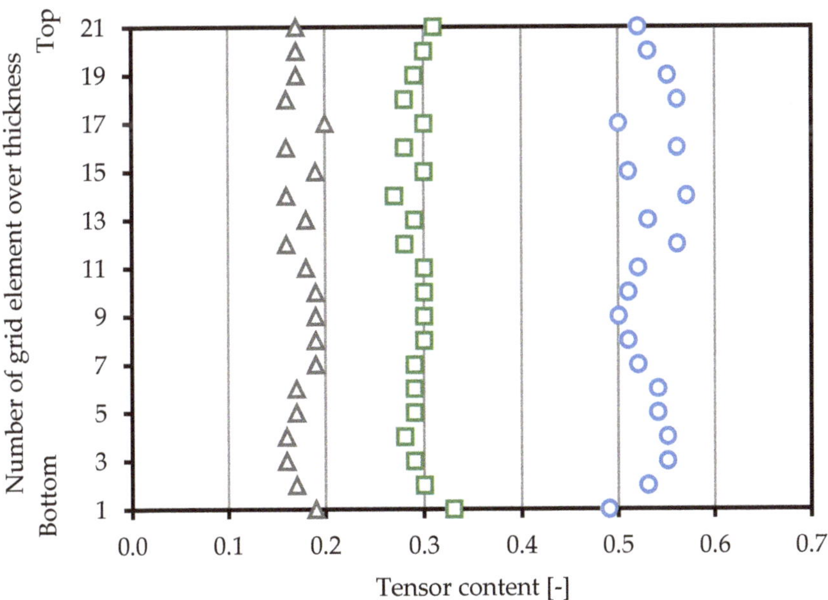

Figure A1. CT2–RS = 288 mm/s, LT = 140 µm, LP = 36 W.

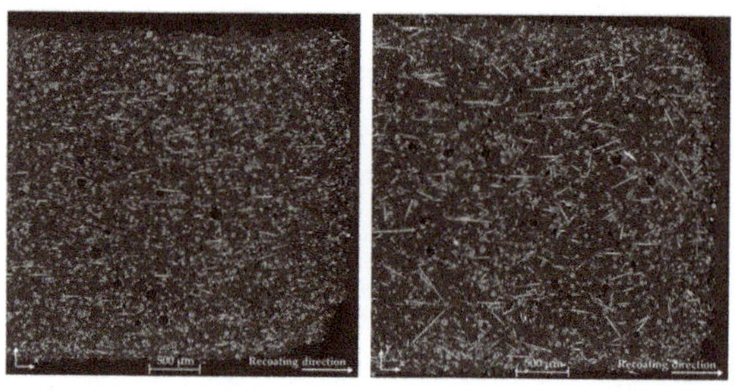

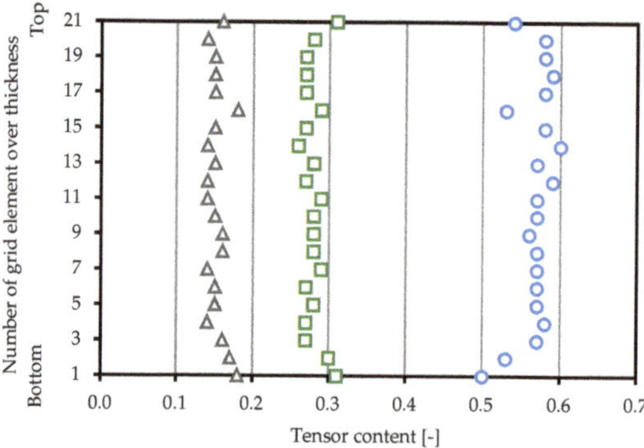

Figure A2. CT3–RS = 288 mm/s, LT = 100 μm, LP = 36 W.

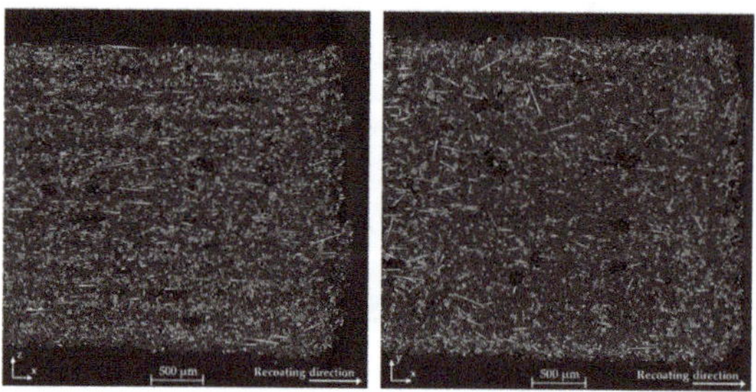

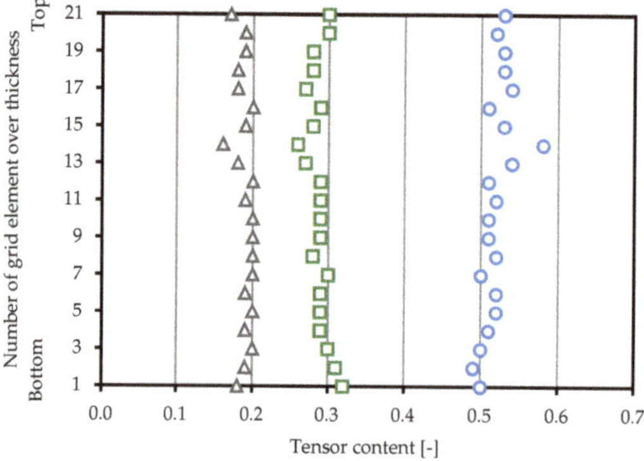

Figure A3. CT4–RS = 288 mm/s, LT = 120 μm, LP = 16 W.

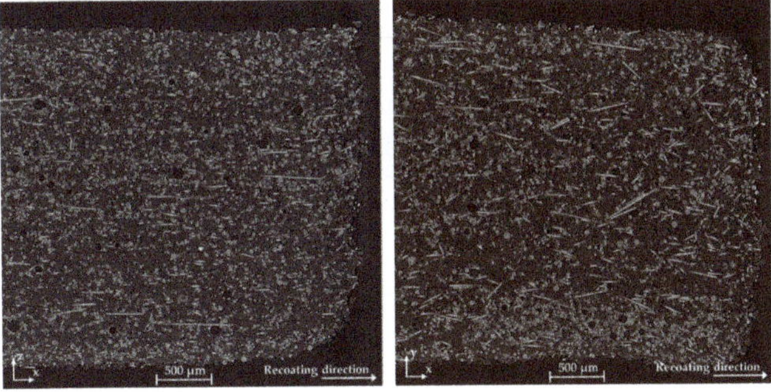
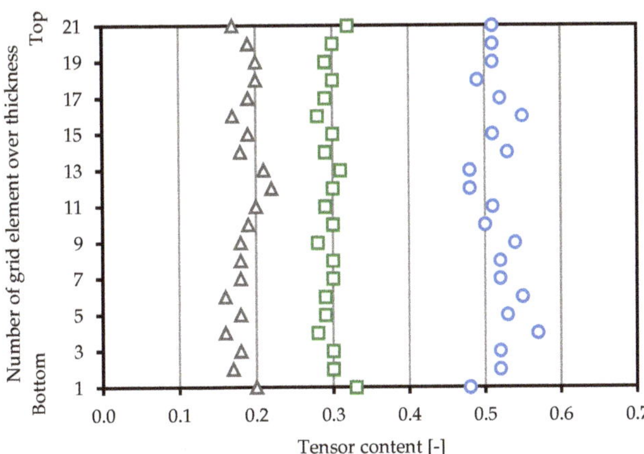

Figure A4. CT5–RS = 168 mm/s, LT = 120 μm, LP = 36 W.

References

1. Bay, R.S.; Tucker, C.L. Fiber orientation in simple injection molding. Part II: Experimental results. *Polym. Compos.* **1992**, *13*, 332–341. [CrossRef]

2. Gandhi, U.N.; Goris, S.; Osswald, T.A.; Song, Y.-Y. *Discontinuous Fiber-Reinforced Composites: Fundamentals and Applications*; Hanser: Munich, Germany, 2020.

3. Colon, J.L.; Heckner, T.; Chrupala, A.; Pollock, J. Experimental study of particle migration in polymer processing. *Polym. Compos.* **2019**, *40*, 2165–2177. [CrossRef]

4. Jansson, A.; Pejryd, L. Characterisation of carbon fibre-reinforced polyamide manufactured by selective laser sintering. *Compos. Struct.* **2016**, *9*, 7–13. [CrossRef]

5. Arai, S.; Tsunoda, S.; Yamaguchi, A.; Ougizawa, T. Effect of anisotropy in the build direction and laser-scanning conditions on characterization of short-glass-fiber-reinforced PBT for laser sintering. *Opt. Laser Technol.* **2019**, *113*, 345–356. [CrossRef]

6. Schmid, M. *Selektives Lasersintern (SLS) Mit Kunststoffen: Technologie, Prozesse und Werkstoffe*; Hanser: Munich, Germany, 2015; p. 178.

7. Beitz, S.; Uerlich, R.; Bokelmann, T.; Diener, A.; Vietor, T.; Kwade, A. Influence of powder deposition on powder bed and specimen properties. *Materials* **2019**, *12*, 297. [CrossRef]

8. Drexler, M.; Greiner, S.; Lexow, M.; Lanzl, L.; Wudy, K.; Drummer, D. Selective laser melting of polymers: Influence of powder coating on mechanical part properties. *J. Polym. Eng.* **2018**, *38*, 667–674. [CrossRef]

9. Heckner, T.; Raisch, S.R.; Huelder, G.; Middendorf, P. The Influence of Laser Power Variation on SLS-Printed PA6 Parts and the Resulting Effect on The Parts' Long-Term Properties. In Proceedings of the ANTEC 2020: The Virtual Edition, Brookfield, CT, USA, 29 March–2 April 2020.

10. Lanzl, L.; Wudy, K.; Drummer, D. The effect of short glass fibers on the process behavior of polyamide 12 during selective laser beam melting. *Polym. Test.* **2020**, *83*, 106313. [CrossRef]
11. Zhu, W.; Yan, C.; Shi, Y.; Wen, S.; Liu, J.; Wei, Q.; Shi, Y. A novel method based on selective laser sintering for preparing high-performance carbon fibres/polyamide12/epoxy ternary composites. *Sci. Rep.* **2016**, *6*, 33780. [CrossRef]
12. Simon, D.; FARSOON Europe GmbH, Stuttgart, Germany. Personal communication, 2020.
13. Wudy, K. Alterungsverhalten Von Polyamid 12 Beim Selektiven Lasersintern. Ph.D. Thesis, University of Erlangen-Nuremberg, Erlangen, Germany, 2017.

Journal of
Composites Science

Article

Insights into the Processing of Recycled Carbon Fibers via Injection Molding Compounding

Jochen Wellekötter [1,*], **Julia Resch** [1,*], **Stephan Baz** [2], **Götz Theo Gresser** [2] **and Christian Bonten** [1]

[1] Institut für Kunststofftechnik (IKT), University of Stuttgart, Pfaffenwaldring 32, 70569 Stuttgart, Germany; christian.bonten@ikt.uni-stuttgart.de

[2] Deutsche Institute für Textil- und Faserforschung Denkendorf (DITF), Körschtalstraße 26, 73770 Denkendorf, Germany; Stephan.Baz@ditf.de (S.B.); goetz.gresser@ditf.de (G.T.G.)

* Correspondence: jochen.wellekoetter@ikt.uni-stuttgart.de (J.W.); julia.resch@ikt.uni-stuttgart.de (J.R.); Tel.: +49-711-685-62864 (J.W.); +49-711-685-62841 (J.R.)

Received: 16 September 2020; Accepted: 20 October 2020; Published: 27 October 2020

Abstract: Although fiber-reinforced plastics combine high strength and stiffness with being lightweight, major difficulties arise with high volume production and the return of manufactured parts back into the cycle of materials at the end of their lifecycles. In a novel approach, structural parts were produced from recycled material while utilizing the so-called injection molding compounding process. Recycled fibers and recycled polyamide matrix material were used by blending carbon and matrix fibers into a sliver before processing. Injection molding was then used to produce long fiber-reinforced parts through a direct fiber feed system. Recycled matrix granules were incorporated into the injection molding process by means of an injection molding compounder to investigate their influences on the mechanical properties of the parts. The findings show that the recycled fibers and matrix perform well in standardized tests, although fiber length and fiber content vary significantly and remain below expectations.

Keywords: carbon fiber recycling; lightweight design; long fiber-reinforced thermoplastics; parameter-optimization; injection molding compounding

---

1. Introduction

Recycling concepts for fiber-reinforced plastic (FRP) materials should be developed early on to improve the materials' ecological impact [1]. The reuse of FRP is a challenge due to the difficulty in the separation of the materials, and mechanical properties gradually deteriorate owing to the shortening of the reinforcing fibers with each recycling cycle. This affects both the matrix material used and the fibers. A reuse in the form of continuous fibers in the typical composite technologies is not possible, and alternative routes need to be established for the shortened fibers. This phenomenon has been extensively researched and reviewed in various articles. Examples and case studies were summarized by Pimenta et al. [2] and applications of recycled fiber composites were reviewed by Pickering et al. [3]. Oliveux et al. [4] gave examples of how recycled discontinuous fibers compare to virgin material. Pickering [5] analyzed and compared mechanical recycling and the use of fibers as filler materials to pyrolysis processes. This knowledge can nowadays be considered state-of-the-art, as seen in [6].

Since the fiber length deteriorates during recycling, discontinuous fiber technologies are particularly well-suited for the recycling of these materials, as can be seen in [2,3]. However, the material supply into, e.g., the injection molding process, still proves difficult. The authors present a novel method in this study. Here, end-of-life carbon fibers and production waste are blended with recycled thermoplastic matrix fibers and processed into a bundle of aligned fibers, or a "sliver." This sliver can then be processed via injection molding. Additionally, the sliver can be further processed into a yarn, which allows for established textile processing methods.

In the case of this study, the sliver was processed in a so-called injection molding compounder (IMC) utilizing a direct fiber feed for fiber incorporation. The injection molding compounder was first introduced by KraussMaffei and enables the direct compounding and injection molding in one process step (single-stage) by combining a twin-screw extruder with an injection molding unit. The IMC and its advantages compared to a two-stage process (separate compounding and injection molding processes) have been proven in various research papers, e.g., by Truckenmüller [7]. Further studies analyzed the use of the IMC, e.g., for the processing of natural fibers [8] or the improvement of resulting fiber length—e.g., see Willems et al. [9] and Wellekötter et al. [10]. It could be shown that the IMC's single-stage process results in longer fibers compared to regular injection molding, thereby improving mechanical properties compared to short-fiber-reinforced thermoplastics [10,11]. The direct fiber feed module draws the fibers into the twin-screw extruder of the IMC near the end of the compounding step, reducing fiber damage and allowing for higher fiber length in the final part.

1.1. Recycling of Carbon Fibers and Fiber-Reinforced Plastics

Studies on the development of the composites market (e.g., [12–14]) indicate that the demand for FRP will increase significantly in the near future. For this reason, it is vital to address the challenges of recycling at an early stage. FRPs cannot be easily reused after their lifecycles, nor can they be easily separated, sorted, and treated. In addition, structural damage, such as delamination, is not necessarily visible from the surface, increasing the amount of FRP waste through the replacement of parts to comply with safety regulations.

Thermal recovery (combustion) and thus the recovery of stored energy is possible, although it is extremely unsatisfactory from an ecological and economic point of view, as shown by Meng et al. [1]. However, the reuse of parts from thermoplastic matrices as fillers (see below) in less stressed plastic parts is well-established (examples are summarized in [2,5].

Any other recycling process requires a high degree of purity of the fibers and matrices. Solvolysis (e.g., [4,15,16] and pyrolysis (e.g., [2,4,17,18] (combustion in the absence of oxygen) are already being applied industrially to recover carbon fibers. In pyrolysis, the matrix material is completely burned and energetically recovered, while the carbon fibers have a higher degradation temperature, remain largely inert, and thus retain their properties, albeit with shortened fibers due to previous processing. The fibers also lose their sizing during this process [2–4,12,18,19].

Since the fibers are not infinitely long anymore, but rather have a fiber length distribution depending on the parts which they incorporated, most of the conventional textile processing methods are not applicable. Further processing is limited to reusing these fibers in either compression molding, as reinforcements in the injection molding process, or as fibers for wet- or dry-laid non-wovens [3,18,20–23].

A promising approach is to align the fibers according to the load path and use their full potential as a pre-aligned non-woven (e.g., [23]). Several groups blended recycled carbon fibers with thermoplastic matrix fibers to improve processability and enable reuse in thermoplastic applications as hybrid non-wovens [22], yarns or tapes [24], and filaments for 3D-printing [25].

1.2. Short Fiber Reinforced Thermoplastics

In injection molding, fillers enable cost reduction, reduce shrinkage, and can even improve processing properties [11,26]. In contrast to fillers, plastics are reinforced to improve the material's mechanical behavior. Reinforcement is achieved with a component that has higher mechanical strength and stiffness compared to the matrix. A sufficiently large length–thickness ratio (l/d) of the reinforcing component is required for effective reinforcement. Discontinuous fibers are often used for this purpose [11].

In addition to a high fiber–matrix adhesion, the fiber length defines the mechanical properties, as illustrated by Thomason [27] and shown in Figure 1. Fiber length influences stiffness, strength, and impact resistance. In general, the longer the fibers in a matrix, the better their mechanical properties [11,27].

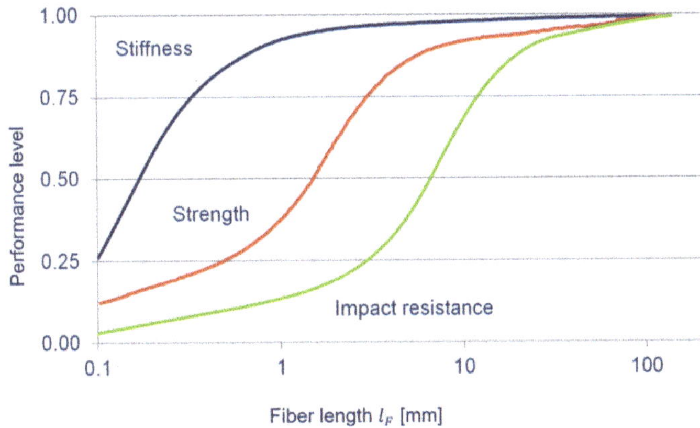

Figure 1. Influences of fiber length on stiffness, tensile strength, and impact strength.

It is, therefore, evident that the longest possible fiber length is desirable for recycled fibers during injection molding to optimize the properties of the new part. In this study, test specimens were produced from different material combinations in a direct processing method (single-stage) [11]. Compared to a regular two-step process, the one-step process allows for longer fibers and improved fiber qualities [11]. This is due to the elimination of repeated melting, the reduction of the fiber path in the extruder, and the lack of a pelletizing step [11]. Both short fibers and endless roving can be introduced into the already fully-plasticized melt via a direct-fiber-feed module. In addition, inexpensive raw materials can be used instead of cost-intensive fiber-reinforced pellets, in combination with large proportions of recycled material.

1.3. Recycling of Plastics

The recyclability of plastics is determined by their atomic and structural properties, intermolecular forces, and degree of cross-linking, and by different fillers and reinforcing materials [11,28]. Plastics can be recycled in various ways. Material recycling is the reuse of thermoplastic waste by remelting the material and thus creating new products, retaining the macromolecular structure of the polymer. This process is carried out on thermoplastics that have few or no impurities. It is also possible to filter out mixed and contaminated plastics by separation processes before recycling them [29]. If the processing is too expensive because of high levels of impurities or mixing, or if the quality is insufficient after too many recycling cycles, raw-material recycling can be performed. In this case, the macromolecular structure of the polymer is broken down, and the raw material is used for other products. If both methods fail, thermal recycling is an option, i.e., burning the plastic to generate energy [30,31].

Various factors determine the recyclability of thermoplastics and can influence the quality of the products. In addition to possible contamination by dust, paint residues, or mixing with other polymers, a recycling-related degradation of the polymer chains must be taken into account. Multiple uses and processing can lead to thermal, mechanical, hydrolytic, and oxidative degradation. Consequently, recycled plastics have different mechanical properties compared to virgin material. The change in molecular structure also influences the flow behavior of the polymer melt, possibly causing discoloration, burns, or bubbles in the product [32,33]. Important factors in thermal decomposition are the processing temperature and residence time. The presence of water can cause hydrolytic decomposition, especially in polycondensation products, such as polyamides. Furthermore, an oxidative decomposition process can occur during processing in an oxygen-rich environment [32,33].

In this study, a polyamide was used as a matrix material. A fraction of recycled material was added for all experiments. The recycling of the polyamide has been extensively analyzed, and a short summary of key insights is given below.

In 2003, the mechanical behavior of different fractions of recycled material (15, 30, 50, and 100%) of polyamide 6 (PA6) was investigated by Maspoch et al. [30]. The PA6 used for recycling came from fiber production waste and contained 20% glass particles and 10% glass fibers. The findings show that a small proportion of only 15% recycled material significantly reduced the Young's modulus, the maximum tensile stress, and the elongation at break. Similar behavior was shown for bending properties. Furthermore, reductions in the maximum tensile stress and Young's modulus were observed after three recycling cycles.

In 2013, Crespo et al. [34] investigated the effects of recycling PA6. Virgin plastic was recycled five times and analyzed after each cycle on injection-molded test specimens. The test results showed almost constant values for tensile strength and elongation at break. After the fifth cycle, a decrease of less than 2% in tensile strength and a decrease of 1% in elongation at break was observed. In comparison, the impact strength was significantly reduced even after the first recycling cycle. After the fifth cycle, the impact strength was reduced by more than 30% [34].

In another study, Su et al. [35] also investigated the effects of recycled PA6 on the mechanical properties of injection-molded parts; 16-fold recycling of the plastic was performed. A slight increase in yield stress was observed after each recycling cycle. Similarly to the study by Crespo et al., the elongation at break remained constant until it decreased after the 13th cycle. The flexural strength and modulus increased steadily. The impact strength steadily decreased, similarly to the results from Crespo et al., attributed to the shortening of the molecular chains and the broader distribution of the chain lengths. The originally soft and tough PA6 became hard and brittle with the large number of recycling cycles [35].

Recent studies [36] on the recycling of cast PA6 waste showed that the behavior of the material could be adjusted by compounding. Cast PA6 waste was converted into an extrudable polyamide with suitable additives. Rheological and mechanical analyses revealed that the lubricant leads to higher viscosity and impact strength. Recyclates modified with an oxidizing agent resulted in higher degrees of crystallinity due to the reduced molecular weight and increased stiffness and tensile strength in the final material. Due to the shorter polymer chains, the recyclates had lower viscosities than comparable virgin grades. The material could, therefore, be processed by injection molding.

The authors of this paper further analyzed the recyclability of polyamide by comparing a single-stage and a two-stage process [10]. In the single-stage process, significantly longer fibers and thus better mechanical properties were achieved using direct feed module. The investigations showed that longer fibers could be reached at lower fiber contents. However, the influence of the recycled matrix fraction was negligible.

Pimenta [2] gives a detailed overview and a list of further studies about the incorporation of recycled glass and carbon fibers into the injection molding process (without using recycled matrix material). These studies are, therefore, not be presented here.

1.4. Aim

It is evident that the behavior of recycled material cannot yet be fully predicted. The use of an IMC for the recycling of glass fibers has been investigated in a previous study by the authors. There, little influence of the recycling on the mechanical properties could be observed. The authors have not found any further research with respect to recycled (and thus discontinuous) carbon fibers and a single-stage IMC process.

Questions remain: How might recycling carbon fibers be introduced into recycled polyamide grades? How does this process compare to the use of virgin glass fibers? It is assumed that the processing of recycled carbon fibers in an IMC leads to increased fiber length and mechanical properties compared to established processes, such as injection molding. Therefore, recycled carbon fibers were introduced into the direct fiber feed module of the IMC and mixed with a recycled polyamide matrix. The fiber content, fiber length, and mechanical behavior were then analyzed to show whether the use of a sliver and an IMC is a feasible approach for the recycling of carbon fibers.

2. Experimental

The production of a test specimen in the IMC comprises several process steps. First, a fibrous matrix material is created from recycled material (Section 2.1). These fibers are mixed with recycled carbon fibers (Section 2.2) and processed into a sliver (Section 2.2.3). This sliver is then used as reinforcement for injection molded parts by utilizing a direct fiber feed module of an IMC (Section 2.3).

2.1. Plastics Recycling

For the production of various recycling mixtures, a fraction of the plastic is recycled and mixed with virgin material. An industrial-grade PA6 from BASF SE, Ludwigshafen, Germany, with the trade name Ultramid® B3S, was used for all experiments. To exclude a premature degradation reaction of the moisture-sensitive polymer chains, the material was dried for five to six hours at 80 °C. The pre-dried granulate was then processed on an Allrounder 520S 1600-400 injection molding machine from Arburg GmbH, Loßburg, Germany, to form type 1A tension rods in accordance with ISO 527-2.

In the next step, the produced parts were cut into granules with a granulator type FZK 280 from Fellner and Ziegler GmbH, Frankfurt am Main, Germany. The processed recyclate was mixed with virgin material according to the mixing ratios to be tested.

In this study, the different recycling mixtures are labeled based on new material content and recycled material content. Recycling contents of 25% and 75% were evaluated. Additionally, each mixture was analyzed with double-recycled material. Double recycling is indicated by a squared recycling content value. For example, a label of "mixture 25%recy^2" identifies a mixture with 75% new material and 25% twice-recycled material.

For the production of the hybrid sliver, a fibrous matrix was required in addition to the carbon fibers. To obtain a part made of 100% recycled material, a multifilament made out of 24 single filaments and a fineness of 100 dtex (110f24) was spun from recycled-polyamide-6 (rPA6) granules on a semi-industrial melt spinning machine. To protect the filaments during further processing and enable stable winding, the sizing SYNTEX 242 from Schill + Seilacher GmbH, Böblingen, Germany, was applied to the multifilament.

The filaments were post-stretched to a single fiber fineness of 4.2 dtex and then crimped in a texturing process with a circular knitting and heat fixation machine from Maschinenfabrik HARRY LUCAS GmbH, Nürnberg, Germany.

The textured multifilament yarns were then cut to a staple fiber length of 61 mm using an NM-150 cutting converter from Oerlikon Neumag, Neumünster, Germany. The length was selected so that it could be produced with the cutting converter and to match the carbon fibers in terms of length distribution. Uniform fiber lengths are advantageous for textile processing. The manufacturing steps are shown in Figure 2.

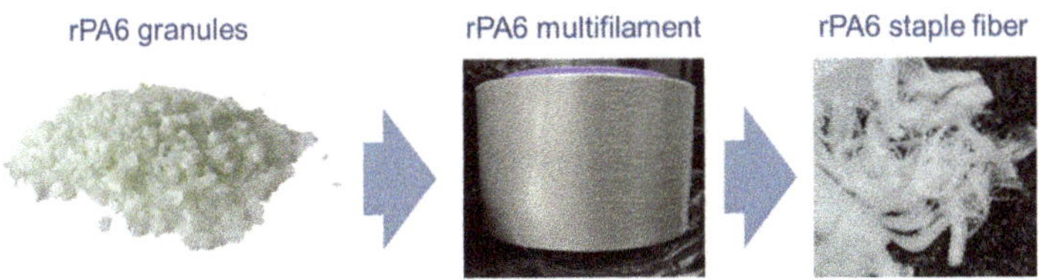

Figure 2. Manufacturing steps of rPA6 staple fibers from rPA6 granules.

2.2. Carbon Fiber Recycling

The formation of a hybrid recycling sliver containing recycled carbon fibers and recycled polyamide matrix fibers spans three processes: opening, blending, and texturizing. The opened fibers are processed to a sliver by means of a card. The pre-opened and blended carbon and matrix fibers are fed into the card with a defined area weight by a feeding chute. The card processes the fiber flocks to single fibers and orients them. Upon the delivery of the card, the formed card web is then merged into a sliver. A detailed overview of the sliver formation is given in Section 2.2.3. First, fiber selection and sizing will be investigated briefly.

2.2.1. Selection of Fibers

In principle, recycled carbon fibers can be divided into two categories. The first category describes carbon fibers that are present in the form of production waste (e.g., spool remnants or edge trimmings, roving sections; Table 1, right column) [5]. These carbon fibers have a sizing adapted to the original processing and use. Foreign fibers, e.g., from sewing threads from scrim production, may also be present. The advantage of carbon fibers in the first category is that the sizing is evenly distributed on the fibers, which allows for easier textile processing of the carbon fibers compared to that of fibers without sizing (e.g., less fiber loss). Disadvantages include the presence of partially contained foreign fibers and the unsuitable sizing of the carbon fibers for new applications (e.g., carbon fibers with epoxy sizing to be used in thermoplastic composites).

Table 1. Comparison of the recycled fibers.

Type	Carbiso C60/90 Pyrolyzed and Cut Rovings		Carbiso CT60/90 Cut Rovings	
Picture				
Fiber Fineness	Sample: 10 kg batch:	0.81 dtex 0.70 dtex	Sample: 10 kg batch:	0.83 dtex 0.82 dtex
Sizing Amount	Sample: 10 kg batch:	0.0% 0.0%	Sample: 10 kg batch:	1.4% 0.8%
Tenacity	Sample: 10 kg batch:	3140 MPa 3620 MPa	Sample: 10 kg batch:	3100 MPa 3830 MPa
Mean Fiber Length	Sample: 10 kg batch:	46.5 mm 101.0 mm	Sample: 10 kg batch:	47.5 mm 41.0 mm

The second category describes carbon fibers that have been dissolved out of the matrix by pyrolysis or solvolysis. Possible sources of carbon fibers in this category are parts at the end of their lifecycles, defective parts, and prepreg waste. The thermal or chemical removal of the matrix results in pure carbon fibers without sizing (Table 1, left column). Without sizing, the carbon fibers no longer adhere to each other, which is beneficial for mixing with matrix fibers. The disadvantage of this property is that, in textile processing, there is a significant amount of fiber-dust formation caused by fiber breakage. An advantage is that, theoretically, both processing and fiber–matrix adhesion can be facilitated by applying a sizing adapted to the new application. However, there is currently no available sizing or sizing application process for recycled carbon fibers, and the sizing has to be applied manually.

ELG Carbon Fiber, Coseley, England, offers both original roving sections (production waste) and pyrolyzed carbon fibers. Table 1 shows the results of the investigation of the carbon fibers. There is no significant difference in the fineness-related maximum tensile strength of the carbon fibers before and after pyrolysis.

Table 1 also shows that the use of recycled material leads to fluctuations in material properties. Some carbon fibers have lower individual strengths. In addition, there are large irregularities in the average fiber length and the fiber length distribution within a batch. According to the manufacturer, the fiber length should be in between 60 and 90 mm. This is the case for pyrolyzed carbon fibers. For the non-pyrolyzed production waste, the average fiber length is 47.5 mm in the best case, well below the manufacturer's specification. The measured value of the average fiber length for the carbon fibers used for sliver production is 41 mm.

2.2.2. Sizing

Sizing plays an important role in the recycling and reuse of carbon fibers. Presently, thermoset matrix materials dominate the market, so that a sizing specifically characterized for these materials is often used but not well-suited for thermoplastics. Production waste and residues, in particular, are usually not pre-treated but directly processed, which can lead to considerable difficulties during recycling and a lack of fiber–matrix adhesion in the final part [18].

Fibers from pyrolysis behave even more critically since the sizing is removed in the pyrolysis process [18]. Generally, the "bare" carbon fibers exhibit good fiber–matrix adhesion with PA6 but are difficult to process with the methods used in this study.

To improve the quality and handling of the fibers recovered by pyrolysis, preliminary sizing trials were carried out by the authors. Different sizing compositions were compared. The fibers were also processed without sizing. Samples were produced to analyze the resulting mechanical properties.

In the initial opening and processing trials of the carbon fiber variants, the need for sizing to minimize fiber damage was demonstrated. Significant reductions in fiber length by sizing carbon fibers could be achieved in the fiber preparation process and in fiber opening and sliver production.

The effects of four different sizing systems on the processing of recycling carbon fibers were compared: two industry-grade sizings (Stantex S 6256 PC from Pulcra Chemicals, Geretsried, Germany and Silastol CF1/CF2 from Schill + Seilacher GmbH, Böblingen, Germany), unsized, and original fibers, i.e., roving sections from production waste. Non-woven fabrics with defined carbon-fiber content were produced in a carding process, and the weight yield was recorded as an indicator of the processability of the fibers. Low carbon-fiber content indicates a high loss of fibers due to fiber breakage and entanglement, suggesting worse processing quality.

As shown in Table 2, sizing is necessary for a carding process; otherwise, a product, or "card web," cannot be produced (weight loss without sizing: 100%). If the fibers are sized manually, however, the losses are kept within limits. Manually sized fibers performed better compared to roving sections. Fibers from rovings need to be separated first, resulting in fiber loss, unlike the "fluffy" structure of fibers from pyrolysis [2–4,18].

Table 2. Carbon fiber weight yield after the carding process.

Sizing	Mass of PA6 in g	Mass of Carbon Fibers in g	Mass after Processing in g	Loss in %
Silastol CF1/CF2	222	235	331	28
Stantex S 6256 PC	222	235	349	24
Roving Sections	222	235	188	59
Unsized	222	235	0	100

After processing, the card web was doubled and compacted to a non-woven-like structure. Two sample plates for each sizing were then produced: the non-wovens were heated to two different temperatures and pressed using a hydraulic parallel press. Microscopic images were then analyzed to determine the fiber–matrix adhesion. An example is given in Figure 3. The examination of the fiber–matrix adhesion was conducted with a scanning electron microscope (SEM). For evaluation, fracture surfaces were generated at temperatures of −196 °C. All tested sizings showed similar results. Since both commercial sizing types showed a comparable fiber–matrix adhesion and resulted in low fiber losses, one was randomly chosen for further tests in this study. Additionally, roving sections were used for sliver formation trails.

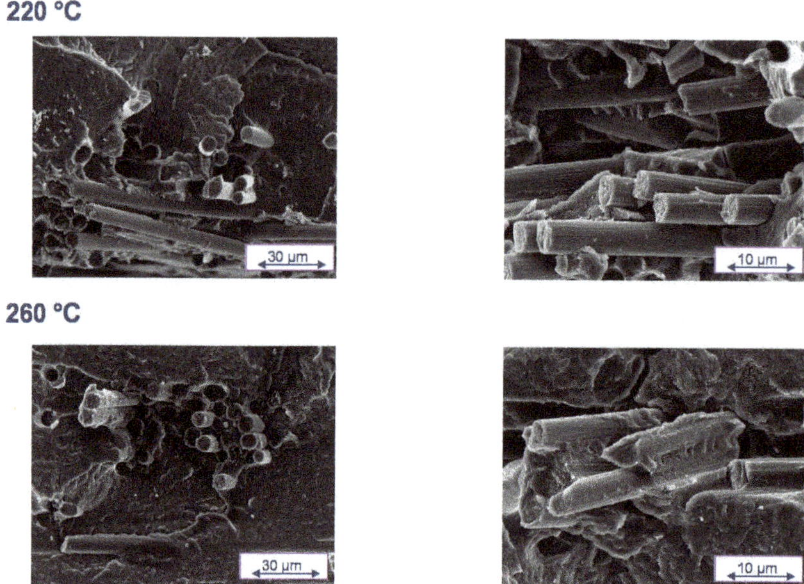

Figure 3. Microscopic images of the sizing Silastol CF1/CF2 after processing at 220 and 260 °C.

The processing of pyrolyzed carbon fibers without sizing proved to be not effective (see Section 1). However, the manual application of an appropriate sizing proved possible. Nonetheless, the processability could not be sufficiently improved, owing to a relatively uneven application. Therefore, roving sections were initially selected as the carbon fiber source for the sliver production.

2.2.3. Sliver Production

A sliver is a bundle of fibers that is usually spun into a yarn. In this study the sliver is used as a means to incorporate fibers into the injection molding process via the direct fiber feed module of an IMC. The production steps needed to produce the sliver will be presented in the following.

The first step toward realigning the recycled carbon fibers into a sliver and obtaining an infinite, homogeneous, semi-finished product is the preparation (mixing and opening) of the raw material. The second technological step involves machines that ideally dissolve and orient the fibers, which is called the carding process. Roller cards, like the one in this study, were originally used for processing wool. Wool fibers have low strength and have always been an expensive raw material. For these reasons, processing must be as gentle as possible without major fiber losses. Due to its processing properties, wool fibers are the most comparable textile fibers to carbon fibers.

Figure 4 shows a schematic cross-section of the textile-process chain to produce a sliver out of carbon fibers. The blue line illustrates the material flow through the machine. The general principle of the textile conversion of randomly oriented fibers into an aligned semi-finished product like the sliver is to transport, comb, and draw single fibers with metal pins, or "metallic card clothing." The size of the pins decreases as the process progresses to gradually separate and orient the fibers. The mechanical work required for orientation is applied step by step and more gently than if individual fibers are pulled from a large collection at the beginning of the process. Combing processes are generated through different surface speeds of the working elements.

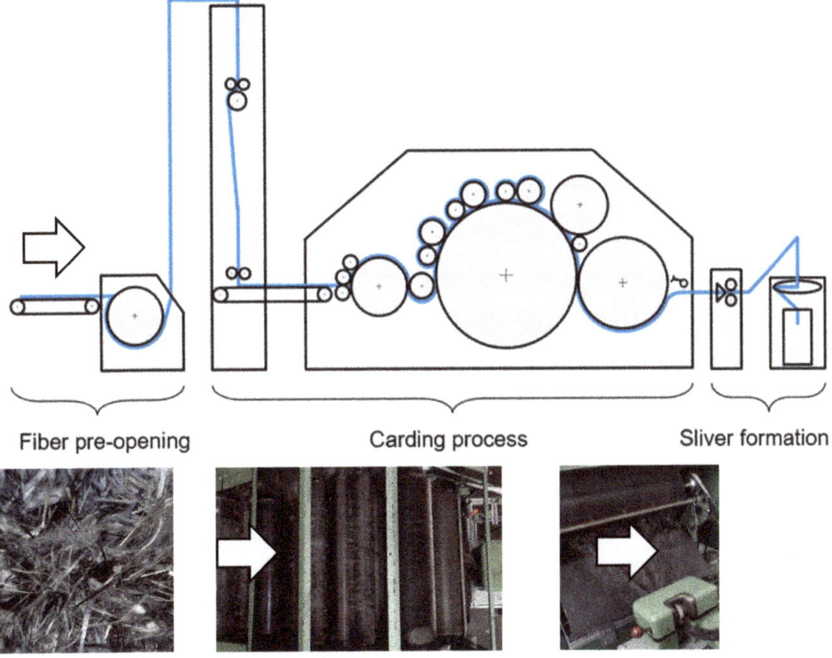

Figure 4. Schematic illustration of the textile process steps used in this study for sliver production (**top**) and pictures of the material feed (**left**), the roller card (**middle**), and the sliver formation (**right**).

In addition to the orientation of the fibers, the carbon fibers were mixed with the thermoplastic matrix fibers (rPA6) to achieve a homogeneous blend within the sliver. The carding process separates and orients the fibers via mechanical forces, which leads to the breaking of the carbon fibers. At the same time, this separation process results in homogeneous mixing, with the matrix and carbon fibers lying directly next to each other, depending on the fiber diameters of the mixing components.

The process starts with a coarse blend of carbon and rPA6 fibers on an opener from the company Dilo Group, Eberbach, Germany (fiber pre-opening in Figure 4). The processability of the pre-opened fiber blend on the roller card resulted in almost trouble-free further processing with acceptable carbon fiber reduction when the fiber blend was opened twice (two passages, Figure 5). A fiber volume content of 55% for the sliver was specified as the mixing ratio. To compensate for possible losses during processing, the fiber flocks were mixed with a volume ratio of 65% carbon fibers to 35% rPA6.

(**a**) material feeding chute

(**b**) 1st opening step

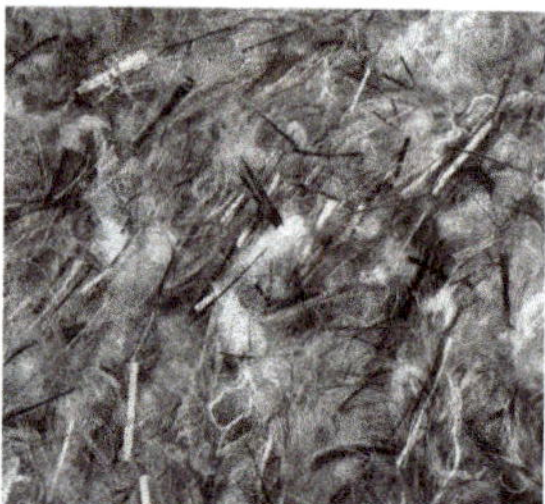
(**c**) 2nd opening after feeding

Figure 5. Feeding (**a**) first (**b**) and second (**c**) opening and mixing process of carbon (black) and rPA6 (white) fibers.

The pre-opened and mixed fibers are further opened and mixed in the card's feeding chute Type FBK from Trützschler GmbH, Mönchengladbach, Germany. The feeding chute is followed by the carding machine of the company Maschinenfabrik Memmingen, Memmingen, Germany. The output of the card, the card web, is compacted through a cone and stored in a can by means of a can coiler. The sliver take-up and can coiler have been adapted to the card by the authors.

The resulting sliver is shown in Figure 6. On the right side, the oriented carbon fibers within the sliver and the homogenous blend of black carbon and white matrix fibers (grey) can be seen.

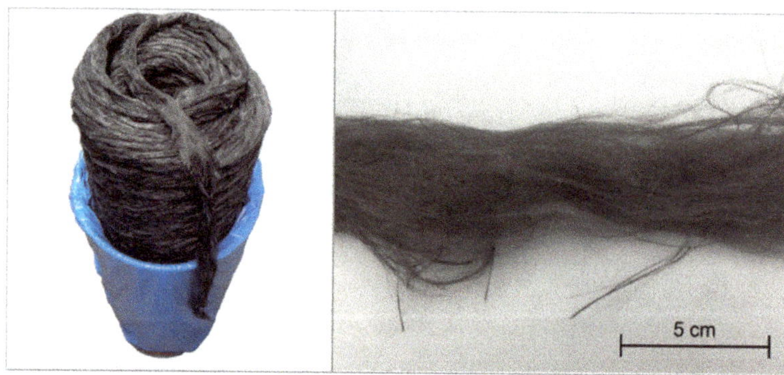

Figure 6. Sliver coil (**left**) and texture of the recycled staple fiber sliver (**right**).

2.3. Production of Test Specimens

To fabricate the test specimen, recycled carbon fiber sliver and polyamide matrix granules were processed via injection molding compounding on an IMC 200-1400 C2, KraussMaffei Group, Munich. The fiber content can be adjusted by the number and fineness of the fibers fed into the direct feed module of the twin-screw extruder of the IMC and by the screw speed n and the mass throughput $\dot{m}$ of the compounder. The theoretical fiber content φ is calculated according to Equation (1), as described in the documentation of the IMC provided by KraussMaffei:

$$\varphi = \frac{x{\cdot}SDF{\cdot}n{\cdot}k}{(x{\cdot}SDF{\cdot}n{\cdot}k) + \dot{m}}; \ IPF = d_f{\cdot}U{\cdot}60, \ U = d_{si}{\cdot}(1+k){\cdot}\pi, \tag{1}$$

where x corresponds to the number of rovings (in the case of this study, slivers) used in the feed module, *IPF* is the initial throughput factor, k is a correction factor in calculating the actual screw circumference U from the inner screw diameter d_{si}, and d_f is the fiber mass per kilometer in tex (g/1000 m).

To produce test specimens with a fiber content of 20% in the IMC, for example, the screw speed was set to 120 min^{-1}, the mass throughput to 4 kg/h, and rovings with a fineness of 2.4 g/m were used. By increasing the screw speed to 160 min^{-1} and reducing the mass throughput to 3 kg/h, a fiber content of 30% was achieved.

To further compare the results for recycled carbon fibers, test specimens with glass fiber rovings from Johns Manville, Denver, CO, USA, were also produced. The roving type E-glass PR 440 2400 871 (StarRov®LFTPlus) was adapted for processing with polyamide matrices with a silane sizing.

2.4. Characterization of Test Specimens

The characterization of fiber-reinforced plastics was carried out with classical methods to determine mechanical properties. The tensile tests, according to ISO 527, were performed by a universal testing machine type 1455 from Zwick GmbH and Co. KG, Ulm, Germany on the injection-molded basic specimens of type 1A. The bending properties were determined on the same machine, according to ISO 178. Seven to ten test specimens per mixing ratio were tested and statistically evaluated.

Tests such as the determination of fiber content (ISO 1172 method A), fiber length (ISO 22314), and fiber–matrix adhesion were also carried out. Due to the limited resources, complexity, and time consumption involved in carbon fiber length determination in an injection molded part using a solvolysis method, only two test specimens were measured: specimen one with a fiber content of 27 wt-% ($\dot{m}$ = 3 kg/h, n = 300 min^{-1}) and specimen two with a fiber content of 31 wt-% ($\dot{m}$ = 3 kg/h, n = 350 min^{-1}). A specimen made of virgin fiber roving with a fiber content of 26 wt-% was also measured for comparison.

The single fiber length of the sliver was determined according to the standards DIN 53,803 and DIN 53808-1. For this purpose, 150 individual fibers were randomly and carefully removed from the fiber sample by hand and measured under a magnifying glass using a ruler.

2.5. Summary of the Production Steps

In the following section, the test evaluation is presented. Below is a short summary of the production process. The production cycle basically contains two steps:

- A staple fiber hybrid sliver, containing recycled carbon fibers (product scraps) and recycled PA6 fibers, is created and analyzed. Since product scraps are used, no additional sizing is applied to the fibers, and no information about the previous sizing is available.
- An IMC then produces recycling fiber-reinforced parts for evaluation. Virgin PA6 is mixed with recycled PA6 in various ratios. The sliver is added to the matrix via a direct feed module.

3. Results

3.1. Sliver Characterization

Throughout the textile process, the fiber length and mixing ratio were measured to continuously assess the individual process steps and changes in the material. During the production of the sliver, mechanical work was applied to open and align the fibers, resulting in a shortening of the carbon fibers, as shown in Figure 7. Despite ongoing optimization of all textile process steps, the carbon fibers were shortened by 20% from the raw material to the sliver regarding the median fiber length.

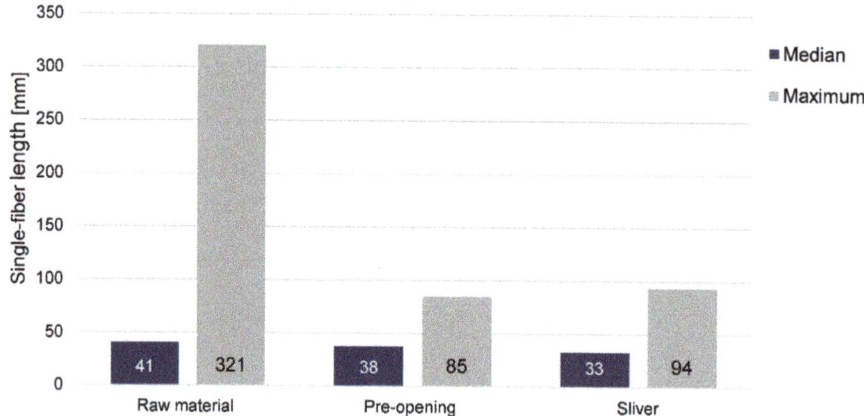

Figure 7. Change of the average and maximum carbon fiber length through the sliver production process.

To measure the single-fiber length, the examined carbon fibers were laid out on black velvet, covered by a glass plate, and placed under a magnifying glass with illumination. Using tweezers and fingers, the carbon fibers were carefully removed from the fiber agglomeration. The sliver in particular, had to be delicately dissolved by hand to remove the carbon fibers without damage. For this reason, the resulting single-fiber length is questionable. The tendency of a fiber to shorten after each processing step, however, was assumed to be correct.

Even slight mechanical processing of the fibers during pre-opening resulted in shortening, especially for the longest carbon fibers. A further reduction of the carbon fiber length was due to the mechanical processing of the fibers in the carding machine. Overall, the processing of carbon fibers results in a more uniform fiber length distribution, which can be beneficial for further processing and lead to stable process conditions.

Figure 8 shows the change in the mixing ratio over the sliver production process. The development of fiber dust in the carding process can be detected by measuring a change in the mixing ratio. While the pre-opening, which included measurement inaccuracies due to the sample size, did not lead to a change in the mixing ratio, the mechanical processing of the carbon fibers in the carding machine caused fiber breakage, and thus, fiber dust, decreasing the carbon fiber fraction. The further reduction of the carbon content was due to the removal of fiber dust from the sliver. Overall, a reduction of the carbon content by 7% resulted during manufacturing.

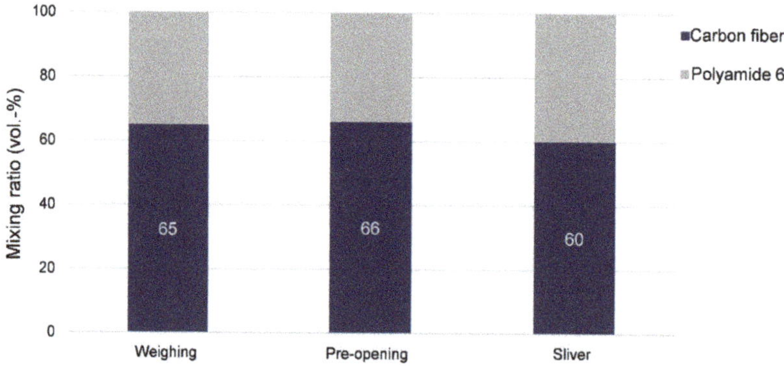

Figure 8. Change of the mixing ratio over the sliver production process.

3.2. Preliminary Investigations of Fiber Content in Relation to the Process Settings

In a previous study, comparatively strong results were achieved when processing glass fibers in a single-stage process via the direct fiber feed module [10], and the fiber content was predicted successfully. Hitherto, carbon fibers had never been processed in a single-stage process using a direct fiber feed module, so this study investigated the transferability of the results to the processing of recycled carbon fibers. Therefore, the settings of speed and mass throughput at the IMC for a direct fiber feed of carbon fibers had to be determined in preliminary tests. To ensure comparability with previous tests of glass-fiber-reinforced recycled polyamide [10], carbon fiber contents of 20 wt-% and 30 wt-% were targeted.

Contrary to the use of virgin carbon fiber rovings, it was not yet known whether the processing settings of the recycled carbon fibers via the staple-fiber sliver could reliably be determined with the calculation model for the direct fiber feed module. Due to its hybrid structure, the carbon fiber content of the sliver is only roughly 60 wt-%. In addition, the low consolidated structure (low tex value) of the sliver presented difficulty. Hence, a high sliver throughput was needed, requiring high screw speeds, which could increase fiber slippage and thus reduce the fiber content. Therefore, three setting approaches for the sliver were attempted in the preliminary tests. The selected test settings are shown in Table 3.

Table 3. Settings of the preliminary tests.

Type of Fiber	Mass throughput $\dot{m}$ (kg/h)	Screw Speed n (rpm)	Calculated Fiber Content (wt-%)
Reinforced Carbon Fiber	4	160	30
	4	100	25
	5	100	20
Recycled Staple Fiber Sliver	3	200	-
	3	300	-
	3	400	-

The results of the fiber content measurements of the preliminary tests in Figure 9 show that the calculation model is also applicable to the processing of (virgin) carbon fiber rovings, in general. For the sliver, in contrast, the mass throughput of $\dot{m}$ = 3 kg/h and a screw speed of n = 300 min^{-1} yielded a fiber content of approximately 20 wt-%, which is lower than predicted. Since the screw speed of n = 400 min^{-1} induced a fiber content that was too high, the speed for the further investigation was set to n = 350 min^{-1} for a mass throughput of $\dot{m}$ = 3 kg/h.

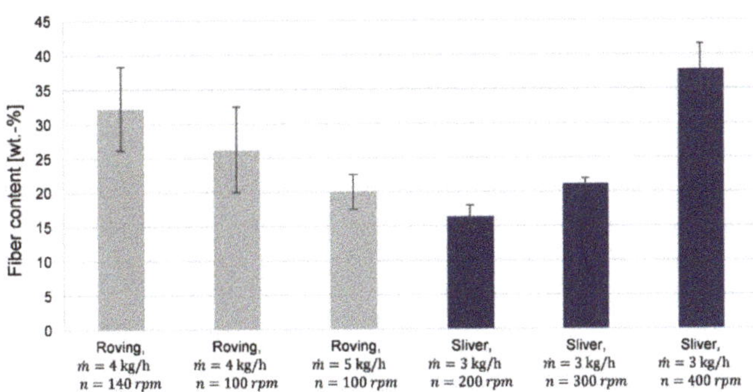

Figure 9. Fiber content in preliminary tests with carbon fiber roving and recycled staple fiber sliver.

In general, the deviations of the measured values were quite large. This could have been due to the unsteady sliver formation process. High fluctuations were observed during the single-stage process with the glass fibers as well, which could be attributed to the IMC [10].

3.3. Fiber Content

Figure 10 shows the averaged fiber contents of the different mixtures from the single-stage process. When processing the sliver, the calculated settings did not lead to the desired results. Nearly all mixtures exhibited higher fiber content values than the calculated values according to Equation (1). In particular, for the calculated fiber content of 20 wt-%, all measured values were above 30 wt-%. In some cases, this discrepancy was even larger at higher speeds. Overall, the fiber contents showed high deviations and inconsistent behavior.

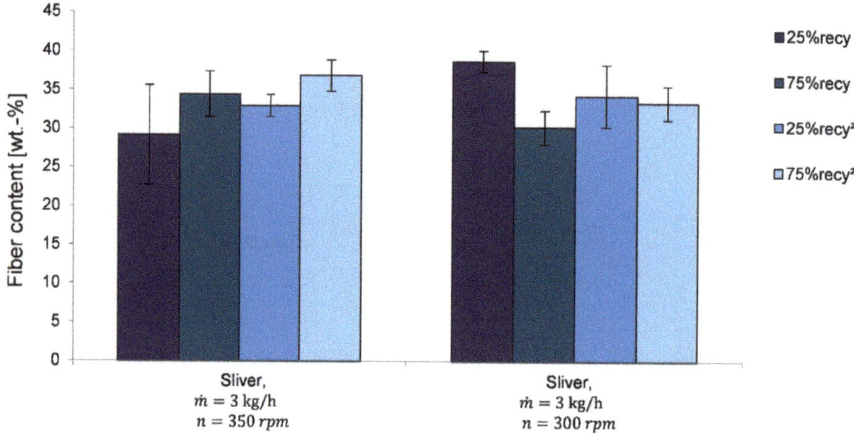

Figure 10. Averaged fiber content of the recycled staple fiber sliver with varying recycling content of the matrix.

The deviation of fiber content was due to the sliver formation process. Further, the feeding of the sliver into the IMC led to difficulties during processing. The sliver did not have a homogeneous width (see Figure 6) and was stretched irregularly when entering the extruder because of its unconsolidated texture. As a result, the sliver was not drawn in continuously and sometimes even tore during processing, resulting in an uneven fiber distribution in the melt. Further, varying process parameters can additionally cause small fluctuations in the fiber content. For example, friction occurs during fiber draw-in at the extruder, which counteracts the fiber draw-in force and can thus have effects on fiber retention and length. The friction is influenced by the roving and roving guide type, bobbin winding, screw geometry, and physical properties of the melt.

An analysis of the recycling content in the matrix material shows no significant influence on the fiber content. The scatter of the measured fiber content values was large. Therefore, the influence of process fluctuations on fiber content is assumed to be strong. In previous work with glass fibers [10], it was also shown that there is no clear correlation between fiber content and recycling rate. Only in the case of twice-recycled material could a dependency be determined. The values obtained here, however, show only small deviations with no statistical significance, even for twice-recycled material.

In summary, the settings calculated according to Equation (1) in the single-stage process with the sliver did not lead to the desired results. The calculation model for the direct fiber feed in a single-stage process cannot, therefore, be applied to the staple fiber sliver made of recycled carbon fibers.

3.4. Fiber Length

The fiber lengths of two test specimens were measured: specimen one with a fiber content of 27 wt-% ($\dot{m}$ = 3 kg/h, n = 300 min^{-1}) and specimen two with a fiber content of 31 wt-% ($\dot{m}$ = 3 kg/h, n = 350 min^{-1}). The results were compared to a specimen made of virgin fiber roving with a fiber content of 26 wt-%. Analyses of the weight-averaged fiber lengths their distributions were based on the ISO 22,314 standard.

Table 4 shows the average fiber lengths of the different mixtures of the single-stage process. Overall, fiber lengths between 200 and 330 μm were obtained. When comparing the different types of carbon fibers used with an analogous adjustment of the fiber content, longer fibers were observed in the test specimens from rovings. The average fiber length values were about 35% higher.

Table 4. Settings of the preliminary tests.

Type of Fiber	Fiber Content (wt-%)	Averaged Fiber Length (mm)
Roving	26	0.33
Sliver	27	0.20
Sliver	31	0.32
Glass Fiber	23	1.06
Glass Fiber	31	0.74

When comparing the fiber contents of the two test specimens made of sliver, the higher fiber content of 31 wt-% yielded significantly longer fibers compared to the lower fiber content of 27 wt-%. The comparison of the fiber length distribution also shows that a fiber content of 31 wt-% leads to significantly longer fibers of up to 5 mm. At a fiber content of 27 wt-%, however, a maximum fiber length of 0.8 mm was observed.

Investigations with glass fibers utilizing a direct fiber feed module in the single-stage process, on the other hand, have shown that a higher fiber content increases the fiber degradation [10]. This effect is explained by the associated higher screw speed and the resulting increase in shear forces in the melt. The reduction of the mass flow rate led to an increased shear energy input, and thus, additional stress to the fibers. Additionally, an increase in the material residence time in the mixing zone could lead to a further reduction in fiber length. Likewise, as the fiber content increases, the fiber length decreases owing to the intensive fiber–fiber interaction [10]. Nonetheless, findings in the literature are contradictory in terms of the influence of the screw speed on fiber length. Higher screw speeds can also result in longer fibers because of the reduced melt viscosity and dwell time, as described in [37,38].

The fibers of the (virgin) glass-fiber-reinforced test specimens produced in the single-stage process were significantly longer (an average of 0.74 mm) than the fibers of the carbon-fiber-reinforced test specimens made of staple fiber sliver (Table 4). The shorter fiber lengths of the recycled carbon fiber test specimens can be attributed to the carbon fibers being thinner and more brittle than glass fibers. It has also been observed in the literature that when comparing short-fiber-reinforced composites with a fiber content of 30 wt-%, the glass-fiber-reinforced test specimens have, on average, significantly longer fibers [39]. Dependence of the recycling fraction and fiber length cannot be determined owing to the small number of samples investigated.

In summary, the average fiber length of the carbon fibers was 0.3 mm. Shorter fibers were measured at a lower fiber content of 27 wt-%, whereas a higher fiber content of 31 wt-% resulted in longer fibers. The influence of the recycled content of the matrix on carbon fiber length could not be determined due to the small number of samples measured.

Since only representative samples were evaluated, the actual fiber length of each test specimen is unknown. The results are intended to test the possibility of fiber length measurement to this extent and to show a trend in fiber length and fiber-length distribution.

3.5. Mechanical Properties

Figure 11 shows the average Young's moduli of the test specimens from the single-stage process. The test specimens with a fiber content of 27 wt-% had similar values to those of the test specimens with a fiber content of 30 wt-%, roughly 25,000 MPa. This suggests an overall good performance of the material compared to the recycling routes summarized by Pimenta [2], especially considering the advantages of the injection molding process compared to bulk molding compound compression or compression molding of non-wovens, which are other solutions for discontinuous fiber processing described in the literature. Pimenta stated that Young's moduli of 20 GPa, 5–30 GPa, and 14–16 GPa were reached for bulk molding compounds, non-wovens, and injection molded parts, respectively [2].

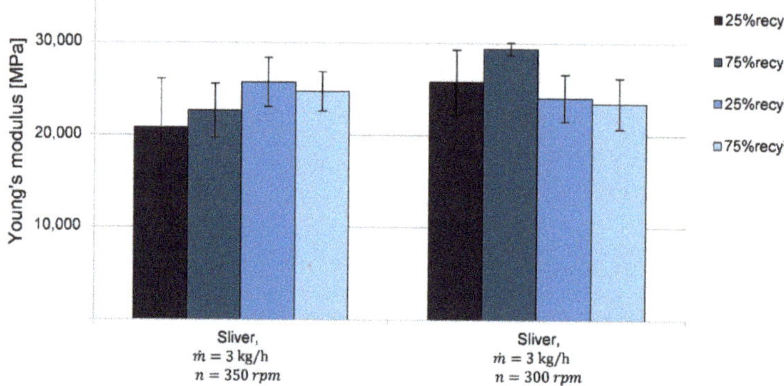

Figure 11. Young's moduli of specimens produced in the single-stage process with varying recycling content of the matrix.

When the results of the mechanical test are compared with those of the fiber content measurements presented above, a certain correlation can be established. Figure 12 shows the average Young's moduli of the test specimens and their average fiber contents. Overall, parallels can be seen between Young's modulus and fiber content, as expected.

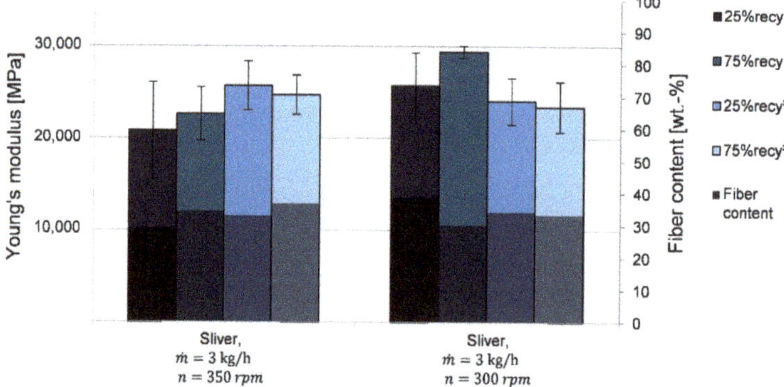

Figure 12. Young's moduli of specimens produced in the single-stage process with varying recycling content and corresponding fiber content of the matrix.

However, the dependence of Young's modulus on the recycled content of the matrix cannot be determined. The double reuse of PA6 does not lead to any significant effects. This means that the recycling-related impairments of the plastic have no apparent effect on the tensile modulus. Instead, the material parameter is primarily determined by the fiber content.

The standard deviations of Young's modulus are very high, up to 13%, due to processing instabilities caused by the IMC and the resulting scattering of fiber content within a batch. As previously explained, there were irregularities in the feeding of sliver, which led to fluctuations of the fiber content in the melt.

When Young's moduli of the carbon-fiber-reinforced recycled parts tested are compared to those of the glass fiber-reinforced plastic parts produced under almost identical conditions, an influence of the type of reinforcing fiber on the part can be clearly established. In Figure 13, glass-fiber-reinforced parts produced in the same process exhibit Young's moduli of approximately 10,000 MPa, nearly half of that of recycled carbon-fiber-reinforced parts. The use of carbon fibers, therefore, leads to significantly higher stiffness of the part, even though the length of the recycled carbon fibers is significantly shorter than the length of the glass fibers.

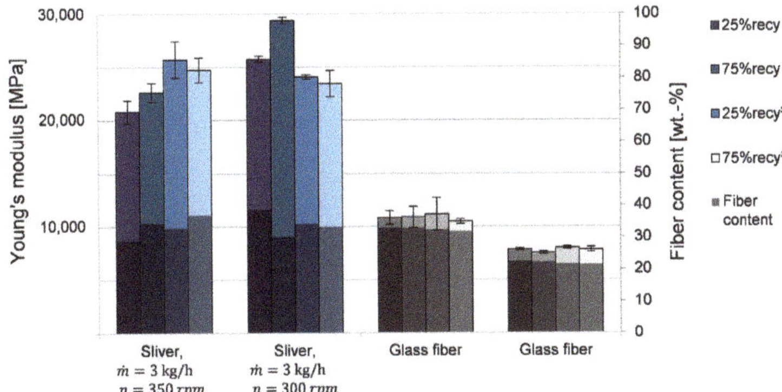

Figure 13. Young's moduli of specimens produced in the single-stage process with varying recycling content and corresponding fiber content of the matrix compared to parts produced with virgin glass fiber rovings and similar fiber content.

To further evaluate the mechanical properties, the tensile strength was determined. The results are shown in Figure 14. As expected, the tensile strength is also dependent on the fiber content. Similar to the modulus of elasticity, a correlation with the fiber content can be observed for the tensile strength as well.

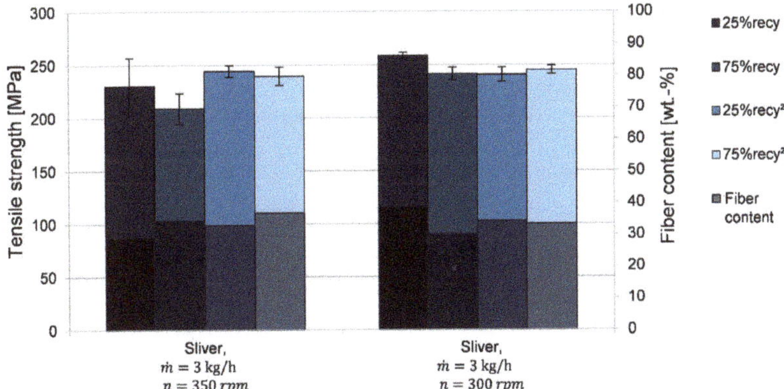

Figure 14. Young's moduli of specimens produced in the single-stage process with varying recycling content and corresponding fiber content of the matrix.

For the comparison of the two different fiber types, the average tensile strengths of the glass fiber specimens and the sliver specimens are shown in Figure 15. As with Young's modulus, the test specimens from the recycled carbon fiber sliver showed significantly higher tensile strengths compared to the glass fiber specimens, confirming previous results.

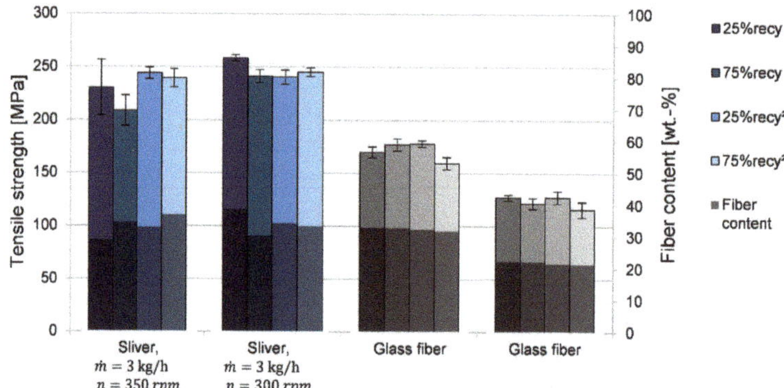

Figure 15. Tensile strengths of specimens produced in the single-stage process with varying recycling content and corresponding fiber content of the matrix compared to parts produced with virgin glass fiber rovings and similar fiber content.

The analyses of the elongation at break, the impact strength, and the flexural modulus and strength showed similar results and will not be presented in this study to avoid redundancy.

4. Summary and Outlook

In this study, a novel method to recycle carbon fibers efficiently while preserving fiber length was investigated. A hybrid sliver with 60 vol-% carbon fiber content was produced from recycled carbon and PA6 fibers. A shortening of the carbon fiber of about 50% resulted after the entire process. At the same time, a narrowing of the fiber length distribution could be observed. The sliver was then used for the production of test specimens on an IMC utilizing a direct fiber feed module.

The feeding of the staple fiber sliver into the melt was challenging. The sliver stretched unevenly and sometimes tore. The predicted fiber content could not be achieved in the single-stage process. In fact, the fiber contents were higher than the calculated values and had high fluctuations. The fiber content measurements do not show a clear dependency on the recycled content of the matrix.

A fiber length measurement was performed on three random samples. The results show an increasing fiber length with increasing fiber content. Fibers of parts made of virgin carbon fiber rovings were longer than those of the recycled carbon fiber sliver.

Tensile tests were carried out to characterize the mechanical properties. Overall, very good mechanical characteristics could be achieved. Correlations between the actual fiber content and the mechanical properties were determined. The higher the fiber content, the better the mechanical properties. Large deviations of the measured values were observed. Furthermore, it can be stated that the recycling-related impairment of the matrix has no effect on the modulus of elasticity, tensile strength, flexural strength, or impact strength.

When comparing the recycled carbon-fiber-reinforced test specimens with the glass fiber test specimens, a clear dependence of the mechanical characteristics on the type of reinforcement fiber was determined. The sliver test specimens all achieved significantly higher results, although the recycled carbon fibers were significantly shorter than the glass fibers.

The results show that desirable mechanical properties can be achieved in principle when recycled carbon fibers are processed in the form of a staple fiber sliver. However, it is questionable whether the processing of staple fiber sliver allows a reproducible adjustment of the fiber content, and thus, predictability of the properties. The loose structure of the sliver causes difficulties with drawing the fibers in the process evenly. The use of the hybrid yarn could simplify the drawing of the recycled fibers, as the strength of the yarn is considerably higher. Thus, a yarn spinning process was developed to further process the sliver before injection molding. This further reduces the fiber content but improves the fiber distribution. Fewer process fluctuations might be reached this way.

Author Contributions: Conceptualization, J.W. and S.B.; methodology, J.W. and J.R.; validation, J.R.; investigation, J.W. and J.R.; resources, G.T.G. and C.B.; writing—original draft preparation, J.W. and J.R.; writing—review and editing, S.B., G.T.G. and C.B.; visualization, J.R. and S.B.; supervision, G.T.G. and C.B.; project administration, J.W. and S.B.; funding acquisition, G.T.G. and C.B.; All authors have read and agreed to the published version of the manuscript.

Funding: This research was funded by the Baden-Württemberg-Stiftung, grant number MAT0005.

Conflicts of Interest: The authors declare no conflict of interest.

References

1. Meng, F.; Olivetti, E.; Zhao, Y.; Chang, J.C.; Pickering, S.J.; McKechnie, J. Comparing Life Cycle Energy and Global Warming Potential of Carbon Fiber Composite Recycling Technologies and Waste Management Options. *ACS Sustain. Chem. Eng.* **2018**, *6*, 9854–9865. [CrossRef]

2. Pimenta, S.; Pinho, S.T. Recycling carbon fibre reinforced polymers for structural applications: Technology review and market outlook. *Waste Manag.* **2011**, *31*, 378–392. [CrossRef] [PubMed]

3. Pickering, S.J.; Liu, Z.; Turner, T.; Wong, K. Applications for carbon fibre recovered from composites. *Iop Conf. Ser. Mater. Sci. Eng.* **2016**, *139*, 012005. [CrossRef]

4. Oliveux, G.; Dandy, L.O.; Leeke, G.A. Current status of recycling of fibre reinforced polymers: Review of technologies, reuse and resulting properties. *Prog. Mater. Sci.* **2015**, *72*, 61–99. [CrossRef]

5. Pickering, S. Recycling technologies for thermoset composite materials—current status. *Compos. Part A Appl. Sci. Manuf.* **2006**, *37*, 1206–1215. [CrossRef]

6. Schürmann, H. *Konstruieren mit Faser-Kunststoff-Verbunden*; Springer Science and Business Media LLC: Berlin/Heidelberg, Germany, 2005.

7. Truckenmüller, F. Direktverarbeitung von Endlosfasern auf Spritzgießmaschinen. Möglichkeiten und Grenzen. Dissertation, Universität Stuttgart, Stuttgart, Germany, 1996.

8. Widmayer, S.; Fritz, H.-G.; Bonten, C. Spritzgießcompoundieren: Feuchte vereinzelt Naturfasern. *Kunststoffe Munchen.* **2012**, *12*, 71–74.

9. Willems, F.; Bonten, C. Influence of processing on the fiber length degradation in fiber reinforced plastic parts. In Proceedings of the 34th International Conference of the Polymer Processing Society, Gratz, Austria, 14 December 2015; AIP Publishing: Melville, NY, USA, 2016; Volume 1779, p. 020003.

10. Wellekötter, J.; Christian, B. Comparison of fiber-reinforced recycled polyamide in single stage and two stage injection molding. *Polym. Compos.* **2018**, *40*, 1731–1739. [CrossRef]

11. Bonten, C. *Plastics Technology—Introduction and Fundamentals*; Hanser: Munich, Germany, 2019.

12. Witten, E.; Kraus, T.; Kühnel, M. Composites-Marktbericht 2016. Available online: https://elib.dlr.de/106030/ (accessed on 15 September 2020).

13. Gardiner, G. Closing the CFRP lifecycle loop. Recycled carbon fiber update. *Compos. World* **2014**, 28–33.

14. Lefeuvre, A.; Garnier, S.; Jacquemin, L.; Pillain, B.; Sonnemann, G. Anticipating in-use stocks of carbon fiber reinforced polymers and related waste flows generated by the commercial aeronautical sector until 2050. *Resour. Conserv. Recycl.* **2017**, *125*, 264–272. [CrossRef]

15. Pillain, B.; Loubet, P.; Pestalozzi, F.; Woidasky, J.; Erriguible, A.; Aymonier, C.; Sonnemann, G. Positioning supercritical solvolysis among innovative recycling and current waste management scenarios for carbon fiber reinforced plastics thanks to comparative life cycle assessment. *J. Supercrit. Fluids* **2019**, *154*, 104607. [CrossRef]

16. Rybicka, J.; Tiwari, A.; Leeke, G.A. Technology readiness level assessment of composites recycling technologies. *J. Clean. Prod.* **2016**, *112*, 1001–1012. [CrossRef]

17. Witik, R.A.; Teuscher, R.; Michaud, V.; Ludwig, C.; Manson, J.-A.E. Carbon fibre reinforced composite waste: An environmental assessment of recycling, energy recovery and landfilling. *Compos. Part. A Appl. Sci. Manuf.* **2013**, *49*, 89–99. [CrossRef]

18. Naqvi, S.; Prabhakara, H.M.; Bramer, E.; Dierkes, W.; Akkerman, R.; Brem, G. A critical review on recycling of end-of-life carbon fibre/glass fibre reinforced composites waste using pyrolysis towards a circular economy. *Resour. Conserv. Recycl.* **2018**, *136*, 118–129. [CrossRef]

19. Wong, K.; Rudd, C.; Pickering, S.; Sandinge, A. Composites recycling solutions for the aviation industry. *Sci. China Ser. E Technol. Sci.* **2017**, *60*, 1291–1300. [CrossRef]

20. Hofmann, M.; Gulich, B. Verarbeitung von rezyklierten Carbonfasern für die Herstellung von Verbundbauteilen. *Light. Des.* **2013**, *6*, 20–23. [CrossRef]

21. Jacob, A. Composites can be recycled. *Reinf. Plast.* **2011**, *55*, 45–46. [CrossRef]

22. Hengstermann, M.; Raithel, N.; Abdkader, A.; Hasan, M.; Cherif, C. Development of new hybrid yarn construction from recycled carbon fibers for high performance composites. Part-I: Basic processing of hybrid carbon fiber/polyamide 6 yarn spinning from virgin carbon fiber staple fibers. *Text. Res. J.* **2015**, *86*, 1307–1317. [CrossRef]

23. Longana, M.L.; Ong, N.; Yu, H.; Potter, K.D. Multiple closed loop recycling of carbon fibre composites with the HiPerDiF (High Performance Discontinuous Fibre) method. *Compos. Struct.* **2016**, *153*, 271–277. [CrossRef]

24. Reichert, O.; Ausheyks, L.; Baz, S.; Hehl, J.; Gresser, G.T. Innovative rC Staple Fiber Tapes - New Potentials for CF Recyclates in CFRP through Highly Oriented Carbon Staple Fiber Structures. *Key Eng. Mater.* **2019**, *809*, 509–514. [CrossRef]

25. Tian, X.; Liu, T.; Wang, Q.; Dilmurat, A.; Li, D.; Ziegmann, G. Recycling and remanufacturing of 3D printed continuous carbon fiber reinforced PLA composites. *J. Clean. Prod.* **2017**, *142*, 1609–1618. [CrossRef]

26. Osswald, T.A.; Menges, G. *Materials Science of Polymers for Engineers*, 3rd ed.; Hanser: München, Germany, 2012.

27. Thomason, J. The influence of fibre length and concentration on the properties of glass fibre reinforced polypropylene: 5. Injection moulded long and short fibre PP. *Compos. Part. A Appl. Sci. Manuf.* **2002**, *33*, 1641–1652. [CrossRef]

28. Martens, H. *Recyclingtechnik. Fachbuch für Lehre und Praxis*; Spektrum: Heidelberg, Germany, 2011.

29. Johannaber, F.; Michaeli, W. *Handbuch Spritzgießen*, 2nd ed.; Hanser: München, Germany, 2004.

30. Maspoch, M.L.; Ferrando, H.E.; Velasco, J.I. Characterisation of filled and recycled PA6. *Macromol. Symp.* **2003**, *194*, 295–304. [CrossRef]

31. Henning, F.; Moeller, E. *Handbuch Leichtbau. Methoden, Werkstoffe, Fertigung*; Hanser: München, Germany, 2011.

32. Hopmann, C.; Michaeli, W. *Einführung in die Kunststoffverarbeitung*, 7th ed.; Hanser: München, Germany, 2015.

33. Wolters, L. (Ed.) *Kunststoff-Recycling*; Hanser: München, Germany, 1997.

34. Crespo, J.; Parres, F.; Peydro, M.; Navarro, R. Study of rheological, thermal, and mechanical behavior of reprocessed polyamide 6. *Polym. Eng. Sci.* **2012**, *53*, 679–688. [CrossRef]

35. Su, K.-H.; Lin, J.-H.; Lin, C.-C. Influence of reprocessing on the mechanical properties and structure of polyamide 6. *J. Mater. Process. Technol.* **2007**, *192–193*, 532–538. [CrossRef]

36. Formisano, B.; Göttermann, S.; Bonten, C. Recycling of cast polyamide waste on a twin-screw-extruder. In Proceedings of the 34th International Conference of the Polymer Processing Society, Lyon, France, 25–29 July 2016; AIP Publishing: Melville, NY, USA, 2016; Volume 140002.

37. Moritzer, E.; Heiderich, G.; Hirsch, A. Fiber length reduction during injection molding. In Proceedings of the AIP Conference 2055, Dresden, Germany, 27–29 June 2017; AIP Publishing: Melville, NY, USA, 2019.

38. Cieslinski, M. Using a Sliding Plate Rheometer to Obtain Material Parameters for Simulating Long Fiber Orientation in Injection Molded Composites. Ph.D. Thesis, Virginia Polytechnic Institute and State University, Blacksburg, VA, USA, 2015.

39. Sarasua, J.R.; Remiro, P.M.; Pouyet, J. The mechanical behaviour of PEEK short fibre composites. *J. Mater. Sci.* **1995**, *30*, 3501–3508. [CrossRef]

Publisher's Note: MDPI stays neutral with regard to jurisdictional claims in published maps and institutional affiliations.

Journal of
Composites Science

Article

Characterization of Mechanical Performance of Composites Fabricated Using Innovative Carbon Fiber Wet Laid Process

Hicham Ghossein [1,2,*], Ahmed Arabi Hassen [3], Seokpum Kim [4], Jesse Ault [5] and Uday K. Vaidya [1,4]

1   Department of Mechanical, Aerospace and Biomedical Engineering, Fibers and Composites Manufacturing Facility (FCMF), University of Tennessee, Knoxville, TN 37916, USA; Uvaidya@utk.edu
2   Endeavor Composites, Inc., Knoxville, TN 37932, USA
3   Materials Science and Technology Division (MSTD), Oak Ridge National Laboratory (ORNL), Oak Ridge, TN 37830, USA; hassenaa@ornl.gov
4   Energy and Transportation Science Division (ETSD), Oak Ridge National Laboratory (ORNL), Oak Ridge, TN 37830, USA; kimsp@ornl.gov
5   School of Engineering, Brown University, Providence, RI 02912, USA; jesse_ault@brown.edu
*   Correspondence: Hghossein@endeavorcomposites.com

Received: 28 July 2020; Accepted: 18 August 2020; Published: 22 August 2020

Abstract: Recent innovation in production of optimized nonwoven wet laid (WL) carbon fiber (CF) mats raised the question of optimal translation of the performance and isotropy into composites formed through these dry preforms. This work explores the mechanical behavior of composites produced from WL-CF mats in conjunction with the microstructure predicted through Object Oriented Finite Element Analysis (OOF). The mats used for the composites were prepared in two dispersion regimes using 25.4 mm long CF. The mixing regimes discussed in the author's previous work, are identified as Method 1 for the traditional processing regime and Method 2 for the innovative regime that provided optimal nonwoven WL-CF mats. Composite panels from Method 2 mats showed a normalized tensile strength increase of 52% over those from Method 1 panels. Reproducibility analysis of composites made from Method 2 mats demonstrated a standard deviation of 2% in fiber weight content, 2% in tensile modulus and 9% in tensile strength, while composites made from Method 1 mats demonstrated a standard deviation of 5% in fiber weight content, 5% in tensile modulus and 17% in tensile strength. Systematic study of the microstructure and its analysis through OOF confirmed the isotropy translation of mats produced through method 2 to the composites. This study validated the hypothesis that optimal nonwoven mats lead to a well-balanced composite with optimal performance and that non-optimal nonwoven mats do not pack into a well-balanced composite.

Keywords: wet laid; isotropic; tensile; carbon fiber; discontinuous

1. Introduction

Recent environmental issues related to global climate change and greenhouse gas emissions have prompted automotive manufacturers to focus on the development of lightweight and fuel efficient vehicles [1]. Fiber reinforced composites possess the advantage of high strength-to-weight and stiffness-to-weight ratios, light weight, low fatigue susceptibility and superior damping capacity [2–4].

The authors in their previous work [5] explored the production of nonwoven CF wet laid (WL) mats through two methods of fiber dispersion. Method 1 used a shear mixer to spread the fiber bundles, but it resulted in an unequal fiber distribution in the mats with somewhat inconsistent reproducibility. Method 2 used an innovative mixer based on chaotic advection theory which provided a fully balanced fiber distribution and consistent reproducibility of the mats. Researchers [6–10] have shown that porous nonwoven fabrics like the nonwoven WL-CF mats possess high specific surface area, light weight and ease of processing into complex geometries. This study investigates the mechanical properties of composite plates produced from nonwoven CF-WL mats made through Method 1 and Method 2. The experimental mechanical properties of both types of composites are provided within this paper. Further, a microstructure based finite element analysis (FEA) has been conducted on the mats made by each of the two methods to determine the effect on the mechanical properties of the final composite.

The effect of microstructure on mechanical properties of a composite is well explored. Straumit et al. [11] used X-ray computed tomography to quantify the internal structure of textile composites using an automated voxel model. Wan et al. [12] investigated the tensile and compressive properties of chopped carbon fiber tapes with respect to the changes in microstructure based on tape length and molding pressure. The authors found that increasing structural integrality improved the composite mechanical properties. In a second study, Wan et al. [13] analyzed the microstructural differences in CF composites using X-ray micro-CT. This method provided information on the morphology of the composite which was used to predict its mechanical properties. Tseng et al. [14] conducted numerical prediction of fiber orientation to predict mechanical properties for short/long glass and carbon fiber reinforced composites. Their model came within a 25% variance to the experimental data. Feraboli et al. [15] compared performance of different microstructures of various materials like recycled CF fabric/epoxy composites, twill laminates and sheet molding compounds (SMC). Similar mechanical properties were exhibited by the twill laminates and SMC under similar microstructural conditions. Caba et al. [16] characterized the fiber-fiber interactions in carbon mat thermoplastics (CMT) produced through the WL technique. Their study established a foundation for understanding the relation between the fiber volume fraction and the mechanical behavior of the composites. Evans et al. [17] proposed directed fiber compounding (DFC),producing a material similar to SMC via an automated spray deposition process of CF. They reported tensile stiffness and strength values of 36 GPa and 320 MPa for isotropic materials at 50% fiber volume. Selezneva et al. [18] investigated the mechanical properties of randomly oriented strand thermoplastic composites in an effort to quantify the effect of strand size and found that properties are dependent on the strand length. Amaro et al. [19] stated that the elastic modulus, tensile strength, and impact resistance of fiber composites increase as fiber length increases. Thomason [20] investigated the influence of fiber length and concentration on the properties of reinforced composites, and reported that above a critical fiber length (l_c), the full reinforcing potential of the reinforcement is realized due to higher fiber aspect ratio. The nonwoven WL-CF mats used in this study are produced using a 25.4 cm long CF. In order to avoid the complexity and vast data produced with X-Ray tomography to analyze the microstructure, this study refers to simpler image analysis techniques with higher precision.

Langer et al. [21] introduced an image based finite element analysis software, a novel numerical approach called object oriented finite element (OOF). OOF is a desktop software application for studying the relationship between the microstructure of a material and its overall mechanical properties using finite element models based on real or simulated micrographs. Reid et al. [22] discussed the mathematical approach and operational method for OOF and OOF2. This novel numerical approach has been used by researchers to provide fundamental insight on expectations of mechanical properties of fiber reinforced polymeric composites [23–27]. Goel et al. [28] compared experimental results for Young's modulus of long fiber thermoplastic (LFT) against predictions based on mathematical models and OOF. They reported that the closest prediction to experimental results was achieved through the OOF software.

Building on this approach, this study (a) validated the isotropy of composites made from the nonwoven WL-CF mats produced through the two discussed methods from the author's previous work [5], and (b) evaluated the composites mechanical performance with respect to theoretical prediction from Halpin–Tsai equations [29] for oriented discontinuous reinforcements.

2. Materials and Methods

This study used un-sized chopped Zoltek™ PX35 Type 02 CF with a length of 25.4 mm, average diameter of 7 μm, specific gravity of 1.81 g/cm^3, tensile strength of 4137 MPa, and tensile modulus of 242 GPa. The matrix used was a West System epoxy 105/206 mix with a specific gravity of 1.18 g/cm^3, tensile strength of 50.33 MPa and tensile modulus of 3.17 GPa.

2.1. Sample Preparations and Experimental Setup

Two different sets of CF/epoxy plates were prepared using the vacuum assisted resin transfer method (VARTM) technique. Three 306 × 306 mm plates were produced for each set, with every plate using the same five layers laminate of nonwoven WL-CF mats. The first set was processed using Method 1 mats; the second was processed with Method 2 mats. The mats contain randomly distributed fibers as seen in Figure 1 [5]. All plates were prepared on a flat glass mold surface with a nylon vacuum bag. The setup was a de-bulked under vacuum for 30 min, the resin was degassed for 10 min and the infusion took another 10 min until all fibers were fully wetted. After resin cure, samples were collected from each plate following the layout in the schematic seen in Figure 2. The samples' distribution allowed a statistical representation from the vacuum side of the plate, the center of the plate and the resin infusion side of the plate. The three sites for sample collection allow the exploration of the effect of resin to fiber distribution on the mechanical properties of the composite. The average mechanical properties of each set were considered in the study, in order to examine the reproducibility of the composite and to understand the difference between composites made from Method 1 mats and those made using Method 2 mats.

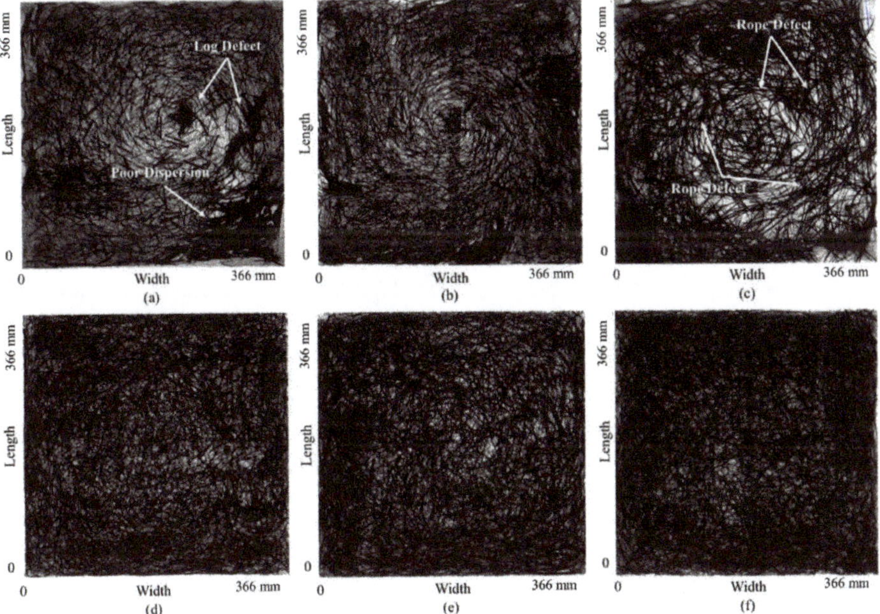

Figure 1. Back light scatter images of nonwoven wet laid carbon fiber (WL-CF) mats prepares using: (**a**) Method 1 at 10 min fiber dispersion; (**b**) Method 1 at 20 min fiber dispersion; (**c**) Method 1 at 30 min fiber dispersion; (**d**) Method 2 at 10 min fiber dispersion; (**e**) Method 2 at 20 min fiber dispersion; (**f**) Method 2 at 30 min fiber dispersion. All mats used in this study are prepared with 20 min fiber dispersion. Adopted from [5].

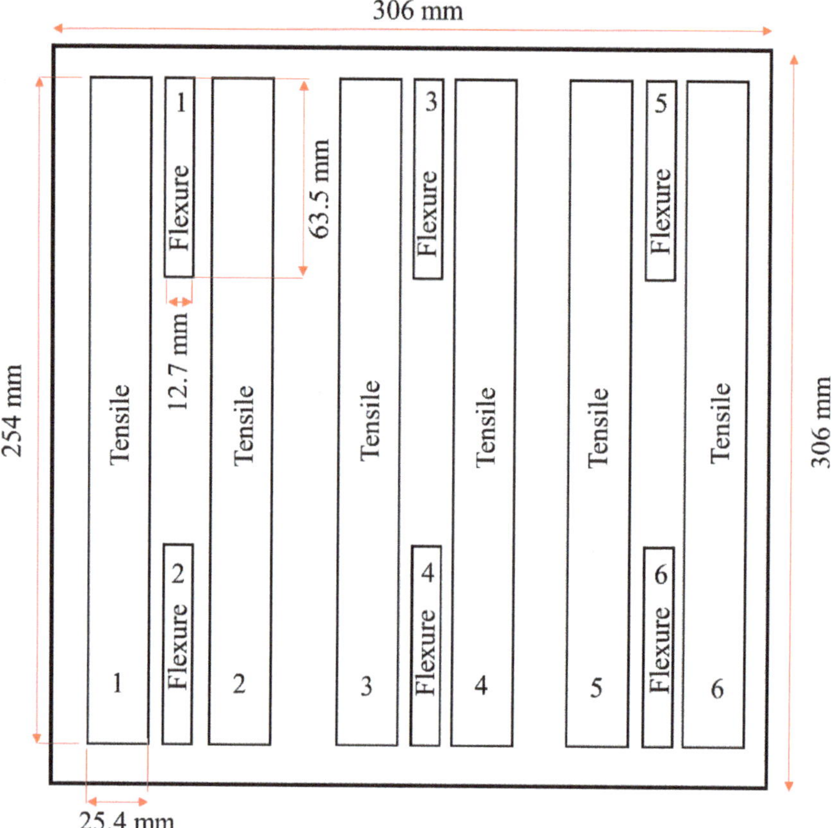

Figure 2. Location of tensile and flex samples machined from the WL nonwoven carbon fiber composite plates.

All tensile samples were tested based on ASTM D5083, all flexure samples were tested based on ASTM D790 and Inter Laminar Shear Stress (ILSS) samples were tested based on ASTM 2344. Furthermore, samples of 6.45 square cm were taken from each tensile sample for a burn off test in order to confirm the fiber weight ratio to resin distribution. All samples were weighed before and after matrix burn-off for comparison. The matrix was subjected to burn-off using a Thermo Fisher Scientific 1100 °C box furnace CF51800 series at 450 °C for three hours. Another set of 6.45 square cm samples were selected from each tensile sample for void calculations per ASTM D2734. Scanning electron microscopy (SEM) analysis was conducted to observe the break surface of the tensile samples.

2.2. Halpin-Tsai Theory and Relevance

The elastic modulus of isotropic reinforced composites is calculated theoretically by the Halpin–Tsai equations for oriented reinforcements and the Rule of Mixtures (ROM) [30]. Halpin [29] argued that it is possible to construct a material having isotropic mechanical properties from layers or plies of another or similar material. Mallick [31] stated that the Halpin–Tsai method is used to calculate the longitudinal and transverse properties of aligned discontinuous reinforcement composites, which can in turn be used to calculate the modulus of randomly oriented reinforcement composites. This study uses the Halpin–Tsai calculations to compute theoretical values for the composites under investigation by assuming the following conditions: (1) fiber cross section is circular; (2) fibers are arranged in a square array; (3) fibers are uniformly distributed throughout the matrix; (4) perfect bonding exists between the fibers and the matrix; (5) matrix is free of voids.

For these conditions, the tensile modulus of randomly oriented discontinuous fiber reinforced composites is calculated as:

$$E_{random} = \frac{3}{8}E_L + \frac{5}{8}E_T \tag{1}$$

where E_L and E_T are the empirical longitudinal and transverse moduli, respectively, for a unidirectional discontinuous fiber reinforced composite and they are calculated as follows:

$$E_L = \frac{1 + 2\left(\frac{l_f}{d_f}\right)\eta_L v_f}{1 - \eta_L v_f} E_m \tag{2}$$

$$E_T = \frac{1 + 2\eta_T v_f}{1 - \eta_T v_f} E_m \tag{3}$$

η_L and η_T are calculated as follows:

$$\eta_L = \frac{\left(\frac{E_f}{E_m}\right) - 1}{\left(\frac{E_f}{E_m}\right) + 2\left(\frac{l_f}{d_f}\right)} \tag{4}$$

$$\eta_T = \frac{\left(\frac{E_f}{E_m}\right) - 1}{\left(\frac{E_f}{E_m}\right) + 2} \tag{5}$$

where E_m and E_f are the matrix tensile modulus and the fiber tensile modulus, respectively; and l_f is the fiber length, d_f is the fiber diameter and v_f is the fiber volume of the composite.

As for the theoretical strength, failure is predicted when the maximum tensile stress in the laminate equals the strength averaged over all possible fiber orientation angles, known as the Hahn's approach [32]. Failure is predicted when the maximum tensile stress in the laminate equals the following strength averaged over all possible fiber orientation angles:

$$S_r = \frac{4}{\pi} \sqrt{S_L S_T} \tag{6}$$

where S_r is the strength of the random fiber laminate, S_L and S_T are, respectively, the empirical longitudinal and transverse strength of a 0° laminate of continuous fibers calculated through the rule of mixture using fiber and matrix properties provided by the manufacturer.

2.3. OOF Analysis Approach with Assumption of Composites Isotropy and Its Validation

Composite microstructure has a direct influence on its mechanical properties. In order to quantify microstructural features, both morphological and material properties must be characterized. Image processing is a robust technique for the determination of morphological features.

The OOF2 software follows an adaptive meshing algorithm and specifies properties to the grid units based on pixel assignment to the material (black = Matrix, white = Fiber). Figure 3 displays polished surfaces of samples produced using a single infused nonwoven WL-CF mat from Method 1 and Method 2, showing the microstructure distribution in each samples. The use of a single mat played an important role in preventing any bias of fiber packing from multiple layers in a composite panel. A clear qualitative difference is apparent in the microstructures of the two methods. Method 1 demonstrates a higher fiber agglomeration while Method 2 has a wider fiber distribution. In both cases, a validation of the isotropic nature of the composites is required. Such validation is performed by measuring the Young's modulus in multiple directions of load application this can be done experimentally or through computer simulation and analysis of the microstructure. Isotropy is defined by having equal Young's modulus for all load directions or a ratio of 1 when dividing the values of Young's modulus obtained from two different directions of load mounting. In this study, the microstructure performance of both methods, in different load mounting directions is evaluated through computer-simulation analysis. Using OOF2 an adaptive mesh is conformed on sample sets from each method, the mesh is then subjected to a simulated deformation through ABAQUS 2018 – HF5 to calculate the Young's modulus values for a tensile load in two principal directions.

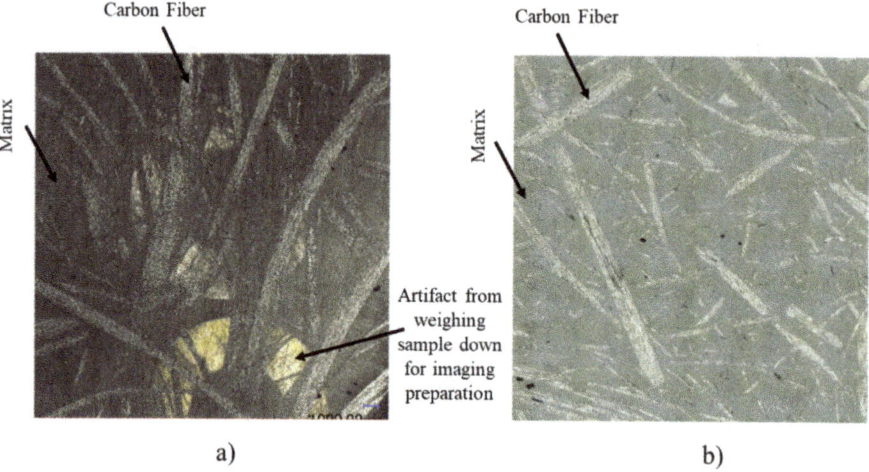

Figure 3. Polished samples showing the microstructure obtained in composite; (**a**) Micrographs for samples obtained by Method 1, and (**b**) Micrographs for samples obtained by Method 2.

The first step in this approach is to select images of a representative microstructure. Single mats prepared in similar methods to the mats shown in Figure 1b, and Figure 1e were infused with epoxy resin by the same methodology of composite preparation out of several stacked layers. Three samples per method were taken from the median of the infused mats at an area of 18 × 18 mm each. They were mounted in epoxy resin for ease of polishing, polished to reveal the microstructure and imaged at a magnification of 100×. Note that the representative image of the microstructure is somewhat subjective, thus the need for several samples for higher degree of fidelity. The samples had an average area fraction of ~33% which is representative of the overall volume fraction of fibers in the composite.

Before the OOF software is able to analyze the images and generates FEA meshes, some image processing filters need to be adopted. The constituents within the microstructure are separated into distinct grey levels through color thresholding and application of a blurring filter and contrast adjustment as seen in Figure 4. This process is similar to the approach by Goel et al. [27].

Figures 4 and 5 show the processed images with their corresponding meshes generated by OOF2 and simulated through ABAQUS for Method 1 and Method 2 sample sets, respectively. Figure 6 shows a representative microstructure and process analysis to assign a conforming mesh for the FEA processing. The area fraction of the samples is close to 33% with some variations. The samples from Method 1 showed higher variance in area fraction which reflected on the FEA results, while samples from Method 2 had higher consistency. This reflects the optimal fiber distribution of Method 2 versus the non-optimal fiber distribution of Method 1 as discussed in [5].

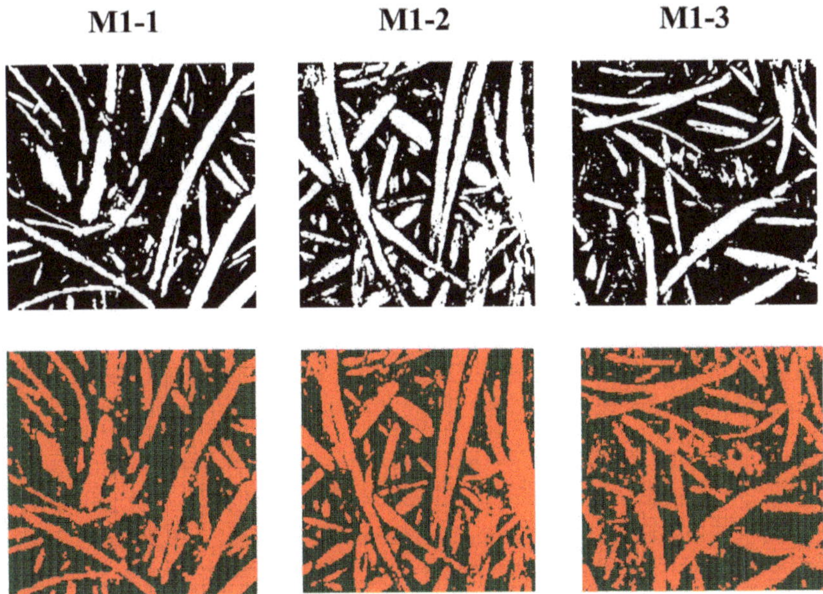

Figure 4. Method 1 modified images and their respective generated finite element analysis (FEA) meshes. M1-1 stands for Method 1-Sample 1.

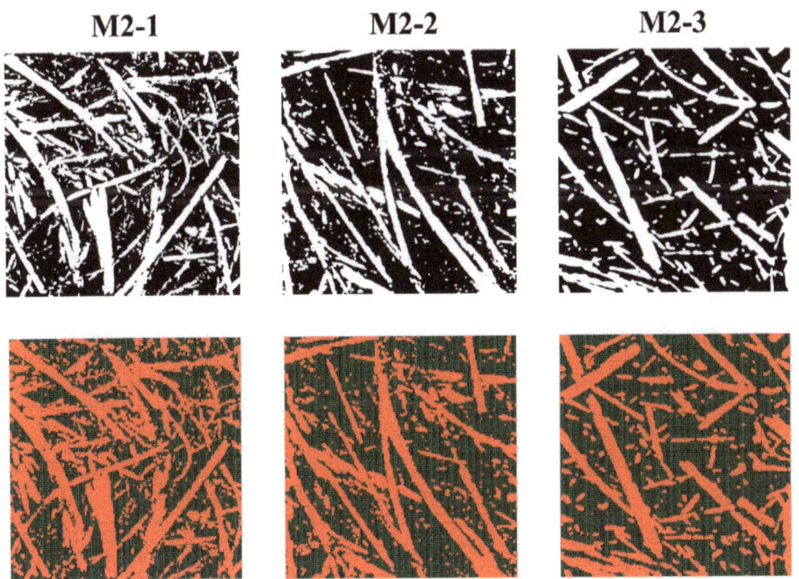

Figure 5. Method 2 modified images and their respective generated FEA meshes. M2-1 stands for Method 2-Sample 1.

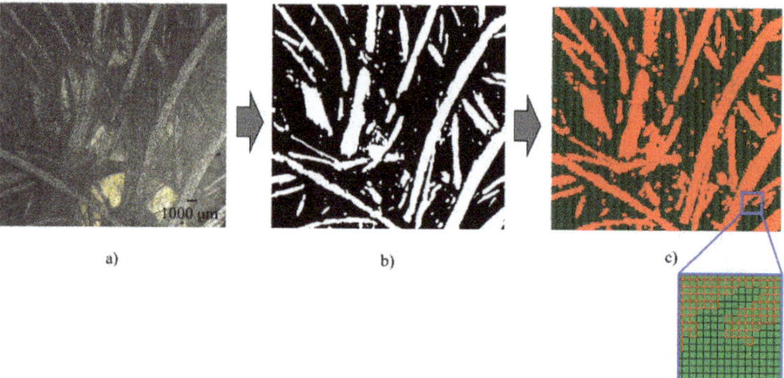

Figure 6. Image of representative microstructure and process analysis to assign a conforming mesh; (**a**) Original image collected through optical microscopy with a magnification of 100× showing the carbon fiber distribution from mixing Method 1, (**b**) Processed image after transferring to black and white and application of color threshold and blur and contrast filters the areal fiber fraction is ~33%, and (**c**) Conforming mesh to the microstructure based on color separation.

The next step was to assign the material properties to respective pixel groups. The properties of the material can be found in the 1st paragraph of the Materials and Methods section of this paper. Fibers and matrix are both modeled as elastic materials since they are examined within the linear elastic regime of the stress-strain curve. The simulation was performed on all samples collected for Method 1 set and Method 2 set by applying load in two principal directions (x axis and y axis considering the 2D plane of the field as a Cartesian coordinate system). If the mats are isotropic, they should yield comparable properties in both directions x and y as is hypothesized to result from the WL system.

The following boundary conditions, as seen in Figure 7, were used: (a) the displacement of the left edge was fully constrained (i.e., set to zero). The displacement of the lower left node in both x and y directions was also set to zero. A force in the x direction was applied to the right of the microstructure. (b) Displacement of the lower edge was set to zero. The displacement of the lower left node in both x and y directions was also set to zero. A force in the y direction was applied to the top of the microstructure.

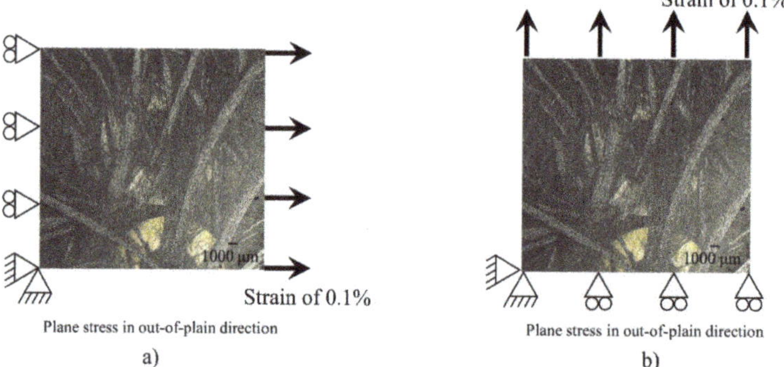

Figure 7. Boundary and loading condition for one of the original images collected through optical microscopy with a magnification of 100×. This sample areal fiber fraction is ~33%; (**a**) Left boundary fixed, lower left and bottom node fixed and constant force applied from the right hand side to cause a strain of 0.1% in the x axis direction in a Cartesian coordinate system, and (**b**) Lower boundary fixed, lower left and bottom node fixed and constant force applied from the top side to cause a strain of 0.1% in the y axis direction in a Cartesian coordinate system.

3. Results and Discussion

3.1. OOF Analysis Results for Isotropy Validation

The goal of this computational approach is to validate the isotropy of the WL mats. In general, the mats were assumed to be isotropic in nature as the WL system generates random orientation for the fibers in the mats. Isotropy is defined by having equal properties in multiple direction, in a way that the ratio of one property such as Young's modulus, measured in two orthogonal directions, should yield a value of 1. Table 1 displays the results of the FEA simulation and the ratio of Young's modulus calculated by the FEA analysis performed in ABAQUS 2018 – HF5 for the two principal directions of the 2D plane of the images. From these ratios, Method 2 samples are seen to have higher isotropy over those produced using Method 1. Figure 8 shows an example of the FEA simulation, where the fibers in the loading direction are highlighted in red as the primary stress bearing elements, while the fibers in normal direction to the load are indicated in the lower stress colder colors based on the Von-mises scale.

Table 1. Tensile Modulus in GPa as predicted by the OOF based FEA study.

Sample	Dir-X	Dir-Y	Ratio (x/y)	Sample	Dir-X	Dir-Y	Ratio (x/y)
M11	9.4	13.2	0.7	M21	15.0	16.5	0.9
M12	11.1	33.6	0.3	M22	9.2	11.8	0.8
M13	15.2	13.0	1.2	M23	9.5	8.4	1.1

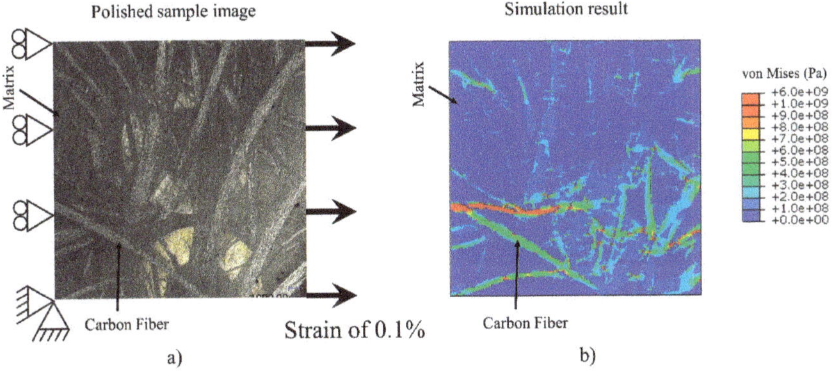

Figure 8. FEA simulation example showing (**a**) the original microstructure microscope image with the loading direction and (**b**) the fibers in the loading direction, bear the load and highlighted in warm colors. While the fibers that are perpendicular to the load direction remain in cold colors as they do not contribute to carrying the load.

3.2. Experimental Results

Table 2 summarizes the average tensile properties with respect to the weight fractions for Method 1 and Method 2 plates, respectively. 90% of the tensile samples failed in AGM (angled, gage, middle) mode and 10% in LGM (lateral, gage, middle). Table 3 summarizes normalized average tensile properties for each method in respect to unified weight fraction of 33%. The normalized values were calculated based on a linearized fit of the properties based on neighboring weight fraction. It must be noted that such fit will not be valid for difference of weight fraction above 5% between the considered samples for normalization. The linearized fitting equation is [33]:

$$Normalized\ value = test\ value * \frac{Chosen\ weight\ fraction}{weight\ fraction\ of\ test\ specimen} \tag{7}$$

Table 2. Experimental vs. theoretical tensile modulus (E) and ultimate tensile strength (σ) for individual plates. Where the nomenclature Mi Pj, indicates the Method number through i and the plate number through j.

Plate	W_f%	$E_{Experimental}$ (GPa)	STDE (GPa)	$E_{Theoretical}$ (GPa)	% Out of Theoretical	$\sigma_{Experimental}$ (MPa)	STDE (MPa)	$\sigma_{Theoretical}$ (MPa)	% Out of Theoretical
M1 P1	35	22	2	28	77	128	31	230	56
M1 P2	39	20	3	32	64	106	21	239	44
M1 P3	37	20	7	30	66	149	36	234	64
M2 P1	30	19	3	24	79	168	17	217	77
M2 P2	29	19	3	23	83	141	19	214	66
M2 P3	28	20	2	23	88	152	15	211	72

Table 3. Normalized average tensile properties for each method based on unified, normalized, fiber weight fraction of 33%.

Method	W_f% Average	$E_{Experimental}$ (GPa)	$E_{Theoretical}$ (GPa)	% Out of Theoretical	$\sigma_{Experimental}$ (MPa)	$\sigma_{Theoretical}$ (MPa)	% Out of Theoretical
M1	33	19	27	71	114	225	51
M2	33	22	27	83	173	225	77

Figure 9 compares the average experimental tensile Young's modulus of each plate to the theoretical modulus calculated using Halpin–Tsai equations [29] with respect to their fiber volume fraction. Since each set contained three plates, the average values presented in Figure 9 form a cluster distribution of data points. Analysis of the data cluster distribution presents higher control and repeatability of fiber weight content with composites made using Method 2 mats, with the three plates having 28%, 29% and 30% fiber volume ratios. The composites made with Method 1 mats had larger fiber weight content distribution at 35%, 39% and 37% for plates M1-P1, M1-P2, and M1-P3, respectively. This is attributed to the imbalance of fiber distribution in each layer, resulting in unequal fiber weight fractions in the various locations of the plates. This is reflected in the low standard deviation of 3% in Method 2 for fiber weight content, while for Method 1 the fiber weight content standard deviation was above 10%.

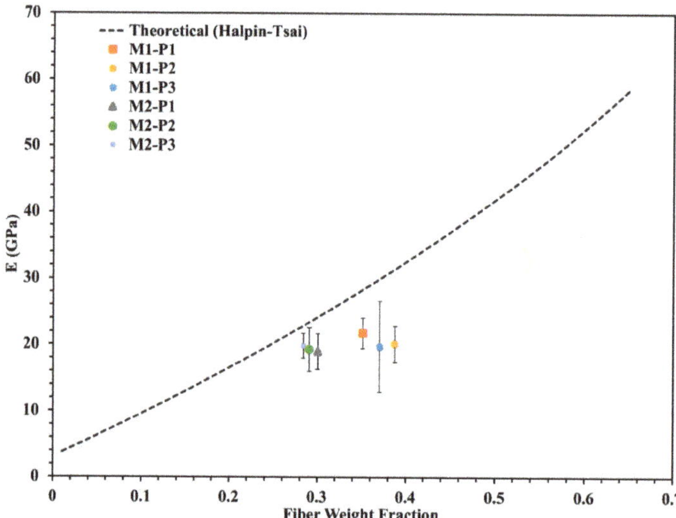

Figure 9. Average tensile modulus versus fiber weight fraction of each plate in both sets of Method 1 and Method 2. The blue line indicates the theoretical value calculated using the Halpin-Tsai equations.

The improvement of fiber distribution had direct correlation with an increase of Young's modulus experimental value obtained through Method 2 that was only 17% below the calculated theoretical value for a composite at the same fiber weight ratio. The imbalanced fiber distribution in Method 1 mats lead to average experimental values that are 32% below the calculated theoretical one for a composite of the same fiber weight content.

Figure 10 represents the tensile strength comparison between theoretical and experimental values for the plates of both sets. In Method 1, plates had a standard deviation of 21, 31 and 36 MPa (valued at 20%, 24.2% and 24%, respectively). Method 2 plates showed narrower standard deviation at 15, 17 and 19 MPa (valued at 10%, 14% and 9.8%). The cluster of the average tensile strength from samples of Method 1 showed a wider spread due to variability in weight content, and an average difference of 46% from theoretical value. The average cluster for Method 2 was closer to the theoretical value, with only a 28% difference. Such differences can be attributed to voids in the samples. The microstructure dependency plays an important role toward the deviation from theory (which assumes perfect isotropic material). Still, samples produced through Method 2 mats demonstrated a narrower standard deviation that did not surpass 10% between themselves, while the standard deviation for samples produced through Method 1 mats was higher than 20% when comparing between all three plates. These results were in accordance with the findings of the tensile modulus for individual samples collected from plates produced using mats from both methods.

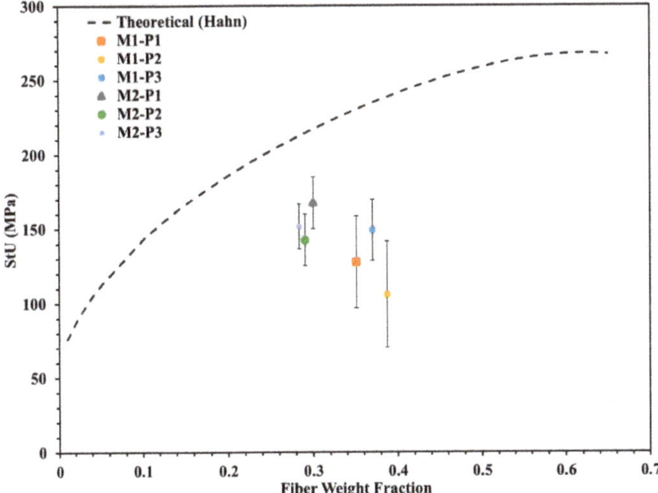

Figure 10. Average tensile strength vs. the fiber fraction weight of each of the plates from the sets of each method. The blue curve represents the theoretical value calculated based on the Hahn's equation.

Figure 11 shows the normalized tensile values with respect to a unified fiber weight fraction of 33% as presented in Table 3. This bar graph clarifies the improved performance of Method 2 over Method 1 by 16% in Young's modulus value and 52% in ultimate tensile strength. Optimal fiber distribution in Method 2 resulted in closing the gap with the theoretical values as well, with 10% improvement toward theoretical Young's modulus based on the Halpin–Tsai prediction, and a 26% improvement in tensile strength based on the Hahn's approach [32].

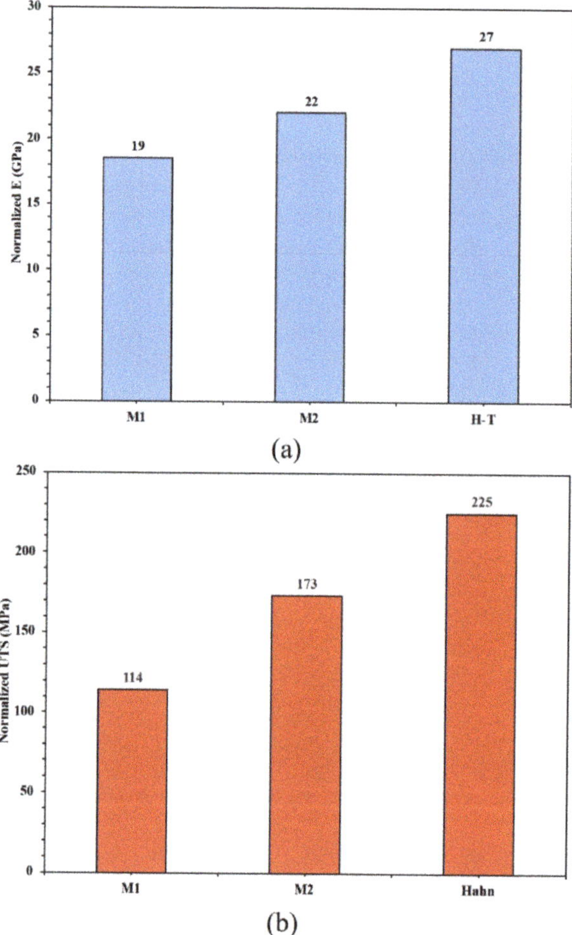

Figure 11. Normalized tensile properties for each method in respect to a unified fiber weight fraction of 33%. Method 2 shows significant improvement in performance over Method 1 both in; (**a**) Tensile modulus by 16%, and (**b**) Tensile strength by 52%.

Figure 12 shows SEM images of the fracture surface of the tensile samples. It is clear from Figure 12a that there is significant fiber grouping due to lack of proper dispersion in Method 1 and significant fiber pullout. On the other hand, Figure 12b presents a clean break surface with more fiber distribution due to innovation of fiber dispersion in Method 2.

Table 4 summarizes the average flexural properties, weight fractions, and void content for the Method 1 and Method 2 plates, respectively. Table 5 summarizes normalized flexural properties with respect to a unified fiber weight fraction of 33% for the plates.

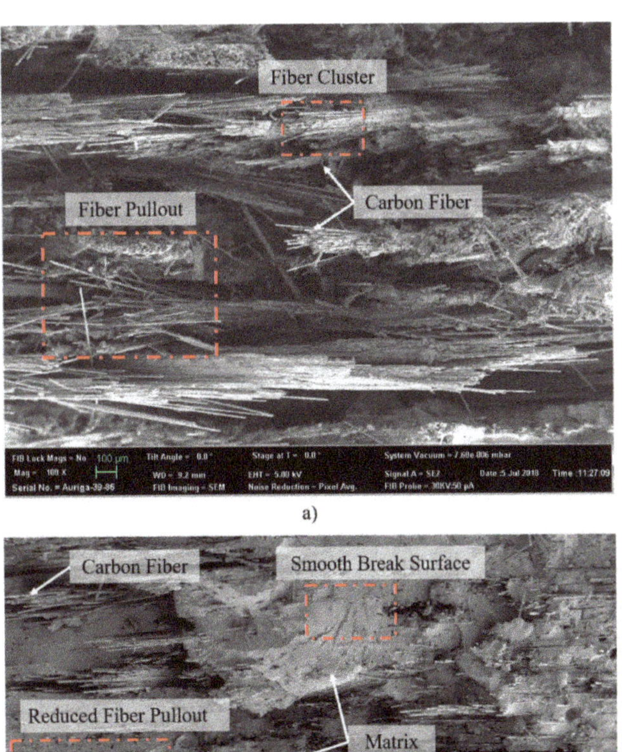

Figure 12. SEM images for selected break surface showing fiber random orientation and dispersion level for; (**a**) Method 1 sample break surface showing high density of bundle fibers, a sign of non-optimal dispersion, and (**b**) Method 2 sample break surface showing a clean area, a sign of optimal fiber dispersion.

Table 4. Experimental flexural results for the CF/Epoxy plates produced with WL-CF mats from Method 1 and Method 2 of nonwoven production.

Plate	$W_f\%$	Flex Modulus (GPa)	STDE (GPa)	STDE (%)	Flex Strength (MPa)	STDE (MPa)	STDE (%)
M1 P1	35	14	6	46	252	85	34
M1 P2	39	13	3	25	242	43	18
M1 P3	37	13	3	22	241	43	18
M2 P1	30	12	1	6	243	23	10
M2 P2	29	12	2	17	239	38	16
M2 P3	28	12	1	11	234	22	10

Table 5. Normalized average flexural values for the plates from each method at a unified fiber weight fraction of 33%.

Method	W_f%	Flex Modulus (GPa)	Flex Strength (MPa)
M1	33	12	219
M2	33	14	270

Figure 13 shows the variance of the flexural modulus and strength in each method and a comparison between the normalized results of both methods. The normalized data for a normalized fiber weight fraction of 33% shows a significant increase in flexural modulus and strength of 17% and 23%, respectively, in Method 2 over Method 1. The other significant difference lies in the standard deviation for the individual plates where Method 2 plates have a variance of 6%, 17% and 11% while Method 1 plates have a variance of 46%, 25% and 22%.

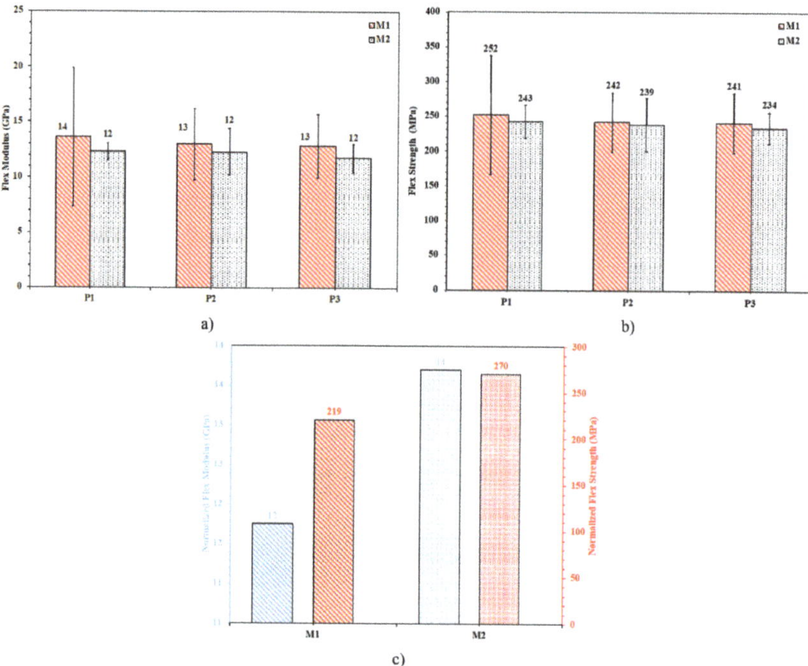

Figure 13. Flexural data for; (**a**) Flex modulus for Method 1 and Method 2 plates, (**b**) Flex strength for Method 1 and Method 2 plates, (**c**) Normalized average flex modulus and strength for Method 1 Set versus Method 2 set for a unified fiber weight fraction of 33%, the superiority of Method 2 is reflected by the positive slope of the trendline between the calculated values.

Table 6 and Figure 14 summarizes the average ILSS for each individual plate from both sets and Table 7 shows the average of each set. Figure 15 highlights the performance in ILSS for both methods, with a bar graph showing superiority of Method 2 over Method 1 by 15 MPa. It is hypothesized that with optimal filament dispersion, increased fiber surface area causes the improvement of the ILSS.

Table 6. Average ILSS of each individual plate from both sets.

Plate	$W_f\%$	ILSS (MPa)	STDE (MPa)	STDE%
M1-P1	35	16.44	5.62	34
M1-P2	39	21.39	3.46	16
M1-P3	37	23.24	5.63	24
M2-P1	30	31.92	2.41	8
M2-P2	29	30.87	2.05	7
M2-P3	28	25.91	1.58	6

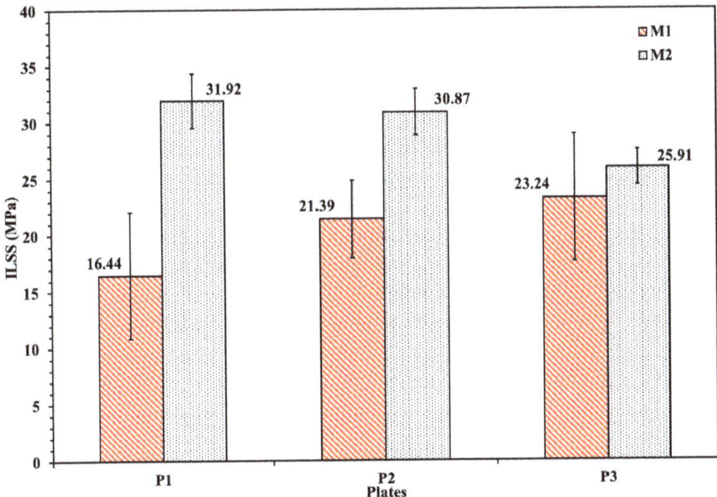

Figure 14. Average ILSS of each plate from both samples sets fabricated by Method 1 and Method 2; despite the superiority of fiber weight fraction in Method 1 plates, it is clear that Method 2 plates are outperforming Method 1 plates by values of 10%, to 90% with higher consistency in standard deviation.

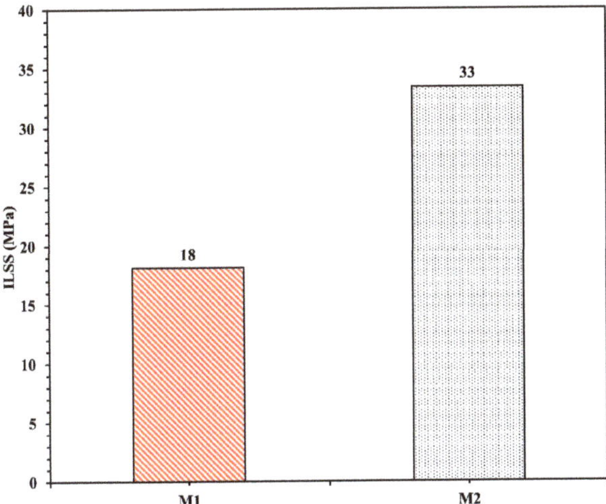

Figure 15. Normalized average ILSS of each set showing the improvement of 83% caused by optimal fiber dispersion in Method 2 when compared to Method 1.

Table 7. Normalized average ILSS values of each set for a unified fiber weight fraction of 33%.

Plates	$W_f\%$	Normalized ILSS (MPa)
M1	33	18
M2	33	33

4. Conclusions

Two sets of composites were made from WL-CF nonwoven mats and were consolidated and tested for tensile, flex and ILSS. The mats of each set were prepared through the two mixing regimes, Method 1 and Method 2 presented in [5]. The composites' isotropy was validated by computational method through OOF. OOF demonstrated higher isotropy in composites made through Method 2 versus those made through Method 1, thanks to the optimal fiber dispersion achieved in Method 2. The tensile properties were analyzed and compared against the Halpin–Tsai predictions for Young's modulus and the Hahn's approach for ultimate tensile strength. The optimal fiber dispersion improved the values of Young's modulus by 16% from Method 1 and more closely aligned with the Halpin–Tsai prediction by 10%. The effect of optimal fiber dispersion in Method 2 reflected as well on the flexural properties, which increase by 17% and 23% for modulus and strength, respectively. As for the ILSS, Method 2 showed a superiority of 83%, which may be attributed to the uniform fiber distribution. With these findings it can be concluded that use of WL CF nonwoven mats produced by the innovative Method 2 proposed in [5] should be adopted for composites production. These optimized mats helped bridge the gap between theoretical and experimental values, enabling the design and fabrication of complex geometry parts with long discontinuous fibers through low-cost manufacturing techniques, such as out of autoclave methods and VARTM.

Author Contributions: Conceptualization, H.G., A.A.H. and U.K.V.; methodology, H.G., A.A.H.; software, S.K., J.A.; validation, H.G., A.A.H. and S.K.; formal analysis, H.G.; investigation, H.G.; resources, U.K.V.; data curation, H.G.; writing—original draft preparation, H.G.; writing—review and editing, H.G., A.A.H., S.K., J.A. and U.K.V.; visualization, H.G. and A.A.H; supervision, U.K.V.; project administration, H.G. and U.K.V.; funding acquisition, U.K.V. All authors have read and agreed to the published version of the manuscript.

Funding: The research was funded in part by the Office of Energy Efficiency and Renewable Energy (EERE), U.S. Department of Energy, under Award Number DE-EE0006926. Support from the Institute for Advanced Composites Manufacturing Innovation (IACMI-The Composites Institute is gratefully acknowledged.

Acknowledgments: The authors want to acknowledge and thank Stephen Langer from the Mathematical Software Group, at the National Institute of Standards and Technology (NIST) for providing access to the OOF software. His generous help made this work possible. This research was supported by the DOE Office of Energy Efficiency and Renewable Energy (EERE), Advanced Manufacturing Office and used resources at the Manufacturing Demonstration Facility, a DOE-EERE User Facility at Oak Ridge National Laboratory.

Conflicts of Interest: The authors declare no conflict of interest.

References

1. Pervaiz, M.; Panthapulakkal, S.; KC, B.; Sain, M.; Tjong, J. Emerging Trends in Automotive Lightweighting through Novel Composite Materials. *Mater. Sci. Appl.* **2016**, *7*, 26–38. [CrossRef]
2. Knopp, A.; Scharr, G. Tensile Properties of Z-Pin Reinforced Laminates with Circumferentially Notched Z-Pins. *J. Compos. Sci.* **2020**, *4*, 78. [CrossRef]
3. Chung, D.D.L. Review: Materials for vibration damping. *J. Mater. Sci.* **2001**, *36*, 5733–5737. [CrossRef]
4. Akonda, M.H.; Lawrence, C.A.; Weager, B.M. Recycled carbon fibre-reinforced polypropylene thermoplastic composites. *Compos. Part A Appl. Sci. Manuf.* **2012**, *43*, 79–86. [CrossRef]
5. Ghossein, H.; Hassen, A.A.; Paquit, V.; Love, L.J.; Vaidya, U.K. Innovative Method for Enhancing Carbon Fibers Dispersion in Wet-Laid Nonwovens. *Mater. Today Commun.* **2018**, *17*, 100–108. [CrossRef]
6. Yeole, P.; Ning, H.; Hassen, A.A.; Vaidya, U.K. The Effect of Flocculent, Dispersants, and Binder on Wet-laid Process for Recycled Glass Fiber/PA6 Composite. *Polym. Polym. Compos.* **2018**, *26(3)*, 259–269. [CrossRef]
7. Lu, L.; Xing, D.; Xie, Y.; Teh, K.S.; Zhang, B.; Chen, S.M.; Tang, Y. Electrical conductivity investigation of a nonwoven fabric composed of carbon fibers and polypropylene/polyethylene core/sheath bicomponent fibers. *Mater. Des.* **2016**, *112*, 383–391. [CrossRef]
8. Deng, Q.; Li, X.; Zuo, J.; Ling, A.; Logan, B.E. Power generation using an activated carbon fiber felt cathode in an upflow microbial fuel cell. *J. Power Sources* **2010**, *195*, 1130–1135. [CrossRef]
9. Meng, C.; Liu, C.; Fan, S. Flexible carbon nanotube/polyaniline paper-like films and their enhanced electrochemical properties. *Electrochem. Commun.* **2009**, *11*, 186–189. [CrossRef]
10. Shen, L.; Wang, J.; Xu, G.; Li, H.; Dou, H.; Zhanga, X. NiCo$_2$S$_4$ nanosheets grown on nitrogen-doped carbon foams as an advanced electrode for supercapacitors. *Adv. Energy Mater.* **2015**, *5*, 2–8. [CrossRef]
11. Straumit, I.; Lomov, S.V.; Wevers, M. Quantification of the internal structure and automatic generation of voxel models of textile composites from X-ray computed tomography data. *Compos. Part A Appl. Sci. Manuf.* **2015**, *69*, 150–158. [CrossRef]
12. Wan, Y.; Takahashi, J. Tensile and compressive properties of chopped carbon fiber tapes reinforced thermoplastics with different fiber lengths and molding pressures. *Compos. Part A Appl. Sci. Manuf.* **2016**, *87*, 271–281. [CrossRef]
13. Wan, Y.; Straumit, I.; Takahashi, J.; Lomov, S.V. Micro-CT analysis of internal geometry of chopped carbon fiber tapes reinforced thermoplastics. *Compos. Part A Appl. Sci. Manuf.* **2016**, *91*, 211–221. [CrossRef]
14. Tseng, H.; Chang, R.; Hsu, C. Numerical prediction of fiber orientation and mechanical performance for short/long glass and carbon fiber-reinforced composites. *Compos. Sci. Technol.* **2017**, *144*, 51–56. [CrossRef]
15. Feraboli, P.; Kawakami, H.; Wade, B.; Gasco, F.; DeOto, L.; Masini, A. Recyclability and reutilization of carbon fiber fabric/epoxy composites. *J. Compos. Mater.* **2011**, *46*, 1459–1473. [CrossRef]
16. Caba, A.C.; Loos, A.C.; Batra, R.C. Fiber-fiber interactions in carbon mat thermoplastics. *Compos. Part A Appl. Sci. Manuf.* **2007**, *38*, 469–483. [CrossRef]
17. Evans, A.D.; Qian, C.C.; Turner, T.A.; Harper, L.; Warrior, N.A. Flow characteristics of carbon fibre moulding compounds. *Compos. Part A Appl. Sci. Manuf.* **2016**, *90*, 1–12. [CrossRef]
18. Selezneva, M.; Lessard, L. Characterization of mechanical properties of randomly oriented strand thermoplastic composites. *J. Compos. Mater.* **2016**, *50*, 2833–2851. [CrossRef]
19. Amaro, A.M.; Reis, P.N.B.; Santos, J.B.; Santos, M.J.; Neto, M.A. Effect of the electric current on the impact fatigue strength of CFRP composites. *Compos. Struct.* **2017**, *182*, 191–198. [CrossRef]
20. Thomason, J.L. The influence of fibre length and concentration on the properties of glass fibre reinforced polypropylene: Injection moulded long and short fibre PP. *Compos. Part A Appl. Sci. Manuf.* **2002**, *33*, 1641–1652. [CrossRef]
21. Langer, S.A.; Fuller, E.R.; Carter, W.C. OOF: An image-based finite-element analysis of material microstructures. *Comput. Sci. Eng.* **2001**, *3*, 15–23. [CrossRef]
22. Reid, A.C.E.; Lua, R.C.; García, R.E.; Coffman, V.R.; Langer, S.A. Modelling Microstructures with OOF2. *Int. J. Mater. Prod. Technol.* **2009**, *35*, 361. [CrossRef]
23. Dong, Y.; Bhattacharyya, D. Morphological-image analysis based numerical modelling of organoclay filled nanocomposites. *Mech. Adv. Mater. Struct.* **2010**, *17*, 534–541. [CrossRef]
24. Dong, Y.; Bhattacharyya, D.; Hunter, P.J. Characterisation and Object-Oriented Finite Element Modelling of Polypropylene/Organoclay Nanocomposites. *Key Eng. Mater.* **2007**, *334–335*, 841–844. [CrossRef]

25. Cannillo, V.; Esposito, L.; Pellicelli, G.; Sola, A.; Tucci, A. Steel particles–porcelain stoneware composite tiles: An advanced experimental–computational approach. *J. Eur. Ceram. Soc.* **2010**, *30*, 1775–1783. [CrossRef]
26. Bakshi, S.R.; Patel, R.R.; Agarwal, A. Thermal conductivity of carbon nanotube reinforced aluminum composites: A multi-scale study using object oriented finite element method. *Comput. Mater. Sci.* **2010**, *50*, 419–428. [CrossRef]
27. Coffman, V.R.; Reid, A.C.E.; Langer, S.A.; Dogan, G. OOF3D: An image-based finite element solver for materials science. *Math. Comput. Simul.* **2012**, *82*, 2951–2961. [CrossRef]
28. Goel, A.; Chawla, K.K.; Vaidya, U.K.; Chawla, N.; Koopman, M. Two-dimensional microstructure based modelling of Young's modulus of long fibre thermoplastic composite. *Mater. Sci. Technol.* **2008**, *24*, 864–869. [CrossRef]
29. Affdl, J.C.H.; Kardos, J.L. The Halpin-Tsai equations: A review. *Polym. Eng. Sci.* **1976**, *16*, 344–352. [CrossRef]
30. Shokrieh, M.M.; Moshrefzadeh-Sani, H. On the constant parameters of Halpin-Tsai equation. *Polymer* **2016**, *106*, 14–20. [CrossRef]
31. Mallick, P.K. *Fiber-Reinforced Composites: Materials, Manufacturing, and Design*, 3rd ed.; CRC Taylor & Francis Group: Philadelphia, PA, USA, 2007.
32. Hahn, H.T. On Approximations for Strength of Random Fiber Composites. *J. Compos. Mater.* **1975**, *9*, 316–326. [CrossRef]
33. DOF. *Composite Materials Handbook-MIL 17*, 1st ed.; Taylor & Francis: New York, NY, USA, 1999. [CrossRef]

Journal of
Composites Science

Article

Development of Recyclable and High-Performance In Situ Hybrid TLCP/Glass Fiber Composites

Tianran Chen [1,*], Dana Kazerooni [1], Lin Ju [1], David A. Okonski [2] and Donald G. Baird [1,3]

[1] Macromolecules Innovation Institute, Virginia Tech, Blacksburg, VA 24061, USA; dk4z@vt.edu (D.K.); linj7@vt.edu (L.J.); dbaird@vt.edu (D.G.B.)
[2] GM Global Research & Development Center, Warren, MI 48092, USA; david.a.okonski@gm.com
[3] Department of Chemical Engineering, Virginia Tech, Blacksburg, VA 24061, USA
* Correspondence: tianran@vt.edu

Received: 26 July 2020; Accepted: 19 August 2020; Published: 24 August 2020

Abstract: By combining the concepts of in situ thermotropic liquid crystalline polymer (TLCP) composites and conventional fiber composites, a recyclable and high-performance in situ hybrid polypropylene-based composite was successfully developed. The recycled hybrid composite was prepared by injection molding and grinding processes. Rheological and thermal analyses were utilized to optimize the processing temperature of the injection molding process to reduce the melt viscosity and minimize the degradation of polypropylene. The ideal temperature for blending the hybrid composite was found to be 305 °C. The influence of mechanical recycling on the different combinations of TLCP and glass fiber composites was analyzed. When the weight fraction ratio of TLCP to glass fiber was 2 to 1, the hybrid composite exhibited better processability, improved tensile performance, lower mechanical anisotropy, and greater recyclability compared to the polypropylene reinforced by either glass fiber or TLCP alone.

Keywords: recycling; hybrid composites; polymer-matrix composites (PMCs); thermotropic liquid crystalline polymer; glass fibers

---

1. Introduction

The early constructions of vehicles like automobiles, locomotives, and aircraft were designed using dense metals with high strength capabilities. However, recent advancements in material science have enabled fiber-reinforced composites to replace traditional metals because of higher strength-to-weight ratios [1–3]. For instance, aluminum has traditionally been one of the most common metals used in the aerospace industry, but its usage dropped from 50% in the Boeing 777 aircraft to only 20% in the Boeing 787 [1]. The advantages that fiber reinforced composite materials have over traditional metal materials include: (1) light weight, (2) high stiffness and strength, (3) corrosion resistance, and (4) design flexibility. These attributes have been embraced by the automotive industry, which has increased its use of fiber-reinforced composite materials to improve fuel efficiency and reduce greenhouse gas emissions.

Among all of the different types of reinforcements utilized on a commercial scale, 65% of the revenue generated by the sale of fiber reinforced materials comes from glass fiber [4]. In 2020, the global glass fiber reinforced composite market is expected to grow to about 60 billion dollars [5]. Glass fiber is especially attractive as a reinforcement for composites because of its low cost, superior mechanical and physical properties (e.g., stiffness and strength, impact resistance, stability, and durability). The tensile modulus and the strength of E-glass fiber are around 72 GPa and 3.5 GPa, respectively. This outperforms aluminum with a tensile modulus of 68.9 GPa and tensile strength of 310 MPa [6,7].

Thermotropic liquid crystalline polymers (TLCPs) are another type of reinforcement that is being extensively studied and used in both academia and industry [8,9]. Tremendous efforts have been made toward the development of TLCPs that exhibit high modulus and strength coupled with outstanding melt processability [8,10–12]. The drawn TLCP filaments display a modulus of up to 100 GPa and tensile strength of about 1.5 GPa, which is comparable to the properties of E-glass fiber [13].

Both glass fiber and TLCP have excellent mechanical performance, high strength-to-weight ratio, and chemical resistance, but glass fiber is still a more attractive reinforcement choice over TLCP for three major reasons. First, TLCPs are more expensive than glass fiber. Depending on the grade of TLCP, the cost may range from eight to 12 dollars per pound [14]. Second, glass fiber has a higher tensile strength than TLCPs, especially when the TLCP is not generated using the fiber spinning or strand extrusion process. The full reinforcing potential of the TLCP fiber cannot be achieved under other processing techniques. Finally, glass fiber reinforced composites have lower mechanical anisotropy than their TLCP-filled counterparts. This is primarily due to the TLCP fibrils being created in situ under strong unidirectional elongation and shear flow [15].

One of the advantages of using TLCPs in reinforcing thermoplastic materials is the processability. It is well known that the incorporation of glass fibers into thermoplastics results in a substantial increase in viscosity, which gives rise to the difficulty in processing and high energy consumption [8]. During the processing of a TLCP composite melt, rigid chain TLCP molecules adopt highly oriented states relative to the partially oriented flexible chain molecules displayed by conventional thermoplastics [16]. Therefore, the melt viscosity of TLCP reinforced composites become much lower than that of glass-filled composites, leading to more facile processing. In addition, the surface smoothness of TLCP reinforced composites is greater than that of composites reinforced by glass fiber only [17]. The higher surface smoothness is related to the diameter of TLCP fibrils, which is one order of magnitude less than glass fiber. Additionally, the density of TLCP is around 1.4 g/cm^3, which is about half the density of E-glass fiber (2.58 g/cm^3) [7,18]. The composite parts that utilize TLCP exhibit a lower weight than those of glass fiber, making the TLCP composite an attractive material, specifically for transportation applications. Finally, the recyclability of the TLCP composite has been found to be superior to glass-filled systems [19]. TLCP is able to regenerate a highly oriented molecular structure during the reprocessing while glass fiber would suffer severe fiber breakage during the recycling process.

To capitalize on the advantages of using TLCPs and glass fibers, composites consisting of both TLCPs and glass fiber as reinforcements in thermoplastics have been studied [20–25]. The in situ hybrid composite consists of three components: microscopic TLCP fibrils, macroscopic conventional fibers (e.g., glass and carbon fiber), and the matrix polymer [25]. Bafna et al. [23] reported the use of glass fiber and TLCP reinforcements to enhance the mechanical properties and reduce the mechanical anisotropy of in situ TLCP composites with polyetherimide (PEI) as the matrix material. Tensile and flexural moduli increased and the anisotropy reduced with increasing glass fiber content. Furthermore, the creep performance of PEI composites improved when TLCP and glass fiber were blended together. Another study looked to combine the advantages of short fiber composites and the TLCP composite. He et al. [25] investigated the mechanical, rheological, and morphological properties of hybrid in situ carbon fiber or glass fiber/TLCP composite systems. Improvement in tensile and flexural properties, lower melt viscosity, and more oriented fibers in the flow direction have been observed.

Although composite materials have a variety of advantages, one of the major challenges for fiber reinforced composites is their recyclability. The disposal of composite waste in an environmentally friendly manner is a crucial task to our society. Typically, fiber-reinforced composite materials are very difficult and energy intensive to recycle due to the nature of heterogeneity, technology limitations, high recycling cost, and low quality of recycled products. More restrictive environmental legislation drives the market toward recycling and reusing fiber reinforced composites. There are three major recycling methods to reclaim fiber reinforced composites: (1) thermal process, (2) solvolysis, and (3) mechanical recycling [26–29]. Mechanical recycling has less environmental impact, can recover both fiber and matrix polymer, and requires no use of solvents compared to thermal and solvent methods [30]. Mechanical recycling uses the principles of shredding or crushing the composite part into small particulates and then feeding these into a manufacturing machine to produce recycled parts. The recycled composites have very limited applications. Usually, the recycled composites, acting as "filler", are blended with virgin materials to make a product with similar performance as new "virgin" parts. The incorporation level of the recycled composites is usually no more than 10 wt % to minimize the negative impact from the recycled materials [30].

Mechanical recycling is considered environmentally friendly and cost-effective. However, the application of this method is hindered because the recycling process reduces the performance of subsequent composite parts. Extensive investigations have been carried out on the influence of mechanical recycling on the properties of glass or carbon fiber reinforced composites [31–34]. The mechanical properties of fiber reinforced composites decreased significantly after mechanical recycling, which is mainly due to fiber attrition during the recycling process. The need for developing a recyclable and high-performance composite is becoming extremely urgent.

Previous work has explored the effect of mechanical recycling on the properties of TLCP and glass fiber reinforced polypropylene [19]. The results illustrated that the TLCP composite had superior recyclability relative to that of its glass fiber reinforced counterpart. It is of great interest to determine whether there exists a formulation of TLCP, glass fiber, and matrix polymer that may be mechanically recycled without compromising the mechanical performance of the subsequent composite part. The objective of this work was to utilize glass fiber and TLCP to develop a recyclable and high-performance in situ polypropylene-based hybrid composite. The in situ hybrid composites were mechanically recycled once by injection molding and grinding processes. The processing temperature was optimized through rheological analyses to improve the processability and reduce the thermal degradation of polypropylene. The optimal formulation of glass fiber and TLCP enables the best combination of high recyclability, high mechanical properties, and low mechanical anisotropy of the hybrid composite.

2. Materials and Methods

2.1. Materials

The thermotropic liquid crystalline polymer (TLCP) used in this study, trade name Vectra B950 and made by Celanese (Florence, KY, USA), is an aromatic poly(ester-co-amide) composed of 6-hydroxy-2-naphthoic acid (60 mol%), terephthalic acid (20 mol%), and aminophenol (20 mol%). The melting point of Vectra B950 is around 280 °C [35]. The long glass fiber (GF) reinforced polypropylene was provided by SABIC (Ottawa, IL, USA) as 8.0 mm long pellets with a reinforcement loading of 50 wt %. The diameter of the glass fiber is around 17 μm. The unfilled matrix polypropylene (PP), catalog name Pro-fax 6523, was purchased from LyondellBasell (Houston, TX, USA) and has a melt flow rate of 4.0 g/10 min at 230 °C [36].

2.2. Melt Compounding and Recycling of Hybrid Composites

There were four combinations of the glass fiber and TLCP composite used in this study. These composites were designated as 30GF/PP, 20GF10TLCP/PP, 10GF20TLCP/PP, and 30TLCP/PP. The number before the component represents the weight percent of that particular reinforcement in the composite. For instance, 20GF10TLCP/PP means that the composite consists of 20 wt % glass fiber, 10 wt % TLCP, and 70 wt % polypropylene. The weight fraction of reinforcement in each composite was kept at 30 wt %. To prepare the hybrid composites, the materials (e.g., TLCP, 50 wt % GF/PP, and PP) were dried in a vacuum oven at 80 °C for 24 h. A single screw extruder with a 1-inch diameter screw and L/D ratio of 24 was used for compounding the TLCP and PP at 290 °C. The extrudate was cooled down in a water bath and pelletized. The TLCP/PP and GF/PP pellets were injection molded (BOY 35E) into end-gate plaques to form the pristine hybrid composite. The pristine hybrid composites were shredded by a granulator (Cumberland/John Brown D-99050). The recycled hybrid composites were prepared by injection molding the shredded materials. The injection molding temperature was optimized at 305 °C based on rheological analyses.

2.3. Rheological Measurements of Polypropylene, TLCP, and Hybrid Composite

An ARES-G2 rheometer with 25 mm parallel plates was used to investigate the viscoelastic properties of TLCP at four different experimental temperatures: 290, 300, 305, and 310 °C. All work was completed under a nitrogen atmosphere unless expressly stated as otherwise. Each sample was equilibrated to the experimental temperature for 5 min. First, pure TLCP pellets were loaded directly into the rheometer and then tested using a shear step strain at a strain of 0.5% [37]. The relaxation modulus (G(t)) at each temperature was plotted against time. Further experimentation on the pure TLCP pellets included running a small amplitude oscillatory shear (SAOS) frequency sweep per temperature to obtain the complex viscosity, storage modulus, and loss modulus. Next, virgin PP material was first tested using SAOS rheology under time sweep mode at each temperature. The complex viscosity of PP was tracked as a function of time per temperature. Finally, all the compositions of recycled GF/TLCP/PP hybrid materials were used to measure the complex viscosity for each composition using the SAOS frequency sweep at 305 °C.

2.4. Mechanical Properties

Rectangular bars with dimensions of $75.0 \times 8.0 \times 1.5$ mm were cut from the injection molded end-gated plaques of pristine and recycled hybrid composites in the flow direction and perpendicular to flow direction (transverse direction). For the flow direction, three strips were cut from one plaque from one edge of the plaque to the middle. The tensile properties of this plaque in the flow direction were the average properties of the three rectangular bars. At least five plaques for each composition were tested to obtain the average tensile properties. For the transverse direction, one strip was cut from the middle of the plaque and the tensile properties of each hybrid composite were the average properties of at least five plaques. All tensile properties of each material were measured using an Instron uniaxial tensile tester (Model 4204) with a 5 kN load cell. The tensile strains of the specimen were measured with an extensometer (MTS 634.12). The cross-head speed was set at 1.27 mm/min.

2.5. Differential Scanning Calorimetry (DSC) Characterization

A DSC (TA Instruments Discovery) was used to examine the thermal properties of TLCP. Under a nitrogen atmosphere, a sample was subjected to a heat/cool/heat cycle. The material was first equilibrated at 50 °C for 5 min and then heated up to 320 °C at 10 °C/min. The materials were cooled down to 50 °C at -10 °C/min and then heated back to 320 °C at 10 °C/min. The melting temperature (T_m) was determined by the TA Instruments TRIOS software.

3. Results and Discussion

3.1. Optimization of the Injection Molding Temperature

To generate the in situ TLCP/glass fiber/polypropylene composites, the processing temperature has to be optimized using a series of rheological analyses. One advantage of blending TLCPs with other thermoplastics is the reduction in melt viscosity, which results in an improvement of the polymer blend's processability. TLCPs have lower melt viscosity when compared to many thermoplastics, and to take advantage of this, thermal and rheological properties of TLCPs need to be examined. The TLCP used in this study has a melting point around 280 °C [35]. The rheological properties of TLCP above its melting point are critical in the processing of this material. Transient behaviors of TLCPs are related to the rigid TLCP molecules and domain structure [37,38]. The transient responses of pure TLCP at various test temperatures, following step strain were measured, as shown in Figure 1. The relaxation modulus (G) is plotted against time on the log-log scale. At the testing temperatures below 305 °C, the relaxation modulus decreases gradually over time, producing a long relaxation tail. This is graphically demonstrated by the relaxing curves at 290 and 300 °C. On the other hand, the relaxation modulus (G) sharply dropped when the temperature was equal and above 305 °C. It is speculated that the slow decay of the relaxation modulus at low temperature is due to the presence of unmelted crystals in the TLCP [37]. As the temperature further increased, TLCP crystals change to a liquid phase, causing the long relaxation tail to disappear, as seen by a dramatic drop in the relaxation modulus at the temperatures of 305 and 310 °C.

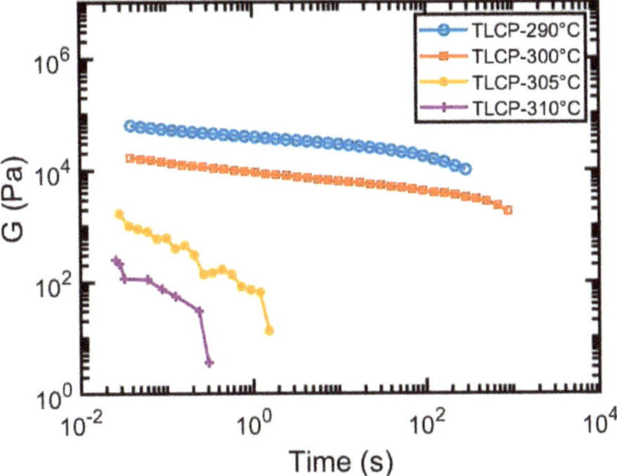

Figure 1. Stress relaxation of pure TLCP following a step shear strain at different test temperatures.

The complex viscosity ($|\eta^*|$) of pure TLCP was measured using a rheometer in the frequency sweep mode. Figure 2 illustrates the complex viscosity of TLCP as a function of frequency for each experimental temperature. At 290 and 300 °C, the $|\eta^*|$ is more than an order of magnitude higher than the complex viscosity of TLCP at 305 and 310 °C, suggesting the TLCP has greater resistance to flow at lower temperatures. This further reinforces the idea that at low temperatures, there may be unmelted TLCP crystals. The TLCP melt exhibits a weak dependence between the complex viscosity and frequency at a temperature of 305 °C and above. On the other hand, the relationship between complex viscosity and frequency at 290 and 300 °C unveils a stronger correlation with the change of angular frequency. TLCPs usually exhibit three distinct regions of shear viscosity in which the first reflects a shear thinning behavior at low shear rates, followed by a Newtonian plateau, and eventually finishes with another shear thinning region at high shear rates [39]. In between the frequency range of 1–500 rad/s, the Newtonian region was not seen for the temperature at 290 and 300 °C, instead, the TLCP showed shear thinning behavior over the entire frequency sweep range. Only the Newtonian plateau and first shear thinning region were observed at 305 °C and 310 °C, which is probably due to the angular frequency not being low enough. Since the complex viscosities of TLCP at 305 and 310 °C almost overlap with each other, the viscosity of TLCP is not very sensitive to the changes in temperature in this temperature region. As the lowest complex viscosity of TLCP at all frequencies was exhibited at the temperature of 305 and 310 °C, the ideal processing temperature of TLCP blends would have to be equal to or higher than 305 °C to reduce the viscosity during processing.

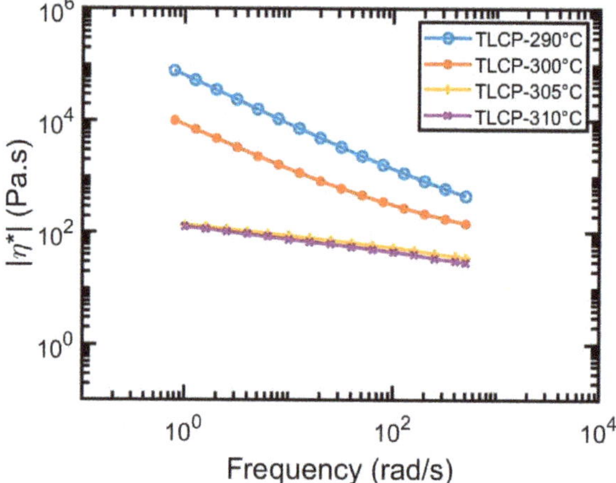

Figure 2. Complex viscosity of TLCP at temperatures from 290 to 310 °C.

The viscoelastic properties of pure TLCP at different temperatures were studied using oscillatory shear rheology. Figure 3 illustrates the change of storage (G′) and loss (G″) moduli of TLCP as the function of angular frequency at varying operational temperatures. Storage modulus and loss modulus measure the stored energy and energy dissipated as heat, respectively. At 290 °C, the storage modulus was higher than the loss modulus over the entire frequency range, suggesting that the TLCP exhibits a solid-like character. After a 10 °C increase, the curves for G′ and G″ crossover at 200 rad/s. The crossover frequency characterizes the transition of a polymer melt from the elastic (solid-like) to the viscous (liquid-like) state. At 305 °C and above, the G″ is higher than G′ at all angular frequencies, indicating that the TLCP behaves like a viscous liquid. The G″ > G′ over the entire frequency range is probably due to the complete melt of TLCP crystals. The transitioning effect observed from the dramatic change of viscoelastic properties of TLCP with an increase in the testing temperature reinforces the notion that the existence of TLCP crystals significantly impacts the rheological behavior of TLCP.

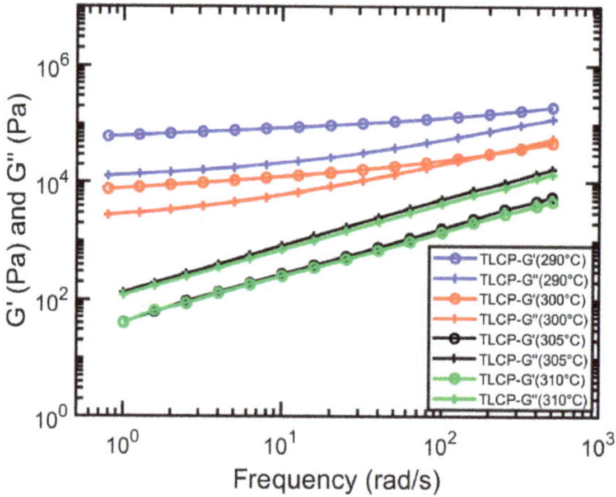

Figure 3. Storage modulus and loss modulus of TLCP at different temperatures above its melting point.

A differential scanning calorimeter (DSC) was utilized to study the melting behavior of pure TLCP and to confirm any residual crystals of TLCP at temperatures above 280 °C. Two distinct transition phases were observed during the melting of TLCP, as shown by the two peaks in the DSC heating scan in Figure 4. The first peak was observed at around 280 °C and the second peak was observed around 297 °C. The different crystal structures develop during the process, leading to the mesophase transition of TLCP [40]. Literature reports state that the melting point of TLCP is around 280 °C [35]. Several studies have successfully melt processed the PP with TLCP between 285 to 300 °C, resulting in improved mechanical performance [35,41,42]. Nevertheless, to utilize the low viscosity effect of TLCP, the processing temperature must be high enough to melt all the TLCP crystals. Thus, the melting endotherm ends around 305 °C, which confirms that the TLCP crystals have completely melted and shows consistency with the results obtained from the rheological analyses.

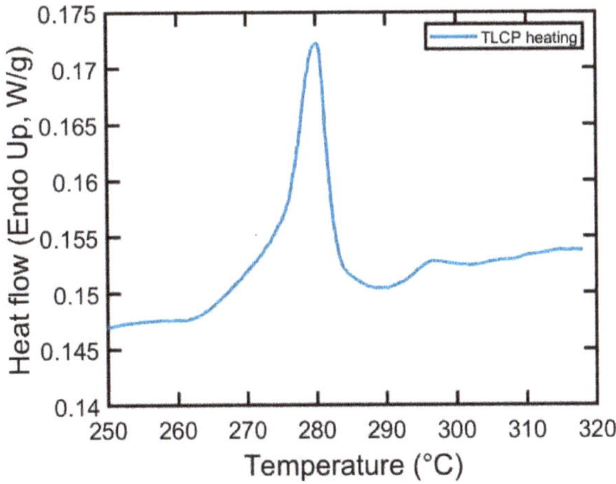

Figure 4. DSC heating scan of the TLCP material with the heating rate of 10 °C/min.

Before processing the in situ hybrid GF/TLCP/PP composite, the PP used as the matrix of the composite needs to be tested to make sure it can withstand the high temperatures for processing. The high processing temperature improves the processability of the polymer blend, but also raises a concern regarding the thermal degradation of the polypropylene. The thermal stability of polypropylene was measured using the isothermal time sweep rheological test. Figure 5 shows the complex viscosity ($|\eta^*|$) of polypropylene as a function of time at different experimental temperatures. The isothermal rheological tests were carried out under a nitrogen atmosphere to simulate the nitrogen purge hopper used during the injection molding process. The overall residence time exposed to high temperature for processing was around 240 s.

During this time, the viscosities of polypropylene decreased by about 14.6% at 290 °C and 42.9% at 310 °C. Polypropylene undergoes severe thermal degradation at 310 °C. The degradation rate of polypropylene at 300 °C and 305 °C in the nitrogen environment was much lower than at 310 °C, where the complex viscosity dropped around 26.7% at 305 °C (Figure 5). Therefore, to achieve better processability and reduce the thermal degradation of the polypropylene, 305 °C was selected as the processing temperature for injection molding of the hybrid composite materials.

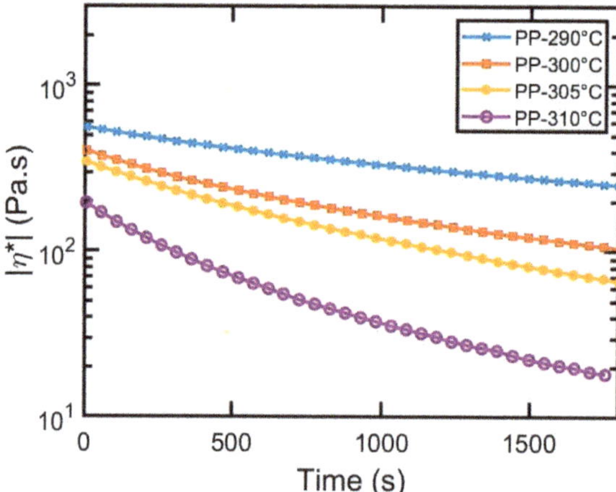

Figure 5. Thermal stability of polypropylene at various temperatures (isothermal time sweep mode).

3.2. Rheology of the In Situ TLCP/GF Hybrid Composites

To characterize the rheological properties of the recycled hybrid composite, small amplitude oscillatory shear (SAOS) frequency sweep tests were carried out at 305 °C. The complex viscosity was plotted against the angular frequency for each formation of the recycled hybrid composite, as indicated in Figure 6. Hybrid composites with varying TLCP and glass fiber compositions displayed dramatic differences in rheological behaviors. At low frequencies, the viscosity increased with increasing glass fiber concentration. When glass fiber was used as the sole reinforcement in the composite, the viscosity jumped two-fold compared to its TLCP counterpart at 305 °C. At low frequencies, the complex viscosity plateau was not observed for the glass-filled hybrid composite, but rather the complex viscosity appreciably increased with the decrease in angular frequency. This response of the composite to low frequency was due to the fiber–fiber and glass–fiber matrix interactions in glass-filled systems [43]. As frequency increases, these interactions are interrupted, and the complex viscosity of glass-filled composite decreases dramatically. On the other hand, TLCP/PP showed overall weak dependence between complex viscosity and angular frequency. The incorporation of glass fibers into the TLCP/PP composite significantly increased the complex viscosity of the hybrid composite. Especially in the low angular frequency, the complex viscosity of the hybrid composite increased exponentially with the addition of glass fiber. The hybrid composite with a higher concentration of TLCP showed significant reduction in viscosity, thereby confirming that TLCP reduced the viscosity of polymer blends [8]. Not only did the TLCP increase the processability of the GF/PP composite by lowering the melt viscosity, but it also mitigated the fiber breakage issue by reducing stresses acting on the fibers by the matrix polymer. In turn, diminishing fiber breakage leads to higher mechanical performance of the composites [44–46].

One concern regarding this rheological test is the thermal degradation of polypropylene at 305 °C, which decreases the complex viscosity over time. This may cause the ill-defined viscosity data at high frequency (rheological test sweeps the frequency from low to high), but this frequency sweep test provides a method to qualify the benefit of a higher TLCP concentration, resulting in the lower viscosity of a hybrid composite.

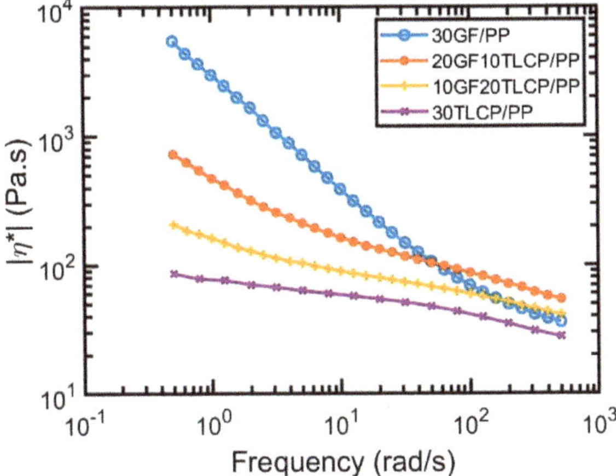

Figure 6. SAOS frequency sweep of hybrid composites at 305 °C in a nitrogen atmosphere.

3.3. Mechanical Properties of the Recycled TLCP/GF Hybrid Composite

The first injection molded samples were defined as pristine materials (blue). The recycled composites (red) were obtained by grinding the pristine materials and re-injection molding into end-gated plaques. The tensile properties of each hybrid composite are depicted in Figures 7–9. The modulus of pristine 30 wt % GF/PP was 5.07 GPa, and the modulus of pristine TLCP/PP was about 3.83 GPa. The lower modulus of the TLCP blend was speculated to have originated from the injection molding process, which is unable to achieve high elongational deformation. The tensile modulus of 26 wt % TLCP/PP is reported to be 13.5 GPa when the sample is prepared by strand extrusion [47]. By replacing 10 wt % of the TLCP with glass fiber, the tensile modulus improved by 7.4%. The tensile modulus of the hybrid composite fell between the 30 wt % GF/PP and TLCP/PP. To determine the effect of mechanical recycling on the hybrid composite materials, the pristine composites were ground and injection molded into end-gated plaques. After recycling, the tensile modulus of the GF/PP composite dropped to 4.87 GPa while the tensile modulus of TLCP/PP remained the same. The major factor for the decrease in the GF/PP composite modulus was the fiber length attrition during the recycling process [34]. The TLCP was able to regenerate the TLCP fibrils during the mold filling process, and the TLCP-filled composites maintained their mechanical performance after mechanical recycling. In addition, the recyclability of a glass-filled composite was improved with the presence of TLCP, as seen with the 20GF10TLCP/PP hybrid composite only losing 4.1% of its modulus after recycling.

The tensile strengths of the pristine and recycled hybrid composites are shown in Figure 8. The tensile strength of pristine GF/PP was much higher than that of TLCP/PP because the tensile strength of the TLCP (~0.5 GPa) was lower than the tensile strength of GF (~3.5 GPa). The same difference in the properties of the composites was also expected to be seen [47,48]. By adding glass fiber into the TLCP blend, the tensile strength was significantly enhanced. Replacing 10 wt % TLCP with glass fiber enabled the improvement of the tensile strength from 36.8 to 50.0 MPa. Mechanical recycling imposes different degrees of impact on the tensile strength of hybrid composites. The tensile strength of 30 wt % GF/PP dropped from 71.5 to 61.7 MPa after recycling, but there was only a 9.6% decrease in the tensile strength of 20GF10TLCP/PP. Just as for the tensile modulus, this was due to the severe fiber length attrition during the blending and grinding processes. The higher content of glass fiber and higher melt viscosity accelerated the fiber breakage during processing. The short glass fiber composite showed lower mechanical properties than the long glass fiber composite [46]. The higher content of TLCP in the hybrid composite lessened the decrease in the mechanical properties of the composites after recycling.

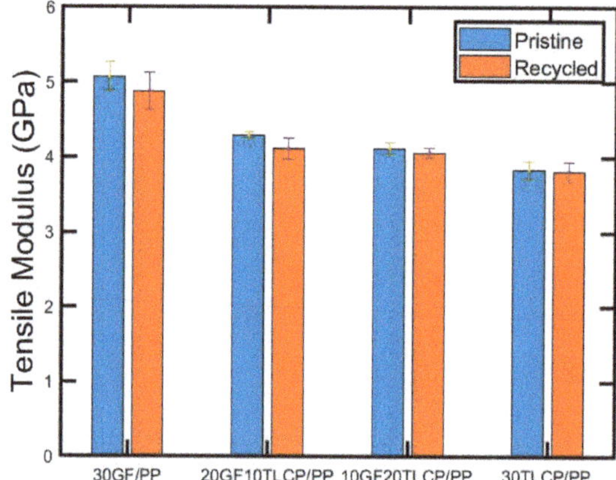

Figure 7. Tensile modulus of pristine and recycled hybrid composites in the flow direction.

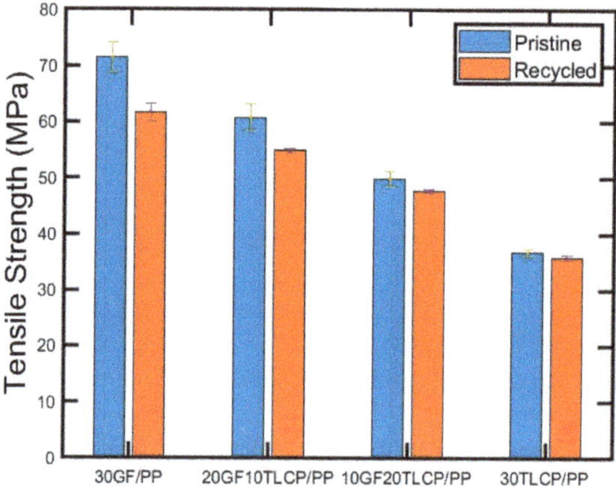

Figure 8. Tensile strength of pristine and recycled hybrid composites in the flow direction.

The mechanical properties of the hybrid composites in the transverse-to-flow direction were measured and are shown in Figure 9. The rectangular bar was cut from injection-molded, end-gated plaques in the direction perpendicular-to-flow in order to obtain the performance of the hybrid composites in the transverse-to-flow direction. Similar to the mechanical behaviors in the flow direction, the tensile properties of TCLP/PP were lower than the GF/PP due to the combining effect of different mechanical properties of the pure materials used and processing methods. The addition of 10 wt % glass fiber improved the tensile modulus of TLCP/PP from 1.8 to 2.5 GPa. In terms of recyclability, the large decrease in the mechanical properties of GF/PP after recycling can be seen in Figure 9. The extent of the decrease in the mechanical properties of the composites after recycling in the transverse-to-flow direction was mitigated by replacing glass fiber with TLCP. The tensile strength of 10GF20TLCP/PP only decreased from 21.4 to 20.2 MPa after mechanical recycling, while the 30 wt % TLCP/PP composite exhibited little or no change in the tensile modulus and strength in the transverse-to-flow direction. Hybrid composite formulations provide the balance between recyclability and mechanical performance.

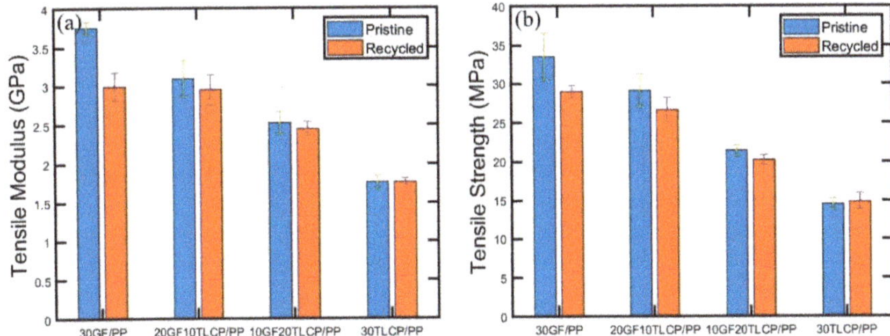

Figure 9. Tensile properties of hybrid composites in the transverse-to-flow direction. (**a**) Tensile modulus; (**b**) tensile strength.

3.4. Mechanical Anisotropy of the Hybrid TLCP/GF Composite

Mechanical anisotropy is an important parameter to evaluate the performance of the composite material. Table 1 illustrates the tensile modulus of the composites in the flow and transverse-to-flow directions as well as the mechanical anisotropy. Mechanical anisotropy is the ratio of the property in the flow direction over the property in the transverse-to-flow direction. Several studies have been published about the mechanical anisotropy of injection-molded composites [23,49,50].

TLCP composites have significantly higher mechanical anisotropy than their glass-filled counterparts, which is caused by TLCP fibrils being generated in the direction of elongational flow developed at the melt front. The transverse-to-flow direction has less TLCP fibrils than the flow direction. With a higher content of TLCP in the blend, the degrees of mechanical anisotropy increased rapidly [50]. This is one of the major limitations when using in situ TLCP composites. For glass-filled composites, the glass fibers can be aligned in both flow and transverse-to-flow directions through flow kinematics during the mold filling process, leading to less mechanical anisotropy. As indicated in Table 1, the modulus of 30 wt % TLCP/PP was 3.83 and 1.76 in the flow and transverse-to-flow directions, respectively. The anisotropy for the TLCP/PP composite was calculated to be 2.2, but the mechanical anisotropy of glass fiber reinforced polypropylene was only 1.35. Therefore, the anisotropy of the TLCP-filled composite can be lessened with the addition of glass fiber, which is demonstrated in Table 1, where we can see that the 10GF20TLCP/PP hybrid composite had its mechanical anisotropy reduced by 26%. The hybrid composites effectively reduced the mechanical anisotropy of the in situ TLCP composite.

Table 1. Mechanical anisotropy of the hybrid composites.

Material	Modulus (GPa) Flow	Modulus (GPa) Transverse	Mechanical Anisotropy
30GF/PP	5.07	3.75	1.35
20GF10TLCP/PP	4.25	3.10	1.37
10GF20TLCP/PP	4.11	2.52	1.63
30TLCP/PP	3.83	1.76	2.2

3.5. Recyclability of the Hybrid TLCP/GF Composite Material

To evaluate the recyclability of each composite material, the mechanical properties of the recycled composite were compared against the pristine counterpart. The normalized values of the tensile properties of the hybrid composites are presented in Figure 10. The normalized values were obtained by dividing the tensile properties of the recycled composites by their pristine material properties. For the tensile properties of the hybrid composites in the flow direction, the normalized values increased with increasing weight fraction of TLCP, suggesting an improvement in the recyclability with the presence of TLCP (Figure 10a). The 30 wt % GF/PP only retained 86% of its tensile strength after recycling, and 10GF20TLCP/PP retained 96% of its tensile strength compared to the pristine composite. The tensile modulus in the flow direction was not significantly impacted by mechanical recycling. The different impact of recycling on tensile modulus and strength is due to the different sensitivity of each property to the change in fiber length [46].

Figure 10b exhibits the normalized tensile properties of the hybrid composite in the transverse-to-flow direction. The properties follow a similar trend as before where properties increased with increasing TLCP concentration. The tensile modulus of GF/PP dropped significantly after recycling, which may be due to the joint influence of fiber breakage and decrease of fiber orientation in the transverse-to-flow direction. In the injection molding process, long fiber polymer blends have a larger core region where fibers are randomly oriented [51,52]. These effects may lead to large drops in the tensile modulus of GF/PP in the transverse-to-flow direction. The normalized values of tensile properties are significantly influenced by the content of glass fiber and TLCP in both the flow and transverse-to-flow directions. The recycling process has a greater impact on the tensile strength of composite materials than their tensile modulus in both the flow and transverse-to-flow directions.

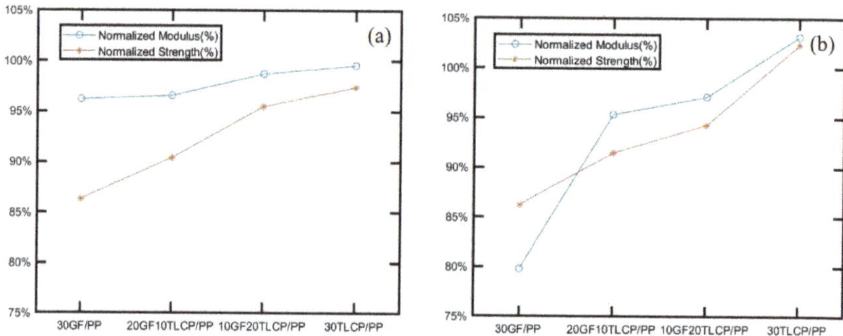

Figure 10. Percentage of mechanical properties of the recycled composite to pristine composite in (**a**) flow direction and (**b**) transverse-to-flow directions.

The knock-down (KD) factor is used to quantify the degree of recyclability of a hybrid composite where a low KD factor means better recyclability. The knock-down factor for a composite is determined by comparing the property of recycled material relative to its pristine material [53]. The KD factor is defined by the following equation:

$$KD = \left(1 - \frac{P_r}{P_v}\right) \times 100(\%) \tag{1}$$

where P_r/P_v is the ratio of recycled material to the pristine material property. Table 2 shows the KD factor of each composite in the flow direction. The KD factor of tensile modulus and strength for glass fiber reinforced polypropylene was 3.8 and 13.7, respectively. The KD factor decreased in value with increasing TLCP concentration, thereby confirming that the higher TLCP weight fraction in a TLCP/GF hybrid composite gives rise to greater recyclability. Composites with a KD factor less than five are typically considered to be within design limits for various applications where the recycled part can be used to replace that made of pristine material [53]. The formulation of 10 wt % glass fiber and 20 wt % TLCP resulted in the creation of a recyclable and high-performance hybrid material. Based on Figure 10, the formulation that contains a ratio of 2 to 1 or higher TLCP to glass fiber will yield a recyclable hybrid composite.

Table 2. The knock down (KD) factor of the in situ hybrid composite.

Material	KD (Modulus)	KD (Strength)
30GF/PP	3.8	13.7
20GF10TLCP/PP	2.4	9.6
10GF20TLCP/PP	1.3	4.5
30TLCP/PP	0.4	2.6

4. Conclusions

It has been found that the combination of TLCP fibrils and glass fiber can result in a high-performance and recyclable hybrid composite. The processing temperature of the injection-molding process was determined by rheological analyses and DSC. To provide a low viscosity blend to ensure that all the TLCP crystals were melted and to reduce excessive thermal degradation of polypropylene, the composites were processed at 305 °C. In situ hybrid composites were successfully generated at the optimized processing temperature. Due to the hybrid nature of the glass fiber and TLCP, the in situ hybrid composite exhibited balanced performance with respect to processability, mechanical properties, and recyclability. The 10 wt % glass fiber and 20 wt % TLCP hybrid composite material was the best formulation of the hybrid composite because it lowered the melt viscosity, thereby increasing its processability while maintaining high tensile properties, lowered mechanical anisotropy, and increased recyclability of the TLCP/GF hybrid composite.

Author Contributions: Conceptualization, T.C.; Methodology, T.C.; Validation, T.C.; Formal analysis T.C.; Investigation T.C., D.K. and L.J.; Writing—original draft preparation, T.C.; Writing—review and editing, D.K., L.J., D.A.O. and D.G.B.; Supervision, D.A.O. and D.G.B.; Funding acquisition, D.A.O. All authors have read and agreed to the published version of the manuscript.

Funding: The funding is from General Motor with Project Number GAC3053.

Acknowledgments: Support from General Motor (GM) is gratefully acknowledged. The authors would like to thank Celanese and SABIC for providing materials.

Conflicts of Interest: The authors declare no conflict of interest.

References

1. Lu, K. The future of metals. *Science* **2010**, *328*, 319–320. [CrossRef] [PubMed]
2. Rajak, D.K.; Pagar, D.D.; Menezes, P.L.; Linul, E. Fiber-reinforced polymer composites: Manufacturing, properties, and applications. *Polymers* **2019**, *11*, 37. [CrossRef]
3. Clyne, T.; Hull, D. *An Introduction to Composite Materials*; Cambridge University Press: Cambridge, UK, 2019.
4. Report, M.A. Fiber Reinforced Polymer (FRP) Composites Market Analysis by Fiber Type. Available online: https://www.grandviewresearch.com/industry-analysis/fiber-reinforced-polymer-frp-composites-market (accessed on 29 June 2020).
5. Report, M.R. GFRP Composites Market by End-Use Industry. Available online: https://www.marketsandmarkets.com/Market-Reports/glass-fiber-reinforced-plastic-composites-market-142751329.html (accessed on 30 June 2020).
6. ASM Aerospace Specification Metals Inc. Aluminum 6061-t6. Available online: http://asm.matweb.com/search/SpecificMaterial.asp?bassnum=MA6061T6 (accessed on 5 June 2020).
7. Prashanth, S.; Subbaya, K.; Nithin, K.; Sachhidananda, S. Fiber reinforced composites—A review. *J. Mater. Sci. Eng.* **2017**, *6*, 1–6.
8. Weiss, R.A.; Wansoo, H.; Nicolais, L. Novel reinforced polymers based on blends of polystyrene and a thermotropic liquid-crystalline polymer. *Polym. Eng. Sci.* **1987**, *27*, 684–691. [CrossRef]
9. Chae, H.G.; Kumar, S. Rigid-rod polymeric fibers. *J. Appl. Polym. Sci.* **2006**, *100*, 791–802. [CrossRef]
10. Wang, X.-J.; Zhou, Q.-F. *Liquid Crystalline Polymers*; World Scientific Publishing Company: Singapore, 2004.
11. Donald, A.M.; Windle, A.H.; Hanna, S. *Liquid Crystalline Polymers*; Cambridge University Press: Cambridge, UK, 2006.
12. Qian, C.; Mansfield, C.D.; Baird, D.G. Extrusion blow molding of polymeric blends based on thermotropic liquid crystalline polymer and high density polyethylene. *Int. Polym. Process.* **2017**, *32*, 112–120. [CrossRef]

13. Kalfon-Cohen, E.; Marom, G.; Wachtel, E.; Pegoretti, A. Characterization of drawn monofilaments of liquid crystalline polymer/carbon nanoparticle composites correlated to nematic order. *Polymer* **2009**, *50*, 1797–1804. [CrossRef]

14. Collier, M.C. Reclamation and Reprocessing of Thermotropic Liquid Crystalline Polymer from Composites of Polypropylene Reinforced with Liquid Crystalline Polymer. Ph.D. Thesis, Virginia Tech, Blacksburg, VA, USA, 1998.

15. Handlos, A.A.; Baird, D.G. Processing and associated properties of in-situ composites based on thermotropic liquid-crystalline polymers and thermoplastics. *J. Macromol. Sci.-Rev. Macromol. Chem. Phys.* **1995**, *C35*, 183–238. [CrossRef]

16. Collyer, A.A. *Liquid Crystal Polymers: From Structures to Applications*; Springer Science & Business Media: Berlin, Germany, 2012; Volume 1.

17. Baird, D.G.; Huang, J. Injection molding of polypropylene reinforced with thermotropic liquid crystalline polymer microfibrils. Part II: Effect of impact toughening. *J. Inject. Mold. Technol.* **2002**, *6*, 107.

18. Williams, D. Applications for thermotropic liquid crystal polymer blends. *Adv. Polym. Technol.* **1990**, *10*, 173–184. [CrossRef]

19. Chen, T.; Mansfield, C.D.; Ju, L.; Baird, D.G. The influence of mechanical recycling on the properties of thermotropic liquid crystalline polymer and long glass fiber reinforced polypropylene. *Compos. Part B Eng.* **2020**, *200*, 108316. [CrossRef]

20. Yu, X.B.; Wei, C.; Xu, D.; Lu, C.H.; Yu, J.H.; Lu, S.R. Wear and mechanical properties of reactive thermotropic liquid crystalline polymer/unsaturated polyester/glass fiber hybrid composites. *J. Appl. Polym. Sci.* **2007**, *103*, 3899–3906. [CrossRef]

21. Yu, X.; Chun, W.; Lu, S.; Yu, J.; Deng, X.; Lu, C. Preparation and mechanical properties of tlcp/up/gf in-situ hybrid composites. *T. Nonferr. Metal. Soc.* **2006**, *16*, s529–s533. [CrossRef]

22. Huang, J.; Baird, D.G. Injection molding of polypropylene reinforced with thermotropic liquid crystalline polymer microfibrils. Part III: Combination of glass and tlcp. *J. Inject. Mold. Technol.* **2002**, *6*, 187.

23. Bafna, S.S.; Desouza, J.P.; Sun, T.; Baird, D.G. Mechanical-properties of in-situ composites based on partially miscible blends of glass-filled polyetherimide and liquid-crystalline polymers. *Polym. Eng. Sci.* **1993**, *33*, 808–818. [CrossRef]

24. He, J.S.; Zhang, H.Z.; Wang, Y.L. In-situ hybrid composites containing reinforcements at two orders of magnitude. *Polymer* **1997**, *38*, 4279–4283. [CrossRef]

25. He, J.S.; Wang, Y.L.; Zhang, H.Z. In situ hybrid composites of thermoplastic poly(ether ether ketone), poly(ether sulfone) and polycarbonate. *Compos. Sci. Technol.* **2000**, *60*, 1919–1930. [CrossRef]

26. Pickering, S.J. Recycling technologies for thermoset composite materials—Current status. *Compos. Part A Appl. Sci. Manuf.* **2006**, *37*, 1206–1215. [CrossRef]

27. Piñero-Hernanz, R.; Dodds, C.; Hyde, J.; García-Serna, J.; Poliakoff, M.; Lester, E.; Cocero, M.J.; Kingman, S.; Pickering, S.; Wong, K.H. Chemical recycling of carbon fibre reinforced composites in nearcritical and supercritical water. *Compos. Part A Appl. S. Compos.* **2008**, *39*, 454–461. [CrossRef]

28. Cunliffe, A.M.; Jones, N.; Williams, P.T. Pyrolysis of composite plastic waste. *Environ. Technol.* **2003**, *24*, 653–663. [CrossRef]

29. Howarth, J.; Mareddy, S.S.; Mativenga, P.T. Energy intensity and environmental analysis of mechanical recycling of carbon fibre composite. *J. Clean. Prod.* **2014**, *81*, 46–50. [CrossRef]

30. Oliveux, G.; Dandy, L.O.; Leeke, G.A. Current status of recycling of fibre reinforced polymers: Review of technologies, reuse and resulting properties. *Prog. Mater. Sci.* **2015**, *72*, 61–99. [CrossRef]

31. Chrysostomou, A.; Hashemi, S. Influence of reprocessing on properties of short fibre-reinforced polycarbonate. *J. Mater. Sci.* **1996**, *31*, 1183–1197. [CrossRef]

32. Kuram, E.; Ozcelik, B.; Yilmaz, F. The influence of recycling number on the mechanical, chemical, thermal and rheological properties of poly(butylene terephthalate)/polycarbonate binary blend and glass-fibre-reinforced composite. *J. Thermoplast. Compos. Mater.* **2016**, *29*, 1443–1457. [CrossRef]

33. Colucci, G.; Ostrovskaya, O.; Frache, A.; Martorana, B.; Badini, C. The effect of mechanical recycling on the microstructure and properties of pa66 composites reinforced with carbon fibers. *J. Appl. Polym. Sci.* **2015**, *132*, 9. [CrossRef]

34. Eriksson, P.A.; Albertsson, A.C.; Boydell, P.; Prautzsch, G.; Manson, J.A.E. Prediction of mechanical properties of recycled fiberglass reinforced polyamide 66. *Polym. Compos.* **1996**, *17*, 830–839. [CrossRef]

35. Postema, A.R.; Fennis, P.J. Preparation and properties of self-reinforced polypropylene liquid crystalline polymer blends. *Polymer* **1997**, *38*, 5557–5564. [CrossRef]

36. Lyondellbasell Pro-Fax 6523 Technical Data Sheet. Available online: https://www.lyondellbasell.com/en/polymers/p/Pro-fax-6523/cca46629-99c9-4596-8390-83b67fd362ff (accessed on 6 June 2020).

37. Done, D.; Baird, D.G. Transient flow of thermotropic liquid-crystalline polymers in step strain experiments. *J. Rheol.* **1990**, *34*, 749–762. [CrossRef]

38. Viola, G.G.; Baird, D.G. Studies on the transient shear-flow behavior of liquid-crystalline polymers. *J. Rheol.* **1986**, *30*, 601–628. [CrossRef]

39. Cocchini, F.; Nobile, M.R.; Acierno, D. Transient and steady rheological behavior of the thermotropic liquid-crystal copolymer 73/27 hba hna. *J. Rheol.* **1991**, *35*, 1171–1189. [CrossRef]

40. Chung, T.S.; Cheng, M.; Pallathadka, P.K.; Goh, S.H. Thermal analysis of vectra b950 liquid crystal polymer. *Polym. Eng. Sci.* **1999**, *39*, 953–962. [CrossRef]

41. Datta, A.; Baird, D.G. Compatibilization of thermoplastic composites based on blends of polypropylene with 2 liquid-crystalline polymers. *Polymer* **1995**, *36*, 505–514. [CrossRef]

42. Qin, Y.; Brydon, D.L.; Mather, R.R.; Wardman, R.H. Fibers from polypropylene and liquid-crystal polymer blends: 3. A comparison of polyblend fibers containing vectra-a900, vectra-b950 and rodrun-lc3000. *Polymer* **1993**, *34*, 3597–3604. [CrossRef]

43. Shenoy, A.V. *Rheology of Filled Polymer Systems*; Springer Science & Business Media: Berlin, Germany, 2013.

44. Vonturkovich, R.; Erwin, L. Fiber fracture in reinforced thermoplastic processing. *Polym. Eng. Sci.* **1983**, *23*, 743–749. [CrossRef]

45. Zhang, G.; Thompson, M.R. Reduced fibre breakage in a glass-fibre reinforced thermoplastic through foaming. *Compos. Sci. Technol.* **2005**, *65*, 2240–2249. [CrossRef]

46. Thomason, J.L. The influence of fibre length and concentration on the properties of glass fibre reinforced polypropylene: 5. Injection moulded long and short fibre pp. *Compos. Part A Appl. Sci. Manuf.* **2002**, *33*, 1641–1652. [CrossRef]

47. Baird, D.G.; Robertson, C.G.; De Souza, J.P. Liquid Crystalline Polymer-Reinforced Thermoplastic Fibers. U.S. Patent 5,834,560, 10 November 1998.

48. Mallick, P.K. *Fiber-Reinforced Composites: Materials, Manufacturing, and Design*; CRC Press: Boca Raton, FL, USA, 2007.

49. Mortazavian, S.; Fatemi, A. Effects of fiber orientation and anisotropy on tensile strength and elastic modulus of short fiber reinforced polymer composites. *Compos. Part B Eng.* **2015**, *72*, 116–129. [CrossRef]

50. Mehta, A.; Isayev, A.I. Rheology, morphology, and mechanical characteristics of poly(etherether ketone)-liquid crystal polymer blends. *Polym. Eng. Sci.* **1991**, *31*, 971–980. [CrossRef]

51. Toll, S.; Andersson, P.O. Microstructure of long-fiber and short-fiber reinforced injection molded polyamide. *Polym. Compos.* **1993**, *14*, 116–125. [CrossRef]

52. Thomason, J.L. Micromechanical parameters from macromechanical measurements on glass reinforced polyamide 6,6. *Compos. Sci. Technol.* **2001**, *61*, 2007–2016. [CrossRef]

53. Eriksson, P.A.; Albertsson, A.C.; Boydell, P.; Manson, J.A.E. Durability of in-plant recycled glass fiber reinforced polyamide 66. *Polym. Eng. Sci.* **1998**, *38*, 348–356. [CrossRef]

Journal of
Composites Science

Article

Strength Prediction Sensitivity of Foamed Recycled Polymer Composite Structures due to the Localized Variability of the Cell Density Distribution

Daniel P. Pulipati and David A. Jack and David A. Jack *

Department of Mechanical Engineering, Baylor University, Waco, TX 76706, USA; Daniel_Pulipati@baylor.edu
* Correspondence: David_Jack@baylor.edu; Tel.: +1-254-710-3347

Received: 30 June 2020; Accepted: 14 July 2020; Published: 16 July 2020

Abstract: The need for novel methods for the reuse of post-industrial/post-consumer polymer solid wastes (PSW) is of increasing societal importance. Unfortunately, this objective is often limited due to material stream variability or insufficient load-carrying capacity of the fabricated goods. This study investigates a large format fiber-reinforced structural member that contains spatially varying material properties, specifically density. The application is focused on the unique features of closed-cell foamed composite structures made from recycled post-industrial/post-consumer PSW composed of High-Density Polyethylene (HDPE) and Glass Fiber Polypropylene (GFPP). The structures in this research are manufactured using a hybrid extrusion process, which involves foaming enabled by chemical blowing agents that form a fully consolidated solid outer shell and a closed-cell core. The cell distribution is inhomogeneous, in size distribution and spatial distribution, leading to significant spatial variations of the local effective stiffness. To understand the correlation between density variations and effective stiffness and strength, a low-cost method using digital imaging is introduced and integrated into a finite element subroutine. The imaging approach includes sectioning the structural member and analyzing the resulting image using various custom imaging processing techniques in the MATLAB environment. The accuracy of the imaging technique was experimentally verified using a Keyence digital microscope, and the error was found to be 3% in any given spatial feature. The processed image is then correlated to a localized density map of the cross-section using a weighted spatial averaging technique, and the local effective material properties of the foamed region are predicted using the presented micromechanical approach. The local stiffness is a function of void density, local fiber orientation, constitutive behavior of both the fiber and the matrix blend, and the non-linear response of the matrix blend. The spatially varying stiffness and nonlinear strength response at each spatial location are then integrated into a finite element subroutine within the COMSOL multiphysics environment, and results are presented for the deflection and internal stress state of the composite structure. Results indicate that the internal microstructural variations have a nominal impact on the bulk deflection profile. Conversely, results show the peak of the internal stress is increased by ∼11% as compared to the uniform core assumption, thus safe designs must consider core density spatial variations in the final product design.

Keywords: composite foams; closed cells; image processing; finite element analysis

---

1. Introduction

Sustainability of polymers has become a significant concern both within the polymer industry and within society, with various researchers focusing on the use of polymer composite structures, specifically plastic lumber, made from various polymer solid wastes (PSW) including Polyethylene and Polypropylene (see e.g., [1–4]). With the intent of reducing waste and also reducing the use of high density wooden structures, themselves a depleting resource, recycled polymer composite structures are being used as crossties (sleepers), marine pilings, short bridges, and other load bearing applications. Researchers have demonstrated the feasibility of these structures as viable load bearing applications, which are being used all over the world [5–7].

The use of fillers, such as glass fibers and natural fibers, are often integrated into the polymer matrix to enhance stiffness, strength, and impact resistance (see e.g., [8,9]). To reduce manufacturing costs, improve light weighting, and allow easy installation of screw fasteners in crossties, chemical blowing agents (CBA) such as azodicarbonamide (ADC) are introduced into the extrusion process [10]. Kord et al. [11] studied the effect of the blowing agent ADC on composites made from High-Density Polyethylene (HDPE), wood flour, and nanoclay. They observed that the increase of ADC content resulted in increased cell size and average cell density, which decreased the overall density of the structure and also reduced the tensile modulus. Therefore, the amount of ADC induced into the composite needs to be precise to achieve an acceptable density spatial distribution to maximize the strength to weight ratio.

When using chemical blowing agents, melt temperature and pressure are key components for nucleation to occur [12]. As PSWs in their molten state are mixed with ADC, the blowing agents undergo a chemical reaction by dissolving into a gas as the pressure in the melt increases. Due to a pressure differential, the nucleation of the individual cells is caused when the gas is released within the mold. The density of the foam varies based on the gas fraction reacted in the molten PSWs, gas lost into the atmosphere, and the rate at which the gas decompresses [12]. As the nucleated cells achieve their maximum growth, the glass fibers in the mold are forced into the dense outer solid shell adding reinforcement to the shell where the maximum stresses are expected [6]. For the expected loading applications, the foamed core does not yield the same stress requirements as the exterior shell. Ruiz et al. [12] observed that when blowing agents are not activated properly, the inactivated ADC particles can form clusters, inducing poor cell morphology and leading to poor internal load transfer.

The cooling rates of the mold have an impact on the formation of cells as well [13]. When the mold is externally cooled using water channels, the shell layer thickness varies based on the total heat transfer between the mold walls and the molten polymer. The molten reinforced polymer is cooled faster in the exterior region limiting the cell growth to the core with a fully densified outer shell. Yousefian et al. [14] studied the Polypropylene/Nano-crystalline cellulose composite and found that by varying mold temperatures between 30 °C and 80 °C, the ratio between the shell and the core could be altered by 5%. Tissandier et al. [15] studied the effects of temperature, blowing agents, and fiber content on composites made from HDPE, flax fiber, and ADC, which were mixed and manufactured using injection molding. This process included applying a temperature gradient (0–60 °C) inside the mold and achieved acceptable microcellular asymmetric structure with high fiber content, blowing agents, and mold temperature. They showed that if a bar is cooled using a gradient, the cell wall thickness on the cooler side is thicker than the warmer side. In their configuration, they concluded that the cold side of the mold controlled the final morphology of the foaming process.

The mold temperature also influences cell size, cell density, and distribution [16]. It was found that as the mold temperature decreases in injection molded HDPE structural foams foamed with ADC, a small but measurable increase is observed in the average cell density and cell size [14]. Cell growth can be limited by increasing the fiber content, which increases the viscosity of the melt and potentially impacting processability [17]. These variations of cell size and its dependence on material properties, namely elastic stiffness, have been studied by Redenbach et al. [18] by using Laguerre tessellations, a weighted mathematical generalization used to model foams using convex cells. They found that as the cell aspect ratio, defined as the ellipsoidal major to a minor axis, increased from spherical, the effective elastic stiffness decreased.

Cell wall thickness also impacts the material properties of foams. Barbier et al. [19] studied Voronoi closed-cell foams (which is a mathematical partitioning of regions in proximity to grid points in space) under a variety of imposed strains caused by wall stretching, and they found a correlation between the density and cell irregularity on the effective elastic and plastic properties. They concluded that under elastic deformation, material properties rely only on relative density. Whereas, under plastic deformation, there is a dependence on both relative density and structural deformation.

To calculate the material properties of closed cell foams, many researchers have used various micromechanical approaches to estimate the effective homogenized medium, specifically, the tensile modulus, as either an isotropic or anisotropic structure (see e.g., [20]). The micromechanical approaches do not account for various combinations of closed cells, including cell sizes, cell shapes, and relative density, to predict the material behavior. Zhang et al. [20] have reviewed various micromechanics models (see e.g., [21–25]) using uniform closed cells by studying HDPE foams made by compression molding techniques. They suggested that the differential scheme [24] and the square power-law suggested by Moore et al. [26] are the most accurate within the cell volume fraction of 0–55%. In the present research, the square-power law will be used. Recently, Lo et al. [27] studied transversely isotropic PVC foams and modeled and predicted the elastic stiffness using unit cell representation for the foam microstructure. They modified the Halpin–Tsai equations by introducing an apparent volume fraction to calculate the material properties and replaced the actual resin volume fraction. Tucker and Liang [28] suggested that the Halpin–Tsai equations are most applicable for long fibers, but for the short fiber the reinforced system employed in the present study, the Tandon and Weng [29] model is more appropriate. The present paper uses the closed form solution of the Tandon and Weng model, implied by Tucker and Liang as derived in the work by Zhang [30], for the fiber reinforced micromechanics modeling.

It has been shown that many structural foams have a spatially varying density gradient that often decreases in density from the outer solid shell to the closed cell foamed core [31]. To understand the complex nature of foams with varying cell sizes, density, and cell size distribution, various studies using images from Scanning Electron Microscopy (SEM) [32], digital microscopy [33,34], and Micro-CT/X-ray [35–37] were used. Sadik et al. [33] studied images obtained from a digital microscope at 5× magnification and used the micrographs to binarize cells and measure the void fraction using ImageJ software. The void fraction was then used to calculate linear elastic material properties of the foamed polymer such as tensile, shear, and flexural modulus. These images were analyzed for the foamed structures by using micrographs and calculating the void fractions. Davari et al. [32] studied chemically linked polyethylene foams and the foam morphology using images from SEM. They concluded that density is the dominant parameter for predicting material properties, but when the density is uniform, there is a small but measurable contribution to the material properties as a function of the cell size distribution.

Recently, Zhu et al. [36] used MATLAB image processing techniques to study aluminum foams by examining images obtained from X-ray computed tomography (μCT) and extended the foam characteristics into a Finite Element Model (FEM). They showed that the image processing techniques can be used to study the porosity of aluminum foams and can reasonably predict the compression behavior. In an attempt to characterize porosity using inexpensive and quick modeling techniques, Yunus et al. [38] studied the porosity of polyurethane foams by using a Canon camera, a black box, and a LaserJet scanner. They then processed the resulting images in a custom MATLAB subroutine. They also validated their results for the porosity measured by each instrument using a stereomicroscope and an SEM. They concluded that the black box results were the most comparable to the results obtained from an SEM.

In a recent work by the authors [39], the composite structures made from recycled materials were studied by integrating the fiber aspect ratio, uniform cell density, orientation state of the fibers, and the constitutive properties of the raw materials. These properties were used as inputs into micromechanics models, and the effective elastic moduli of the solid shell outer core of the composite as well as the foamed core were predicted. Stress values of the raw material were measured at a 0.2% reference strain, and the reference stress of the shell and core formed a nonlinear stress–strain relationship using a modified rule of mixtures and micromechanics models. The predicted elastic moduli and reference stresses were then used in a nonlinear finite element model to predict the flexural deflection response of the composite crosstie under a 4-point bend test using ASTM D6109-13. The results from the model were compared to experimental tests from sixteen different samples, and the finite element results were found to be within the experimental variation.

In this paper, a weighted method is proposed and demonstrated to characterize the density of glass fiber reinforced HDPE/PP (High-Density Polyethylene/Polypropylene) foamed core with cells of varying sizes and distribution using a commercial off-the-shelf (COTS) EOS60D Canon Camera (Canon Inc., Tokyo, Japan). This is achieved by taking 2D images of the cross-section, such as that depicted in Figure 1, using the COTS camera and obtaining a density map by measuring areas of cells contained within the image using image processing techniques implemented in MATLAB (2018a, MathWorks, Natick, MA, USA). The cross-sectional area of the cells within a sample image is tabulated and their results are validated against results obtained using a high-resolution VR-3100 3D digital microscope (Keyence Corporation of America, Itasca, IL, USA). The density map is translated to a point by point mapping of the elastic modulus using a differential scheme. A finite element model is then implemented using a gridding technique to map the spatial varying density results obtained from the imaging results, and the solutions are compared to those from the uniform density method. The von-Mises stress is analyzed for comparison purposes and the results that include local variations in the cell density yield a ~11% higher peak stress as compared to that of the uniform density core. This method will allow for safe design considerations for various applications due to the stresses caused by variations in cell size and distributions.

Figure 1. Example of a cross-section.

2. Methodology

2.1. Materials and Manufacturing

The large-format composite structures discussed in this research are made of a blend of recycled High-Density Polyethylene (HDPE), recycled Glass Fiber Polypropylene (GFPP), ADC (chemical blowing agent), and carbon black. This mixture consists of ~20 % GFPP, ~80% HDPE, ~0.5% ADC, and carbon black by mass. This blend was mixed, melted, and pressurized into a mold cavity. The exterior of the mold was cooled using room temperature water. The cooling affected the outer shell first and allowed the blowing agent to react in the core forming a dense outer core and a foamed inner core, with varying cell size distribution and spatially varying density. Three samples from different crossties were sectioned using an industrial chop saw, and debris from the cutting process was removed. The cross-section of the crosstie was then photographed in a low light environment with the flash from the COTS camera, yielding an image that was approximately 3000 × 2300 pixels. The dimensions of the core for each sample were 0.19 m × 0.14 m (7.38 in × 5.38 in) as shown in Figure 2a.

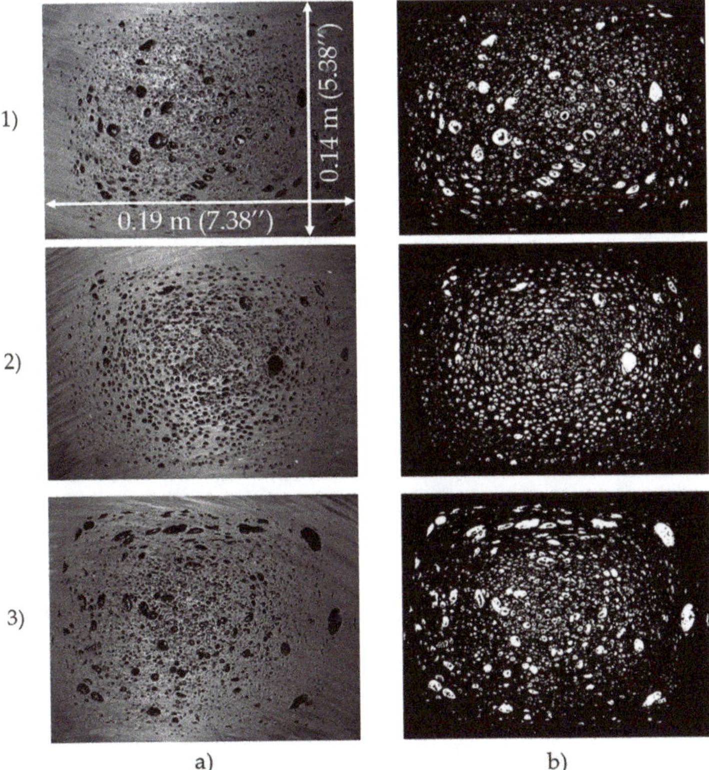

Figure 2. Cross-section of the presented samples. (**a**) Images of sample 1, 2, and 3 from the commercial off-the-shelf (COTS) camera. (**b**) Binary plots of samples 1, 2, and 3.

2.2. Image Processing to Generate Cells

The photographed samples were imported into the MATLAB environment as an array representing the image color density as a function of pixel location. The uploaded images taken from three different cross-sections are shown in Figure 2a. Pixels were converted from a grayscale value to a binary value using color contrast differentiation. It was found that a yellow tint using the automatic enhance option included in the Windows photo editor provided reasonable pre-processing contrast variations in differentiating the solid region from the darker cells. A threshold value of 25% was selected for the images, and the noise within the binary image was removed using a morphological structure operation by ignoring individual false single pixels. As shown in Figure 2b, the white pixels denote cells, and the neighboring solid surface region is denoted in black.

This method then considers the effective area of the cells shown in white pixels and draws a circle around it. The dimensions of this circle are defined as the effective diameter using the automated MATLAB subroutine which is much faster compared to highly sophisticated and manual use of an optical microscope. During image processing, false positives must be identified and properly addressed. As seen in Figure 3a,b, it can be noted that in the picture taken with the COTS camera, there are color contrast variations within the cell cavity itself. The first false positive, termed a donut, is a cell region that contains black pixels resembling a solid shell. For the overall sample size of 7.38 × 5.38 inches, the image is approximately 3000 × 2300 pixels, and the size of each pixel is 63.5 μm. For the composite structure discussed, the radius of the fibers was measured to be 8.56 μm ± 0.57 μm [39] and thus the fibers are not visible within the image. Optical microscopy was performed on the surface of the cells, and the fibers did not protrude from the cell wall. A second false positive, termed a double circle, is a shell region that appears to cover a portion of the cell due to flaws introduced from the saw blade used in the sectioning process.

The first false positives were corrected using a subroutine that automatically takes into account the overall area by circles drawn around each cell, including the black pixels within the cell region, as shown in Figure 3c. The second false-positive requires an additional processing step whereby any captured features that reside within the domain of another feature, such as those highlighted in Figure 3c, were removed from consideration. If the double circles were not removed, the area of the cells within the region would be overestimated. Figure 3d shows the cell distribution requiring substantially less archived data as only the centroid and effective radius is stored and used in subsequent processing. This down-select process allows the homogenization algorithm to process more quickly and retains the effective features of the various cells for micromechanical predictions.

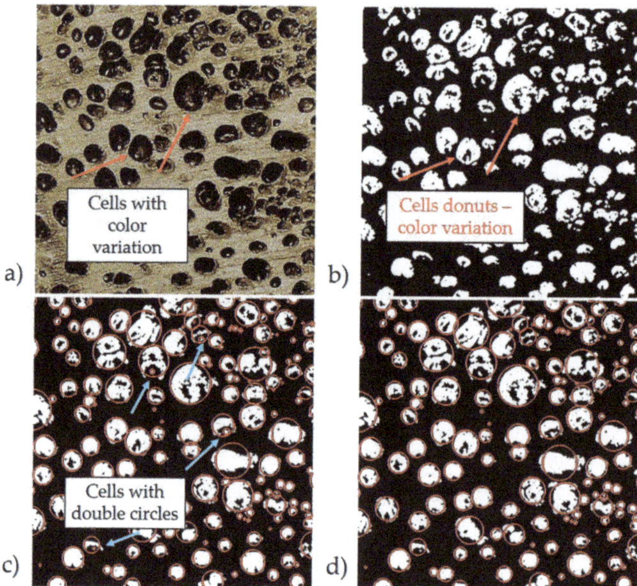

Figure 3. Image processing false positives. (**a**) Image with a yellow tint showing color contrasts within cells. (**b**) Pixel formation of cell donuts based on color contrast. (**c**) Cells with double circles. (**d**) Filtering the image of double circles.

It is important to note that the cross-sections shown are cut at a random plane along the longitudinal axis of the composite. As this cutting process may not necessarily pass through the centroid of the cell, it will tend to underestimate the actual cell spherical equivalent volume. To account for the underestimation, a correction factor is used to estimate the largest radius of the 3D spheroidal cell instead of the chord length as seen in a 2D cell. This correction factor is taken from the work of Pinto et al. [40] and the radius of each cell is adjusted as

$$r_{3D} = 1.273\, r_{2D} \tag{1}$$

2.3. 2D Area Validation

To validate the cell area obtained using image thresholding and the COTS camera, a detailed study was conducted on the subregion of the COTS image shown in Figure 4a for sample 2. This region contains 36 identifiable cells of varying sizes, and each cell is compared to the area measured manually using the high-resolution 3D Keyence 3100 microscope. For example, a single cell was selected and is shown in Figure 4b. Its effective pixel diameter was measured using the image thresholding algorithm discussed above, and the effective area is compared to the area measured using the Keyence digital microscope that includes the non-circular perimeter of the cell in the area calculation. Each of the 36 cells were measured manually using the Keyence digital microscope and their regions are highlighted in Figure 4c. The areas from the aforementioned imaging algorithm, shown in Figure 4d, were then compared to each of manually identified areas from the high-resolution microscopy results.

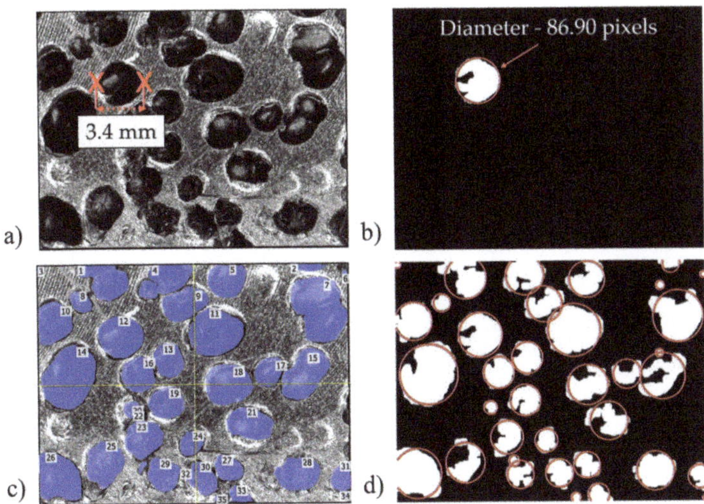

Figure 4. Cell area validation using a Keyence digital microscope. (**a**) Measuring diameter as seen from the Keyence microscope. (**b**) Measuring the effective diameter of the same cell using the new proposed method. (**c**) Manually measuring the area of each cell using the Keyence microscope. (**d**) Automated measuring of the effective diameters of all the cells.

The total area for the region of one cell is highlighted in Figure 4b. The total area for the cells in the image was calculated with image thresholding approach using the image from the COTS camera as 170.69 mm^2, and the total area for all 36 cells was measured using high-resolution microscopy was 166 mm^2, yielding an error of 3%. The error was calculated using the equation,

$$Err = \frac{|A_{Microscopy} - A_{COTS\ Imaging}|}{A_{Microscopy}} \times 100 \tag{2}$$

2.4. Image Mirroring and Density Homogenization

To calculate the homogenized density of the surface for use in the finite element model, a grid of 400 × 250 points was mapped onto the surface of each respective sectioned sample. The sample image was then mirrored about the edges to avoid biasing in any region near the surface and to allow continuous regions for homogenization at all points within the image. A region for a single point was depicted in Figure 5. A fixed pixel radius of $R = 300$ was taken for each of the 3000 × 2300 pixel images, and every cell was identified within the radius R.

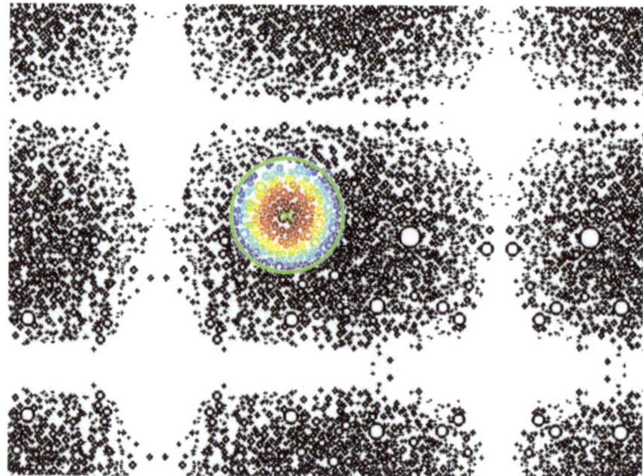

Figure 5. Mirrored image preventing image biasing near the boundaries. A single grid point is highlighted, showing the region of consideration used for local density homogenization.

The cells identified within the circle were averaged to calculate the local density using an exponential decaying weight function. This allowed the areas of the cells closest to the grid point to have the highest contribution to the effective density, whereas the cells farther from the center have less of a contribution to the effective density. This is also denoted by the color variations, as seen in Figure 5. The cells closest to the center are colored red and provide the highest contribution to the density homogenization, whereas the cells farthest from the center are shown in colors that transition to yellow and then to blue as their contribution decreases. The exponential decay function is defined as,

$$\widetilde{\rho}(x,y) = \int_0^{2\pi} \int_0^{\infty} \psi(r,\theta)e^{-(r(x,y)/R)^2}r(x,y)drd\theta \simeq \int_0^{2\pi} \int_0^{3R} \psi(r,\theta)e^{-(r(x,y)/R)^2}r(x,y)drd\theta \tag{3}$$

where r is the local radius from the center of the region of radius R, $\widetilde{\rho}(x,y)$ is the localized density, and $\psi(r,\theta)$ is the probability of finding a cell at r,θ. The upper limit on the radial integral is approximated as being upper bounded by $3R$, as this will take into account 99.9% of the area within the distribution while avoiding unnecessary calculations over all cells captured within the image. As observed, the cell sizes are much smaller than the region R. Therefore, for discrete data sets, Equation (3) can be approximated as,

$$\widetilde{\rho}(x,y) \simeq \sum_{i=1}^{N} A_i e^{-(r(x,y)/R)^2} \tag{4}$$

where A_i is the area of the i_{th} cell and N is the total number of cells in the entire image. Again, Equation (4) as in Equation (3), will disregard cells that have a centroid more than $3R$ from the grid point being analyzed as their contribution will not significantly alter the final solution due to the pre-multiplication of the exponential decaying function. To normalize over the entire area of the image, the density $\rho(x,y)$ is calculated as,

$$\rho(x,y) = \frac{\widetilde{\rho}(x,y) \sum_{i=1}^{N} A_i}{\int \int \widetilde{\rho}(x,y)dxdy} \tag{5}$$

where all the cell areas within the image are summed from $i = 1$ to N and x and y are the grid points selected.

2.5. Material Properties Prediction

Based on the previous research [39], material properties were predicted using only constitutive properties. This was accomplished using the inputs of raw materials, including as elastic moduli and Poisson's ratios of HDPE and GFPP, glass fiber aspect ratio, and foamed core density. Since modeling over each individual fiber in a 2.62 m long composite structure is computationally impractical, Advani and Tucker [41] proposed a model which uses the average orientation of the fibers within a region, termed the orientation tensors. Specifically of interest are the second order a_{ij} and the fourth order a_{ijkl} tensors, defined as,

$$a_{ij} = \int_{\mathbb{S}} p_i p_j \psi(\mathbf{p}) d\mathbb{S}, \quad a_{ijkl} = \int_{\mathbb{S}} p_i p_j p_k p_l \psi(\mathbf{p}) d\mathbb{S} \tag{6}$$

where, $\mathbb{S}$ is the surface of a unit sphere and $\mathbf{p}$ is the unit vector representing an individual fiber. The orientation tensor is symmetric, $a_{ij} = a_{ji}$, and $a_{ijkl} = a_{klij} = a_{jikl} = a_{ilkj} = \cdots$. Based on the property that the integral of the probability density function over the entire unit sphere must be 1, the trace of second-order orientation tensor is similarly equal to 1. For example, the second order orientation tensor of randomly distributed fibers has components of $a_{11} = a_{22} = a_{33} = 1/3$ with all other components being equal to zero.

Using the second-order orientation tensor a_{ij}, the fourth-order orientation a_{ijkl} was approximated using the orthotropic fitted (ORT) closure method [42]. Along with the orientation tensors, the unidirectional composite stiffness tensor $\overline{C}_{ijkl}$ was used in the homogenization approach to calculate the anisotropic stiffness of the composite. The underlying unidirectional stiffness tensor of the composite was calculated using the method given by Tandon and Weng [29], which uses the matrix and fiber material properties and the fiber aspect ratio as shown in Table 1. Within the fiber aspect ratio calculations, the weighted average and number average fiber lengths were calculated and the weight average fiber length was used for the aspect ratio based on the discussion as shown in [39].

Table 1. Micromechanics input material properties, taken from [39].

Material Property	Value
Matrix modulus	1.29 GPa
Fiber modulus	70 GPa
Aspect ratio	60
Volume fraction	3.33%
Poisson's ratio of the fiber	0.23
Poisson's ratio of the matrix	0.45

From the calculated unidirectional stiffness tensor $\overline{C}_{ijkl}$, the effective anisotropic stiffness tensor $< C_{ijkl} >$ was computed using the equation,

$$\begin{aligned}
\left\langle C_{ijkl} \right\rangle =\ & B_1 a_{ijkl} + B_2(a_{ij}\delta_{kl} + a_{kl}\delta_{ij}) + B_3(a_{ik}\delta_{jl} + a_{il}\delta_{jk} + a_{jl}\delta_{ik} \\
& + a_{jk}\delta_{il}) + B_4(\delta_{ij}\delta_{kl}) + B_5(\delta_{ik}\delta_{jl} + \delta_{il}\delta_{jk})
\end{aligned} \tag{7}$$

where δ_{ij} is the Kronecker delta, and the B_i terms are shown in terms of the calculated unidirectional stiffness tensor as (see, e.g., [41]),

$$\begin{aligned}
B_1 &= \overline{C}_{1111} + \overline{C}_{2222} - 2\overline{C}_{1122} - 4\overline{C}_{1212} & B_4 &= \overline{C}_{2233} \\
B_2 &= \overline{C}_{1122} - \overline{C}_{2233} & B_5 &= \tfrac{1}{2}(\overline{C}_{2222} - \overline{C}_{2233}) \\
B_3 &= \overline{C}_{1212} + \tfrac{1}{2}(\overline{C}_{2233} - 2\overline{C}_{2222})
\end{aligned} \tag{8}$$

From the calculated anisotropic stiffness matrix $< C_{ijkl} >$, the fourth-order compliance tensor was calculated using $< S_{ijkl} > = < C_{ijkl} >^{-1}$. Using this method, Young's modulus of the isotropic solid shell region can be predicted using the equation $E_{11} = 1/S_{1111}$ and was calculated as 1.73 GPa.

The Young's modulus of the foamed core depends on varying densities calculated using Equation (5). As the density of the cells locally increases, the Young's modulus decreases. Based on the comparison study of existing micromechanics models discussed by Zhang et al. [20], the empirical power law equation given by Moore [26] was used in the present study and is given as

$$E_f = E_c \left(1 - f\right)^n \tag{9}$$

where E_c is the Young's Modulus of the solid composite, E_f is the Young's modulus of the localized cell region, f is the calculated density, and $n = 2$ is given by the square power law relationship. This calculation was performed over each individual grid point mapped onto the surface of the cross-sectioned image and yielded an array of Young's moduli values of E_f at 400×250 points.

2.6. Finite Element Analysis

In a previous work [39], the aforementioned micromechanical modeling on coupon samples was experimentally validated for the shell and the core region. The earlier work was extended to include a full non-linear finite element solution, with inputs of the constitutive properties of the neat polymers and fibers. The previous modeling work was extended in the present study to allow for spatial variations of the foam density as indicated in Figure 2 using the image processing technique discussed in the previous sections. The Young's modulus grid developed using Equation (9) was exported into the finite element software package COMSOL (Version 5.4, COMSOL Inc., Burlington, MA, USA) using a custom subroutine written in MATLAB that takes the spatial location as inputs and returns the effective local anisotropic stiffness tensor. The finite element domain is that of a 0.178 m × 0.229 m (7 in × 9 in) tie that is 2.62 m (8.6 ft) in length as shown in Table 2, which is placed in a four-point bend testing configuration conforming to the ASTM D6109-13 dimensions.

Table 2. Finite Element Analysis model dimensions.

Property	Dimensions (m)	Dimensions (in)
Overall Length	2.62	103.2
Width	0.23	9
Depth	0.18	7
Span Length	1.52	60

The domain was given a point cloud of over 10 million points, and the spatially varying local stiffness was computed at each point using the results from the cell homogenization process discussed above. The lookup table data set contains the spatial x, y, z location of the point cloud along with the Young's modulus value. This data set was then exported to a single lookup table for later access by a subroutine within COMSOL using a gridded interpolation technique. This pre-processing work was done to avoid a constant shifting between MATLAB and COMSOL that would significantly increase the computational time. The use of a lookup table resulted in reducing the total computational time of the FEA model to the order of several hours, whereas the prior solution process required more computational time than the authors were willing to consider, and solutions were terminated after a week of computation.

To make use of the available computing power and reduce the computation time, symmetry was used to model the four-point bend test. A quadrilateral mesh was used to model this geometry and is shown in Figure 6. This model was then run with 3.5 million degrees of freedom, and the presented results were compared to results obtained at 8 million degrees of freedom, which was the threshold of the computational resources available on the workstation utilized for the study, an Intel i9-7940X CPU at 3.10 GHz, 14 cores, 28 logical processors, and 64 GB memory. The presented results are graphically indistinguishable from those from the higher resolution model.

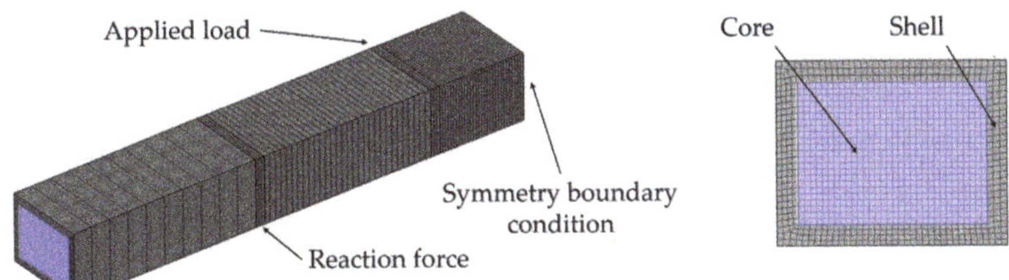

Figure 6. Quadrilateral Mesh for various surfaces of the symmetric FEA model.

The lookup table was read by the non-linear FEA model to study the load-deflection relationship under a four-point bend test that conforms to ASTM-D6109-13. Load-deflection data samples at regular intervals from 0 to 111 kN (0 to 25 kip) were collected to generate results. To account for the overall deflection, the (x, y, z) coordinates of the undeformed structure were mapped to the displaced points as $(x, y, z - w)$, where $w = w(x, y, z)$ represents the z-deflection in COMSOL. This allowed each point to retain its individual spatially varying stiffness behavior due to cell density variations.

To analyze the importance of including the non-linearity of the composite and spatial variations of the density within the core, the material properties from the uniform core structure were used from the identical material system from the authors' previous work [39] and are given in Table 3 for completeness. This model uses the Ramberg–Osgood [43] model, which is a non-linear curve fitting method as given in COMSOL Multiphysics to analyze the stress $\sigma(x)$, and strain $\varepsilon(x)$, based on Youngs's modulus E, a nonlinear reference stress σ_{ref}, and a nonlinear reference strain ε_{ref}, and is shown as,

$$\varepsilon(x) = \frac{\sigma(x)}{E} + \varepsilon_{ref}\left(\frac{\sigma(x)}{\sigma_{ref}}\right)^{n} \tag{10}$$

Table 3. Material inputs for the non-linear FEA model.

Material	Property	Result
Shell	Tensile Modulus	1.73 (GPa)
	Nonlinear Model Reference Stress	12.48 (MPa)
	Nonlinear Model Reference Strain	0.002
Core	Tensile Modulus	0.96 (GPa)
	Nonlinear Model Reference Stress	6.83 (MPa)
	Nonlinear Model Reference Strain	0.002

3. Results

3.1. Cell Generation and Cell Size Distribution

Using the image thresholding discussed previously, along with the automated removal of any false-positives, the image of the cross-section was analyzed, and the results for the three samples shown in Figure 3 are shown in Figure 7a. Notice that each cell region is converted to a circular region with only the center position and radius archived in a look-up table for fast processing. The histogram of the cell effective pixel radius distribution is provided in Figure 7b. Observe that the cell size distribution varies between each of the three samples. Samples 1 and 3 have similar cell size trends, with a number average for the effective radius of 8.4 pixels and 11.04 pixels, respectively, and a weighted average for the effective radius of 17.04 pixels and 22.7 pixels, respectively. Conversely, the cells in sample 2 have a number and weighted average for the effective radius of, respectively, 9.6 pixels and 14.8 pixels. The weighted average of samples 1 and 3 are greater than sample 2 because the effective radius in the samples 1 and 3 have larger cells compared to sample 2 overall.

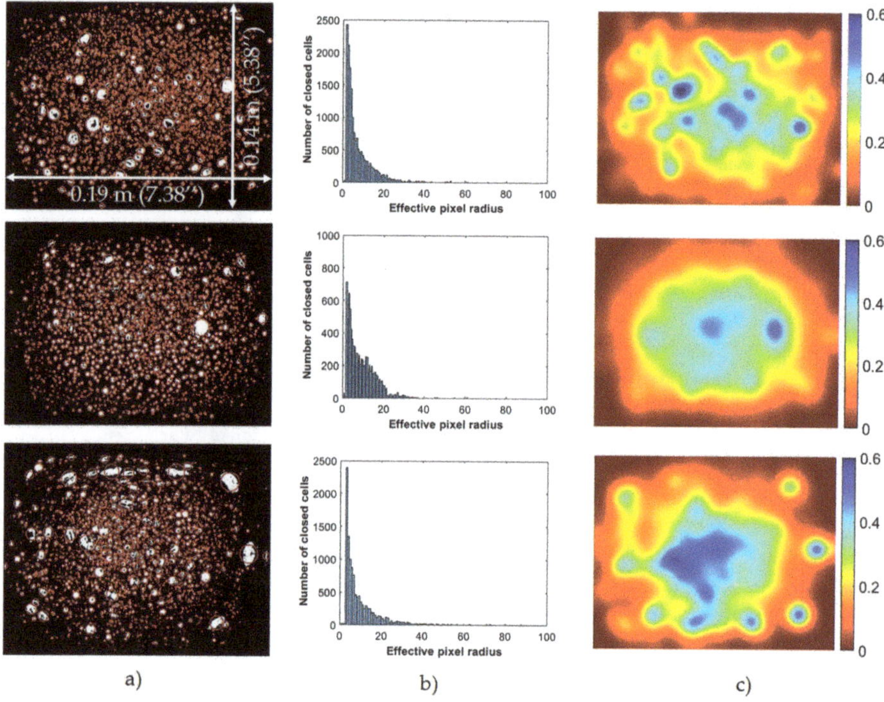

Figure 7. Various stages within the MATLAB image processing technique. (**a**) Cell area identification for samples 1 through 3. (**b**) Histograms of the cell size distribution for samples 1 through 3. (**c**) Spatially varying effective local density for samples 1 through 3 within the foamed core.

Although it is not obvious from visual observations, sample 2 contains fewer cells than sample 1 and 3. However, it is clear from the distribution that there is a more uniform distribution of cell sizes for the cells in sample 2, whereas in samples 1 and 3 there is a higher probability of the smaller cells with $r < 10$ pixels that are more dispersed within throughout the core along with several large cells with $r > 50$ pixels. It is important to note that the overall density of the three samples is similar, with ~14% of the surface area representing a cell, as are the basic manufacturing parameters, but it is clear from Figure 7c that there is some variation in both the size and spatial distribution between the three samples. This variability of cell size and distribution is due to the variability within the use of recycled materials as well as the manufacturing parameters.

The spatially varying density estimation from Equation (5) shows regions of significant density variation in the cross-section for each of the three samples. The point-wise grid with an $R = 300$ pixels gives a homogenized density range from 0–60% relative to that of the fully densified polymer composite. Each of the three samples has substantial variation in the spatial density as observed in Figure 7c. The larger open cells provide more weight in the regional density homogenization, and they affect the regions around them by causing a reduction in the local density. It is worth noting that each sample, although having the same overall density, has a unique spatial distribution of density. Specifically, in sample 1 the cells tend to be well dispersed, whereas in sample 3, there is a clear region in the core with a higher concentration of cells and a slow increase in the density as one moves away from the center region. This dispersion of cells can be observed through the image thresholding approach coupled with the spatially varying density weighting. The spatial varying density obtained in Figure 7c is then saved to a lookup table for later use in a finite element subroutine.

3.2. Finite Element Analysis

The pointwise Young's modulus was calculated from the spatially varying density using Equation (9) using the data shown in Figure 7c. This spatially varying modulus was then read into COMSOL as discussed in Section 2.4 and the results are presented in Figure 8 for each of the three samples. The Young's modulus was predicted for the shell region as E_c = 1.73 GPa [39], where, E_c is the Young's modulus of the HDPE, GFPP composite, and is used for the solid outer shell. Figures 7c and 8 are similar in spatial distribution as the localized density relates to the localized stiffness from Equation (9).

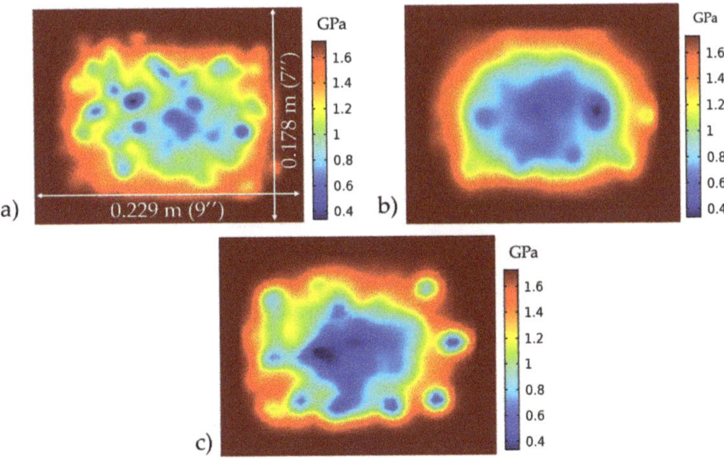

Figure 8. Spatially varying Young's modulus obtained from image analysis for (**a**) sample 1, (**b**) sample 2, (**c**) sample 3.

At the boundary between the foamed core and uniform density shell, a linear blending technique was implemented to allow for a smooth and continuous transition of the material properties within the finite element model cross-section. This blend is defined as

$$E_t = E_s(1 - \alpha) + E_f \alpha \tag{11}$$

where E_f is the foamed core modulus calculated using Equation (9) and E_t is the modulus approximated near the transition region around the shell and the core. The constant α is a normalized scalar that linearly increases from 0 to 1, with 0 corresponding to a point fully within the shell region and 1 corresponding to a point within the foamed core region. For the results presented, a transition region of 0.1 inch was taken with the region centered about the transition between the core and the shell.

Performing the FEA simulation for the model discussed in Section 2.4 with the applied load of 111 kN (25,000 lbs), results were plotted to study the load-deflection and stress behavior of the structure. The deflection of the core structure for sample 3 is shown in Figure 9, and the von-Mises stress on the core is superimposed on the deformed structure.

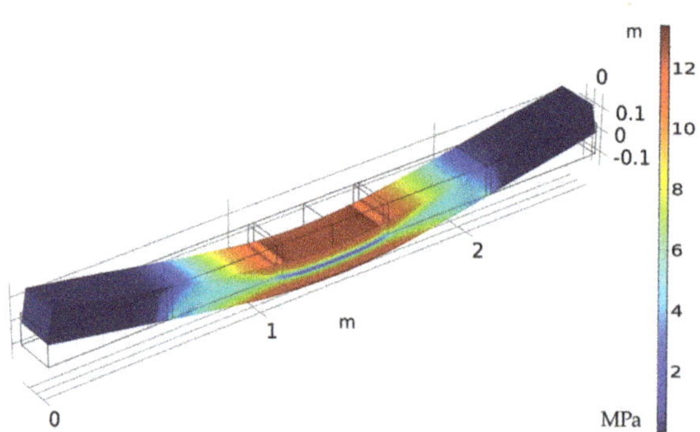

Figure 9. FEA results for core stress under deformation for sample 3.

To compare the load-deflection of the four models, the results from samples 1, 2, and 3 were compared to a model that one would obtain if the core had a uniform distribution. The results are shown in Figure 10a and it is observed that all four scenarios yield equivalent results for the overall load-deflection behavior. Equation (12) is used to convert the load to stress and displacement to strain, and the results are shown in Figure 10b.

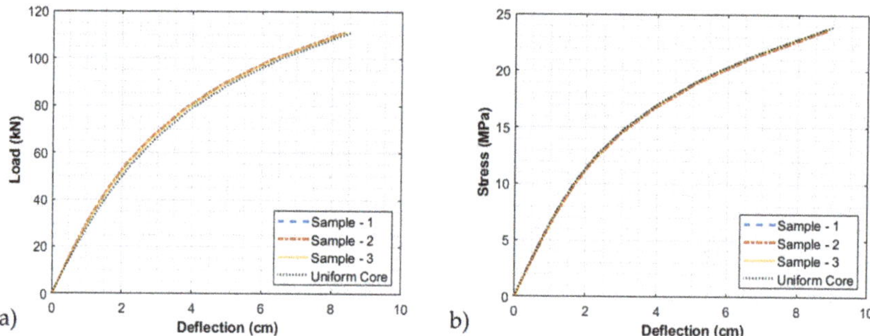

Figure 10. Results from FEA (**a**) comparing load–deflection from various samples and (**b**) comparing stress–deflection from various samples.

The plotted stress, S, value is defined in the ASTM standard D6109 and is plotted against the strain r, defined as [44]

$$S = PL/bd^2$$
$$r = 4.70Dd/L^2 \tag{12}$$

where, L is the total span length, b is the width, d is the depth of the beam as shown in Table 2, P is the applied load, and D is the deflection due to the applied load.

The real benefit of the spatially varying cell density homogenization approach allows one to look carefully at the point-wise stress distribution within the core region. Notice in Figure 11a that for the uniform core with no spatial variation, the internal von-Mises stress varies linearly in y with a zero-stress state at the geometric centroid. This behavior changes quickly for the three samples, with spatial variations in the cell density as shown in Figure 11b–d.

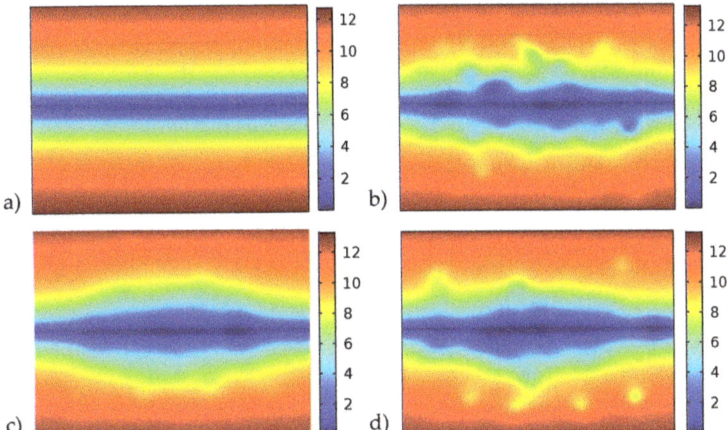

Figure 11. Core von-Mises stress at mid-point for each four-point bend simulation in the units of MPa for (**a**) uniform core, (**b**) sample 1, (**c**) sample 2, and (**d**) sample 3.

Notice that the stress state is no longer a linear function in the vertical dimension, and there are regions where the peak stress is measurably higher within the core than previously. The peak von-Mises stress predicted for sample 3 is a stress value of 13.7 MPa at peak loading, whereas the peak stress in the uniform core is 12.3 MPa for the same loading configuration. There is an ~11% variation in maximum stresses of the samples when compared to the uniform core. Variations in the peak von-Mises stress within the core region at the center-plane of the beam loaded in 4-point bending are shown in Figure 12 as a function of increasing deflection.

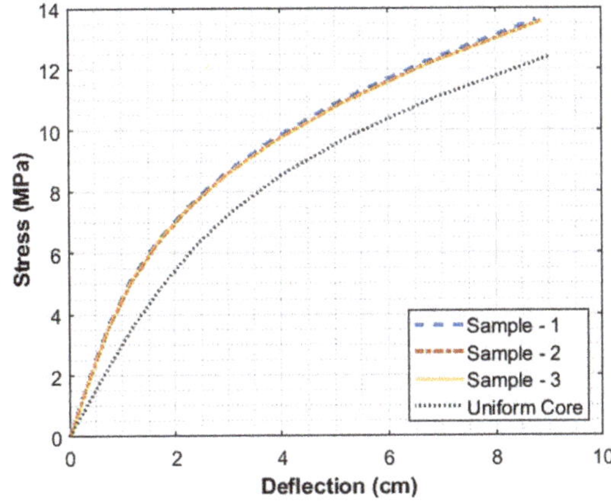

Figure 12. Stress comparison for the various FEA models with image cross-sections and the uniform core.

It is evident that with each of the different cores, the peak von-Mises stress changes as a function of both cross-head displacement and local cell distribution. It is interesting to note in Figure 12 that there is little difference in the results from the three samples with spatially varying density, but there is a difference in the results with and without spatially varying density. Figure 11a shows the von-Mises stresses with a uniform core, while Figure 11b–d show maximum stress with cores from samples 1–3.

Although the stresses with samples 1–3 have higher stresses overall, the main load and stress are distributed within the outer shell. In addition, the distribution has a relatively smooth transition from the shell to the core, even if there is variability within the core, it is carrying a nominal amount of the overall load. The present study focused mainly on the onset of failure. Various researchers have studied advanced methods for tracking crack propagation within the failed composite (see e.g., [45]), and it is worth studying in future works.

4. Conclusions

In this paper, a method to locally estimate the cell density without the full meshing of the individual cells was presented. This local homogenization was performed using images captured using a digital camera, and a procedure was proposed to analyze the correlation of the cell size and cell distributions to the local stiffness. These results were then fed into a finite element model of a structure subjected to four-point bending. Images taken using a digital camera were analyzed using a custom script written in the MATLAB environment, and an exponential function was proposed to establish a weighting method to allow for a smooth continuous function to be used to represent the local cell variations. This weighted method calculates a density of cells based on their distance from the point analyzed. A periodic boundary was used to estimate the density near the boundaries. Once the local densities were calculated, a micromechanics model that included the constitutive properties of the fiber and matrix along with fiber length and volume fraction of fibers was used to calculate the pointwise Young's modulus of the core as a function of cell density.

The spatially varying Young's modulus was then used in a finite element model to analyze the load-deflection of a four-point bend test. These results were compared to a uniform core model previously validated for the macroscopic response to a physical specimen. The macroscopic response for the spatially varying core density and the uniform core density was graphically equivalent. The local von-Mises stress was then analyzed between the two density distribution assumptions, and it was seen that the stresses from the spatially varying cell density resulted in a $\sim$11% higher stress within the core as compared to the uniform core.

The present study identifies that the spatial variation of the cell distribution within the core impacts the as manufactured performance, but it is unclear from the present work the sensitivity of the final performance on subtle differences in the spatial variations. Future work will consider sectioning multiple regions within the composite structure to study the variability in location, cell size, and distribution, and its impact on final part performance. This method also allows quality control procedures to remain in place without needing an extensive need for equipment and personnel. The ease of taking images and analysis of the proposed method can be extended to other handheld devices like a cell phone or a tablet.

Author Contributions: D.A.J. and D.P.P. were instrumental in the conceptualization of this method. D.P.P. implemented the concept and performed the analysis with assistance from D.A.J. D.P.P. developed the finite element model. D.P.P. wrote the original draft and D.A.J. reviewed the manuscript and edited it. D.A.J. supervised all facets of this research. All authors have read and agreed to the published version of the manuscript.

Funding: This research was funded by Axion Structural Innovations.

Acknowledgments: We would also like to thank Axion Structural Innovations for providing the manufactured crosstie cross-section samples.

Conflicts of Interest: The authors declare no conflict of interest.

References

1. Bajracharya, R.; Manalo, A.; Karunasena, W.; Lau, K.T. Characterisation of recycled mixed plastic solid wastes: Coupon and full-scale investigation. *Waste Manag.* **2015**, *48*, 72–80. [CrossRef]
2. Hugo, A.M.; Scelsi, L.; Hodzic, A.; Jones, F.; Dwyer-Joyce, R. Development of recycled polymer composites for structural applications. *Plast. Rubber Compos.* **2011**, *40*, 317–323. [CrossRef]
3. Grigore, M.E. Methods of Recycling, Properties and Applications of Recycled Thermoplastic Polymers. *Recycling* **2017**, *2*, 24. [CrossRef]
4. Momanyi, J.; Herzog, M.; Muchiri, P. Analysis of Thermomechanical Properties of Selected Class of Recycled Thermoplastic Materials Based on Their Applications. *Recycling* **2019**, *4*, 33. [CrossRef]

5. Manalo, A.; Aravinthan, T.; Karunasena, W.; Ticoalu, A. A review of alternative materials for replacing existing timber sleepers. *Compos. Struct.* **2010**, *92*, 603–611. [CrossRef]

6. Lotfy, I.; Farhat, M.; Issa, M.A.; Al-Obaidi, M. Flexural behavior of high-density polyethylene railroad crossties. *Proc. Inst. Mech. Eng. Part F J. Rail Rapid Transit* **2016**, *230*, 813–824. [CrossRef]

7. Zyka, K.; Mohajerani, A. Composite piles: A review. *Constr. Build. Mater.* **2016**, *107*, 394–410. [CrossRef]

8. Jones, R.M. *Mechanics of Composite Materials*; CRC Press: Boca Raton, FL, USA, 2018.

9. Barbeo, E.J. *Introduction to Composite Materials Design*; CRC Press: Boca Raton, FL, USA, 2017.

10. Altan, M. Thermoplastic Foams: Processing, Manufacturing, and Characterization. In *Polymerization*; IntechOpen: Rijeka, Croatia, 2018; Chapter 6.

11. Kord, B.; Varshoei, A.; Chamany, V. Influence of chemical foaming agent on the physical, mechanical, and morphological properties of HDPE/wood flour/nanoclay composites. *J. Reinf. Plast. Compos.* **2011**, *30*, 1115–1124. [CrossRef]

12. Ruiz, J.A.R.; Vincent, M.; Agassant, J.F.; Sadik, T.; Pillon, C.; Carrot, C. Polymer foaming with chemical blowing agents: Experiment and modeling. *Polym. Eng. Sci.* **2015**, *55*, 2018–2029. [CrossRef]

13. Carlos, T.C.; Rubén, G.N.; Denis, R. Effect of Mold Temperature on Morphology and Mechanical Properties of Injection Molded HDPE Structural Foams. *J. Cell. Plast.* **2008**, *44*, 223–237.

14. Yousefian, H.; Rodrigue, D. Nano-crystalline cellulose, chemical blowing agent, and mold temperature effect on morphological, physical/mechanical properties of polypropylene. *J. Appl. Polym. Sci.* **2015**, *132*. [CrossRef]

15. Tissandier, C.; González-Núñez, R.; Rodrigue, D. Asymmetric microcellular composites: Morphological properties. *J. Cell. Plast.* **2014**, *50*, 449–473. [CrossRef]

16. Villamizar, C.A.; Han, C.D. Studies on structural foam processing II. Bubble dynamics in foam injection molding. *Polym. Eng. Sci.* **1978**, *18*, 699–710. [CrossRef]

17. González-Núñez, R. Microcellular Agave Fibre-High Density Polyethylene Composites Produced By Injection Molding. *J. Mater. Sci. Eng.* **2012**, *2*, 667–692.

18. Redenbach, C.; Shklyar, I.; Andrae, H. Laguerre tessellations for elastic stiffness simulations of closed foams with strongly varying cell sizes. *Int. J. Eng. Sci.* **2012**, *50*, 70–78. [CrossRef]

19. Barbier, C.; Michaud, P.; Baillis, D.; Randrianalisoa, J.; Combescure, A. New laws for the tension/compression properties of Voronoi closed-cell polymer foams in relation to their microstructure. *Eur. J. Mech. A/Solids* **2014**, *45*, 110–122. [CrossRef]

20. Zhang, Y.; Rodrigue, D.; Ait-Kadi, A. High Density Polyethylene Foams. II. Elastic Modulus. *J. Appl. Polym. Sci.* **2003**, *90*, 2120–2129. [CrossRef]

21. Mori, T.; Tanaka, K. Average Stress in Matrix and Average Elastic Energy of Materials With Misfitting Inclusions. *Acta Metall.* **1973**, *21*, 571–574. [CrossRef]

22. Weng, G. Some elastic properties of reinforced solids, with special reference to isotropic ones containing spherical inclusions. *Int. J. Eng. Sci.* **1984**, *22*, 845–856. [CrossRef]

23. Farber, J.; Farris, J. Model for prediction of the elastic response of reinforced materials over wide ranges of concentration. *J. Appl. Polym. Sci.* **1987**, *34*, 2093–2104. [CrossRef]

24. McLaughlin, R. A study of the differential scheme for composite materials. *Int. J. Eng. Sci.* **1977**, *15*, 237–244. [CrossRef]

25. Gibson, L.J.; Ashby, M.F. *Cellular Solids: Structure and Properties*; Cambridge University Press: Cambridge, UK, 1997.

26. Moore, D.; Iremonger, M. The Prediction of the Flexural Rigidity of Sandwich Foam Mouldings. *J. Cell. Plast.* **1974**, *10*, 230–236. [CrossRef]

27. Lo, K.H.; Miyase, A.; Wang, S.S. Stiffness predictions for closed-cell PVC foams. *J. Compos. Mater.* **2016**, *51*, 3327–3336. [CrossRef]

28. Tucker, C.L., III; Liang, E. Stiffness Predictions for Unidirectional Short-Fiber Composites: Review and Evaluation. *Compos. Sci. Technol.* **1999**, *59*, 655–671. [CrossRef]

29. Tandon, G.; Weng, G.J. The effect of aspect ratio on the elastic properties of unidirectionally aligned composites. *Polym. Compos.* **1984**, *5*, 327–333. [CrossRef]

30. Zhang, C. Modeling of Flexible Fiber Motion and Prediction of Material Properties. Master's Thesis, Baylor University, Waco, TX, USA, 2011.

31. Barzegari, M.; Rodrigue, D. Tensile Modulus Prediction of Structural Foams using Density Profiles. *Cell. Polym.* **2008**, *27*, 285–301. [CrossRef]

32. Davari, M.; Razavi Aghjeh, M.K.; Seraji, S. Relationship between the cell structure and mechanical properties of chemically crosslinked polyethylene foams. *J. Appl. Polym. Sci.* **2012**, *124*, 2789–2797. [CrossRef]

33. Sadik, T.; Pillon, C.; Carrot, C.; Reglero Ruiz, J.; Vincent, M.; Billon, N. Polypropylene structural foams: Measurements of the core, skin, and overall mechanical properties with evaluation of predictive models. *J. Cell. Plast.* **2016**, *53*, 25–44. [CrossRef]

34. Shen, H.; Oppenheimer, S.; Dunand, D.; Brinson, L. Numerical Modelling of Pore Size Distribution in Foamed Titanium. *Mech. Mater.* **2006**, *38*, 933–944. [CrossRef]

35. Chen, Y.; Das, R.; Battley, M. Effects of cell size and cell wall thickness variations on the stiffness of closed-cell foams. *Int. J. Solids Struct.* **2015**, *52*, 150–164. [CrossRef]

36. Zhu, X.; Ai, S.; Fang, D.; Liu, B.; Lu, X. A novel modeling approach of aluminum foam based on MATLAB image processing. *Comput. Mater. Sci.* **2014**, *82*, 451–456. [CrossRef]

37. Nasrabadi, A.M.; Hedayati, R.; Sadighi, M. Numerical and experimental study of the mechanical response of aluminum foams under compressive loading using CT data. *J. Theor. Appl. Mech.* **2016**, *54*, 1357–1368. [CrossRef]

38. Yunus, S.; Sefa-Ntiri, B.; Anderson, B.; Kumi, F.; Mensah-Amoah, P.; Sonko Sackey, S. Quantitative Pore Characterization of Polyurethane Foam with Cost-Effective Imaging Tools Image Analysis: A Proof-Of-Principle Study. *Polymers* **2019**, *11*, 1879. [CrossRef] [PubMed]

39. Pulipati, D.P.; Jack, D.A. Characterization and model validation for large format chopped fiber, foamed, composite structures made from recycled olefin based polymers. *Polymers* **2020**, *12*, 1371. [CrossRef] [PubMed]

40. Pinto, J.; Solórzano, E.; Rodríguez-Pérez, M.; De Saja, J. Characterization of the cellular structure based on user-interactive image analysis procedures. *J. Cell. Plast.* **2013**, *49*, 554–574. [CrossRef]

41. Advani, S.G.; Tucker, C.L., III. The use of tensors to describe and predict fiber orientation in short fiber composites. *J. Rheol.* **1987**, *31*, 751–784. [CrossRef]

42. VerWeyst, B.E.; Tucker, C.L., III. Fiber suspensions in complex geometries: Flow-orientation coupling. *Can. J. Chem. Eng.* **2002**, *80*, 1093–1106. [CrossRef]

43. Ramberg, W.; Osgood, W.R. *Description of Stress-Strain Curves by three Parameters*; Technical Note No. 902; National Aeronautics and Space Administration: Washington, DC, USA, 1943.

44. Flexural Properties of Unreinforced and Reinforced Plastic Lumber and Related Products. In *Standard Test Methods*; ASTM International: West Conshohocken, PA, USA, 2013.

45. Mehrmashhadi, J.; Chen, Z.; Zhao, J.; Bobaru, F. A stochastically homogenized peridynamic model for intraply fracture in fiber-reinforced composites. *Compos. Sci. Technol.* **2019**, *182*, 107770. [CrossRef]

MDPI
St. Alban-Anlage 66
4052 Basel
Switzerland
Tel. +41 61 683 77 34
Fax +41 61 302 89 18
www.mdpi.com

Journal of Composites Science Editorial Office
E-mail: jcs@mdpi.com
www.mdpi.com/journal/jcs